# Nanodiamonds

## Advanced Material Analysis, Properties and Applications

# Nanodiamonds

## Advanced Material Analysis, Properties and Applications

Edited by
**Jean-Charles Arnault**

ELSEVIER

Elsevier
Radarweg 29, PO Box 211, 1000 AE Amsterdam, Netherlands
The Boulevard, Langford Lane, Kidlington, Oxford OX5 1GB, United Kingdom
50 Hampshire Street, 5th Floor, Cambridge, MA 02139, United States



**Library of Congress Cataloging-in-Publication Data**
A catalog record for this book is available from the Library of Congress

**British Library Cataloguing-in-Publication Data**
A catalogue record for this book is available from the British Library

ISBN: 978-0-323-43029-6

For information on all Elsevier publications
visit our website at https://www.elsevier.com/books-and-journals



*Publisher:* Matthew Deans
*Acquisition Editor:* Simon Holt
*Editorial Project Manager:* Sabrina Webber
*Production Project Manager:* Julie-Ann Stansfield
*Designer:* Greg Harris

Typeset by VTeX

# Contents

# Contributors

**Be-Ming Chang**
Institute of Atomic and Molecular Sciences, Academia Sinica, Taipei, Taiwan

**Huan-Cheng Chang**
Institute of Atomic and Molecular Sciences, Academia Sinica, Taipei, Taiwan

**Sylvie Chevillard**
CEA DRF iRCM, Laboratory of Experimental Cancerology, Fontenay-aux-Roses, France

**Petr Cigler**
Institute of Organic Chemistry and Biochemistry, v.v.i., Academy of Sciences of the Czech Republic, Prague, Czech Republic

**Jozo Delic**
CEA DRF iRCM, Laboratory of Experimental Cancerology, Fontenay-aux-Roses, France

**Valerii Y. Dolmatov**
(FSUE SCTB) "Technolog", Saint-Petersburg, Russia

**Yury Gogotsi**
A.J. Drexel Nanomaterials Institute and Department of Materials Science & Engineering, Drexel University, Philadelphia, PA, United States

**Romain Grall**
CEA DRF iRCM, Laboratory of Experimental Cancerology, Fontenay-aux-Roses, France

**Neeraj Gupta**
Shenyang National Laboratory for Materials Science, Institute of Metal Research, Chinese Academy of Sciences, Shenyang, China

**Jan Havlik**
Institute of Organic Chemistry and Biochemistry, v.v.i., Academy of Sciences of the Czech Republic, Prague, Czech Republic;
Faculty of Science, Charles University, Prague, Czech Republic

**Dean Ho**
Division of Oral Biology and Medicine and Department of Bioengineering, UCLA, Los Angeles, CA, United States

**Alon Hoffman**
Schulich Faculty of Chemistry, Technion, Haifa, Israel

**Anke Krueger**
Institute for Organic Chemistry, Julius-Maximilians-Universität Würzburg, Würzburg, Germany

**Michel Mermoux**
Grenoble Alpes University, LEPMI, Grenoble, France;
CNRS, LEPMI, Grenoble, France

**Shaul Michaelson**
Schulich Faculty of Chemistry, Technion, Haifa, Israel

**Dimitar Mitev**
Australian Center for Research on Separation Science (ACROSS), School of Physical Sciences, University of Tasmania, Hobart, TAS, Australia

**Vadym N. Mochalin**
Department of Chemistry, Missouri University of Science & Technology, Rolla, MO, United States

**Christoph E. Nebel**
Fraunhofer-Institute for Applied Solid State Physics (IAF), Freiburg, Germany

**Jitka Neburkova**
Institute of Organic Chemistry and Biochemistry, v.v.i., Academy of Sciences of the Czech Republic, Prague, Czech Republic;
First Faculty of Medicine, Charles University, Prague, Czech Republic

**Ioannis Neitzel**
A.J. Drexel Nanomaterials Institute and Department of Materials Science & Engineering, Drexel University, Philadelphia, PA, United States

**Pavel N. Nesterenko**
Australian Center for Research on Separation Science (ACROSS), School of Physical Sciences, University of Tasmania, Hobart, TAS, Australia;
ARC Centre of Excellence for Electromaterials Science (ACES), School of Physical Sciences, University of Tasmania, Hobart, TAS, Australia

**Elke Neu**
Saarland University, Saarbrücken, Germany

**Nicholas Nunn**
Adámas Nanotechnologies, Inc., Raleigh, NC, United States

**Vincent Paget**
CEA DRF iRCM, Laboratory of Experimental Cancerology, Fontenay-aux-Roses, France

**Alexander M. Panich**
Department of Physics, Ben-Gurion University of the Negev, Be'er-Sheva, Israel

**Brett Paull**
Australian Center for Research on Separation Science (ACROSS), School of Physical Sciences, University of Tasmania, Hobart, TAS, Australia;
ARC Centre of Excellence for Electromaterials Science (ACES), School of Physical Sciences, University of Tasmania, Hobart, TAS, Australia

**Tristan Petit**
Helmholtz-Zentrum Berlin für Materialien und Energie GmbH, Berlin, Germany

**Goutam Pramanik**
Institute of Organic Chemistry and Biochemistry, v.v.i., Academy of Sciences of the Czech Republic, Prague, Czech Republic

**Helena Raabova**
Institute of Organic Chemistry and Biochemistry, v.v.i., Academy of Sciences of the Czech Republic, Prague, Czech Republic;
Department of Inorganic Chemistry, University of Chemistry and Technology Prague, Prague, Czech Republic

**Bohuslav Rezek**
Faculty of Electrical Engineering, Czech Technical University, Prague, Czech Republic

**Alexander I. Shames**
Department of Physics, Ben-Gurion University of the Negev, Be'er-Sheva, Israel

**Olga Shenderova**
International Technology Center, Raleigh, NC, United States

**Stepan Stehlik**
Institute of Physics, Czech Academy of Sciences, Prague, Czech Republic

**Dangsheng Su**
Shenyang National Laboratory for Materials Science, Institute of Metal Research, Chinese Academy of Sciences, Shenyang, China;
Dalian Institute of Chemical Physics, Chinese Academy of Sciences, Dalian, China

**François Treussart**
Laboratoire Aimé Cotton, Orsay, France

**Stuart Turner**
Aurubis Belgium Research, Development and Innovation, Olen, Belgium

**Jan Vavra**
Institute of Organic Chemistry and Biochemistry, v.v.i., Academy of Sciences of the Czech Republic, Prague, Czech Republic

**Igor I. Vlasov**
General Physics Institute RAS, Moscow, Russia

**Qi Wang**
Shenyang National Laboratory for Materials Science, Institute of Metal Research, Chinese Academy of Sciences, Shenyang, China

**Guodong Wen**
Shenyang National Laboratory for Materials Science, Institute of Metal Research, Chinese Academy of Sciences, Shenyang, China

# About the Editor

**Jean-Charles Arnault** is currently a Research Director at CEA. He obtained his PhD in 1993 at Poitiers University, France. From 1993 to 2007 he was an Assistant Professor at Strasbourg University and obtained his HDR in 2001. His research activities in Strasbourg concerned diamond nucleation on different substrates and interactions between MPCVD plasmas with surfaces using in situ surface analysis methods (electron spectroscopies), and scanning probe microscopies. These studies were completed by electron microscopies. In 2003, he moved to the Physics and Chemistry of Surfaces and Interfaces Laboratory at CEA for a research period focused on diamond on reconstructed cubic silicon carbide surfaces (2003–2007). Since 2007, Jean-Charles Arnault performed his research activities in the Diamond Sensors Laboratory at CEA. His research interests are currently focused on diamond nanoparticles, diamond heteroepitaxy and the growth of boron-doped single crystalline diamond. For diamond nanoparticles, surface modifications by MPCVD plasma or annealing are performed to control their surface terminations for efficient grafting. Intrinsic properties of hydrogenated and surface graphitized nanodiamonds were investigated during the last years especially for bioapplications. In 2008, he initiated diamond heteroepitaxy on iridium at CEA. This topic is now developed in collaboration with a French consortium involving four other groups. Jean-Charles Arnault is regularly involved in the organization of international symposia related to these topics (Spring and Fall MRS, E-MRS). In January 2015, he became an Editor of Diamond and Related Materials. He recently contributed to three chapters of books, two of them concerning diamond nanoparticles.

# Preface

The research field related to nanodiamonds (i.e. diamond nanoparticles) was initiated in Russia during the 1960s. It has seen a strong revival during the last 10 years. This resurgence is mainly explained by interests in nanodiamonds for the fabrication of composites (polymers, lubricants) and for the development of bio-applications like biomarkers or nanocarriers for drug delivery.

The present book aims to provide an up-to-date overview of nanodiamonds properties evidenced by advanced characterization tools. The most promising fields of applications are also included. This book adopts a didactic approach in each chapter by providing basic information useful to students or researchers starting in the field of nanodiamonds.

Nanodiamonds combine many properties of bulk diamond like chemical inertness, wide-band gap electronic properties, excellent thermal conductivity, and outstanding mechanical behavior to those conferred by their high specific surface area, which could reach 400 $m^2/g$ for nanodiamonds produced by detonation synthesis. Before focusing on diamond properties at nanoscale, we need to describe the usefulness of bulk diamond. The first chapter presents bulk electronic properties of diamond.

The properties observed at nanoscale are significantly influenced by factors like the size and the shape of nanodiamonds, the nature and the concentration of impurities, the presence of structural defects, and the surface chemistry of outershells. An extensive characterization of nanodiamonds is then needed to identify the key factors governing their properties. This book evaluates the current knowledge from the production methods of nanodiamonds to their most advanced characterization tools.

Recent developments in synthesis and processing of nanodiamonds are exposed in Chapter 2. First, the methods like detonation shock-wave assisted and laser assisted synthesis are detailed. The alternative crushing of high pressure high temperature (HPHT) or chemical vapor deposition (CVD) bulk diamond materials is also presented. The crushing process and recent HPHT results appear as promising approaches to obtain doped nanodiamonds (with boron, phosphorus or nitrogen) with a controlled level of concentration. The purification and disaggregation procedures are also discussed.

The diamond core of nanodiamonds could be investigated by different advanced techniques which are described in Chapters 3 to 7. Its crystalline structure, the presence of structural defects or $sp^2$ carbon, of chemical impurities could be characterized by High Resolution Transmission Electron Microscopy (HRTEM) and Electron Energy Loss Spectroscopy (EELS) as shown in Chapter 3. Raman spectroscopy appears as a complementary technique which can also probe the crystalline structure and the presence of chemical impurities (see Chapter 4). At the same time, this technique is sensitive to strain effects and could be applied to probe surface chemistry. Inductively Coupled Plasma (ICP) spectroscopy and related techniques allow measuring the concentrations of chemical impurities with detection limits down to $10^{10}$–$10^{-11}$ wt% for certain elements (Chapter 5). Paramagnetic defects like some structural defects or impurities could be probed by Electron Paramagnetic Resonance (EPR) or Nuclear Magnetic Resonance (NMR). This is the topic of Chapter 6. Lastly, photoluminescence properties provided by color centers like nitrogen/vacancy (NV) or silicon/vacancy (SiV) could be added to render nanodiamonds fluorescent. Chapter 7 compares the assets of these two color centers. Such fluorescent nanodiamonds are currently used for bioapplications or for quantum physics and sensing.

As previously underlined, some properties at nanoscale are closely related to the surface chemistry of nanodiamonds. A complete overview of the current issues is provided in Chapter 8 dealing with different surface terminations and grafting of complex moieties. Chapters 9 and 10 present complementary characterizations of their outershells. In particular, the chemical nature of surface groups, their thermal stability, their reactivity toward ambient air, the surface potential of nanodiamonds are investigated using High Resolution (HR) EELS and Kelvin Force Microscopy experiments, in Chapters 9 and 10, respectively.

Nanodiamonds are currently used to elaborate composites with polymers or lubricants (Chapters 14 and 15). In this way, tribological and thermal properties of composites could be tuned. These aspects are very promising for future industrial applications involving nanodiamonds. Another active field concerns bio-applications like drug delivery or labeling, and Chapter 16 is devoted to these developments. For the latter applications, the toxicity of nanodiamonds remains a major issue. State-of-the-art of in vitro and in vivo investigations is provided in Chapter 12. The behavior of nanodiamonds in solvents is essential for most of the previous applications. Chapter 11 focuses on the interactions between nanodiamonds and water molecules, especially probed by advanced X-ray and IR spectroscopy methods.

In addition to well-identified applications, descriptions of promising fields for nanodiamonds like quantum physics or catalysis are included in this book. Sensing applications based on fluorescent nanodiamonds are described in Chapter 17. Chapter 18 concerns the performance of catalysts synthetized with nanodiamonds. Finally, a very recent trend linked to the fabrication of hybrids based on nanodiamonds like core-shell structures with polymers, silica, metals or proteins is discussed in Chapter 13.

The present overview constitutes a precious basis for new scientific and technologic developments. Indeed, the combination of the presented advanced techniques will allow a better understanding of diamond properties at nanoscale in the future. This will then lead to the emergence of novel applications at the frontier of disciplines involving composites or core-shell particles.

# About the Authors

**Be-Ming Chang** obtained her BS degree in Chemistry from Tamkang University and MSc degree from National Taiwan Normal University. Currently, she is working on her PhD degree in the Institute of Atomic and Molecular Sciences of Academia Sinica, Taiwan, under Dr. Huan-Cheng Chang's supervision. Her research work involves in biolabeling applications of fluorescent nanodiamonds and development of hybrid diamond nanoparticles for dual-modality imaging.

**Huan-Cheng Chang** received his BS degree in Agricultural Chemistry from National Taiwan University in 1981 and his PhD degree in Physical Chemistry from Indiana University at Bloomington in 1990, followed by a postdoctoral study at Harvard University. He joined the Institute of Atomic and Molecular Sciences (IAMS) of Academia Sinica in 1994 and is now a Distinguished Research Fellow of IAMS. His research interests are focused on the development of new methods, tools, and technologies based on physical chemistry and applying them to solve problems of biological and medicinal significance.

**Dr. Petr Cigler** is a Chemist with broad experience in synthesis and analysis of nanoparticles, organic molecules, and biomolecules. He received his PhD degree from the Institute of Chemical Technology Prague, Czech Republic in 2008. Then he joined The Scripps Research Institute, La Jolla, United States as a postdoctoral fellow (Prof. M.G. Finn). Since 2013 he is a head of Synthetic Nanochemistry Group, Institute of Organic Chemistry and Biochemistry, AS CR, Prague. He is focused on studies of nanomaterials based either on bioorganic or inorganic cores including fluorescent nanodiamonds, virus-like particles, and plasmonic systems for use in therapeutics, imaging, and diagnostics of diseases.

**Valerii Y. Dolmatov** is a Doctor of Chemical Sciences and a Doctor of Technical Sciences. He has been investigating detonation nanodiamonds since 1984, and is an author of over 270 scientific papers and more than 100 patents. Moreover, Dr. Dolmatov is a Laureate of the Lenin Komsomol Award in Science and Technology (1981), received a Gold Medal of the exhibition of National Economic Achievements of the USSR (1986), was named the best inventor of Leningrad (1990), Prize Laureate of the Ministry of Higher Education of the USSR (1989), holds Honorary Diploma of the Ministry of Justice of Russia for creation and innovation the Russian Patents No. 2348580 and No. 23599902, is Honorary Doctor of Sciences of the Superhard materials Institute (Kiev, Ukraine, 2011), and is the author of 3 patents top 100 patents of RF (Nos. 2348580, 2359902 and 2381241).

**Yury Gogotsi** is a Distinguished University Professor and Trustee Chair of Materials Science and Engineering at Drexel University in Philadelphia, United States. He is the founding Director of the A.J. Drexel Nanomaterials Institute and an Associate Editor of *ACS Nano*. He works on nanostructured carbons and two-dimensional carbides for energy related and biomedical applications. He has co-authored 2 books, more than 400 journal papers and obtained more than 50 patents. He has received numerous national and international awards for his research. He has been recognized as a Highly Cited Researcher by Thomson-Reuters in 2014 and 2015, and elected a Fellow of AAAS, MRS, RSC, ECS, and ACerS, as well as a member of the World Academy of Ceramics.

**Jan Havlik** is a PhD student at the Department of Inorganic Chemistry, Faculty of Science, Charles University in Prague, and has been working in Petr Cigler's Synthetic Nanochemistry Group at the Institute of Organic Chemistry and Biochemistry, AS CR, in Prague, Czech Republic since 2010. He is focused on irradiation, preparation, and surface chemistry of fluorescent nanodiamonds. He is also interested in fields related to STEM, education, science popularization, and science journalism.

**Dean Ho** is currently a Professor of Oral Biology and Medicine, and Bioengineering at the University of California, Los Angeles (UCLA). He is also the Co-Director of the Jane and Jerry Weintraub Center for Reconstructive Biotechnology and member of the Jonsson Comprehensive Cancer Center. He was previously an Associate Professor of Biomedical Engineering and Mechanical Engineering at Northwestern University. His research interests include the development of nanodiamond-based drug delivery and imaging agents, as well as drug development for personalized and precision medicine.

**Prof. Alon Hoffman** is a Full Professor at the Schulich faculty of Chemistry at the Technion and also holds the Joseph Szydlowsky Chair in Science. Prof. Alon completed his DSc in 1987 at the Technion, investigating adsorption processes on single crystal metal alloy surfaces by electron spectroscopy. During 2012–2016 he was Dean of the Schulich Faculty of Chemistry. His research efforts range from basic to fundamental research in surface science and thin films with more than 300 publications. During the last 25 years Prof. Hoffman's research has been related to different aspects associated to the chemical, physical, and electronic properties of diverse carbon surfaces such as graphite, diamond, and amorphous carbon. In particular, he has investigated very extensively the interaction of hydrogen with single and poly crystalline diamond surfaces by high resolution electron energy loss spectroscopy. He has also investigated the different processes involved in the formation – nucleation and growth – of polycrystalline diamond films by chemical vapor deposition onto non-diamond substrates.

**Michel Mermoux** is a Research Director at the CNRS. He works at the Laboratory of Electrochemistry and Physico-chemistry of Materials and Interfaces (LEPMI, UMR 5279). His research activities are focused for several years on the use of Raman spectroscopy and Raman imaging for the analysis of different materials (diamond, carbon materials, semiconductors, oxides, etc.), in correlation with their structures and physicochemical properties. He has authored about 120 publications.

**Vadym N. Mochalin** has received his PhD degree in Physical Chemistry from L.M. Litvinenko Institute of Physical Organic and Coal Chemistry, National Academy of Sciences of Ukraine, and MS degree in Biochemistry from Donetsk National University, Ukraine. In 2005–2015, he worked with Prof. Yury Gogotsi at Drexel University, leading research on nanodiamond. He is now an Associate Professor of Chemistry and Materials Science & Engineering at Missouri University of Science and Technology. His current research interests include synthesis, characterization, purification, chemical modification, computational modeling, and applications of nanodiamond, MXene, nanoonions, nanocarbons, and other materials for composites, energy storage, biomedical applications, and extreme environments.

**Christoph E. Nebel** graduated in Electrical Engineering from the University of Stuttgart where he also got his PhD in 1989. He became a postdoc at the Xerox Research Center, Palo Alto, United States, between 1990 and 1992, funded by the Lynen Program of the Alexander von Humboldt Foundation, Germany. In 1993 he joined the team of Martin Stutzmann at the Walter Schottky Institute of the Technical University Munich where he habilitated in 1998. In 2004 he joined the Diamond Research Center of the National Institute of Advanced Industrial Science and Technology (AIST), Japan, where he was leading the "Bio-Functionalized Device Team" until 2008 when he went back to Germany to become head of the Micro- and Nano-Sensor Department of the Fraunhofer-Institute for Applied Solid State Physics (IAF) in Freiburg. At the center of his scientific activities is diamond, a new and challenging material, which is currently developed as man-made gemstones as well as for power-electronic applications.

**Jitka Neburkova** graduated in 2013 from Biochemistry at the Faculty of Science, Charles University in Prague. Now she is a PhD student at the First Faculty of Medicine, Charles University in Prague and works in Petr Cigler's Synthetic Nanochemistry Group at Institute of Organic Chemistry and Biochemistry, AS CR in Prague, Czech Republic. Her interests cover preparation of nanoparticles, their modifications with biocompatible hydrophilic polymers and other molecules such as targeting moieties or diagnostic probes, and in vitro cell experiments with nanoparticles.

**Elke Neu** studied Physics at the University of Kaiserslautern (Germany), with a focus on quantum and molecular physics. In 2012 she received her PhD at Saarland University (Germany) in Prof. Christoph Becher's group working on diamond color centers as single photon sources. As a postdoc in Prof. Patrick Maletinsky's group at Basel University (Switzerland), she broadened her expertise to scanning probe magnetometry with diamond color centers. In 2015 she has been awarded a Young Investigator Grant ("NanoMatFutur") from German Ministry of Education and Research, enabling her to establish a group investigating applications of diamond sensors in Life Sciences.

**Ioannis Neitzel** received his diploma in Mechanical Engineering from the Bonn Rhein-Sieg University of Applied Sciences in 2008. After receiving a Fulbright scholarship in 2008, he joined the Nanomaterials Group in the department of Materials Science and Engineering, working under supervision of Prof. Yury Gogotsi and Prof. Vadym Mochalin at Drexel University, Philadelphia, United States. During his PhD studies, he worked on the surface functionalization of nanodiamonds for applications in polymer nanocomposites and biomedical applications.

**Prof. Pavel N. Nesterenko** received his PhD (1984) and DSc degrees (2000) in Chemistry from Lomonosov Moscow State University, Moscow, Russia. In 2006 moved to Australia, where he holds a strategic position of New Stars Professor at the School of Physical Sciences of the University of Tasmania, Hobart. He is an author of more than 325 scientific publications including 3 monographs, 10 chapters in books, 26 reviews, 300 regular papers, and 12 patents. Prof. Nesterenko research interests include the development of new advanced adsorbents and technologies for various separation techniques. He is a member of advisory and editorial boards for 7 international journals in the field of Analytical Chemistry and Separation Sciences as well as Editor-in-Chief of Current Chromatography. In 2004 he was awarded E.T.S. Walton professorship of Science Foundation of Ireland. In 2006 Prof. Nesterenko received International Achievement Award for outstanding contributions in the development of Ion Chromatography.

**Nicholas Nunn** graduated from North Carolina State University in the spring of 2014 with a BS degree in Materials Science & Engineering. He also joined Adámas Nanotechnologies in 2012. His primary areas of interest while working at Adámas have been scalable and cost effective means of production, purification, and characterization of both detonation and fluorescent nanodiamonds in commercially viable products. During his time at Adámas, he has authored or co-authored nine peer-reviewed papers.

**Dr. Alexander M. Panich** is currently a Senior Research Scientist at the Department of Physics, Ben-Gurion University of the Negev, Israel. He received his MSc Diploma with Honors in 1971, and PhD Diploma in 1981. He joined Ben-Gurion University of the Negev in 1993. Dr. Panich has published 136 papers in peer-reviewed journals, one monograph, 3 book chapters, and presented 142 papers at international conferences. His current research is mainly focused on investigation of nanomaterials using Nuclear Magnetic Resonance. He is a reviewer of a number of the leading Physical and Materials Science journals.

**Dr. Tristan Petit** obtained his PhD degree in Physics in 2013 from École Normale Supérieure de Cachan for his work on surface modifications of nanodiamonds for biomedical applications conducted at the French Alternative Energies and Atomic Energy Commission (CEA) in Saclay, France. He then joined the Helmholtz-Zentrum Berlin für Materialien und Energie GmbH (HZB), Germany, as an Alexander von

Humboldt postdoctoral fellow. In 2015 he was awarded a Freigeist Fellowship from the Volkswagen foundation to start his junior group at HZB investigating photo- and electrochemical processes at the nanocarbon–water interface using soft X-ray, extreme UV, and infrared spectroscopies.

**Dr. Goutam Pramanik** received his PhD degree in 2013 from the University of Ulm, Germany under the guidance of Prof. Tanja Weil. Currently, he is working in Petr Cigler's research group as a postdoctoral fellow at the Institute of Organic Chemistry and Biochemistry, AS CR, Prague, Czech Republic. His research interests include fundamental issues in the design, synthesis, surface functionalization, and characterization of fluorescent nanoparticles for bioimaging and drug delivery applications.

**Helena Raabova** is a PhD student in Inorganic Chemistry at the University of Chemistry and Technology (UCT) in Prague. She graduated in 2013 from UCT in Inorganic Non-metallic Materials; since 2013 she is a member of Synthetic Nanochemistry Group at the Institute of Organic Chemistry and Biochemistry, AS CR in Prague, Czech Republic, where she focuses on preparation, surface modification, and spectral properties of fluorescent nanodiamonds.

**Bohuslav Rezek** graduated in Physics from the Faculty of Mathematics and Physics at the Charles University in Prague in 1996. In 2001 he obtained his PhD degree at the Academy of Sciences of the Czech Republic (ASCR) for a study of charge transport in silicon thin films with high lateral resolution by using scanning probe techniques. Afterwards he was doing research on diamond and its interfaces in Germany, Switzerland, and Japan. In 2006 he became a research team leader and Purkyne Fellow at the Institute of Physics ASCR. Since 2015 he became the head of Physics Department at the Faculty of Electrical Engineering of the Czech Technical University in Prague. His research is focused on nano-interfaces of semiconductors and organic materials towards opto-electronic and bio-electronic applications.

**Dr. Alexander I. Shames** is a Senior Researcher in charge of the Laboratory of Magnetic Resonance at the Department of Physics, Ben-Gurion University of the Negev. He obtained his MSc degree in Theoretical Physics (*cum laude*) at the Kishinev State University (Moldova) and a PhD degree in Solid State Physics at the Ural State University (Russia). He is the author of over 200 articles in peer reviewed journals and several book chapters. His main scientific interest is magnetic resonance applied to the wide range of objects from novel magnetic materials to nanocarbons including fullerenes, nanoonions, graphenes, and nanodiamonds. He is a co-founder of two hi-tech companies.

**Dr. Olga Shenderova** is the President of Adámas Nanotechnologies, United States. While she started her scientific career in the area of atomistic modeling of carbon nanostructures, within the last decade she has pursued applied research in the fields of nanodiamond particles processing, surface modification, and the development of optical and biological applications of nanodiamonds. She started Adámas in order to accelerate the nanodiamond field by providing cutting edge materials. She has authored over 160 papers, 20 patents/patent applications, and edited 5 books related to nanodiamonds. She received the Nerken Award (2014) for scientific and technological developments of nanodiamond from the American Vacuum Society.

**Stepan Stehlik** graduated in Chemistry from the Faculty of Chemical Technology at the University of Pardubice, Czech Republic, in 2006. He continued there and obtained his PhD degree in Chemistry and Technology of Inorganic Materials in 2010. His research was focused on the synthesis and characterization of silver doped chalcogenide glasses for applications in solid state memories and electrolytes. He joined the group of Bohuslav Rezek at the Academy of Sciences of the Czech Republic in 2011. His research is now focused on modification and size engineering of nanodiamonds as well as on their general characterization with emphasis on scanning probe microscopy techniques. He spent one year (2014–2015) as a SciEx fellow at the University of Basel, Switzerland.

**Dangsheng Su** completed his Physics studies at Jilin University (China) in 1986 and received his PhD degree at the Technical University of Vienna (Austria) in 1991. He moved to the Fritz-Haber-Institut der Max-Planck-Gesellschaft as a postdoc. After a short stay at Hahn-Meitner Institut GmbH and Humboldt Universität zu Berlin, he joined the Fritz-Haber-Institute in 1999, where he worked on nanomaterials in heterogeneous catalysis and electron microscopy. He is now a Professor at the Dalian Institute of Chemical Physics, Chinese Academy of Sciences, Dalian, China. His research interests focus on nanocarbons, nanocatalysis, chemical energy conversion and storage, and electron microscopy. He has published about 400 peer-reviewed papers, several book chapters, and edited several special issues in international journals.

**François Treussart** is a Professor in the Physics Department of Ecole Normale Supérieure de Cachan and the head of the "BioPhotonics" team at Laboratoire Aimé Cotton (Paris Saclay University, Orsay). He was trained in Quantum and Laser Optics. He earned his PhD degree at Laboratoire Kastler Brossel (Paris) by studying high-finesse microcavities for a low threshold laser (thesis supervisor Prof. Serge Haroche). His expertise is in nanophotonics, biophotonics, and single emitters. Since 2003 he studies fluorescent nanodiamond (FND) first to produce single-photon-on-demand and then as bio-imaging labels. With Prof. M. Simonneau (INSERM U894, Paris) he uses FND to probe intraneuronal transport abnormalities induced by genetic risk factors of neurological diseases. He also uses FND as traceable delivery agents of anticancer drug (collaborating with Institut Gustave Roussy, Villejuif).

**Jan Vavra** is an MS student at the University of Chemistry and Technology in Prague. As a member of the Synthetic Nanochemistry Group at the Institute of Organic Chemistry and Biochemistry, AS CR in Prague, Czech Republic, he works on novel approaches to surface coatings of nanodiamonds.

**Igor I. Vlasov** received his MS degree in Physics from the Physics Department of Moscow State University in 1983, and then his PhD degree in Physics from the V.I. Vernadsky Institute of Geochemistry and Analytical Chemistry in 1993. His PhD work was devoted to the development of a new technique of atomic spectroscopy, namely Laser Enhanced Ionization spectroscopy, used for detection of ultra-low concentration of impurities in semiconductors. In 1995 he joined the General Physics Institute (GPI), Russian Academy of Sciences. His scientific interests include developing of advanced carbon nanomaterials for nanophotonics, quantum information processing, biomedicine. He pioneered research works on the search of smallest luminescent nanodiamonds. Now he heads the laboratory of Carbon Nanophotonics.

CHAPTER

# General properties of diamond 1

**Christoph E. Nebel**

*Fraunhofer-Institute for Applied Solid State Physics (IAF), Freiburg, Germany*

**Contents**



## 1.1 INTRODUCTION

Diamond is attracting more and more attention as it is becomes a mature wide-bandgap semiconductor material. The basic properties of diamond have been characterized in detail over a long period of time, and some of the most important parameters are summarized in Table 1.1 in comparison to Si, GaN, and 4H-SiC.[1] Several parameters of diamond are unique like high electron and hole mobilities, both of about 2000 $cm^2/Vs$, thermal conductivity of 25 W/cm·K, electric breakdown field of $10^7$ V/cm, hardness of $10^4$ $kg/mm^2$, and Debye temperature of 1860 K. Diamond is therefore very attractive for a variety of applications due to the discovery of different methods for the production of man-made diamond.

Diamond is nowadays available in single-, poly-, nano-, and ultra-nano-crystalline structure. To fabricate diamond artificially, the high pressure high temperature (HPHT) technique,[2] plasma-discharge-stimulated chemical vapor deposition methods,[3–5] and the hot-filament technique[6] have been optimized during the last 60 years.

The most promising are optical applications (e.g. infrared windows, lenses, ATR units, X-ray windows), thermal applications (heatspreaders, laser submounts, X-ray targets), mechanical applications (cutting tools, scalpels, knives, length gauge tips, wear resistant components, e.g. for textile machines, inserts for dresser tools), electrochemical applications (electrodes, electro-chemical detectors, bio-chemical sensors), radiation sensors (ionizing radiation detectors/dosimeters, fluorescence beam monitors), biolabels, and drug delivery components (diamond nanoparticles, see also Chapter 16). Based on single defects centers in diamond, like the famous nitrogen vacancy center,[7] quantum metrology is a promising field, as is quantum communication and computing (see also Chapter 17). Both can emerge in the near future. Due to the progress of diamond doping, growth, and technology,[8,9] electronic applications

**Table 1.1**

| Properties/Materials | Diamond | Si | 4H-SiC | GaN |
|---|---|---|---|---|
| Bandgap (eV) | 5.47 | 1.12 | 3.26 | 3.5 |
| Electron Mobility ($cm^2$/Vs) bei 300 K | 1,900–2,300 | 1,500 | 900 | 1,250 |
| Hole Mobility ($cm^2$/Vs) bei 300 K | 1,500–2,300 | 600 | 100 | 200 |
| Dielectric Constant | 5.7 | 11.9 | 9.7 | 9.5 |
| Thermal Conductivity (W/cm·K) | 25 | 1.48 | 4.9 | 1.3 |
| Electron Saturation Velocity ($10^7$ cm/s) | 2.7 | 1 | 2.7 | 2.7 |
| Breakdown Field ($10^5$ V/cm) | 100 | 3 | 30 | 30 |
| Debye Temperature (K) | 1,860 | 645 | 1,200 | 608 |
| Hardness (kg/$mm^2$) | 10,000 | 1,000 | 4,000 | |
| Johnson's FoM | 81,000 | 1 | 278 | 215 |
| Baliga's FoM | 25,100 | 1 | 125 | 187 |
| Bipolar Power Switching Product | 1,426,711 | 1 | 748 | 560 |

in power devices are emerging.[10] Here, the missing single-crystalline diamond wafer is a major problem as this is a basic requirement for device fabrication in semiconductor production facilities. However, increasing worldwide activities with respect to hetero-epitaxial growth of single-crystalline diamond on iridium will help to overcome this problem sooner or later.[11–13]

The electronic properties of diamond are determined by the band structure, phonons, intrinsic and extrinsic defects, as well as by dopant atoms. Among natural diamond, boron-doped crystals (named IIb) are available and have been labeled "semiconducting diamond".[14] Boron acts as an acceptor, generating hole-conductivity with an activation energy of 360 meV. Synthetic high pressure high temperature diamond ("HPHT") can contain up to 1000 ppm of boron. Man-made diamond grown by plasma-enhanced CVD-technique can be boron-doped in the regime from $10^{16}$ to $10^{22}$ $cm^{-3}$ by adding gaseous compounds of boron to the plasma.[15] Nitrogen is the dominant extrinsic impurity in diamond.[16] Isolated substitutional nitrogen (P1 center) acts as a deep-donor with an ionization energy of 1.7 eV. In 1997, phosphorus doped CVD-diamond has been grown for the first time by adding phosphine ($PH_3$) to the plasma.[8] The incorporation of phosphorus is varied typically in the range from $10^{16}$ to $10^{20}$ $cm^{-3}$.[17] Spectrally resolved photoconductivity experiments reveal an optical excitation energy of phosphorus of 0.56 eV with a hydrogen-like electronic structure.[18]

## 1.2 PROPERTIES OF INSULATING DIAMOND

In general, no truly intrinsic diamond with a carrier activation energy of about 2.5 eV as expected for the large bandgap of 5.47 eV has been detected up to now. One can estimate the density of free electrons $n$ at 300 K ($k_B T = 25$ meV) in the conduction-

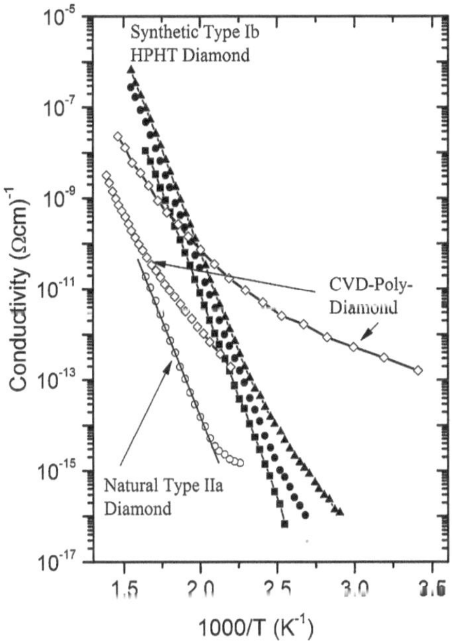


**FIGURE 1.1**

Conductivity properties of natural type IIa, type Ib, and of polycrystalline CVD diamond. The very pure type IIa diamond shows the lowest conductivity which is activated with 1.7 eV.[20] Several Ib-synthetic HPHT diamonds also show 1.7 eV activation energies but they are about three orders of magnitude higher in conductivity.[19] Transport in polycrystalline CVD-diamond is activated with energies increasing with increasing temperature[21]

band by

$$n = N_c \exp\left[-\frac{E_c - E_F}{k_B T}\right] \tag{1.1}$$

where the effective density of states of the conduction band, $N_c$, is assumed to be $\cong 6 \times 10^{19}\ \mathrm{cm}^{-3}$ and $E_c - E_F = 2.5$ eV. This results in an effectively zero ($10^{-19}\ \mathrm{m}^{-3}$) carrier density situation. The electronic properties of diamond will therefore be governed by dopants, defects, extrinsic contaminants, and surface effects even if they are "ultra-low" in density.

A comparison of dark conductivities of very pure diamond (type IIa) with man-made HPHT diamond of type Ib (containing nitrogen),[19,20] and with typical high-quality poly-crystalline PECVD diamond layers[21] is shown in Fig. 1.1. The dark conductivity of type Ib diamonds shows a thermal activation energy of 1.7 eV which is in agreement with the photoexcitation threshold of the P1 center (substitutional nitrogen) of 1.7 eV [22,23] as shown in Fig. 1.2 where the absorption coefficients as functions of photon energies detected for typical type Ib (nitrogen containing HPHT),

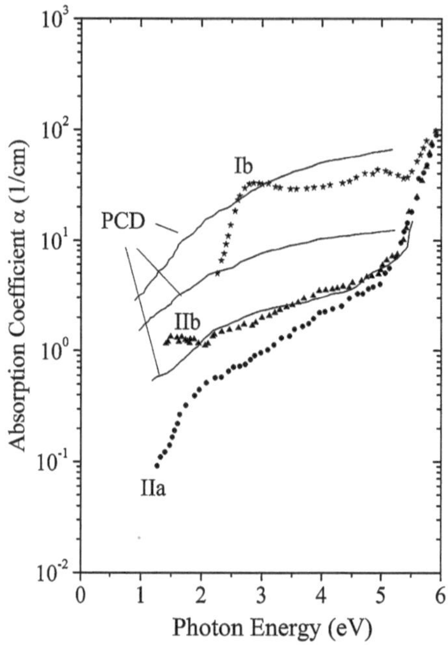


**FIGURE 1.2**

Optical absorption coefficients of type IIa, type Ib, and polycrystalline CVD diamond (PCD) as measured by transmission/reflectivity and photothermal deflection spectroscopy (PDS) experiments

type IIa (ultra-pure diamond), type IIb (natural boron doped diamond) and intrinsic polycrystalline CVD diamond films (PCD) are shown. Based on these data, it is reasonable to assume that the dark conductivity in Ib-diamond is of n-type. The conductivity of type IIa diamond is about three orders of magnitude lower than that of type Ia diamond, but shows the same dark conductivity activation energy of 1.7 eV as for Ib diamond. The purity of IIa diamond is reflected by the very low absorption coefficient in the bandgap regime. We assume that the activation energy of 1.7 eV in type IIa arises from a small amount of substitutional nitrogen atoms, which pin the Fermi level in the bandgap 1.7 eV below the conduction band minimum and cause the n-type nature. It is, however, not possible to generalize this conclusion. Redfield,[24] for example, detected by photo-Hall experiments in several highly resistive layers (comparable to type IIa diamond) a positive photo-Hall-voltage which indicates that some intrinsic diamonds can be of p-type. In the end, it will depend on the extrinsic contamination density of diamond which is either nitrogen or boron, or both.

Fig. 1.1 also shows typical dark conductivities of high-quality intrinsic CVD-polycrystalline diamond layers.[21] No singly activated behavior can be observed. In the temperature regime from 500 to 650 K, the activation energy $E_{act}$ is decreasing by about 1 eV towards lower temperatures. The transport properties of polycrystalline diamond are dominated by conduction in the grains and in grain-boundaries which can be decorated with graphitic states.[25,26] The presence of $sp^2$ states at grain boundaries is causing the continuous absorption spectra shown in Fig. 1.2 and has been discussed in detail in the literature.[25–27]

### Hole mobility, drift velocity, and effective mass

Mobilities of charge carriers in semiconductors are usually measured by the Hall effect technique. This method can, however, only been applied on conductive semiconductors. For highly resistive layers, time-of-flight (ToF) experiments can be applied.[28,29] They are based on a pulsed photoexcitation where electron–hole pairs are generated, for example, by a pulsed laser illumination. Using strongly absorbing light, carriers will be generated close to a contact, and they can be separated by an external applied electric field. Depending on the electric field and the generation of carriers, either electrons or holes are propagating, and their drift velocity and related mobility can be detected as functions of temperature and applied electric field.

Fig. 1.3 summarizes published drift[29–31] and Hall mobilities[32,33] of holes, as detected for highly insulating and natural semiconducting diamond films. The data collected by the different research teams and at very different times show the same tendency. The quality of diamonds (defects, extrinsic contaminations, and dopands) could be slightly different, thereby causing variations as shown. The main features of the shown mobilities are:

1. Below 300 K, the hole drift mobility exhibits a typical $T^{-3/2}$ temperature dependence, expected for acoustic phonon scattering. For higher temperatures, the slopes become steeper, showing a $T^{-s}$ dependence with $s = 2.7$–$2.9$,[25] indicating the onset of intervalley phonon scattering.[22]
2. As shown in Fig. 1.3, for intrinsic diamond, the hole mobility at 290 K is in the range of 1500–2300 $cm^2/Vs$. With increasing electric field strength, the drift velocity of holes exhibits a sublinear increase[30] as shown in Fig. 1.4, which is characteristic for covalent semiconductors.[37] For $T \leq 300$ K, an anisotropic drift velocity is observed, with $v_{d<100>} \geq v_{d<110>}$. Reggiani et al.[34] attribute this result to the warped and nonparabolic features of the valence band. A saturated hole drift velocity of $1.1(\pm 0.1) \times 10^7$ cm/s is detected for electric fields $\geq$25 kV/cm and temperatures $T < 300$ K.
   The limiting drift velocity in a semiconductor with dominant carrier scattering by optical phonons is given by[38]

$$v_{sat} = \left[\frac{8E_{opt}}{3\pi m^*}\right]^{1/2} \tag{1.2}$$

   where $E_{opt}$ is the energy of the optical phonon, and $m^*$ the density of states effective hole mass. Reggiani et al.[34] calculate from these considerations the heavy- (hh) and light- (lh) hole masses to be $m_{hh} = 1.1m_o$ and $m_{lh} = 0.3m_o$, respectively (here $m_o$ is the electron rest mass). A detailed discussion will be given below.

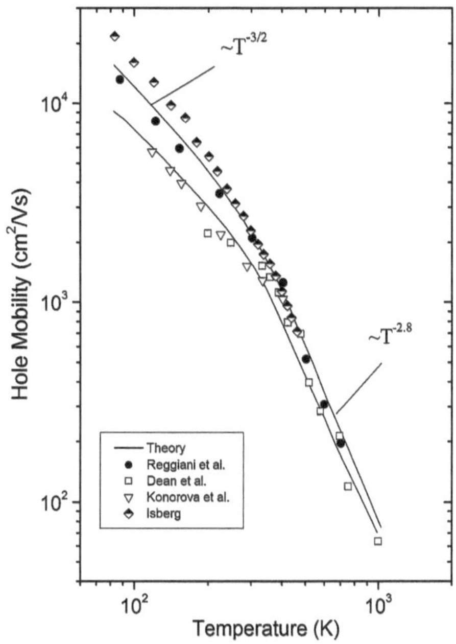


**FIGURE 1.3**

Hole mobilities as a function of temperature in natural diamond. The data have been determined by the Hall effect (Dean[32], Konorova[33]) and time-of-flight experiments (Reggiani[30], Isberg[29]). The continuous curve refers to theoretical calculations by Reggiani et al.[30]

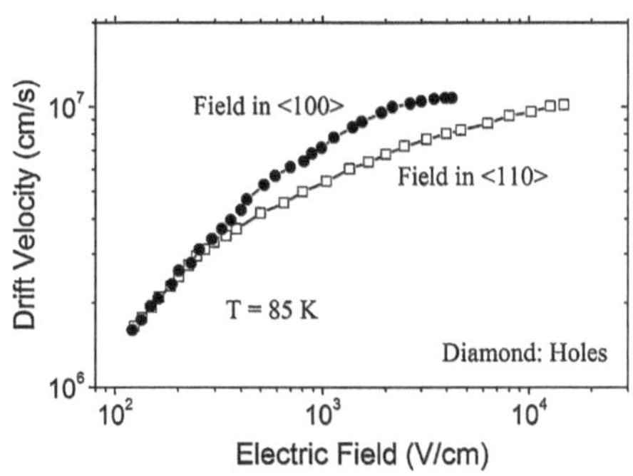


**FIGURE 1.4**

Hole drift velocity as a function of electric field measured at a temperature of 85 K in natural diamond.[30] A characteristic anisotropy $v_{<100>} > v_{<110>}$ is detected

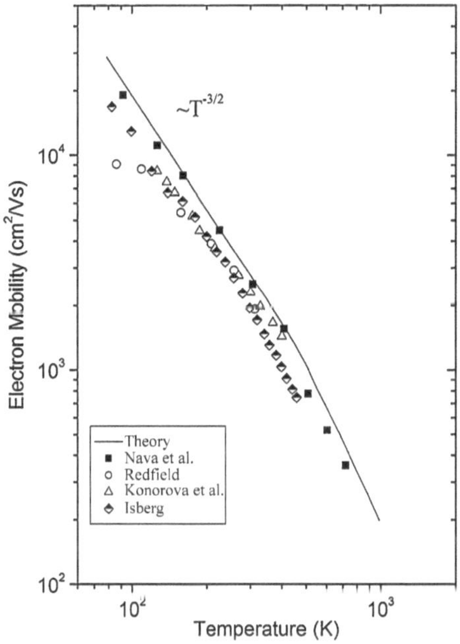


**FIGURE 1.5**

Electron mobility as a function of temperature in natural diamond. The data have been published by Redfield[24] (Hall effect), Konorova[33] (Hall effect), Nava[39], and Isberg[29] (time-of-flight). The solid line refers to theoretical calculations of Nava et al.[39]

## Electron mobility, drift velocity, and effective mass

To characterize electron mobilities, time-of-flight and photo-Hall experiments have been applied on the same samples as shown for holes in Fig. 1.3. The results are summarized in Fig. 1.5.[24,31,33,39] The main features of electron transport have been discussed by Nava et al.[39] and Isberg et al.[31] and can be summarized as follows:

1. As shown in Fig. 1.5, the electron mobility at 300 K is in the regime from 1900 to 2500 $cm^2/Vs$. As for holes, at temperatures <300 K, a typical $T^{-3/2}$ dependence is found (acoustic phonon scattering) with a highest mobility at 92 K of 19000 $cm^2/Vs$. Above 300 K, the slopes becomes steeper, which has been be attributed to phonon induced intervalley scattering.

The electron drift velocity versus electric field (see Fig. 1.6) graph shows also an anisotropy with $v_{<100>}$ larger than $v_{<110>}$. This anisotropy is characteristic for a multi-valley band structure with ellipsoidal constant energy surfaces elongated along equivalent <100> directions. The saturation drift velocity is $1.5(\pm 0.1) \times 10^7$ cm/s for all temperatures below 300 K.

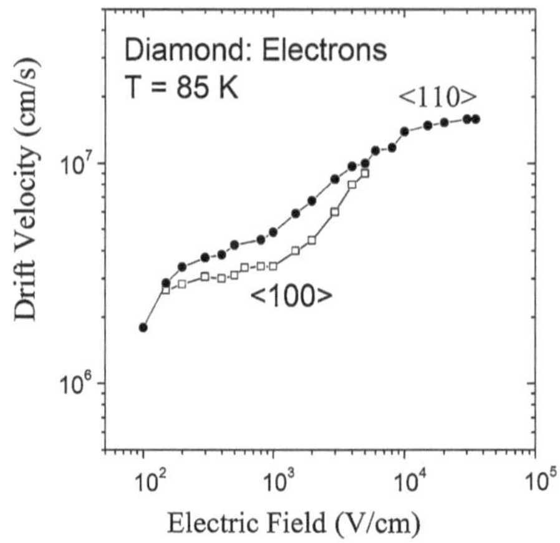


**FIGURE 1.6**

Electron drift velocity as a function of electric field in natural diamond[39] detected at 85 K. The closed and open circles refer to experiments performed along <110> and <100> crystallographic directions

Based on drift mobility data, Nava et al.[39] have estimated the transverse (or $m_\perp$) and longitudinal (or $m_\parallel$) effective mass of electrons in diamond to be $m_t = 0.36m_o$ and $m_l = 1.4m_o$, which is in reasonable agreement with the results from band structure calculations[40] of Willatzen and Cardona who applied linear muffin-tin-orbital and $k \cdot p$ calculations of the effective masses and band structure of semiconducting diamond. They deduced that $m_t$ (or $m_\perp$) $= 0.34m_o$ and $m_l$ (or $m_\parallel$) $= 1.5m_o$.

## 1.3 PROPERTIES OF DOPED DIAMOND

At the very early stage of electronic characterization by Austin and Wolfe in 1956,[41] boron-doped diamond named "natural semiconducting diamond" was investigated by conductivity experiments, which revealed an activation energy of 0.37 eV. At that time aluminum was assumed to be the doping element, which was later corrected to boron. Collins and Lightowlers[42] in 1968 used photothermal ionization spectroscopy to reveal the hydrogen-like nature of the acceptor spectrum with an ionization energy threshold of 0.373 eV.

Nearly 30 years later, in 1997, phosphorus doping of diamond has been demonstrated for the first time generating n-type properties.[8,44] Again, Hall-effect[8,43] and photoionization spectroscopy[18,44] were applied to identify the thermal activation energy of phosphorus to be 0.6 eV as well as its hydrogen-like nature.[45,46] Fig. 1.7 shows a comparison of phosphorus- (n-type) and boron-doped (p-type) diamond containing $5 \times 10^{17}$ cm$^{-3}$ boron and $8 \times 10^{17}$ cm$^{-3}$ phosphorus (data from [17]). Due to the difference in activation energies of 0.37 eV for boron and 0.6 eV for phospho-

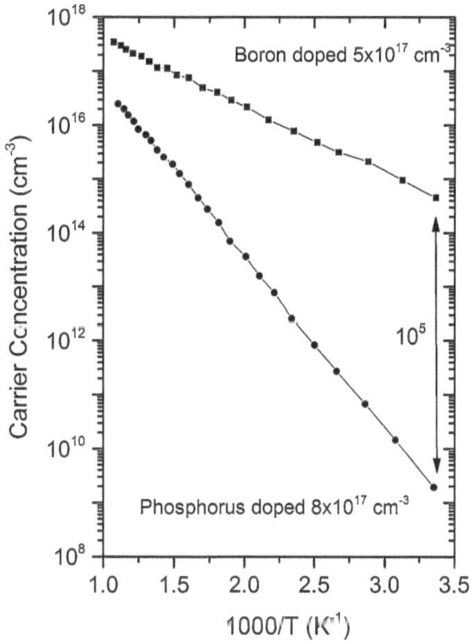


**FIGURE 1.7**

Temperature dependence of carrier concentrations obtained for p-type boron-doped and n-type phosphorus-doped diamond films grown on Ib (001)-oriented substrates. Modified from: Kato, H. Conductivity and Impurity Doping on Single Crystalline Diamond. In Comprehensive Hard Materials, Vol. 3, Sarine, V. K., Nebel, C. E., Eds.; Elsevier, ISBN 978-0-444-63385-9, 2014[17]

rus (both with respect to the conduction band minimum), the carrier concentration difference at room temperature is 5 orders of magnitude.

## Electronic properties of p-type diamond

For a non-degenerate p-type semiconductor (equivalently, for an n-type material) containing an acceptor density, $N_A$, and a density of compensating donors, $N_D$, the density-of-holes $p$ at a given temperature $T$ in the valence band can be calculated using[47]

$$\frac{p(p+N_D)}{(N_A-N_D-p)}=\frac{2}{\beta}\left\{\frac{2\pi m^* k_B T}{h^2}\right\}^{3/2}\exp\left[-\frac{E_A}{k_B T}\right] \tag{1.3}$$

where $\beta$ is the spin degeneracy of acceptors, $m^*$ is the effective density of state hole mass, $E_A$ is the acceptor ionization energy, $k_B$ is the Boltzmann constant, and $T$ the temperature.

## Hole and electron effective density of state masses

To use Eq. (1.3), we need to know the effective density of state hole mass, which can be calculated from the band-structure of diamond. The band-structure of diamond has been discussed by Willatzen and Cardona[40] in 1994 using ab initio calculations based on the linear muffin-tin-orbital method in local-density approximation, which is well known to be one of the most accurate methods to determine the band structure, and so the effective masses, in semiconductors. They showed that the three uppermost valence-bands have their maximum localized at the Γ point of the Brillouin zone. The two upper bands named heavy-hole (hh) and light-hole (lh) bands are degenerated at the Γ point and energetically positioned at a higher energy than the third hole band named split-off band (so). They calculated 13 meV energy separation which is close to the experimentally detected value of 6 meV.[30,34,48] The three valence bands are considered to be parabolic. For temperatures higher than 70 K (>6 meV), the three bands can be considered as degenerate at the Γ point of the Brillouin zone.

## Hole effective masses

The values for the heavy-hole (hh), light-hole (lh) and the split-off (so) masses from Willatzen and Cardona are[40]:

$$
\begin{aligned}
m_{\mathrm{hh}}^{*100} &= 0.427m_{\mathrm{o}},\\
m_{\mathrm{hh}}^{*110} &= 0.69m_{\mathrm{o}},\\
m_{\mathrm{lh}}^{*100} &= 0.366m_{\mathrm{o}},\\
m_{\mathrm{lh}}^{*110} &= 0.276m_{\mathrm{o}},\\
m_{\mathrm{so}}^{*} &= 0.394m_{\mathrm{o}}
\end{aligned}
$$

where $m_{\mathrm{o}}$ is the electron rest-mass.

The corresponding value for the density of state mass of the heavy-hole (hh) is given by[49]

$$m_{\mathrm{hh}}^{*} = \left\{\left(m_{\mathrm{hh}}^{*110}\right)^2 m_{\mathrm{hh}}^{*100}\right\}^{1/3} = 0.588m_{\mathrm{o}}.$$

The density of state light-hole (lh) mass is given by

$$m_{\mathrm{lh}}^{*} = \left\{\left(m_{\mathrm{lh}}^{*110}\right)^2 m_{\mathrm{lh}}^{*100}\right\}^{1/3} = 0.303m_{\mathrm{o}}.$$

The total density of valence band state effective mass $m^*$ can be calculated by[49]

$$m^* = \left\{m_{\mathrm{lh}}^{*\frac{3}{2}} + m_{\mathrm{hh}}^{*\frac{3}{2}} + m_{\mathrm{so}}^{*\frac{3}{2}}\right\}^{\frac{2}{3}} = 0.908m_{\mathrm{o}}. \tag{1.4}$$

## Free hole density for $p << N_A$ and $N_D$

At high temperatures, $p$ approaches the saturation value $N_A - N_D$, and at low temperatures, where $p << N_A, N_D$, Eq. (1.3) results in[51]

$$p \approx \left(\frac{N_A}{N_D} - 1\right)\frac{2}{\beta}\left\{\frac{2\pi m^* k_B T}{h^2}\right\}^{3/2} \exp\left[-\frac{E_A}{k_B T}\right]. \tag{1.5}$$

When an impurity level is created by splitting of states from the conduction or valence band with multiple or degenerate extrema, the impurity level spin degeneracy $\beta$ will be larger than 2. For acceptor levels introduced by group III impurities in Si and Ge, $\beta = 4$, because the heavy-hole and light-hole bands are degenerate at $\boldsymbol{k} = 0$ and the spin-orbit splitting energy $\boldsymbol{\Delta} = 44$ meV in Si and 295 meV in Ge is much larger than $k_B T$, so that in general the split-off band is well-separated and thermally occupied only to a negligible degree.[47,52] In diamond, however, spin-orbit coupling is week, $\boldsymbol{\Delta} = 6$ meV,[30,34,48] so that a threefold band degeneracy with $\beta = 6$ at temperatures above about 70 K can be expected. Below 70 K, a twofold band degeneracy with $\beta = 4$ is a better approximation.

The concentration $p$ of holes is generally obtained from the Hall effect data as

$$p = \frac{r_H}{\mathrm{R\,e}} \tag{1.6}$$

where R is the Hall coefficient, $r_H$ is the Hall factor, and e the elementary charge. For acoustic deformation potential scattering, $r_H = 1.18$ while for ionized impurity scattering $r_H = 1.93$.[53] Most Hall data in the literature have been analyzed by assuming $\beta = 2$ and $r_H = 1.18$. The discrepancy with the correct values discussed above makes relatively little difference for a quantitative analysis, since effectively $(m^*)^{3/2}/\beta$ in Eqs. (1.3) and (1.5) is used as an adjustable parameter to obtain the best fits to the experimental data.

Fig. 1.8 shows a plot of the hole density versus reciprocal temperature for a natural type IIb diamond,[35] showing the predicted features given by Eqs. (1.3) and (1.5). The evaluation results in a boron density of $N_A \cong 8 \times 10^{16}$ cm$^{-3}$, a compensating donor density of $N_D \cong 10^{15}$ cm$^{-3}$, and a boron ionization energy of $E_A \cong 368$ meV, which is in good agreement with the photoionization energy of 373 meV.[42]

### Free hole density for $p >> N_D$

If the acceptor density $N_A$ is significantly larger than the compensating donor density $N_D$ ("nearly uncompensated semiconductor") and for the case where the free hole density in the valence band $p >> N_D$ (which is the case at sufficiently high temperatures), the temperature dependent variation of $p$ is given by[47]:

$$p(T) \cong \left\{\frac{N_V}{\beta}(N_A - N_D)\right\}^{1/2} \exp\left(-\frac{E_A}{2k_B T}\right) \tag{1.7}$$

where $N_V$ is the effective density of valence band states, which varies with temperature as $N_V = N_{Vo} T^{3/2}$.

Fig. 1.9 shows typical hole densities measured as a function of temperature by Hall experiments on samples with different boron doping densities varying from $6.4 \times 10^{16}$ to $1.3 \times 10^{18}$ cm$^{-3}$ up to $10^{20}$ cm$^{-3}$ (data from Teraji et al.[54]). The resistivity of the highest doped sample reaches 0.05 Ω cm at 650 K which is about comparable to values of silicon with a doping density of ca. $10^{18}$ cm$^{-3}$ at the same temperature. The hole concentration shows thermally activated behavior and is increasing with

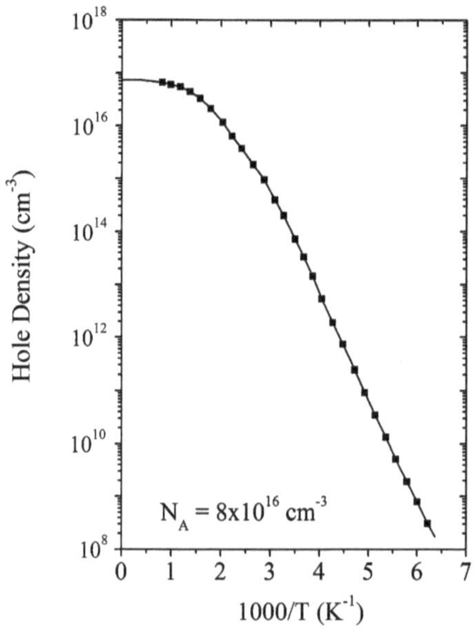


**FIGURE 1.8**

Hole concentration in the valence band as a function of reciprocal temperature.[35] By calculating the fits of Eqs. (1.3) and (1.5) to the data, the authors reveal an acceptor density of $8 \times 10^{16}$ cm$^{-3}$ and the compensating donor density of about $10^{15}$ cm$^{-3}$

increasing doping density. From these data, the boron acceptor activation energy $E_A$ is about 0.36 eV for doping densities below $10^{18}$ cm$^{-3}$. At higher doping densities, $E_A$ decreases gradually and becomes 0.29 eV at ca. $10^{20}$ cm$^{-3}$. It is interesting to note that the activation energy of doped natural conducting diamond is slightly larger, namely 0.368 eV,[35] than of the CVD layers with 0.36 eV activation energy. This may arise by energy level broadening of the excited states of boron due to a constraint that this level is larger in CVD than in natural diamonds.[55]

Hall mobilities measured on the same samples are shown in Fig. 1.10 as a function of temperature. Highest hole mobilities are detected on the sample with the lowest doping density. For a doping density of $6.4 \times 10^{16}$ cm$^{-3}$, it rises from 1620 cm$^2$/Vs at 290 K to 2750 cm$^2$/Vs at 215 K. The Hall mobility $\mu$ is decreasing towards high temperature due to scattering with optical phonons, while at lower temperature scattering with ionized impurities (compensating defects, ionized acceptors) is dominant.

A comparison with hole mobility data measured on undoped diamond film with photoexcitation techniques shows reasonable agreement. The low temperature regime $T < 280$ K is dominated by scattering of holes with ionized impurities in doped films, while this is negligible in undoped diamond. Here the dominant scattering mechanism is acoustic phonon scattering ($T^{-3/2}$). For higher temperatures, the slope becomes steeper, indicating the onset of intervalley phonon scattering in undoped

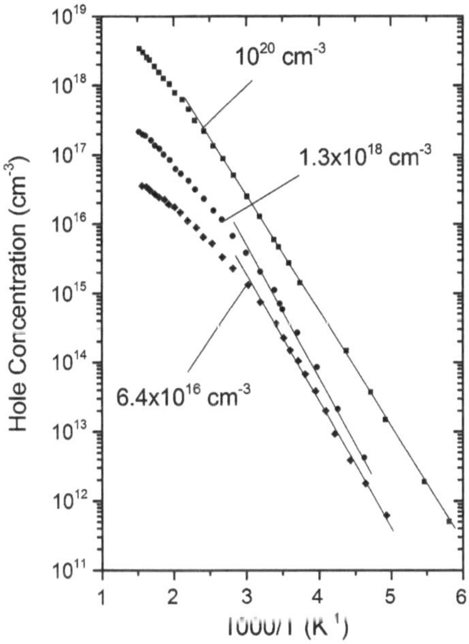


**FIGURE 1.9**

Hole concentration plotted as a function of inverse of temperature for B-doped diamond films grown using high-power MPCVD.[55] The hole concentration shows a thermally activated behavior. The boron acceptor activation energy $E_A$ is 0.36 eV up to $N_A$ of ca. $10^{18}$ cm$^{-3}$; above this density, $E_A$ decreases gradually to become 0.29 eV at $N_A$ of $10^{20}$ cm$^{-3}$

diamond.[29,49] At temperatures >500 K, the scattering with optical phonons becomes dominant in diamond.

## Hole conductivity

The typical dependence of the conductivity on reciprocal temperature for different boron concentrations is shown in Fig. 1.11.[19] The measured electrical conductivity can be described phenomenologically by a combination of various transport mechanisms:

$$\sigma = \sum^{i} \sigma_i . \tag{1.8}$$

In the high temperature regime, holes are thermally excited from acceptor states into the valence band. In this regime, $\sigma$ is thermally activated with $E_{act} = 370$ meV, the acceptor ionization energy. At lower temperatures or higher doping densities, holes

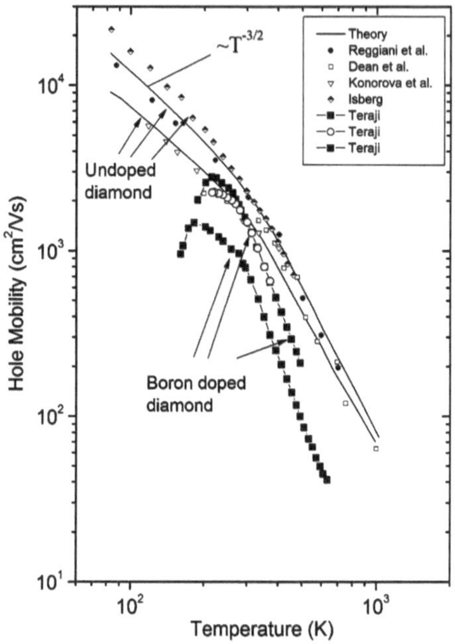


**FIGURE 1.10**

Hall mobilities of boron doped CVD diamond as functions of temperature are shown as detected by Hall-effect measurements and compared to mobilities measured on insulating diamond by photoexcitation techniques. High Hall mobilities of 1620 $cm^2/Vs$ at 290 K and of 2750 $cm^2/Vs$ at 215 K were detected on the $6.4 \times 10^{16}$ $cm^{-3}$ doped layer. The Hall mobilities show a steeper variation as a function of temperature than the data from undoped diamond

propagate via variable range hopping first discussed by Mott[56] where

$$\sigma = \sigma_{hop} \exp\left\{-\left(\frac{T_o}{T}\right)^{1/4}\right\} \tag{1.9}$$

with $T_o$ given by

$$T_o = \frac{18.1}{k_B D(E_F)\alpha^3}. \tag{1.10}$$

$D(E_F)$ is the density-of-states at the Fermi level $E_F$ and $\alpha$ is the effective Bohr radius of boron acceptor levels. The conductivity pre-factor $\sigma_{hop}$ can be calculated by:

$$\sigma_{hop} = e^2 D(E_F) R^2 \nu_{ph} \tag{1.11}$$

where $R$ is the average hopping distance and $\nu_{ph}$ the corresponding exchange frequency (Raman frequency $4 \times 10^{13}$ 1/s). Based on these equations, reasonable fits

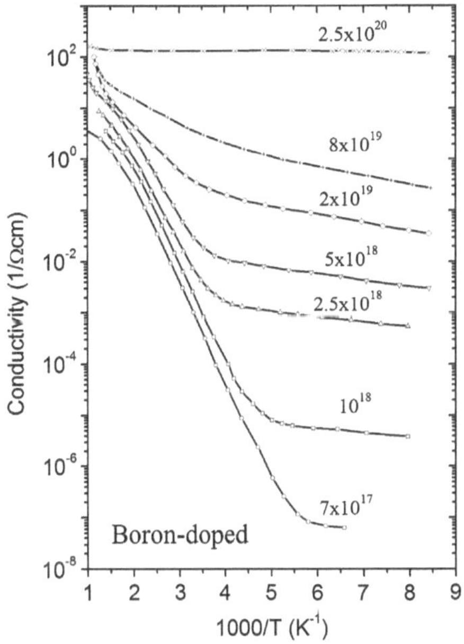


**FIGURE 1.11**

Conductivity data of boron doped diamond.[19] The acceptor density is indicated in the plot in units of $cm^{-3}$. The hopping regime at low temperature is followed by a thermally activated regime where $E_{act} = 370$ meV. With increasing doping, the onset of hopping shifts towards higher temperatures

have been achieved for the low temperature data (see, for example, [19,57,58]) which result in a Bohr radius $\alpha$ of about 6–8 Å. At room temperature, hopping dominates transport at doping levels $\geq 5 \times 10^{19}$ $cm^{-3}$. Here the activation energy drops rapidly towards zero (Fig. 1.12), in conjunction with a decrease of the resistivity over three orders of magnitude (Fig. 1.13). Also shown in Fig. 1.12 is the evolution predicted by the model of Pearson and Bardeen.[59] At acceptor densities exceeding $3 \times 10^{20}$ $cm^{-3}$, a metal–insulator transition takes place.[14,19] Such heavily doped samples show an increasing resistivity with increasing temperature. The metallic resistivity in boron-doped diamond is about $10^{-3}$ to $10^{-2}$ Ω cm.

## Electronic properties of n-type diamond

### The free electron density

The variation of the electron density as a function of temperature as detected by Hall effect measurements on (111)-oriented CVD diamond films with phosphorus densities of $2.3 \times 10^{16}$, $8.9 \times 10^{16}$, and $2 \times 10^{18}$ $cm^{-3}$ is shown in Fig. 1.14.[60] Between 300 and 870 K, these films show negative Hall coefficients, which indicates n-type conduction. The free electron density in the conduction band is exponentially activated

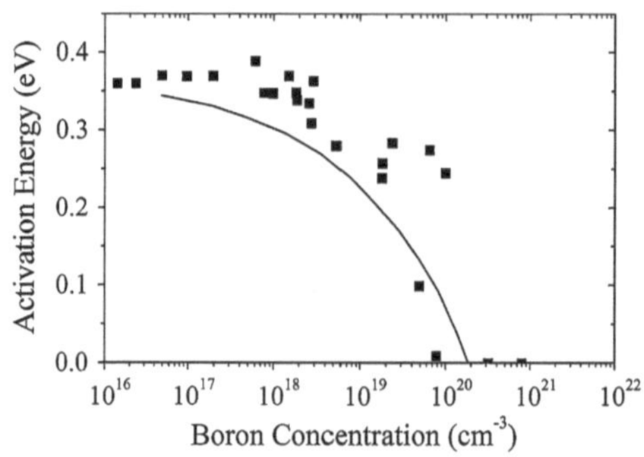


**FIGURE 1.12**

Conductivity activation energies at 300 K as a function of boron content. At about $[B] > 3 \times 10^{18}$ $cm^{-3}$, the activation energy starts to decrease and finally vanishes for $[B] > 2 \times 10^{20}$ $cm^{-3}$. The theoretical dependence according to the model of Pearson and Bardeen (solid curve) is given for comparison (solid line)[59]

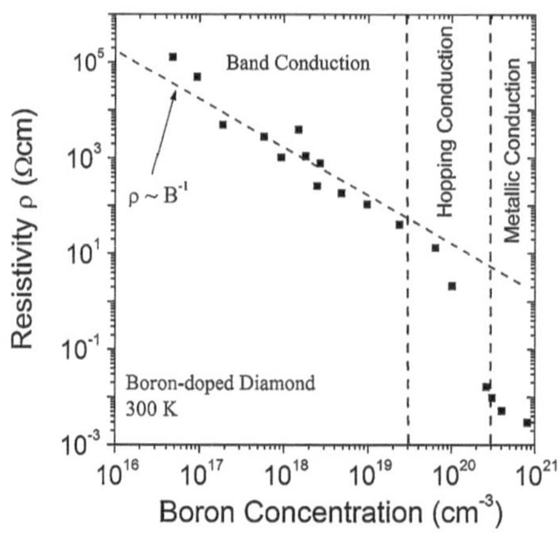


**FIGURE 1.13**

Variation of the resistivity of boron-doped diamond at 300 K with boron-doping levels. Below about $2 \times 10^{19}$ $cm^{-3}$, the decrease is inversely proportional to the acceptor density. In the range from $2 \times 10^{19}$ to $3 \times 10^{20}$ $cm^{-3}$, hopping becomes dominant. For higher concentrations, metallic conductivity is observed

with 0.57 eV in the temperature regime from 300 to 500 K. To evaluate the most important parameters, Eq. (1.12) has been used, where $N_D$ is the phosphorus density, $N_A$ is the compensating acceptor density, $n$ is the free electron density in the conduction band, $N_C$ is the effective conduction band density of states, $g$ is the degeneration factor of donors, $E_D$ is the activation energy of donors, $k_B$ is the Boltzmann constant,

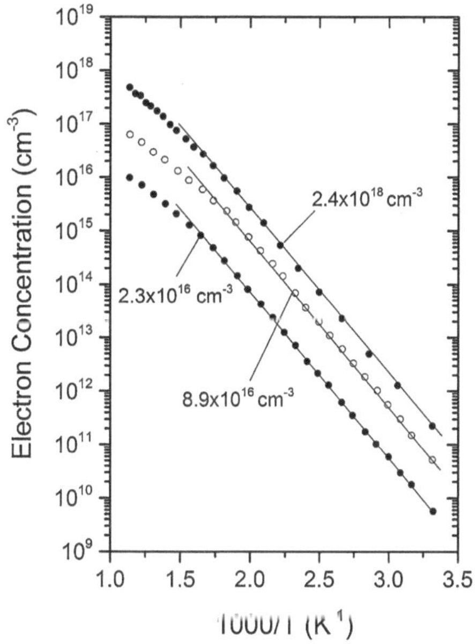


**FIGURE 1.14**

Temperature dependence of the carrier concentrations of P-doped diamond films with various P concentrations is shown.[60] The activation energy of these layers is about 0.57 eV. A compensation ratio of about 0.13 is detected for the film with $8.9 \times 10^{16}$ $cm^{-3}$ donors incorporated

and $T$ is the temperature:

$$\frac{n(n+N_A)}{N_D - N_A - n} = \frac{N_C}{g}\exp\left[-\frac{E_D}{k_B T}\right]. \quad (1.12)$$

For phosphorus-doped films with P concentration of $7 \times 10^{16}$ $cm^{-3}$, best fits to the thermal variation of electrons have been achieved with an activation energy $E_D$ of 0.57 eV, doping density $N_D$ of $6.8 \times 10^{16}$ $cm^{-3}$, and compensating defect density $N_A$ of $8.8 \times 10^{15}$ $cm^{-3}$. This results in a compensation ratio $N_A/N_D$ of 0.13. With decreasing P concentration below $10^{16}$ $cm^{-3}$, the phosphorus doped films become fully compensated and highly resistive. It indicates that the concentration of compensating defects $N_A$ in (111)-oriented CVD diamond is of the order of $10^{15}$ $cm^{-3}$.

## Electron effective mass

For the discussion of electron data, we need to consider six equivalent minima of the conduction band which are located in the Brillouin zone at $2\pi/a(0.742, 0, 0)$ close to the X point. We have to consider two effective masses for the principal directions $m_{\parallel}$

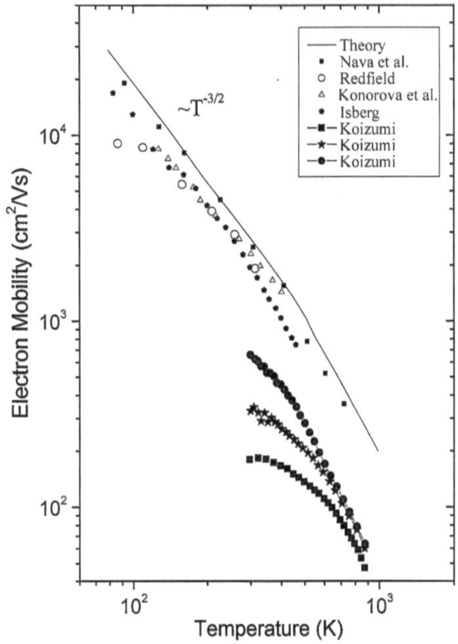


**FIGURE 1.15**

Hall mobilities of different phosphorus-doped diamond films, grown by plasma enhanced CVD[60] are shown in comparison to mobility data measured by photoexcitation in insulating layers. The Hall mobility at RT increases from 410 to 660 $cm^2/Vs$ with decreasing P concentrations from $5 \times 10^{17}$ to $7 \times 10^{16}$ $cm^{-3}$. The Hall mobility for the P-doped film with $7 \times 10^{16}$ $cm^{-3}$ decreases with increasing temperature as $T^{-1.4}$ up to 450 K. This indicates acoustic phonon scattering. At temperatures above 450 K, the Hall mobility is proportional to $T^{-2.6}$ which is attributed to various scattering mechanisms such as intervalley scattering and acoustic phonon scattering

in (100) direction (longitudinal) and $m_\perp$ (transversal) in (010) direction. According to [40], the effective masses are $m_\parallel = 1.5m_o$ and $m_\perp = 0.341m_o$.

Pernot and coworkers[49,50] deduce an effective Hall mobility mass for electrons as

$$m_H^* = \frac{2m_\parallel + m_\perp}{2 + m_\parallel / m_\perp}. \tag{1.13}$$

## Electron Hall mobilities

Fig. 1.15 summarizes typical Hall mobilities of phosphorus-doped films as functions of temperature.[60] The Hall mobility depends on the doping density. At 300 K, it is decreasing from 660 $cm^2/Vs$ for a phosphorus density of $7 \times 10^{16}$ $cm^{-3}$, to 330 $cm^2/Vs$ for $10^{18}$ $cm^{-3}$, to 180 $cm^2/Vs$ for $3 \times 10^{18}$ $cm^{-3}$. A detailed discussion of scattering mechanisms by Pernot and co-workers[51,60] shows that the mobilities at RT are dominated by intravalley acoustic scattering. This arises due to the low density of compensating centers, making ionized and neutral impurity scattering not very effective. At higher temperatures $T > 700$ K intervalley scattering becomes dominant.

Fig. 1.15 shows also electron mobilities detected by time-of-flight and photo-Hall experiments on insulating diamond (see Fig. 1.5). The $T^{-3/2}$ temperature dependence at low temperature has been attributed to acoustical phonon scattering and seems to dominate the mobility in the temperature regime between 80 and 400 K. Pernot[50] concludes that the difference in absolute values cannot be described with a classical model using only one deformation potential related to the conduction band of diamond. He suggests a major difference between experiments performed in thermodynamic equilibrium (Hall experiments) and experiments which are based on photoexcitation.

### n-type conductivity

Finally, typical temperature dependent conductivities $\sigma(T)$ of the n-type diamond films are shown in Fig. 1.16.[61] At lower doping densities ($3 \times 10^{18}$ cm$^{-3}$), the conductivity is singly activated and can be described by

$$\sigma(T) = \sigma_o \exp^{-\frac{E_{act1}}{k_B T}} \tag{1.14}$$

where $\sigma_o$ is the pre-factor of the conductivity, $E_{act1}$ is the activation energy for band conduction, $T$ is the temperature, and $k_B$ is Boltzmann constant.

The pre-factor of conductivity is

$$\sigma_o = q n_o \mu \tag{1.15}$$

where $q$ is the elementary charge, $n_o$ is the density of free electrons in the conduction-band, and $\mu$ is the mobility of electrons. This results in 0.56 eV activation energy and indicates thermal activation of electrons from the phosphorus-doping level to the conduction band in reasonable agreement with Hall effect measurements, which reveal 0.58 eV activation energy. The activation energies $E_{act1}$ measured at high temperatures ($T > 500$ K) are decreasing from 0.56 eV at doping levels $3 \times 10^{18}$ cm$^{-3}$ and $4 \times 10^{19}$ cm$^{-3}$ to about 0.46 eV at $10^{20}$ cm$^{-3}$ due to the temperature-induced decrease of the electron mobility towards higher temperatures.

For higher doping densities, $\sigma(T)$ can be described by a combination of extended state transport ($E_{act1}$) and thermally stimulated hopping transport governed by thermally activated ($E_{act2}$) tunneling transitions ($\alpha R$):

$$\sigma(T) = \sigma_0 \exp^{(-\frac{E_{act1}}{kT})} + \sigma_1 \exp^{(-2\alpha R - \frac{E_{act2}}{kT})}. \tag{1.16}$$

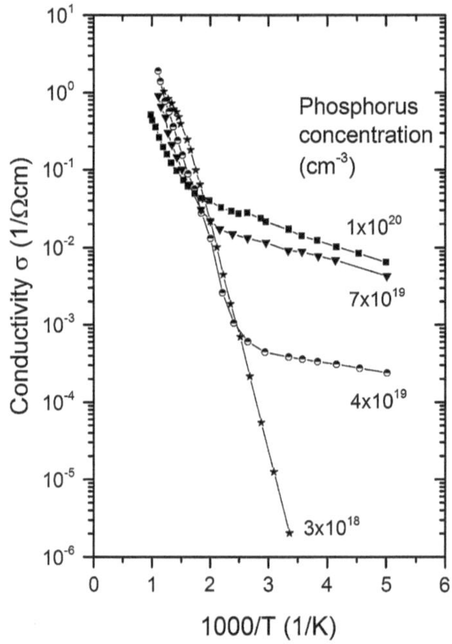


**FIGURE 1.16**

Temperature dependence of the conductivity for P-doped n-type diamond films measured between 200 and 1000 K.[61] The doping densities are indicated. The properties are those of semiconductors with two transport regimes. At lower temperatures the transport is dominated by a thermally activated tunneling transport, whereas at higher temperatures the conductivity is dominated by free electrons in the conduction band

The hopping transport of electrons moving thermally assisted from occupied to nearby unoccupied state is governed the transition probability $P$ which is

$$P(T) = \nu_{\mathrm{r}} \exp^{\left(-2aR-\frac{E_{\mathrm{act2}}}{kT}\right)} \tag{1.17}$$

where $\nu_{\mathrm{r}}$ is the Raman frequency $4 \times 10^{13}$ 1/s, $\alpha$ is the localization length of an electron in a donor state, and $R$ is the distance to the nearest unoccupied state which can also show an energy barrier of $E_{\mathrm{act2}}$ between both states. The discussion of these data by Matsumoto and coworkers[61] reveal a Bohr radius of phosphorus in diamond to be from 3 to 4 Å. This is about half the value of the localization length of holes in the boron level. Evaluating the slope at low temperatures reveals activation energies $E_{\mathrm{act2}}$ of 22 meV for $4 \times 10^{19}$ cm$^{-3}$, 37 meV for $7 \times 10^{19}$ cm$^{-3}$, and 48 meV for the highest doping level of $10^{20}$ cm$^{-3}$. This slight increase with increasing doping is attributed to an increase in Coulomb repulsion. Obviously, the phosphorus-doping level ground-state becomes energetically broader with increasing phosphorus doping.

Due to the Coulomb repulsion effect, a direct correlation with the measured energies is not possible.

## 1.4 SUMMARY

Undoped diamond is an insulator with highest electron and hole mobilities. The conductivity at 300 K is significantly smaller than $10^{-17}$ $\Omega$ cm, showing typical conductivity activation energies of 1.7 eV. This is attributed to small amounts of substitutional nitrogen which dominates the Fermi level. The electron and hole mobilities are very high (>20.000 $cm^2/Vs$ at $T = 80$ K) and are limited by acoustic phonon scattering. With phosphorus and boron doping, diamond becomes semiconducting with an n-type donor activation energy (phosphorus) of 600 meV and an acceptor activation energy (boron) of 370 meV. Both dopands are hydrogen-like in nature. Due to the deep donor levels, conductivity at room temperature is limited, electronic application will therefore be in high-temperature devices. As both electrons and holes show nearly the same mobilities, true bipolar properties could be expected, however, due to the very different doping levels, low temperature devices will be governed by holes with increasing contribution of electrons towards higher temperatures. The effect of the temperature on the carrier mobility can be split into two regimes: at low temperature, acoustic phonon scattering limits the mobility, while at high temperature, intervalley phonon scattering seems to dominate the propagation. Effective masses of carriers have been introduced together with the saturation velocity of electrons and holes. The effect of temperature on the carrier activation from donors and acceptors as well as on transport via hopping propagation has been shown in combination with some basic physical models and descriptions. The results confirm the superior electronic properties of diamond, which makes it very promising for future electronic applications.

## REFERENCES


1. Madelung, O. *Data in Science and Technology, Semiconductors Group IV Elements and III–V Compounds;* Springer, ISBN 3-540-53150-5, 1991.
2. Bundy, F. P.; Hall, H. T.; Strong, H. M.; Wentorf, R. H.; Diamonds, M.-M. *Nature* **1955**, *176* (4471), 51–55.
3. Kiffer, A. D.; Eversole, W. G. *Confidential Memorandum: Synthesis of Diamond From Carbon Monoxide;* Tonawanda Laboratories, Linde Air Products Co, 1956.
4. Angus, J. C. *Diamond Synthesis at Low Pressure*, Internal 3M Report, 1960.
5. Angus, J. C.; Will, H. A.; Stanko, W. S. Growth of Diamond Seed Crystals by Vapor Deposition. *J. Appl. Phys.* **1968**, *39*, 2915–2922.
6. Matsumoto, S.; Sato, Y.; Kamo, M.; Setaka, N. Vapor Deposition of Diamond Particles From Methane. *Jpn. J. Appl. Phys.* **1982**, *21*, L183–L185.

7. Gruber, A.; Drabenstedt, A.; Tietz, C.; Fleury, L.; Wrachtrup, J. Scanning Confocal Optical Microscopy and Magnetic Resonance on Single Defect Centers. *Science* **1997**, *276* (5321), 2012–2014.
8. Koizumi, S.; Kamo, M.; Sato, Y.; Ozaki, H.; Inuzuka, T. Growth and Characterization of Phosphorous Doped {111} Homoepitaxial Diamond Thin Films. *APL* **1997**, *71* (8), 1065–1067.
9. Kato, H.; Yamasaki, S.; Okushi, H. n-Type Doping of (001)-Oriented Single-Crystalline Diamond by Phosphorus. *Appl. Phys. Lett.* **2005**, *86* (22), 222111.
10. Yamasaki, S.; Makino, T.; Takeuchi, D.; Ogura, M.; Kato, H.; Matsumoto, T.; Iwasaki, T.; Hatano, M.; Suzuki, M.; Koizumi, S.; Ohashi, H.; Okushi, H. Potential of Diamond Power Devices. In *Proceedings of the 25th International Symposium on Power Semiconductor Devices & ICs;* Kanazawa, 2013, pp. 307–310, 6.2.
11. Schreck, M.; Roll, H.; Stritzker, B. Diamond/Ir/SrTiO3: A Material Combination for Improved Heteroepitaxial Diamond Films. *Appl. Phys. Lett.* **1999**, *74* (5), 650–652.
12. Ohtsuka, K.; Suzuki, K.; Sawabe, A.; Inuzuka, T. Epitaxial Growth of Diamond on Iridium. *Jpn. J. Appl. Phys. Part 2–Lett.* **1996**, *35* (8B), L1072–L1074.
13. Dai, Z.; Bednarski-Meinke, C.; Loloee, R.; Golding, B. Epitaxial (100) Iridium on A-Plane Sapphire: A System for Wafer-Scale Diamond Heteroepitaxy. *Appl. Phys. Lett.* **2003**, *82* (22), 3847–3849.
14. Williams, A. W.; Lightowler, E. C.; Collins, A. T. Impurity Conduction in Synthetic Semiconducting Diamond. *J. Phys. Part C Solid State Phys.* **1970**, *3* (8), 1727.
15. Nebel, C. E.; Stutzmann, M. Transport Properties of Diamond: Carrier Mobility and Resistivity. In *Properties, Growth and Applications of Diamond;* Nazare, M. H., Neves, A. J., Eds., EMIS Datareviews Series, Vol. 26; Inspec Publication, 2001, pp. 40–52.
16. Newton, M. E. Neutral and Ionized Single Substitutional Nitrogen in Diamond. In *Properties Growth and Applications of Diamond;* Nazare, M. H., Neves, A. J., Eds., EMIS Datareviews Series, Vol. 26; Inspec Publication, 2001, pp. 136–141.
17. Kato, H. Conductivity and Impurity Doping on Single Crystalline Diamond. In Comprehensive Hard Materials, Vol. 3, Sarine, V. K., Nebel, C. E., Eds.; Elsevier, ISBN 978-0-444-63385-9, 2014.
18. Nesladek, M.; Meykens, K.; Haenen, K.; Stals, L. M.; Teraji, T.; Koizumi, S. Low-Temperature Spectroscopic Study of n-Type Diamond. *Phys. Rev. B* **1999**, *59* (23), 14852–14855.
19. Borst, T. H.; Weis, O. Boron-Doped Homoepitaxial Diamond Layers: Fabrication, Characterization, and Electronic Applications. *Phys. Status Solidi A* **1996**, *154*, 423–444.
20. Vandersande, J. W.; Zoltan, L. D. High Temperature Electrical Conductivity Measurements of Natural Diamond and Diamond Films. *Surf. Coat. Technol.* **1991**, *47*, 392–400.
21. Nebel, C. E. Electronic Properties of CVD Diamond. *Semicond. Sci. Technol.* **2003**, *18*, S1–S11, Inst. of Physics Publishing, Guest Editor: R. Jackamn.
22. Dyer, H. B.; Raal, F. A.; du Preez, L.; Loubser, J. H. N. Optical Absorption Features Associated With Paramagnetic Nitrogen in Diamond. *Philos. Mag.* **1965**, *11*, 763–774.
23. Farrer, R. G. On the Substitutional Nitrogen Donor in Diamond. *Solid State Commun.* **1969**, *7*, 685–688.
24. Redfield, A. G. Electronic Hall Effect in Diamond. *Phys. Rev.* **1954**, *94*, 526–537.
25. Nesladek, M.; Vanecek, M.; Stals, L. M. Defect-Induced Optical Absorption in CVD Diamond Films. *Phys. Status Solidi A* **1996**, *154*, 283–303.

26. Rohrer, E.; Nebel, C. E.; Stutzmann, M.; Flöter, A.; Zachai, R.; Jiang, X.; Klages, C.-P. Photoconductivity of Undoped, Nitrogen- and Boron-Doped CVD- and Synthetic Diamond. *Diam. Relat. Mater.* **1998**, *7*, 879–883.
27. Rohrer, E.; Graeff, C. F. O.; Janssen, R.; Nebel, C. E.; Stutzmann, M.; Güttler, H.; Zachai, R. Nitrogen-Related Dopant and Defect States in CVD Diamond. *Phys. Rev. B* **1996**, *54* (11), 7874–7880.
28. Nebel, C. E.; Münz, J.; Stutzmann, M.; Zachai, R.; Güttler, H. Electronic Properties of CVD and Synthetic Diamond. *Phys. Rev. B* **1997**, *55* (15), 9786–9791.
29. Isberg, J.; Hammersberg, J.; Johansson, E.; Wikstrom, T.; Twitchen, D. J.; Whitehead, A. J.; Coe, S. E.; Scarsbrook, G. A. High Carrier Mobility in Single-Crystal Plasma-Deposited Diamond. *Science* **2002**, *297* (5587), 1670–1672.
30. Reggiani, L.; Bosi, S.; Canali, C.; Nava, F.; Kozlov, S. F. Hole-Drift Velocity in Natural Diamond. *Phys. Rev. B* **1981**, *23* (6), 3050–3057.
31. Gabrysch, M.; Majdi, S.; Twitchen, D. J.; Isberg, J. Electron and Hole Drift Velocity in Chemical Vapor Deposition Diamond. *J. Appl. Phys.* **2011**, *109*, 063719.
32. Dean, P. J.; Lightowlers, E. C.; Wight, D. R. Intrinsic and Extrinsic Recombination Radiation From Natural and Synthetic Aluminum-Doped Diamond. *Phys. Rev.* **1965**, *140* (1A), 352–368.
33. Vavilov, V. S.; Konorova, E. A. Semiconducting Diamonds. *Sov. Phys. Usp.* **1976**, *19* (4), 302–316.
34. Reggiani, L.; Waechter, D.; Zukotynski, S. Hall-Coefficient Factor and Inverse Valence-Band Parameters of Holes in Natural Diamond. *Phys. Rev. B* **1983**, *28* (6), 3550–3555.
35. Collins, A. T.; Williams, A. W. S. The Nature of the Acceptor Centre in Semiconducting Diamond. *J. Phys. C Solid State Phys.* **1971**, *4*, 1789–1800.
36. Collins, A. T.; Lightowlers, E. C. Electrical Properties. In *The Properties of Diamond;* Field, J. E., Ed.; Academic Press: London, ISBN 0-12-255350-0, 1979, pp. 79–105.
37. Jacoboni, C.; Reggiani, L. Bulk Hot-Electron Properties of Cubic Semiconductors. *Adv. Phys.* **1979**, *28*, 493–553.
38. Sze, S. M. *Physics of Semiconductor Devices;* John Wiley & Sons, ISBN 0-471-05661-8, 1981, p. 44.
39. Nava, F.; Canali, C.; Jacoboni, C.; Reggiani, L.; Kozlov, S. F. Electron Effective Masses and Lattice Scattering in Natural Diamond. *Solid State Commun.* **1980**, *33*, 475–477.
40. Willatzen, M.; Cardona, M. Linear Muffin-Tin-Orbital and k · p Calculations of Effective Masses and Band Structure of Semiconducting Diamond. *Phys. Rev. B* **1994**, *50* (24), 18054–18059.
41. Austin, I. G.; Wolfe, R. Electrical and Optical Properties of a Semiconducting Diamond. *Proc. Phys. Soc.* **1956**, *69B*, 329–338.
42. Collins, A. T.; Lightowlers, E. C. Photothermal Ionization and Photon-Induced Tunneling in the Acceptor Photoconductivity Spectrum of Semiconducting Diamond. *Phys. Rev.* **1968**, *171* (3), 843–855.
43. Koizumi, S.; Kamo, M.; Sato, Y.; Mita, S.; Sawabe, A.; Reznik, A.; Uzan-Saguy, C.; Kalish, R. Growth and Characterization of Phosphorus Doped n-Type Diamond Thin Films. *Diam. Relat. Mater.* **1998**, *7*, 540–544.
44. Gheeraert, E.; Koizumi, S.; Teraji, T.; Kanda, H.; Nesladek, M. Electronic States of Phosphorus in Diamond. *Diam. Relat. Mater.* **2000**, *9*, 948–951.
45. Gheeraert, E.; Koizumi, S.; Teraji, T.; Kanda, H. Electronic Transitions of Electrons Bound to Phosphorus Donors in Diamond. *Solid State Commun.* **2000**, *113*, 577–580.

46. Gheeraert, E.; Koizumi, S.; Teraji, T.; Kanda, H.; Nesladek, M. Electronic States of Boron and Phosphorus in Diamond. *Phys. Status Solidi A* **1999**, *174*, 39–51.
47. Blakemore, J. S. *Semiconductor Statistics;* Pergamon: Oxford, UK, 1962.
48. Rauch, C. J. Millimeter Cyclotron Resonance Experiments in Diamond. *Phys. Rev. Lett.* **1961**, *7*, 83–84.
49. Pernot, J. *Electrical Transport in Doped Diamond, Diplôme d'habilitation à diriger des recherches;* Université Joseph Fourier: Grenoble, 2010, p. 23.
50. Pernot, J.; Zawadzki, W.; Contreras, S.; Robert, J. L.; Neyret, E.; Di Cioccio, L. Electrical Transport in n-Type 4H Silicon Carbide. *J. Appl. Phys.* **2001**, *90*, 1869–1878.
51. Collins, A. T. The Electrical and Optical Properties of Thin Film Diamond. *Ceram. Int.* **1996**, *22*, 321–327.
52. Sze, S. M. *The Physics of Semiconductors Devices.* 2nd edition; Wiley: New York, USA, ISBN 0-471-05661-8, 1981.
53. Seeger, K. *Semiconductor Physics, An Introduction;* Springer: Berlin, ISBN 3-540-65786, 1973.
54. Teraji, T.; Wada, H.; Yamamoto, M.; Arima, K.; Ito, T. Highly Efficient Doping of Boron Into High-Quality Homoepitaxial Diamond Films. *Diam. Relat. Mater.* **2006**, *15*, 602–606.
55. Nebel, C. E.; Rohrer, E.; Stutzmann, M. Long Living Excited States in Boron Doped Diamond. *J. Appl. Phys.* **2001**, *89*, 2237–2240.
56. Mott, N. F. Conduction in Non-crystalline Materials, Localized States in a Pseudogap and Near Extremities of Conduction and Valence Bands. *Philos. Mag.* **1969**, *19*, 835–852.
57. Massarani, B.; Bourgoin, J. C.; Chrenko, R. M. Hopping Conduction in Semiconducting Diamond. *Phys. Rev. B* **1978**, *17*, 1758–1769.
58. Tsay, Y. F.; Ananthanarayanan, K. P.; Gielisse, P. J.; Mitra, S. S. Electrical Conductivity of Heavily Doped Diamond. *J. Appl. Phys.* **1972**, *43*, 3677–3682.
59. Pearson, G. L.; Bardeen, J. Electrical Properties of Pure Silicon and Silicon Alloys Containing Boron and Phosphorus. *Phys. Rev.* **1949**, *75*, 865–883.
60. Koizumi, S. n-Type Doping of Diamond. In *Physics and Applications of CVD Diamond;* Koizumi, S., Nebel, C. E., Nesladek, M., Eds.; Wiley, ISBN 978-3-527-40801-6, 2008, pp. 237–256.
61. Matsumoto, T.; Kato, H.; Makino, T.; Ogura, M.; Takeuchi, D.; Okushi, H.; Yamasaki, S. Carrier Transport in Homoepitaxial Diamond Films With Heavy Phosphorus Doping. *J. Appl. Phys.* **2014**, *53*, 05FP05.

CHAPTER

# Production and purification of nanodiamonds 2

**Olga Shenderova*, Nicholas Nunn†**

*International Technology Center, Raleigh, NC, United States* Adámas Nanotechnologies, Inc., Raleigh, NC, United States†*

**Contents**



## 2.1 INTRODUCTION

Numerous methods of production of nanodiamond (ND) particles have been invented up until now, including the detonation technique,[1] laser ablation,[2] high-energy ball milling of diamond microcrystals grown at high static pressure and high-temperature (HPHT),[3] chemical vapor deposition (CVD),[4] microplasma-assisted ND formation from ethanol vapor at atmospheric pressure,[5] chlorination of carbides,[6] ion irradiation of graphite,[7] electron irradiation of carbon onions,[8] and ultrasound cavitation.[9] The first three of these methods have been used to produce commercial nanodiamond particles. Nanodiamonds of cosmic origin have been detected in the protoplanetary disks of certain types of stars,[10,11] although the origins of these cosmic sources are still under investigation.

The most widely used of these types of NDs are those of detonation synthesis (DNDs), produced by detonation of carbon-containing explosives, and HPHT nanodiamond particles, obtained by milling of micron-size HPHT particles. While the large scale production of these types of ND particles make them both readily available, due to differences in their structure they are used in different applications. The major differences between DND and HPHT ND are the size of primary particles (monocrystallites) and the state and content of nitrogen impurities in the core of the particles (see Chapter 3). Moreover, due to the difference in the methods of synthesis (bottom up vs top down approaches), the DND and HPHT NDs have distinct morphological differences. DNDs are advantageous due to the small size of primary particles produced (4–5 nm) immediately at the synthesis stage, and the established industrial scale production of ton quantities annually.[12,13] As-produced DNDs form tight, inseparable by sonication aggregates of primary particles. However, due to recent advances in DND deaggregation by milling,[14] colloidal solutions of primary ND particles 4–5 nm in diameter with spherical morphologies have become readily available. Monocrystalline ND particles obtained by processing of HPHT synthetic diamond are available with the smallest average particle size around 10–20 nm; these particles exhibit irregular shapes. One of the identifying characteristics of HPHT NDs is the presence and structure of nitrogen impurities. HPHT NDs of type Ib have nitrogen impurities primarily in the substitutional state (Ns) at the 100–200 ppm level. Type Ia NDs produced by milling of natural diamond exhibit nitrogen primarily in the aggregated state (A centers, or N pairs, for example) up to the 3000 ppm level. High-energy irradiation of type Ib NDs to form vacancies followed by annealing causes the formation of nitrogen-vacancy (NV) color centers with red emission[15] (see Chapter 7), while type Ia NDs exhibit green luminescence originating from formation of NVN (H3) centers after irradiation and annealing.[16] NDs synthesized from explosives have not been the preferred candidates for imaging applications based on NV centers since the high N content (up to 2–3 wt%) exists primarily as aggregates of N which render DNDs optically inactive even with irradiation and annealing.[17]

The kilogram quantity availability of single-digit ND particles produced by detonation synthesis has opened broad prospective applications of NDs in composites, lubricants, and as drug delivery vehicles – topics discussed in the subsequent chapters of this book (see Chapters 14, 15, and 16). For HPHT NDs, a central focus of many research groups remains the controlled production of nitrogen-vacancy centers in nanoscale diamond for applications as fluorescent biomarkers with high photostability, magnetic sensors with nanoscale resolution,[18] or as efficient single pho-

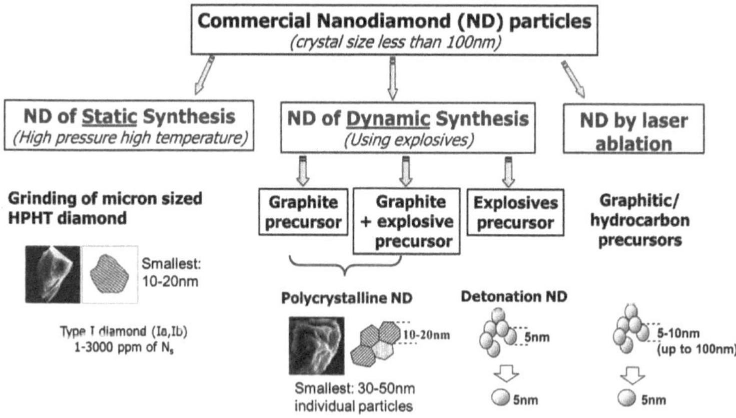


**FIGURE 2.1**

Summary of types of commercial nanodiamond

ton emitters for quantum information processing.[19] The synthesis of NDs of a few nanometers in size with specific color centers remains an important goal, and perhaps new methods of nanodiamond particle synthesis are needed to address this goal. Additionally, the synthesis of boron-doped nanodiamonds, which are conductive and can be used in electroanalysis, electrochemical double-layer capacitors, and batteries, is an important goal.

Here we present a brief survey of the different classes of nanodiamond particles and review some of the well-established and novel methods of synthesis. Modifications of well-established methods, such as detonation synthesis, that allow for production of ND particles with new properties, for example, with smaller primary particle sizes[20] and issues related to doping of NDs and control of the doping process will be discussed. Important processing steps, including purification of ND from $sp^2$ carbon and metallic impurities as well as deagglomeration and fractionation of ND particles, will be also surveyed. We will conclude by summarizing current challenges in the fields of production and processing of nanodiamond particles and the outlook on the future of the field.

## 2.2 TYPES OF NANODIAMONDS AND METHODS OF NANODIAMOND SYNTHESIS

Commercially available ND particles based on their method of synthesis are summarized in Fig. 2.1. Major classes of commercial NDs are related to "static" or "dynamic" synthesis of diamond material, where the former is performed at a high temperature under high static pressure maintained for minutes or hours, and the latter involves a highly non-equilibrium process wherein the high temperature and high pressure conditions exist within a fraction of a microsecond. In order to be commercially viable, particles of diamond produced in static HPHT synthesis using commercial HPHT presses must be on the order of several hundreds of microns or larger in size. Nanodiamonds are obtained by crushing these micron-sized particles. Nanoparticles processed from HPHT synthetic diamond or from natural diamond powders are monocrystalline (Fig. 2.1) and are commercially available with the smallest average particle size around 20 nm (produced, for example, by Microdiamant AG). While ND of static synthesis is by an order of magnitude more expensive than ND of dynamic synthesis, it is a primary material for production of photoluminescent ND particles, where the price of the starting material is an insignificant fraction of the overall production cost.

A few subgroups of nanodiamond are produced via dynamic synthesis, harnessing the energy of explosives and different carbon-containing precursor materials. Three classes of dynamically synthesized NDs manufactured on an industrial scale are summarized in Fig. 2.1 and are as follows: (i) NDs produced by shock compression of graphite mixed with metals[21] (developed by DuPont de Nemours); (ii) NDs produced from a mixture of a carbon precursor and explosives[22]; and (iii) NDs produced from the carbon contained in high-energy explosives themselves (detonation ND).[12,13] Depending on the precursor material used in dynamic synthesis, NDs with primary crystal sizes between 3–25 nm can be synthesized. Based on the carbon precursor, diamond particles can be polycrystalline or consist of "clusters" of primary particles (Fig. 2.1). Polycrystalline nanodiamond powder is processed from micron-sized polycrystalline diamond particles obtained by shock compression of graphite or by detonation of a mixture of graphite and explosives. Polycrystalline particles consist of diamond grains ~10–20 nm in size,[17] forming monolithic particles without pores. The finest polycrystalline diamond fraction produced by micronization and subsequent grading has an average size of individual particles of ~25 nm when dispersed in a variety of solvents. Aggregates of 3–6 nm primary particles of detonation ND encompass the size range from 10 nm up to several hundred nanometers. Contrary to the polycrystalline diamond particles, aggregates of detonation ND contain nanometer-sized pores within clusters of primary particles.[23] Slurries of completely disaggregated 4–6 nm individual DND particles are commercially available.

While the first publications on laser irradiation of graphite precursors in liquids appeared more than a decade ago,[24] ND produced using this method became commercially available only recently (Ray Techniques Inc.). Primary particle size, as reported by the vendor, ranges between 2 and 10 nm,[25,26] and aggregate sizes in water suspensions are about 100 nm. At the moment, this type of ND is an order of magnitude more expensive than detonation ND. The appealing features of NDs produced by laser-assisted synthesis are their high purity, possibility to vary primary particle size by varying laser radiation parameters, and perspectives of controlled doping.

According to their sizes, ND particles can be tentatively categorized into three groups[13]: 1–2 nm molecular forms (diamondoids[21]), single-digit particles with sizes below 10 nm, and nanodiamond particles with sizes above 10 nm. The average size of primary detonation ND particles produced by most vendors lies in the "single-digit" range of 3–6 nm. NDs produced by laser ablation of carbon precursors are also within the single-digit size range. Single-digit HPHT ND has been also recently produced.[27] Details on the methods of synthesis of other single-digit nanodiamond particles are discussed in Section 2.4. The class of nanocrystalline diamond with particles exceeding 10 nm includes both monocrystalline and polycrystalline particles. As discussed above (Fig. 2.1), the monocrystalline forms include ND particles processed from HPHT synthetic diamond, while polycrystalline nanodiamond powder is processed from micron-sized polycrystalline diamond particles obtained by shock compression of graphite or by detonation of a mixture of graphite and explosives. The diamondoid molecule class includes higher and lower diamondoids.[28,29] Higher diamondoids are highly rigid, well-defined, hydrogen-terminated diamond species.[28] With more than 3 crystal diamond cages, higher diamondoids are intermediate in size to the adamantane molecule – the smallest species of H-terminated cubic diamond containing only 10 carbon atoms – and DND particles with sizes more than 3 nm as described above. Higher diamondoids are extracted from petroleum as diamond molecules in the form of nanometer-sized rods, helices, discs, pyramids, etc. So far it has not been possible to synthesize higher diamondoids except anti-tetramantane, a tetramantane isomer.[29] By comparison, lower diamondoids (adamantane, diadamantane, and triadamantane), extracted from crude oil much earlier than the larger members of the diamondoid series, are currently available in kilogram quantities[29] and can be synthesized.

The sections below provide more details on the manufacturing of ND of dynamic and static synthesis, followed by a brief survey of methods of synthesis of ND particles produced at a laboratory scale.

## 2.2.1 NANODIAMOND PARTICLES PRODUCED BY DETONATION SHOCK-WAVE-ASSISTED SYNTHESIS

We briefly survey methods of synthesis and characteristics (crystal size, morphology) of the three classes of dynamically synthesized NDs manufactured on an industrial scale. The history of the discovery of dynamically synthesized NDs is very fascinating and can be found in [30,31].

### *2.2.1.1 Nanodiamond produced from high energy explosives*

An in-depth elaboration of the mechanism of detonation ND synthesis was provided by Danilenko in several recent publications.[32–34] The approach is based on analysis of the *nanocarbon* phase diagram in combination with detonation parameters (pressure and temperature) for high-energy explosives. At the nanoscale, the carbon phase diagram must also include cluster size as a third parameter (in addition to pressure and temperature) because the Gibbs free energy depends on the surface energy, which leads to changes in the phase diagram[33–35] (Fig. 2.2A). The liquid carbon region is shifted to lower temperatures for nanocarbon, and the region of nanodiamond stability is slightly shifted to higher pressures (Fig. 2.2A). The diagram suggests that

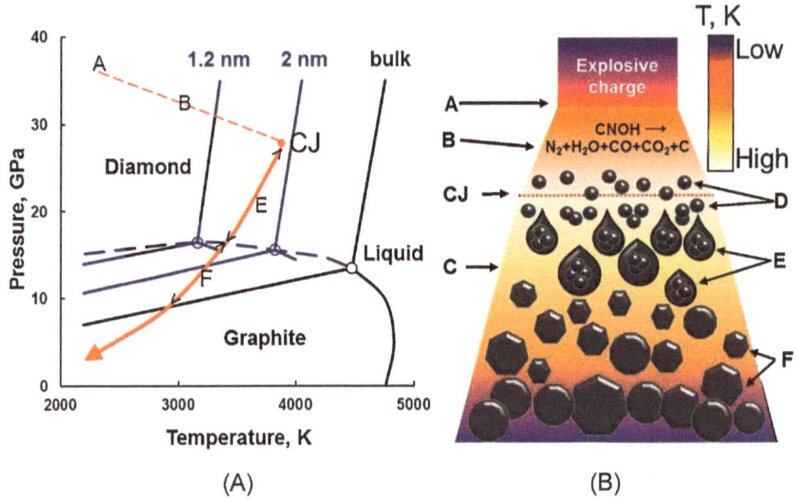


**FIGURE 2.2**

Detonation synthesis of nanodiamond: A The phase diagrams for bulk and nanoscale (blue lines) carbon. Red lines demonstrate a pressure–temperature trajectory in a detonation wave during synthesis of detonation nanodiamond (for the composition B, RDX/TNT 64/36%). B Schematic of the detonation wave propagation showing A the front of the shock wave caused by the explosion; B the zone of chemical reaction; the Chapman–Jouguet plane; C the expanding detonation products; D the formation of carbon nanoclusters; E the coagulation into liquid nanodroplets; and F the crystallization, growth and agglomeration of nanodiamonds. Reprinted by permission from Mochalin et al., The Properties and Applications of Nanodiamonds. Nat. Nanotechnol. 2012, 7, 11

carbon particles, on the order of $10^3$–$10^4$ atoms, can exist in the liquid state at lower temperatures than bulk carbon (Fig. 2.2A).

The suggested mechanism of DND formation is as follows[32–34]: The initial shock wave created by detonation compresses the high-explosive material, heating it and causing chemical decomposition, releasing enormous amounts of energy in a fraction of a microsecond. The detonation wave includes (Fig. 2.2B)[36] the front of the shock wave caused by the explosion (plane A in Fig. 2.2B and point A in Fig. 2.2A), a zone of detonation-induced decomposition of explosive molecules (a zone of chemical reaction, zone B), the Chapman–Jouguet (CJ) plane (where P and T correspond to point CJ in Fig. 2.2A, corresponding to the conditions when reaction and energy release are essentially complete), and a zone of expanding detonation products (zone C). The zones of chemical reaction and expanding detonation products are shown for the so-called composition B explosive in the corresponding phase diagram (Fig. 2.2A, red lines). The decomposition reaction of explosive molecules and formation of free carbon (for an explosive with a negative oxygen balance) proceeds according to the formula

$$CNOH \rightarrow N_2 + H_2O + CO + CO_2 + C.$$

It can be seen in Fig. 2.2A that the P–T values at the CJ point do not reach the region of liquid carbon for a bulk carbon phase diagram, while conditions for the existence of liquid carbon can be achieved on the phase diagram for nanoscale carbon. Importantly, the CJ point of several powerful explosives is located in the region of liquid carbon in the nanocarbon phase diagrams.[32] Pressures and temperatures at the CJ point are not high enough to produce liquid from bulk carbon, but they are high enough to produce liquid carbon at the nanoscale. Thus, it is suggested that nanodiamond is formed by homogeneous nucleation in the volume of the supersaturated carbon vapor via condensation and crystallization of liquid carbon (Fig. 2.2B). After decomposition of the explosive molecules (during ~1 ns), primary carbon clusters (<2 nm in size) are formed in the zone of a chemical reaction (during ~10 ns) (zone D, Fig. 2.2B), followed by their coagulation into carbon droplets >2 nm in size along the isentrope of expanding detonation products above 16.5 GPa (this stage lasts ~$10^2$–$10^3$ ns) (zone E in Figs. 2.2a and 2.2b). Finally, nanodiamond particles are formed by crystallization of the liquid carbon droplets along the isentrope of expanding detonation products in the region 9–16.5 GPa (zone F in Figs 2.2A (outlined by arrows), and 2.2b). The duration of this stage is longer than $10^3$ ns. When the pressure drops below the diamond–graphite equilibrium line (~9 GPa), the growth of diamond is replaced by the formation of graphite. High pressure high temperature conditions favorable for formation of diamond last only a fraction of a microsecond, restricting the growth of diamond crystallites to a size of a few nanometers only. Based on X-ray diffraction (XRD) analysis, the size distribution of nanodiamond crystallites ranges between 2 and 10 nm with a sharp peak at 4–5 nm. Small-angle X-ray scattering (SAXS) and high resolution transmission electron microscopy (HRTEM) analysis also indicates the presence of a small fraction of monocrystals with sizes ~10–20 nm.[17,37] After crystallization, the ND particles continue to move along with gaseous products of detonation in highly turbulent conditions, colliding with each other and the walls of a detonation chamber and forming tight aggregates of primary particles. Tight DND aggregates formed during synthesis drastically differ from many other types of nanoparticles that agglomerate due to electrostatic or van der Waals forces when in solution or during drying from a solution.

From a practical viewpoint, the conversion of the carbon-containing explosive compounds into diamond occurs by firing the explosives in a detonation chamber (Fig. 2.3A) with a typical volume of 2–5 $m^3$ (for 0.1–1 kg charges (Fig. 2.3B)).[38] The resultant product is a mixture of diamond particles with a primary particle size of 4–5 nm, other carbon allotropes, and metallic impurities.[38,39] A wide variety of explosive materials may be used. A typical explosive mixture is TNT (2-methyl-1,3,5-trinitrobenzene) and RDX (1,3,5-trinitroperhydro-1,3,5-triazine) in a proportion ranging from 40% to 60% TNT. The explosion takes place in an inert medium of either gas ($N_2$, $CO_2$, Ar, or other gases) or water (ice), called "dry" or "wet" synthesis, respectively. The medium acts as a coolant.

**FIGURE 2.3**

Photos of detonation chamber with volume ~2 $m^3$ for detonation of 1 kg charge (A) and a detonation charge consisting of a mixture of TNT/RDX (B). Photo courtesy of Dr. V. Dolmatov, Diamond Center, Russia

The commercial product of detonation synthesis, called detonation soot, contains 30 to 80 wt% diamond phase depending on the detonation conditions. The detonation soot yield is 4% to 10% of the weight of the explosive charge, depending on the cooling media (gas versus water or ice).[38,39] The yield of diamond resulting from detonation of 100 g charges of different composition in $N_2$ atmosphere at different pressure of $N_2$ was thoroughly investigated by Kuznetsov et al.[37] Using dark-field imaging, unique information related to the distribution of primary ND particles and their "primary" aggregates directly within the matrix of detonation soot was obtained (Fig. 2.4). Dark-field images of the graphite lattice with (0002) Bragg reflections indicate the scarce presence of nanographite particles (Fig. 2.5B) while (111) diamond reflections (Fig. 2.5C) indicate the distribution of ND particles within the amorphous matrix of soot. ND aggregates of few tens of nanometers are produced in abundance during synthesis under $N_2$ (~10 atm.) and TNT/RDX with a 50/50 charge composition (Fig. 2.5C). Fig. 2.5D demonstrates significant increase of ND fraction in the soot for TNT/RDX composition 50/50 in comparison with the composition 80/20 due to the higher power of the mixture and longer duration of the synthesis process in the diamond region in the phase diagram. Interestingly, based on the XRD data, it was observed that the size of ND primary particles can be varied by modifying the charge composition (Fig. 2.5d) or pressure of the gas cooling media (Fig. 2.5E). Fig. 2.5E

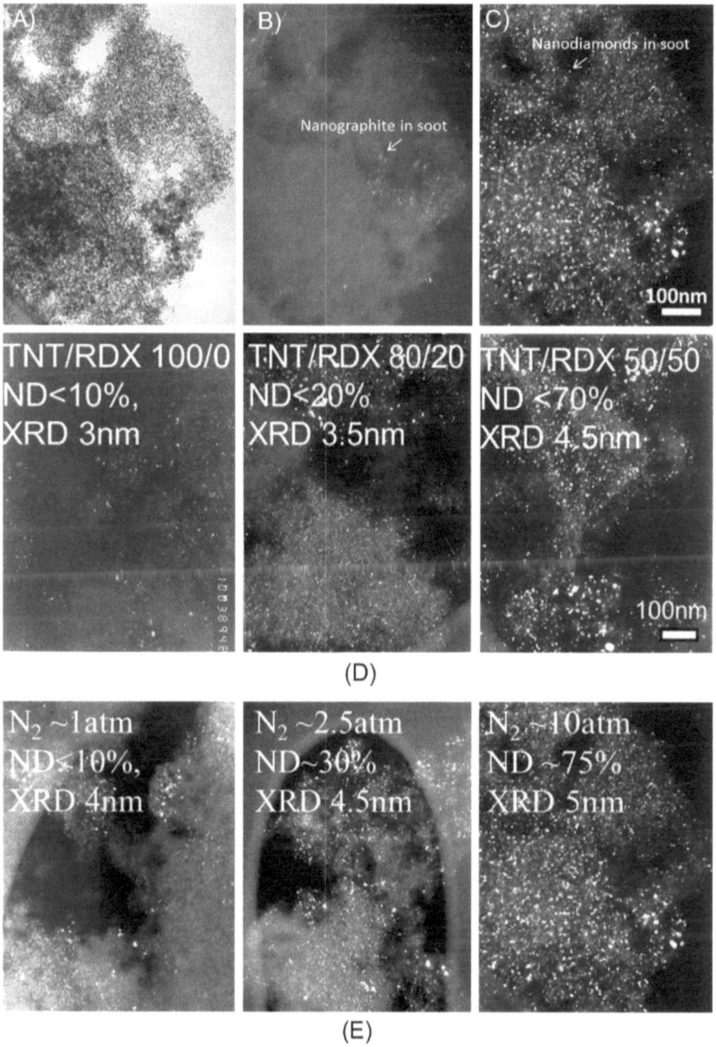


**FIGURE 2.4**

Bright-field (A) and dark-field (B,C) TEM micrographs of the same fragment of soot obtained by detonation of TNT/RDX mixture in $N_2$ cooling media at $N_2$ pressure ~10 atm. Dark-field images indicate the distribution of nanographite (B) and ND (C) particles within the amorphous matrix of soot. Dark-field TEM micrographs of the soot demonstrating the distribution of ND in the soot for different synthesis parameters: (D) detonation in the $N_2$ atmosphere under the pressure ~2.5 atm for different compositions of the charge; (E) detonation of the charge TNT/RDX 50/50 for different pressures of the $N_2$ cooling media. Courtesy of Dr. V. Kuznetsov, BIC, Novosibirsk, Russia. Adapted from: Kuznetsov, V. L., et al. Chem. Phys. Lett. 1994, 222, 343[37]

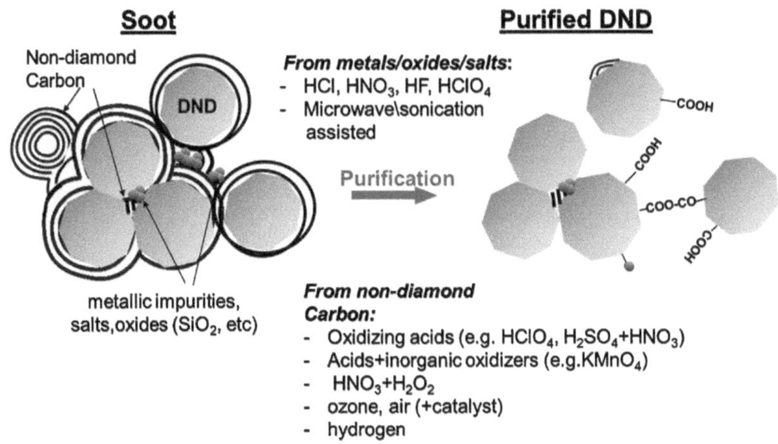


**FIGURE 2.5**

Tentative scheme of major structural components of detonation soot (left) and purified commercial DND product (right). Non-diamond carbon of detonation soot and metallic, incombustible impurities, may be located externally to the tight DND aggregates and be removed during the purification process or can be confined within tight aggregates and remain inaccessible to oxidizing media

illustrates a significant increase of the yield of ND in soot with an increase of the pressure of the gas cooling media.

Further details on the synthesis of DND can be found in books [12,13,32,39,40] and reviews [38,41–44]. Recently, DND synthesis was revisited by several groups to address new demands for this material. Major focuses for novel approaches for DND synthesis are as follows: (i) further reduction of the primary particle size; (ii) control of the level of aggregation at the synthesis stage; (iii) increased $sp^3$ carbon content in detonation soot (via reduction of non-carbon elements and $sp^2$ carbon on DND particle surfaces), and (iv) doping of DND during synthesis, including control of the amount of substitutional nitrogen in the diamond core.

In principle, X-ray diffraction (XRD) data on the size distribution of primary DND particles within detonation soot, which were reported in a limited number of publications,[32,45] include 2 nm peaks. It was reported by Dolmatov[45] that the novel method of using reducing agents (for example, urea and ammonia) in water cooling media allows for the preservation of the DND fraction with a primary particle size less than 3 nm along with a 'typical' fraction of 5 nm primary particles. This method also provides the benefit of increased DND yield from soot (by 2–5 wt%) as well as the overall yield of soot itself (by 100%). It is proposed that the presence of readily oxidizable reducing agents among the detonation products helps to reduce graphitization of DND. It is also reported that the presence of urotropine significantly reduces the amount of incombustible impurities in DND (down to 0.1–0.3 wt%) since urotropine is a strong complexing agent and captures impurities (metals Fe, Ni, Cr,

Cu, Ti, etc.) to form soluble complexes which are easy to remove by subsequent chemical purification, e.g., treatment with nitric acid.

Recently, Pichot et al. investigated the role of the microstructure of the explosive, particularly the size of TNT particles in the mixture with a higher energy explosive, on the size of produced ND primary particles.[20,46] Using the same explosive compositions and charge densities, smaller nanodiamonds (~4.2 nm diameter) were obtained from nanostructured explosives (~100 nm TNT particles in a mixture of RDX/TNT).[46] Microstructured explosives lead to the formation of NDs with a mean diameter of 6.2 nm. In a subsequent study, using nanostructured octolite (octogen/TNT)) with TNT particles size ~40 nm, ND particles with diameters as small as ~2.8 nm were synthesized – confirmed from HRTEM analysis of more than 1000 particles. The authors attributed this phenomenon to a higher nucleation density of NDs in the nanostructured explosives.

Another question relates to the possibility of optimizing the detonation process in order to produce mostly isolated primary particles and only small size aggregates in the detonation soot. This would significantly reduce the cost of the final DND product. In fact, it has been recognized that the dry DND synthesis results in smaller primary DND particle sizes and smaller average aggregate sizes compared to wet synthesis. Another factor influencing the aggregation of DNDs during synthesis is the mass of the charge and the ratio between masses of the charge and wet cooling media used.[47]

There is also a direction of research toward control of ND elemental surface composition starting from the synthesis step. Using a hydrogen-free explosive during detonation, benzotrifuroxan (BTF, C6N6O6), Batsanov et al. produced hydrogen-free ND particles using perchloric acid for purification of ND from graphitic soot.[48] RDX and TNT explosives contain 2.2% and 2.7% of hydrogen, respectively, which, in the high-pressure/high-temperature zone, can contaminate not only the surface, but also the interior of nanoparticles. Particles of hydrogen-free ND were larger (~20 nm) than typical DND due to higher detonation temperature of BTF and the corresponding longer growth time of the diamond phase.

Tailoring DND electronic and optical properties for specific applications can possibly be accomplished at the stage of synthesis by development of different combinations of explosive materials containing solids which are known dopants in diamond, as well as non-traditional cooling media (both gaseous and liquid) containing additives to alter the DND composition (both bulk defects/doping content and surface groups). The possibility to control the N content by a proper choice of the precursor material[17] is discussed in Section 2.5. Of particular importance would be synthesis of boron-doped nanodiamonds, which are electrically conducting.

#### 2.2.1.2 Nanodiamond produced from carbon and carbon/explosive mixture

Under suitable conditions, shock waves produced by explosives in confined volumes for sufficient durations can create high pressure (~20–200 GPa) and high temperature (>2000 K) that will result in partial conversion of graphite into nanometer-sized diamond grains (~20 nm); these grains compact into micron-sized, polycrystalline particles. Copper is mixed with graphite in this process to provide fast heat dissipation to avoid transformation of the diamond back to graphite.[50] The diamond powder with characteristic sizes of particles in the sub-micron range can be processed from micron-sized polycrystalline diamond particles obtained by shock synthesis. Polycrystalline diamond particles are tougher than monocrystalline diamond microparticles (natural or produced by HPHT) and are widely used in fine polishing applications.

NDs produced by another dynamic method – detonation of a mixture of a carbon precursor (graphite or carbon black) and explosives – are much less known in the community. Tatcii[22] recently summarized the data on the production of DALAN ND from three types of precursors with different density: (i) a mixture of RDX and carbon black (charge density 1.34–1.39 g/cm$^3$), (ii) RDX and graphite (charge density 1.61–1.67 g/cm$^3$), (iii) and gunpowder/carbon/RDX (charge density 1.67–1.75 g/cm$^3$). The concentration of carbon precursor materials in the charge mixture can be varied between 5–35%. The detonation pressure measured by dynamic methods (in the Chapman–Jouguet plane) was about 27.5–14.5 GPa. It was concluded that the diamond powders synthesized from graphite consists of 1–3 μm particles of lamellar shape with different phase compositions (diamond, lonsdaleite, and graphite) and different grain sizes. The particles of diamond powder synthesized from carbon black are aggregates consisting of 20–80 nm grains of cubic phase diamond exhibiting either a round or polyhedral shape. The particles of diamond extracted from the solid detonation products of the gunpowder/carbon black/RDX mixture are single-phase particles having a homogeneous grain structure with a grain size of mainly 1–3 nm. These smaller grains form spherule shaped aggregates. These findings demonstrate the key role of the carbon precursor material in the phase composition and the primary particle size of the resulting products. Currently, polycrystalline diamond produced from graphite (soot)/RDX mixture is a valuable commercial product for polishing applications. While micron-sized particles (1–10 μm) with narrow size distribution are of highest demand, the fractions of polycrystalline diamond with sub-100 nm particle sizes are also utilized. Details of the comparative analysis of the structures of different types of ND particles produced by a shock wave assisted synthesis can be found in [17].

## 2.2.2 HPHT NANODIAMONDS

As was discussed above, monocrystalline ND particles processed from HPHT synthetic diamond as well as from natural diamond powders are commercially available with the smallest average particle size being around 20 nm. The diamond nanoparticles obtained by ball milling of micron-sized HPHT or natural diamond particles typically have sharp edges and look in HRTEM (see Chapter 3) like shattered glass particles.[51] Remarkably, ND particles with rounded shapes and sizes below 10 nm were recently obtained by a group of researches starting with HPHT diamond.[3] The reported method is an important breakthrough in the production of single-digit ND containing NV centers, as it was demonstrated by the authors through processing of type Ib diamond. As small as 1–2 nm individual HPHT particles had been obtained by centrifugation from 18 nm ND purified from $sp^2$ carbon.[27]

Besides the top down approach described above, there are laboratory studies of HPHT conversion of exotic carbon forms to nanodiamond. The HPHT process requires approximately 6 GPa of pressure and 1500 °C to convert graphite powder into diamond in the presence of a catalyst. The use of exotic precursor materials such as fullerenes[52] and carbon nanotubes[53] allows much lower temperatures and externally applied pressures as compared to the use of graphite in order to synthesize diamond. For example, the transformation of buckyballs to diamond at high static pressure can be done at room temperature and does not require a catalyst at all.[52] Carbon nanotubes have been converted to diamond at 4.5 GPa and 1300 °C using NiMnCo catalyst.[53] Based upon HRTEM observations, the authors suggest that under HPHT conditions, the tubular structures collapse and broken graphitic shells curl up and close into spheroidal networks to eliminate the dangling bonds at the edges.

Davydov et al.[54] studied high-pressure–high-temperature-induced transformations of naphthalene, octafluoronaphthalene, and their binary mixtures, and observed formation of nano- and micron-size diamond fractions in the products of transformations of hydrocarbons and fluorocarbons. The observation of large amount of nano-size diamonds was attributed to the specifics of carbonization of fluorocarbon compounds under pressure. The authors observed formation of micron-size particles of graphite simultaneously with a significant amount of closed two- to five-layered carbon nanoparticles of 5–15-nm size, acting as precursors for the formation of nano-size diamond fractions. The obtained results open up a new avenue for catalyst-free synthesis of nano/micron-size fractions of pure and doped diamonds.

The growth of nanosized particles by HPHT method is too expensive for commodity products; however, it is an appealing approach when a controlled doping of ND particles is a goal. This approach will be discussed in Section 2.5 on doping of ND particles.

### 2.2.3 LASER-ASSISTED SYNTHESIS OF NANODIAMOND

Generation of nanoparticles by laser ablation in liquids is a relatively new approach that has been actively explored during the last decade.[76] It has a number of important advantages over conventional multi-step chemical synthesis methods that introduce contamination from the intermediate reactants and/or produce agglomerated structures with degraded functionality. The direct laser ablation eliminates the need for chemical precursors and enables generation of "clean" nanoparticles. Due to the highly non-equilibrium nature of laser ablation process, the structure and composition of nanoparticles produced in laser ablation can be very far from the equilibrium, opening an exciting range of opportunities for generation of novel nanostructures with unique properties.

The most recent review of earlier works in the field is available in [31]. In general, publications indicate that a wide variety of experimental conditions (laser parameters, target, and type of liquid) are favorable for ND formation by pulsed laser ablation (PLA). The mechanism of ND formation by PLA can be explained as follows.[56] A pulsed laser beam of high intensity is focused onto the surface of a solid target immersed in a liquid. The sequence of processes in pulse laser ablation results in an explosive decomposition of the overheated surface region of the absorbing target, ejection of clusters and vapor into the surrounding liquid, and mixing of the ejected ablation plume with the liquid. The liquid is brought to boiling or super-critical states and initiates bubble formation. These bubbles collapse and result in the nucleation, growth, and solidification of the nanoparticles under conditions of fast quenching. The formation of the diamond phase is attributed to the extreme temperature/pressure conditions created by collapse of the bubbles. During collapse of the cavitation bubbles, the temperature can reach more than a thousand degrees, while the pressure can reach several GPa. The final size, structure, and composition of the produced nanoparticles dispersed in the liquid are defined by the heat and mass transfer occurring under highly non-equilibrium, quickly evolving conditions. This approach, however, was not implemented in practice due to the low production rate of nanodiamond particles.

A new method of a laser-assisted synthesis of ND particles, named light hydrodynamic pulse, was suggested by Zousman et al.[25,26] The approach is based on the treatment of a multi-component solid target, containing a carbon non-diamond source, by a laser beam focused in a transparent liquid at some predetermined distance from the target surface, rather than at the surface of the target itself like in earlier experiments. A YAG solid-state laser with a wavelength of 1064 nm was used in the experiments. An acoustic shock wave created by plasma impacts the surface of the composite material containing carbon black and hydrocarbons and creates the temperature–pressure conditions sufficient for production of ND particles. Depending on the parameters of laser radiation, NDs with an average size of 4–5 nm size or 250–300 nm can be obtained by this method as confirmed by transmission electron microscopy.[25,26] The size of NDs were controlled by the laser pulse width, while the target, type of liquid, distance of the laser focus from the substrate were kept unchanged. The approach has led to a considerable increase in productivity.

### 2.2.4 OTHER METHODS OF SYNTHESIS OF NANODIAMOND

In this section we briefly outline other methods of ND synthesis which have been demonstrated in the laboratory. Details on the methods can be found in [31].

Temperatures and pressures required for the synthesis of diamond material can be achieved by different means, including ultrasonic cavitation as a result of the rapid collapse of cavitation bubbles generated in a suitable liquid medium. Depending upon the cavitation conditions, the pressure and temperature inside the bubble at the moment of collapse may reach extremely high instantaneous values $\sim 10^5$–$10^6$ bar and 1000 K, respectively.[57] Cavitation-assisted diamond synthesis was recently achieved by Galimov et al.[9] The system for creation of hydrodynamic cavitation bubbles was based on the so-called Venturi tube (a tube consisting of sections with different diameters) connecting two reservoirs and a series of shock waves produced in the system. Benzene was used as a source of carbon and the cavitation media. Cavitation destruction of benzene resulted in production of particles consisting of aggregates of nanocrystallites 10–30 nm in size. Besides diamond and graphite, electron diffraction patterns revealed that the samples contained n-diamond.[58] New diamond (n-diamond) had been proposed as a new carbon allotrope by Hirai and Kondo[59] in 1991, who developed a procedure involving rapid cooling during the shock compression of graphite sheets. The n-diamond's electron diffraction pattern matches that of cubic diamond (space group Fd3m) apart from additional reflections that are forbidden for diamond, indexed as {200}, {222}, and {420}. At the present, the structure of n-diamond is not unambiguously resolved, and some controversies still exist.

Khachatryan et al. reported the formation of monocrystalline micron-sized diamond particles using a combination of two ultrasonic horns aligned toward each other.[60] Micron-sized diamond crystals (5–10 μm) were produced by this ultrasonic cavitation synthesis method from a suspension of powdered graphite (100–200 μm) in an organic liquid. The major factor influencing the diamond yield in the cavitation process was the composition of the cavitation medium. The diamond yield was about 10% of the initial graphite weight. A series of aromatic C–H–O oligomers with low saturated vapor pressure and high boiling temperatures were used as cavitation fluids.

Frenklach and coworkers studied the nucleation and growth of nanodiamond powder directly from the vapor phase in a substrate-free low-pressure microwave-plasma CVD reactor.[95] The ~50 nm particles were collected downstream of the reaction zone on a filter within the tubular flow reactor. A mixture of diamond polytypes were observed in the powder. The same group also studied the effects of heteroatom addition on the nucleation of solid carbon in a low-pressure plasma reactor.[4] The addition of diborane ($B_2H_6$) resulted in substantial production of diamond particles, 5 to 450 nm in diameter, under the same conditions that show no diamond formation without the presence of diborane. The authors do not report if boron was incorporated within the lattice. Studies of diamond nucleation directly from an activated gas phase have important implications in revealing mechanisms of interstellar dust formation.

Another example of homogeneous diamond nucleation in the gas phase is laser-induced decomposition of $C_2H_4$ at low pressures and temperatures that results in diamond powder formation with grain diameters of 6–18 μm.[62] According to the authors, the high purity homogeneously nucleated diamond nanoparticles had spherical and faceted morphology.

Ting et al. developed a method of nanodiamond particle synthesis with 25–50 nm diameters using a microwave plasma torch system operating at near atmospheric pressure.[63] Gas mixtures of $CH_4$ with Ar or $N_2$ were used for ND particle synthesis, with Ar or $N_2$ serving as catalysts. The larger flow rate of Ar and $N_2$ or the greater pressure in the reaction chamber produced larger nanoparticles.

The formation of nanodiamonds at near-ambient conditions using a novel atmospheric pressure microscale plasma (microplasma) process was recently demonstrating by Kumar et al.[5] Particles were homogeneously nucleated in a gas phase by dissociating ethanol vapor, rapidly quenched to limit the size of the particles to the single-digit size range, exhibiting a uniform diameter ~3 nm. By adding $H_2$ gas, the non-diamond phase was etched and the diamond phase was stabilized, leading to a highly purified as-grown nanodiamond product. Microplasma process provides a unique way to grow free standing ND particles significantly smaller than currently available material.

Another group of methods of nanodiamond formation include direct transformation of carbon solids to nanodiamond. Recent experiments have shown that heavy ion or electron irradiation induces the nucleation of diamond crystallites inside concentric nested carbon fullerenes.[8,64] High energy electron irradiation (1.2 MeV) was successfully used to convert the cores of concentric-shell graphitic onions into nanometer-size diamonds at irradiation temperatures above 900 K.[8] Irradiation with $Ne^+$ (3 MeV) at temperatures between 700 and 1100 °C converted graphitic carbon soot into nanometer-size diamonds.[64] Again, the diamonds were found to nucleate in the cores of graphitic onions that developed under irradiation. The increased diamond yield as compared to e-beam irradiation is explained by the higher displacement cross-section, higher energy transfer, and higher total beam current on the specimen. Other carbon materials can also be transformed to nanodiamond by using MeV electron or ion beams. ND nucleation occurred inside graphite under ion irradiation at ambient temperature when implanted with $Kr^+$ ions (350 MeV).[7] The residue of the ion-irradiated graphite was found to contain nanodiamonds with an average diameter of 7.5 nm.

Selective etching of carbides is an attractive technique for the synthesis of various carbon structures including nanocrystalline diamond.[6] Carbon produced by extraction of metals from carbides is called carbide derived carbon. During extraction of silicon from silicon carbide or metal carbide using chlorine-containing gases at ambient pressure and temperatures not exceeding 1000 °C,[6] nanocrystalline diamond with an average crystallite size of 5 nm was formed.

NDs synthesized by traditional large scale methods typically require an additional step, namely purification from non-diamond carbon and metallic impurities, as discussed in the next section.

## 2.3 NANODIAMOND PURIFICATION FROM SP$^2$ CARBON AND METAL IMPURITIES

The most challenging task is purification of detonation ND.[12,38–44] The impurity content of nanodiamonds produced by detonation synthesis is typically much higher when compared with other artificial nanodiamonds. The product of detonation synthesis, called "detonation soot," contains 30–75 wt% of the diamond phase (depending on the detonation conditions), graphitic carbon, and incombustible impurities (metals and metal oxides, 1–8 wt%) (Fig. 2.5). The metal impurities originate from the igniter used to initiate detonation, the steel walls of the detonation chamber, admixtures in coolants, and contamination from reagents used in sp$^2$ carbon oxidation process. Mitev et al. developed an efficient screening of elemental impurities in commercial detonation nanodiamond using sector field inductively coupled plasma-mass spectrometry (ICP-MS)[65] and this is a topic of Chapter 5 of this book. It was concluded that the contaminant profile for each respective DND can also be considered as a fingerprint, characteristic for every producer and technology used.

The primary goals of purification are twofold: (i) the removal of non-diamond carbon forms (NDCFs) and (ii) the removal of inorganic impurities arising from metal and metal-oxide contaminants, as discussed in a recent excellent review by Aleksenskii.[66] The removal of NDCFs typically revolves around the use of selective oxidation, where NDCFs are much more reactive than diamond.[66] The removal of metal-oxides involves the use of acids (such as HCl, HBr, and HF) or alkalis (NaOH) for dissolution.[66]

The impurities can be inside the nanodiamond aggregates or attached to their outer surface (Fig. 2.5); the ND aggregates should be disintegrated in order to remove the trapped impurities. Post synthesis processing includes purification of detonation soot from metallic impurities and non-diamond carbon using strong oxidizers and is typically performed in conjunction with detonation soot synthesis by the same vendor. After typical purification steps, powders of DND can be considered a composite consisting of different forms of carbon (~80% to 89%), nitrogen (~2% to 3%), hydrogen (~0.5% to 1.5%), oxygen (up to ~10%), and an incombustible residue (~0.5% to 8%).[38,39] The carbon phase consists of a mixture of diamond (90% to 99%) and non-diamond carbon (1% to 10%).

In general, methods of DND purification as well as DND purity vary from vendor to vendor. For DND purification from detonation soot, mechanical and chemical methods are used. After mechanically removing process admixtures (by sieving), the soot is subjected, for example, to thermal oxidation with nitric acid under pressure to separate the diamond phase.[38] In this method, metals are dissolved, and non-diamond carbon is oxidized simultaneously. Other "classical" purification methods, based upon the use of liquid oxidizers for the removal of non-diamond species, include sulfuric acid, mixtures of sulfuric and nitric acids, potassium dichromate in sulfuric acid, as well as other schemes such as ozone purification.[44,68,69,73] Schrand et al.[42] reported on comparative purification of DND using HCl, $HCl/HNO_3$, HF/HCl, and $H_2O_2$–NaOH. The most efficient purification in this series was achieved using treatments with HF followed by HCl; an incombustible impurity content as low as 0.2 wt% (from 1.4 wt% in starting DND) was achieved. Also, while the initial sample had a very low colloidal stability, the colloidal stability of the samples after deep purification with HF/HCl was significantly improved, and the average aggregate size decreased by more than twofold. The success is attributed to the ability of HF to dissolve metal oxides. For the oxidation of $sp^2$ carbons, the purification schemes include $KOH/KNO_3$, $Na_2O_2$, $CrO_3/H_2SO_4$, $HNO_3/H_2O_2$ under pressure, mixtures of concentrated sulfuric and perchloric acids, and other approaches.[12] To remove non-carbon impurities, the chemically purified product is subjected, in some cases, to an additional purification process using ion-exchange and membrane technologies. Laboratory scale microwave-assisted purification of NDs in acids was developed by Mitev et al.[67] and achieved trace element impurity amounts more than 100 times lower than those measured in original commercial samples. Microwave-assisted treatment with metal chelating agents (ethylenediaminetetraacetic acid (EDTA) and 2,6-pyridinedicarboxylic (DPA) acid) provided a further decrease of the metal impurities in DND.[67]

Currently, the majority of DND vendors use strong liquid oxidizers at elevated temperatures and pressures. However, liquid-phase purification is both hazardous and costly, contributing up to 40% of the product cost.[42] In addition, the expense of waste pre-treatment and disposal, which is already high, is expected to increase as governmental policy on environmental protection becomes tighter. Alternatively, DND can be very effectively purified from non-diamond carbon in an environmentally friendly manner by a gas phase treatment using ozone at elevated temperatures[68] to eliminate the need for the use of corrosive liquid oxidizers. Ozone oxidation is also more efficient for $sp^2$ carbon removal. Several efforts have attempted to purify DND by oxidation of detonation soot with air at elevated temperatures.[70–73] Osswald et al.[73] demonstrated that for DND with a high content of non-diamond carbon, the optimal temperature for the heat treatment in air within several hours is 400 to 430 °C. Chiganov purified DNDs from soot through thermal oxidation in air, using boric anhydride in order to selectively oxidize the non-diamond carbon.[73] Surface reduction in a hydrogen atmosphere has also been attempted as a purification technique, but non-diamond carbon was not completely removed by this method.[49]

Following purification, the next step in the production process involves the disaggregation of DND particles or the crushing of HPHT particles down to the nanoscale. Common techniques for disaggregation, crushing, and milling are described in the next section.

## 2.4 DISAGGREGATION, CRUSHING, AND FRACTIONATION OF NANODIAMOND

Realizing several useful properties of nanodiamonds usually requires some form of disaggregation or crushing and subsequent size fractionation. In order to avoid confusion and focus the scope of this discussion, we would like to define a few important terms.

*Dis/Deaggregation* – The disintegration of the $sp^2$/hybrid $sp^{2-3}$ aggregate network binding primary particles detonation or laser synthesized diamond via chemical, mechanical, or chemo-mechanical means.

*Crushing* – The fracture of HPHT or natural diamond along susceptible crystallographic planes using primarily mechanical means.

*Dispersion* – The act of homogeneously distributing ND particles within a solid or liquid matrix such as a polymer or a liquid solvent.

*Fractionation* – The *isolation* of specific particle size ranges following disaggregation or crushing (typically achieved via centrifugation).

Indeed, it is possible to write exhaustive manuscripts for each of these individual topics; however, our ensuing discussion will be primarily limited to techniques used for disaggregation, crushing, and fractionation. We will cover the topic of dispersion only as it related to the typical "process" of transferring diamond from a raw (industrially produced) state to a commercially viable product. An important aspect to keep in mind, while we outline some of these common techniques, is that it is often a pitfall of many in the nanodiamond community to make generalized assumptions about all nanodiamonds based on the results from a single source. Such assumptions could not be further from reality, as diamonds are very sensitive to the synthesis and subsequent purification performed by manufacturers. For example, different batches of DND produced by the same supplier may behave completely differently. Although the infrastructure for industrial scale production has been in place for over a decade, standard practices meant to ensure consistency have not been established, yet. Therefore, the viability of these techniques must be evaluated by the producer or researcher with a focus on their specific material, while the results of others should be used as a guideline, rather than a recipe, for development.

## 2.4.1 DISAGGREGATION TECHNIQUES

### *2.4.1.1 Ball milling*

High energy ball milling (or bead milling) is perhaps the most commonly used technique for deagglomeration due to the lower cost relative to other methods, high efficiency, ease of use, and its ability to process large batches of material at a time. The operating principle of ball milling is quite simple; milling media (ranging in sizes from 50 μm to >10 mm) such as zirconia, tungsten carbide, stainless steel, etc., are mixed with the diamond which is then rotated at high speed within a sealed container. The mechanism of action of the milling media depends on the size of milling media used. In the case of smaller milling media (less that about 1 mm), high shear forces generated during rotation break apart diamond aggregates. On the contrary, larger milling media (>1 mm) generate high impact forced which "crush" larger diamond particles to smaller particles via cleavage along susceptible crystallographic planes, primarily the {111} family[51]; however, there are still open questions regarding the ideal brittle fracture planes.[75,76] Ball milling can be performed in both the wet and dry states, where the wet state is performed in the presence of a solvent and is referred to as "colloidal grinding." Colloidal grinding is typically used for DND disaggregation and was first demonstrated as a useful means of producing 4–5 nm primary particle hydrosols in 2004.[77] However, as exhibited by Pentecost et al.,[78] the use of dry milling can be effective for DND aggregates when used in the presence of a load transfer media (described below). One of the major drawbacks of ball milling is the associated high levels of contamination and heat that arises from the process.[14] For example, milling with micron size $SiO_2$ caused an increase in the $SiO_2$ content from 0.05% to 2.5%.[77] These contaminants can cause undesirable performance issues when the milled material is used for various applications.

Diamond has the unfortunate reputation of being an "extreme" material, and many novices may mistakenly assume that "extreme" conditions (high speeds and extensive milling times) are required to have success. Unfortunately, it is often the case that extensive milling can actually have the reverse effect, and eventually will start to cause aggregation (perhaps due to heat buildup with colloidal grinding and cold-welding with dry milling),[79] not to mention the increased amounts of contaminants and lattice distortion that would naturally arise with longer milling times. Therefore, a lot of trial and error is required for optimizing the time required for milling.

#### 2.4.1.2 Jet milling

A less used technique is jet mill, and the reason for this is obvious. Jet milling is not very effective for reducing the size below 1 μm; therefore, it is often used in conjunction with ball milling to produce nanoscale diamond. Unlike ball milling, jet milling does not rely on the use of any additional milling media, but instead uses the impact of the particles with themselves while circulating under a high pressure gas flow. Jet milling has the benefit of inducing little to no additional contamination, but, as stated above, it is limited primarily to the micronization of larger diamond. Boudou et al. demonstrated the use of jet milling in conjunction with ball milling to produce high yields of fluorescent nanodiamonds; however, it should be stressed that nanomilling was performed with ball milling, and the jet milling procedure only yielded particles close to the 1–2 μm range.[3] Atomization techniques were attempted on DND prior to the application of bead milling to little or no effect.[14]

#### 2.4.1.3 Thermal and chemical treatments

A number of thermal and chemical treatments have been shown to be effective in reducing particle sizes of DND, HPHT, and natural diamonds. The general idea with most of these approaches is the removal of non-diamond carbon phases that are present on the surfaces of HPHT diamonds and aggregates of DND and where the use of these techniques can lead to disaggregation and size reduction as well as the removal of metallic impurities which can bridge particles together. Thermal oxidation in air and reduction in hydrogen have both been shown to be effective in reducing particle size by a number of different groups. Reduction deagglomeration in hydrogen at 500 °C for 5 hours was demonstrated by Williams et al. to yield stable suspensions of 4–5 nm primary particles of DND.[74] Etzold and coworkers demonstrated a cyclic technique of reduction and oxidation for size reduction.[80] In their approach, they sought to control the oxidation step by splitting the process into two independent steps – a chemisorption and a high temperature desorption step – to develop a layer-by-layer (LbL) oxidation process. A low temperature chemisorption process (oxidation) was performed at 350 °C followed by a subsequent desorption step at 600 °C in an $H_2$/Ar atmosphere. A size reduction of primary particles from 5.2 to 4.7 nm (as measured via SAXS) was achieved by this approach. It has also been demonstrated that oxidation in air at ~430 °C can be useful for enriching DND with $sp^3$ carbon but the process can actually increase the average primary particle size from 4.7 to 4.8 nm (measured by XRD).[81] As this work demonstrates, care must be taken to avoid excessive oxidation times and temperatures. While shorter oxidation times (2 hours) removed large amounts of undesirable amorphous carbon and non-diamond carbon phases, longer oxidation times (42 hours) led to selective oxidation of smaller crystals, thereby shifting the size distribution to larger values. Air oxidation has been demonstrated to reduce sizes in HPHT diamond.[27,82] Treatment of nitrogen-vacancy containing diamond showed an average size reduction of about 40 nm following a 2 h air treatment versus a 5 h air treatment at 600 °C, and the etch rates of diamond ranged from <1 nm/h at 500 °C to about 10 nm/h at 600 °C. Therefore, this could be a useful method of obtaining small fluorescent nanodiamonds; however, the caveat is that such etching does remove the desired defect centers,[82] so a costly irradiation step with high-energy particles to induce defect formation would be required prior to employing this method.

Wet chemical treatments in strong oxidizing acids such as mixtures of $HNO_3$:$H_2SO_4$ have also been investigated as a means to control particle size; however, while these acids are useful in purifying the $sp^3$ phase from the non-diamond phases, they are not useful for dramatically reducing crystal sizes. It has also been demonstrated that the use of wet oxidizers led to the agglomeration of NDs.[14] The use of HCl has, however, been shown to be a useful means of disaggregation particularly for DND due to the removal of metallic impurities which can bridge particles together (see Fig. 2.5).[83]

#### *2.4.1.4* ***Ultrasonic disaggregation and dispersion***

A popular technique for dispersion of nanoparticles (including nanodiamonds) is the use of ultrasonic cavitation to promote dispersion. While ultrasonic cavitation (typically with the use of a horn-type sonotrode) is a very powerful dispersion technique, it has also been shown to help reduce the sizes of nanoparticles and NDs in suspension and may even induce chemical modification via sonochemistry.[84] The combination of ultrasonic treatment and subsequent fractionation has been demonstrated to be useful for extracting small sizes of polycrystalline DND.[85] We would argue that ultrasonic cavitation treatment is an indispensable step in the extraction of small ND particles since centrifugation is highly ineffective for media in which NDs are poorly dispersed.[42] Nevertheless, the use of ultrasound is typically not efficient enough to extract high yields of small particles, and is most effective when used in conjunction with other disaggregation techniques.

#### *2.4.1.5* ***Specialty techniques & recent advancements***

A number of unique approaches have been developed over the years and continue to be of interest. We highlight some of the more effective of these novel techniques below.

##### ***Bead assisted sonic disintegration (BASD)***

This technique combines the use of ultrasonics with typical milling media and was first reported by Ozawa et al.[86]; in their typical setup, approximately 5 ml of a solvent was mixed with 7.0 g of 50 μm zirconia beads and 50–500 mg of ~200 nm ND powder with 200 nm aggregates. Approximately 10 nm size particles were achieved after 120 minutes of sonication with a 400 W horn-type sonotrode. The technique was used by the same group to perform functionalization reactions, where 4–5 nm primary particles were achieved using BASD.[87] While this method has been shown to be effective, it is obvious that a large amount of contamination would be introduced into the material from both the Ti sonotrode and the zirconia beads, and these materials would require harsh chemical treatments to be removed.[14] Additionally, scale up is unlikely without a specialized system, since the abrasive nature of the milling media would severely damage expensive ultrasonic treatment equipment; however, it can function successfully at the laboratory scale.

#### *Matrix assisted milling/crushing*

The use of an additional solid matrix for load transfer to assist in disaggregation during milling or other processing has also been shown with some success. Pentecost et al.[78] showed that dry milling of high grade DND powder with stainless steel media in the presence of salt or sucrose could disaggregate the aggregates to primary 4–5 nm particles with just 1–1.5 hours of milling. This is a very appealing approach since it eliminates the need for costly zirconia beads (which are traditionally used), and the contaminants (sucrose or sugar) can be easily washed from the material without harsh chemicals. Any contamination from steel can subsequently be removed with hydrochloric acid (which is arguably a much safer alternative compared to other acid moieties). The major drawback from this method is the extensive amount of heat that can be generated from grinding in the dry state with fine particulates. In the case of the above work, the instrumentation had to be maintained at 0 °C throughout the entire milling process.

Kanygin et al.[88] demonstrated a method of forge-rolling an ND/PS composite to disaggregate DNDs; however, this method, while useful for distributing NDs within a solid matrix, is likely not suitable for obtaining the often sought hydrosols of primary 4–5 nm particles of DNDs. Nevertheless, this method does demonstrate a fundamental benefit to using a solid matrix to assist with load transfer.

Chang and coworkers demonstrated a unique approach to crush HPHT diamond from the 30 to sub-20 nm size range using a hydraulic press.[93] In this approach, a mixture of ~30 nm fluorescent nanodiamond and NaF (1:10) were pressed under a pressure of 10 tons using a hydraulic oil press. It was possible to extract a ~60% yield of sub-20 nm particles without causing extensive lattice distortion.

### 2.4.2 FRACTIONATION

We would like to consider fractionation as a separate topic to disaggregation since, typically, disaggregation is a step performed prior to fractionation. It is rare that DND aggregates can be milled to a nearly 100% yield of 4–5 nm primary particles, and obtaining such a narrow size distribution with HPHT or natural diamond is even more unlikely. In other words, no disaggregation technique will be completely effective in yielding a single, narrow size distribution, so some form of fractionation will always be required. Although there are examples where high yields of 4–5 nm primary particles can be achieved with only milling,[77] such procedures require extensive amounts of milling and negatively affect the final product due to contamination and lattice distortion. Moreover, it has been demonstrated that 4–5 nm primary particles can be achieved via ultracentrifugation,[89] so it is not necessary to achieve high yields of primary particles with only milling. Fractionation is typically performed by dispersing the ND within a solvent (typically using ultrasonic cavitation), and subsequent centrifugation to obtain selected particular size fractions.[42] It is a requirement that suspensions of ND exhibit high colloidal stability, or they are nearly impossible to separate by size.[12] Centrifugation is such a common technique at both the laboratory and industrial scale, so we do not wish to belabor details regarding centrifugation schemes; it suffices to say that trial and error are again a necessary component to accomplish successful fractionation of small particles.

One alternative technique involving centrifugation is the use of density gradient centrifugation which has recently been shown to be an effective technique to extract small particles at the gram scale.[90]

## 2.5 DOPING OF NANODIAMOND PARTICLES

Controlled doping of ND particles with foreign atoms remains an important goal. Nitrogen, silicon, and transitional metals incorporated into the diamond lattice provide the possibility to produce optically active centers in the particles (see Chapter 7). Doping of ND with boron is a long standing goal for applications of electrically conductive NDs in electrodes and low resistivity seeds for growth of CVD boron-doped diamond films. Moreover, the combination of donors and acceptors within the diamond lattice gives a rise to phosphorescent diamonds[91] with new applications, for example, in bioimaging.

Nitrogen is the most common impurity in diamond, incorporated into the crystal lattice as isolated substitutional nitrogen (Ns) atoms or two nearest-neighbor substitutional nitrogen atoms, among other numerous N-containing defects. Shenderova et al. investigated the effect of different types of detonation synthesis of ND particles on N content, N location in the ND core, and its ability to form optically active centers through the creation of complexes with vacancies.[17] In ND synthesis using explosives, N is an integral part of the high energy explosive compounds. It can also be present as an unintended impurity or a dopant introduced into graphite (or other carbon-containing materials) in the methods of dynamic synthesis involving bulk carbon precursors for ND production. The nitrogen content of ND samples measured using combustion analysis indicated that the N concentration in the ND produced from a mixture of TNT/RDX explosives is higher than 2–2.5 wt%, from TNT/HNS (HNS is hexanitrostilbene) lower than 1%, and in ND from graphite and graphite/RDX mixtures lower than 0.5%. Thus, variations in the type of explosives can provide variation in the N content in the final ND product. It is possible that the N–N units of the RDX molecule contribute to enhanced N aggregation in the ND core, if they are not completely destroyed during detonation and are incorporated into the ND core as neighboring nitrogen atoms. Substituting HNS for RDX, which contains less N in the molecule, significantly reduces the N content.[14] Varying the type of explosives can be used together with the addition of other carbon precursor materials (for example, a combination of graphite and HNS) to achieve the desired N content. An interesting implementation of HPHT treatment was reported by Baranov et al.[92] Sintering experiments were carried out on detonation ND with average particle sizes 4.5 nm at temperatures 1500–1700 °C and pressure 6–7 GPa. According to the results of the X-ray diffraction analysis, the average size of the primary particles became 8.5 nm after sintering. When the sintering temperature is above 800 °C vacancies start to diffuse. The presence of a significant amount of NV centers was detected in 10 micron aggregates using pulsed EPR (see Chapter 6). Formation of the NV centers can be possibly attributed to the combination of elimination of surface paramagnetic centers and vacancy diffusion resulting from the high temperature of the process.

The typical density of atomically dispersed nitrogen in most commercially available type Ib diamonds synthesized by HPHT methods is 100–150 ppm. Diamond particles with as high N content as up to 400 ppm are also commercially available, though they are of lower quality.[93] An important issue to emphasize is that Ns atoms are heterogeneously distributed in the raw standard industrial HPHT synthetic diamond. Nitrogen spatial distribution is strongly dependent on the internal structure of the diamond crystal and its growth conditions,[94] including the local growth rate. Presently, the impurity concentration in synthetic diamond crystals decreases with increasing radius.[94] There are several reasons for this. For instance, in the HPHT process used to synthesize diamond crystals, the concentration of nitrogen in the melt decreases with time because the growing diamond takes nitrogen from the molten metallic catalyst/solvent through which carbon is transported from the graphite feedstock to the growing diamond crystals during the HPHT process. Another reason is that the growth rate of the diamond crystal is decreasing with increasing radius, and the impurity incorporation decreases with decreasing growth rate. Further, the impurity concentration in the diamond crystal depends on the growth orientation – the crystallographic direction in which growth takes place. In some cases, the cubic or other growth orientation of the diamond are completely free of nitrogen.[94] As a result, the fluorescent nanodiamond processed from a fluorescent HPHT microdiamond exhibits a heterogeneous NV center spatial distribution. Thus, methods of homogeneous Ns distribution in HPHT microdiamond need to be developed instead of relying on the commercial HPHT NDs optimized for the abrasion industry. Also, the decreasing impurity concentration with increasing radius (i.e., negative concentration gradient) causes tangential tensile stresses on the surface of the diamond.[94] Internal stresses might influence vacancy and NV diffusion within a crystal during irradiation and annealing. Thus, there is still a lot of research required before acceptable HPHT diamond precursors will be available for production of fluorescent ND with uniform Ns (and NV) distribution.

Another important dopant is Si with a goal of production photoluminescent SiV centers emitting in the near red spectral range. SiV centers have been produced in diamonds either by the addition of Si during diamond growth by chemical vapor deposition (CVD),[95,96] or by implantation of silicon ions into pristine diamond.[97] Nanodiamonds containing SiV centers were produced by milling of chemical vapor deposited (CVD) diamond films containing SiV[96] or sonic disintegration of the CVD diamond film containing SiV to generate isolated free particles.[97] Irradiation with electrons was not required; the Si defect typically is embedded into a CVD film accompanied by a vacancy. Zhang et al. experimentally demonstrated enhancement of $(SiV)^-$ CL intensity in NDs by introducing nitrogen gas during diamond growth.[98] Isolated nitrogen (N) impurities in diamond act as electron donors that favor the formation of $(SiV)^-$ centers. HPHT synthesis of nano- and microdiamonds with various contents of luminescent silicon-vacancy (SiV) and nitrogen-vacancy (NV) centers has been demonstrated by Davydov et al.[99] using mixtures of hydrocarbon, fluorocarbon, and organic silicon compounds without catalyst metals.

The CVD technique was recently used for the production of electrically conductive nanodiamond particulates by first CVD growth of boron-doped nanodiamond films followed by grinding of the films to the state of diamond nanoparticles.[100] This approach, however, is hardly scalable. Kondo et al. reported a method based on solid-state diffusion of B initially distributed on ND surface into core of ND particles.[101] A mixture of ND/B powder was heat treated in $H_2$ atmosphere at 900 °C resulting in an increase in the ND conductivity from $2 \times 10^{-6}$ to $2.7 \times 10^{-3}$ S cm$^{-1}$. Related methods for production of boron doped diamond based on the solid-state diffusion include rapid thermal diffusion,[102] field-enhanced diffusion,[103,104] and field-enhanced diffusion with optical activation.[105,106]

Boron-doped HPHT diamond ground to nanoparticles, based on our experience, does not provide the required electrical conductivity, since the B doping level is too low (below 800 ppm). Recently, Ekimov et al.[107] succeeded in direct synthesis of heavily boron-doped ($\sim$1%) nanodiamond particles based on HPHT treatment (at a pressure of 8–9 GPa in a temperature range of 1200–2000 K) of a one-component precursor containing the dopant atom(s), the organoboron compound 9-borabicyclo[3,3,1] nonane dimer (9BBN), $C_{16}H_{30}B_2$. Heavily boron-doped diamond nanoparticles with a size below 10 nm have been produced. The production capacity of the method is 100 mg per 5 min cycle. The authors conclude that a key issue in the direct synthesis of doped nanodiamond is the use of one-component organic compound containing hydrocarbon in $sp^3$ configuration and dopant atoms.

## 2.6 CONCLUSION AND FUTURE OUTLOOK

Among the hundreds of nanomaterials, diamond nanoparticles occupy a special niche due to the combination of their outstanding mechanical performance, chemical resistance, biocompatibility, and unique optical and electronic properties. They have emerged as a key platform for many nanoscience and nanotechnology developments.[36] To realize their great potential, precise control of the structure and physical properties is required. Specifically, control of the size of primary ND particles, their surface structure, and the formation of functional crystallographic defects in the ND core is required. Size is important in applications involving phenomena where the physical dimensions of a nanoparticle are critical, for example, the penetration of biological membranes (see Chapter 16). Smaller size also translates to larger specific surface area, which is imperative for increased uploading of drugs and bioligands or formation of a denser network of bonds between ND nanofillers and the surrounding polymer matrix. The surface structure of NDs plays a critical role in the practical implementation of NDs, including: the formation of colloidal suspensions, incorporation into a matrix, and interaction with biomolecules. Controlled incorporation of dopant elements into the diamond lattice opens perspectives for production of luminescent or electrically conductive ND particles.

Control of the key structural properties must start from the ND synthesis, which must be modified according to the new demands, rather than trying to adapt current commercial ND particles produced by "old" technologies with other goals in mind. This requirement is vital for successful implementation of advanced NDs in real world applications. The two major methods of synthesis of commercial NDs remain detonation of carbon-containing explosives and grinding of micron-sized diamond powders manufactured by static HPHT synthesis. Each of the two classes of ND particles has disadvantages. While possessing a single-digit size of primary particles, DND suffers from strong agglomeration of the primary particles during synthesis and, most importantly, a high concentration of N in the particle's core derived from the N contained in explosives. HPHT ND suffers from the irregular shape of ND particles, a highly labor intensive process to mill them to single digit size particles and, most importantly, the non-uniform distribution of dopant elements resulting from their non-uniform distribution in starting micron-sized particles. The shortcomings of HPHT diamond are to be expected; however, this type of diamond is well established for use in the abrasive industry where little care is given to such considerations.

New laboratory-scale developments in ND synthesis discussed in this chapter demonstrate that some of the drawbacks of the current commercial NDs can be overcome and control of size, surface structure, and lattice doping can be implemented at the stage of ND synthesis. In the synthesis of DND, new trends include control of primary particle size,[20] agglomeration state of ND,[49] and the level of N incorporation into the lattice.[17] Synthesis of ND by the HPHT method using organic precursors became possible,[54,107] including control of doping elements.[107] HPHT synthesis of ND free from metal catalysts is also an important direction in terms of particle purity from encapsulated metal inclusions. Various methods of production of ND particles as small as ~3 nm in diameter have emerged.[5] Of particular interest are NDs produced by laser assisted synthesis,[25] which offers the possibility of controlling the primary particle size and, in principle, has great potential for controlled doping.

Advances in the methods of separation and fractionation of DND[77] have led to the availability of 50, 20, and 5 nm particle sizes (average agglomerate size) versus typical 200–300 nm agglomerates, which were not readily available even 5 years ago. Production volumes of these smaller size fractions opens up new markets and expands existing ones as the material meets the demand for specific uses. New technologies are providing the means to increase the production of small size fractions in larger volumes.

Compared to many other nanomaterials, ND produced by detonation of explosives is relatively expensive since it has not experienced a scale-up to large volume production. As in any market, price is a factor which influences expansion of the materials use. The production method for the initial soot has largely remained the same for the past 30 years. Modern methods of increased production of DND by using larger detonation charges have become possible.[49] Other methods of ND synthesis, such as laser-assisted synthesis, are intended to be brought up to large scale production. HPHT ND remains an order of magnitude more expensive than DND.

If a real commercial demand for novel ND takes place as a result of the scientific activity in the nearest future, industrial production of this material will be in order. However, additional innovations in the scale up of the new emerging methods or modified "old" methods of ND synthesis and processing will be needed to satisfy modern requirements for material with specific properties and at reasonable cost, particularly in light of competition with other nanomaterials.

## REFERENCES


1. Volkov, K. V.; Danilenko, V. V.; Elin, V. I. Diamond Synthesis from Detonation Carbon. *Fiz. Goren Vzryva* **1990**, *26*, 123.
2. Yang, G. W.; Wang, J. B.; Liu, Q. X. Preparation of Nano-Crystalline Diamonds Using Pulsed Laser Induced Reactive Quenching. *J. Phys. Condens. Matter* **1998**, *10*, 7923.
3. Boudou, J. P.; Curmi, P. A.; Jelezko, F.; Wrachtrup, J.; Aubert, P.; Sennour, M.; Balasubramanian, G.; Reuter, R.; Thorel, A.; Gaffet, E. High Yield Fabrication of Fluorescent Nanodiamond. *Nanotechnology* **2009**, *20*, 235602.
4. Frenklach, M.; Howard, W.; Huang, D.; Yuan, J.; Spear, K. E.; Koba, R. Induced Nucleation of Diamond Powder. *Appl. Phys. Lett.* **1991**, *59*, 546.
5. Kumar, A.; Lin, P. A.; Xue, A.; Hao, B.; Yap, Y. K.; Sankaran, R. M. Formation of Nanodiamond at Near Ambient Conditions Via Microplasma Dissociation of Ethanol Vapour. *Nat. Commun.* **2013**, 4220.
6. Welz, S.; Gogotsi, Y.; McNallan, M. J. Nucleation, Growth, and Graphitization of Diamond Nanocrystals During Chlorination of Carbides. *J. Appl. Phys.* **2003**, *93*, 4207.
7. Daulton, T. L.; Kirk, M. A.; Lewis, R. S.; Rehn, L. E. Production of Nanodiamonds by High-Energy Ion Irradiation of Graphite at Room Temperature. *Nucl. Instrum. Methods B* **2001**, *175*, 12.
8. Banhart, F.; Ajayan, P. M. Carbon Onions as Nanoscopic Pressure Cells for Diamond Formation. *Nature* **1996**, *382*, 433.
9. Galimov, E.; et al. Experimental Corroboration of the Synthesis of Diamond in the Cavitation Process. *Dokl. Phys.* **2004**, *49*, 150.
10. Guillois, O.; Ledoux, G.; Reynaud, C. Diamond Infrared Emission Bands in Circumstellar Media. *Astrophys. J.* **1999**, *521*, L133.

11. Goto, M.; et al. Spatially Resolved 3 μm Spectroscopy of Elias 1: Origin of Diamonds in Protoplanetary Disks. *Astrophys. J.* **2009**, *693*, 610.
12. Vul, A., Shenderova, O., Eds., *Detonation Nanodiamonds – Science and Applications*; Pan Stanford Publishing, 2014.
13. Shenderova, O., Gruen, D., Eds., *Ultrananocrystalline Diamond;* 2nd edn.; Elsevier, 2012.
14. Osawa, E. Recent Progress and Perspective in Single-Digit Nanodiamond. *Diam. Relat. Mater.* **2007**, *16*, 2018.
15. Chang, Y. R.; et al. Mass Production and Dynamic Imaging of Fluorescent Nanodiamonds. *Nat. Nanotechnol.* **2008**, *3*, 284.
16. Laporte, G.; Psaltis, D. STED Imaging of Green Fluorescent Nanodiamonds Containing Nitrogen-Vacancy-Nitrogen Centers. *Biomed. Opt. Express* **2016**, 7.
17. Shenderova, O. A.; Vlasov, I. I.; Turner, S.; Tendeloo, G. V.; Orlinskii, S. B.; Shiryaev, A. A.; Khomich, A. A.; Sulyanov, S. N.; Jelezko, F.; Wrachtrup, J. Nitrogen Control in Nanodiamond Produced by Detonation Shock-Wave-Assisted Synthesis. *J. Phys. Chem. C* **2011**, *115*, 14014.
18. Balasubramanian, G.; Chan, I. Y.; Kolesov, R.; et al. Nanoscale Imaging Magnetometry with Diamond Spins Under Ambient Conditions. *Nature* **2008**, *455*, 648.
19. Jelezko, F.; Gaebel, T.; Popa, I.; Gruber, A.; Wrachtrup, J. Observation of Coherent Oscillations in a Single Electron Spin. *Phys. Rev. Lett.* **2004**, *92*, 076401.
20. Pichot, V.; Comet, M.; Risse, B.; Spitzer, D. Detonation of Nanosized Explosive: New Mechanistic Model for Nanodiamond Formation. *Diam. Relat. Mater.* **2015**, *54*, 59.
21. DeCarli, P. S.; Jamieson, J. C. Formation of Diamond by Explosive Shock. *Science* **1961**, *133*, 1821.
22. Tatsii, V. F.; Bochko, A. V.; Oleinik, G. S. Structure and Properties of Dalan Detonation Diamonds. *Combust. Explos. Shock Waves* **2009**, *45*, 95.
23. Korobov, M.; Volkov, A.; Avramenko, N. Improving the Dispersity of Detonation Nanodiamond: Differential Scanning Calorimetry as a New Method of Controlling the Aggregation State of Nanodiamond Powders. *Nanoscale* **2013**, *5*, 1529.
24. Ogale, S. B.; Malshe, A. P.; Kanetkar, S. M.; Kshirsagar, S. T. Formation of Diamond Particulates by Pulsed Ruby Laser Irradiation of Graphite Immersed in Benzene. *Solid State Commun.* **1992**, *84*, 371.
25. Zousman, B.; Levinson, O. Pure Nanodiamond Produced by Laser-Assisted Technique. In Williams, O., Ed., *Nanodiamond;* The Royal Society of Chemistry, Vol. 31, 2014.
26. Baidakova, M. V.; Kukushkina, Y.; Sitnikova, A.; et al. Structure of Nanodiamonds Prepared by Laser Synthesis. *Phys. Solid State* **2013**, *55*, 1747.
27. Stehlik, S.; et al. Size and Purity Control of HPHT Nanodiamonds Down to 1 nm. *J. Phys. Chem. C* **2015**, *119*, 27708.
28. Dahl, J. E.; Liu, S. G.; Carlson, M. K. Isolation and Structure of Higher Diamondoids, Nanometer-Sized Diamond Molecules. *Science* **2003**, *299*, 93.
29. Schwertferger, H.; Fokin, A. A.; Schreiner, P. R. Diamonds Are a Chemist's Best Friend: Diamondoid Chemistry Beyond Adamantane. *Angew. Chem., Int. Ed. Engl.* **2008**, *47*, 1022.
30. Danilenko, V. V. On the Discovery of Detonation Nanodiamond. In Shenderova, O. A., Gruen, D., Eds., *Ultrananocrystalline Diamond*; William Andrew Publishing: Norwich, NY, 2006, p. 335.
31. Danilenko, V. V.; Shenderova, O. Advances in Synthesis of Nanodiamond Particles. In Sarin, V., Ed., *Comprehensive Hard Materials*; Elsevier, 2014.

32. Danilenko, V. V. Synthesis and Sintering of Diamond by Explosion. Explosion Physics, Technique, Technology; Energoatomizdat, 2010 (In Russian).
33. Danilenko, V. V. Features of Carbon Condensation in a Detonation Wave and the Conditions for Optimal Synthesis of Nanodiamond. *J. Superhard Mater.* **2006**, *28*, 7.
34. Danilenko, V. V. In Gruen, D. M., Shenderova, O. A., Vul, A. Y., Eds., *Proceedings of the NATO Advanced Research Workshop on Synthesis, Properties, and Applications of Ultrananocrystalline Diamond*; Springer: Netherlands, 2005, p. 181.
35. Viecelli, J. A.; Bastea, S.; Glosli, J. N.; Ree, F. H. Phase Transformations of Nanometer Size Carbon Particles in Shocked Hydrocarbons and Explosives. *J. Chem. Phys.* **2001**, *115*, 2730.
36. Mochalin, V. N.; Shenderova, O.; Ho, D.; Gogotsi, Y. The Properties and Applications of Nanodiamonds. *Nat. Nanotechnol.* **2012**, *7*, 11.
37. Kuznetsov, V. L.; Chuvilin, A. L.; Butenko, Y. V.; Malkov, I. Y.; Titov, V. M. Onion-Like Carbon from Ultra-Disperse Diamond. *Chem. Phys. Lett.* **1994**, *222*, 343.
38. Dolmatov, V. Y. Detonation Nanodiamonds: Synthesis, Structure, Properties and Applications. *Russ. Chem. Rev.* **2007**, *76*, 375.
39. Dolmatov, V. Y. In *Detonation Nanodiamonds*; St. Petersburg, 2011, p. 530.
40. Shenderova, O. A., Gruen, D. M., Eds., *Ultrananocrystalline Diamond*; William Andrew Publishing: Norwich, NY, 2006.
41. Shenderova, O. A., Zhirnov, V. V.; Brenner, D. W. Carbon Nanostructures. *Crit. Rev. Solid State* **2002**, *27*, 227.
42. Schrand, A. M.; Hens, S. A.; Shenderova, O. A. Nanodiamond Particles: Properties and Perspectives for Bioapplications. *Crit. Rev. Solid State* **2009**, *34*, 18.
43. Baidakova, M.; Vul, A. New Prospects and Frontiers of Nanodiamond Clusters. *J. Phys. D, Appl. Phys.* **2007**, *40*, 6300.
44. Dolmatov, V. Y. Detonation Synthesis Ultradispersed Diamonds: Properties and Applications. *Russ. Chem. Rev.* **2001**, *70*, 607.
45. Yur'ev, G. S.; Dolmatov, V. Y. X-Ray Diffraction Study of Detonation Nanodiamonds. *J. Superhard Mater.* **2010**, *30*, 29.
46. Pichot, V.; Risse, B.; Schnell, F.; Mory, J.; Spitzer, D. Understanding Ultrafine Nanodiamond Formation Using Nanostructure Explosives. *Sci. Rep.* **2013**, *3*, 2159.
47. Danilenko, V. V. Assessment of the Influence of Particle Size and Degree of Cohesion on Their Thermal Stability. *J. Superhard Mater.* **2009**, *31*, 218.
48. Batsanov, S. S.; Osavchuk, A. N.; Naumov, S. P.; Efimov, A. E.; Mendis, B. G.; Spperley, D. C.; Batsanov, A. S. *Propellants Explos. Pyrotech.* **2015**, *40*, 39.
49. Danilenko, V. V. Nanodiamonds: Problems and Prospects. *J. Superhard Mater.* **2009**, *32*, 301.
50. Cowan, G. R.; Woodbury, N. J.; Dunnington, B. W.; et al. US Patent 3401019, 1968.
51. Shenderova, O. A.; McGuire, G. E. Science and Engineering of Nanodiamond Particle Surfaces for Biological Applications. *Biointerphases* **2015**, 10.
52. Nunez-Regueira, M.; Monceau, P.; Hodeau, J. L. Crushing C60 to Diamond and Room Temperature. *Nature* **1992**, *355*, 237.
53. Cao, L. M.; Gao, C. X.; Sun, H. P.; et al. Synthesis of Diamond from Carbon Nanotubes Under High Pressure and High Temperature. *Carbon* **2001**, *39*, 311.
54. Davydov, V. A.; Rakhmanina, A. V.; Agafonov, V.; Khabashesku, V. N. On the Nature of Simultaneous Formation of Nano- and Micron-Size Diamond Fractions Under Pressure-Temperature-Induced Transformations of Binary Mixtures of Hydrocarbon and Fluorocarbon Compounds. *Carbon* **2015**, *90*, 231.

55. Yang, L.; May, P. W.; Yin, L.; Smith, J. A.; Rosser, K. N. Growth of Diamond Nanocrystals by Pulsed Laser Ablation of Graphite in Liquid. *Diam. Relat. Mater.* **2007**, *16*, 725.
56. Pearce, S. R. J.; Henleya, S. J.; Claeyssensa, F.; Maya, P. W.; Hallamb, K. R.; Smith, J. A.; Rosser, K. N. Production of Nanocrystalline Diamond by Laser Ablation at the Solid Liquid Interface. *Diam. Relat. Mater.* **2004**, *13*, 661.
57. Brennen, C. E. *Cavitation and Bubble Dynamics;* Oxford University Press: Oxford, 1995.
58. Voropaev, S.; Galimov, E.; Aronin, A. Structural Investigations of Carbon Nanostructures Produced by Hydrodynamical Cavitation Technique. *Fuller. Nanotub. Carbon Nanostructures* **2012**.
59. Hirai, H.; Kondo, K. J. Modified Phases of Diamond Under Shock Compression and Rapid Quenching. *Science* **1991**, *253*, 772.
60. Khachatryan, A. K.; Aloyan, S.; May, P. W.; et al. Graphite-to-Diamond Transformation Induced by Ultrasonic Cavitation. *Diam. Relat. Mater.* **2008**, *17*, 931.
61. Frenklach, M.; Kematrick, R.; Huang, D.; et al. Homogeneous Nucleation of Diamond Powder in the Gas Phase. *J. Appl. Phys.* **1989**, *66*, 395.
62. Buerki, P. R.; Leutwyler, S. Homogenous Nucleation of Diamond Powder by $CO_2$ Laser Driven Reactions. *J. Appl. Phys.* **1991**, *69*, 3739.
63. Ting, C. C.; Young, T.; Jwo, C. S. Fabrication of Diamond Nanopowder Using Microwave Plasma Torch Technique. *Int. J. Add. Manufact. Tech.* **2007**, *34*, 316.
64. Wesolowski, P.; Lyutovich, Y.; Banhart, F.; et al. Formation of Diamond in Carbon Onions Under MeV Ion Irradiation. *Appl. Phys. Lett.* **1997**, *71*, 1948.
65. Mitev, D.; Townsend, A.; Paull, B.; Nesterenko, P. Screening of Elemental Impurities in Commercial Detonation Nanodiamond Using Sector Field Inductively Coupled Plasma Mass Spectrometry. *J. Mater. Sci.* **2014**, *49*, 3573.
66. Aleksenskii, A. E. Technology of Preparation of Detonation Nanodiamond. In Shenderova, O., Vul, A., Eds., *Detonation Nanodiamonds – Science and Applications*; Pan Stanford Publishing, 2014, p. 37.
67. Mitev, D.; Townsend, A.; Paull, B.; Nesterenko, P. Microwave-Assisted Purification of Detonation Nanodiamond. *Diam. Relat. Mater.* **2014**, *48*, 37.
68. Shenderova, O. A.; Koscheev, A.; Zaripov, N.; Petrov, I.; Skryabin, Y.; Detkov, P.; et al. Surface Chemistry and Properties of Ozone-Purified Detonation Nanodiamonds. *J. Phys. Chem. C* **2011**, *115*, 14014.
69. Larionova, I. S.; Molostov, I. N.; Kulagina, L. S.; Komarov, V. F. Russia Patent 2168462, 1999.
70. Mitev, D.; Dimitrova, R.; Spassova, M.; Minchev, C.; Stavrev, S. Surface Peculiarities of Detonation Nanodiamonds in Dependence of Fabrication and Purification Methods. *Diam. Relat. Mater.* **2007**, *16*, 776.
71. Shenderova, O.; Petrov, I.; Walsh, J.; Grichko, V.; Grishko, V.; Tyler, T.; et al. Modification of Detonation Nanodiamond by Heat Treatment in Air. *Diam. Relat. Mater.* **2006**, *15*, 1799.
72. Osswald, S.; Yushin, G.; Mochalin, V.; Kucheyev, S. O.; Gogotsi, Y. Control of $sp^2/sp^3$ Carbon Ratio and Surface Chemistry of Nanodiamond Powders by Selective Oxidation in Air. *J. Am. Chem. Soc.* **2006**, *128*, 11635.
73. Chiganov, A. S. Selective Inhibition of the Oxidation of Nanodiamonds for Their Cleaning. *Phys. Solid State* **2004**, *46*, 620.
74. Williams, O. A.; Hees, J.; Dieker, C.; Jager, W.; Kirste, L.; Nebel, C. E. *ACS Nano* **2010**, *4*, 4824.
75. Brenner, D.; Shenderova, O. *Philos. Trans. R. Soc. A* **2015**, *373*, 20140139.

76. Shenderova, O.; Brenner, D.; Omeltchenko, A.; Su, X.; Yang, L. *Phys. Rev. B* **2000**, *61*, 3877.
77. Krueger, A.; et al. Unusually Tight Aggregation in Detonation Nanodiamond: Identification and Disintegration. *Carbon* **2005**, *43*, 1722.
78. Pentecost, A.; et al. Deaggregation of Nanodiamond Powders Using Salt- and Sugar-Assisted Milling. *Appl. Mater. Interfaces* **2010**, *2*, 3289.
79. Suryanarayana, C. Mechanical Alloying and Milling. *Prog. Mater. Sci.* **2001**, *46*, 1.
80. Etzold, B. J. M.; et al. Layer-by-Layer Oxidation for Decreasing the Size of Detonation Nanodiamond. *Chem. Mater.* **2014**, *26*, 3479.
81. Osswald, S.; et al. Increase of Nanodiamond Crystal Size by Selective Oxidation. *Diam. Relat. Mater.* **2008**, *17*, 1122.
82. Gaebel, T.; et al. Size-Reduction of Nanodiamond Via Air Oxidation. *Diam. Relat. Mater.* **2011**, *21*, 28.
83. Aleksenskii, A. E.; Eydelman, E.; Vul, A. Deagglomeration of Detonation Nanodiamonds. *Nanosci. Nanotechnol. Lett.* **2011**, *3*, 68.
84. Uchida, T.; et al. Disaggregation and Surface Modification of Nano-Size Diamond by Ultrasound Exposure: Relationship Among Acoustic Intensity, Disaggregation, and Surface Modification. *Electron. Commun. Jpn.* **2007**, *90*, 93.
85. Petrov, I.; et al. Polydispersed Detonation Nanodiamond and Approaches for Its Fractioning. *NSTI* **2006**, *1*, 150.
86. Ozawa, M.; et al. Preparation and Behavior of Brownish, Clear Nanodiamond Colloids. *Adv. Mater.* **2007**, *19*, 1201.
87. Liang, Y.; Ozawa, M.; Krueger, A. A General Procedure to Functionalize Agglomerating Nanoparticles Demonstrated on Nanodiamond. *ACS Nano* **2009**, *3*, 2288.
88. Kanygin, M. A.; et al. Polymer-Assisted Forge-Rolling Disaggregation of Detonation Nanodiamonds and Onion-Like Carbon. *Int. J. Nanotechnol.* **2015**, *12*, 182.
89. Morita, Y.; Takimoto, T.; Yamanaka, H.; Kumekawa, K.; Morino, S.; Aonuma, S.; Kimura, T.; Komatsu, N. A Facile and Scalable Process for Size-Controlled Separation of Nanodiamond Particles as Small as 4 nm. *Small* **2008**, *4*, 2154.
90. Peng, W.; et al. Gram-scale fractionation of nanodiamonds by density gradient ultracentrifugation. *Nanoscale* **2013**, *5*, 5017.
91. Gaillou, E.; Post, J.; Byrne, K.; Butler, J. Study of the Blue Moon Diamond. *Gems. Gemol.* **2014**, *50*, 280.
92. Baranov, P. G.; et al. Enormously High Concentration of Fluorescent Nitrogen-Vacancy Centers Fabricated by Sintering of Detonation Nanodiamonds. *Small* **2011**, *7*, 1533.
93. Su, L.; Fang, C.; Chang, Y.; Chen, K.; Yu, Y.; Hsu, J.; et al. Creation of High Density Ensembles of Nitrogen Vacancy Centers in Nitrogen-Rich Type Ib Nanodiamonds. *Nanotechnology* **2013**, *14*, 315702.
94. Evelyn, M. P.; et al. US Patent 6,887,144.
95. Vlasov, I. I.; Barnard, A. S.; Ralchenko, V. G.; et al. *Adv. Mater.* **2009**, *21*, 808.
96. Neu, E.; Arend, C.; Gross, E.; Guldner, F.; et al. *Appl. Phys. Lett.* **2011**, *98*, 243107.
97. Wang, C.; Kurtsiefer, C.; Weinfurter, H.; Burchard, B. Single Photon Emission from SiV Centres in Diamond Produced by Ion Implantation. *J. Phys. B* **2006**, *37*, 37.
98. Zhang, H.; et al. Silicon-Vacancy Color Centers in Nanodiamonds: Cathodoluminescence Imaging Markers in the Near Infrared. *Small* **2014**, *10*, 1908.
99. Davydov, V. A.; Rakhmanina, A. V.; Lyapin, S. G.; llichev, I. D.; Boldyrev, K. N.; Shiryaev, A. A.; Agafonoy, V. N. Production of Nano- and Microdiamonds with Si–V

and N–V Luminescent Centers at High Pressures in Systems Based on Mixture of Hydrocarbon and Fluorocarbon Compounds. *JETP Lett.* **2014**, *99*, 585.
100. Heyer, S.; Janssen, W.; Turner, S.; Lu, Y. G.; Yean, W. S.; Verbeeck, J.; Haenen, K.; Krueger, A. Toward Deep Blue Nano Hope Diamonds: Heavily Boron-Doped Diamond Nanoparticles. *ACS Nano* **2014**, *8*, 5757.
101. Kondo, T.; Okada, N.; Yamaguchi, Y.; Urai, J.; Aikawa, T.; Yuasa, M. Boron-Doped Nanodiamond Powder Prepared by Solid-State Diffusion Method. *Chem. Lett.* **2015**, *44*, 627.
102. Krutko, O. B.; Kodel, P. B.; Wu, R. L. C.; Fries-Carr, S. J.; Heidger, S.; Weimer, J. P-Type Polycrystalline Diamond Layers by Rapid Thermal Diffusion of Boron. *Appl. Phys. Lett.* **2000**, *76*, 849.
103. Sung, T.; Popovici, G.; Prelas, M. A.; Wilson, R. G.; Loyalka, S. K. Boron Diffusion Into Diamond Under Electric Bias. *J. Mater. Res.* **1997**, *12*, 1169.
104. Suarez, A.; Prelas, M. A.; Ghosh, T. K.; Tompson, R. V.; Loyalka, S. K.; Miller, W. H.; Viswanath, D. S. *J. Wide Bandgap Mater.* **2002**, *10*, 15.
105. Mendez, A. E.; Prelas, M. A.; Glascock, M.; Ghosh, T. K. A Novel Method for the Diffusion of Boron in 60–80 micron Size Natural Type Iia Powder. *Mater. Res. Soc. Symp. Proc.* **2006**, *929*, 0929-II05-03.
106. La-Torre-Riveros, L.; Guzman-Blas, R.; Mendez-Torres, A. E.; Prelas, M.; Tryk, D. A.; Cabrera, C. R. Nanoparticles as a Support for Pt and PtRu Catalysts for Direction Methanol Fuel Cells. *ACS Appl. Mater. Interfaces* **2012**, *4*, 1134.
107. Ekimov, E.; Kudryavstev, O.; Khomich, O.; Lebedev, O.; Dolenko, T.; Vlasov, I. I. High-Pressure Synthesis of Boron-Doped Ultrasmall Diamonds from an Organic Compound. *Adv. Mater.* **2015**, *27*, 5518.

## ACKNOWLEDGEMENT

Gary McGuire from International Technology Center is acknowledged for his valuable comments.

CHAPTER

# Structure, shape, defects and impurities in nanodiamonds investigated by HRTEM and STEM-EELS

3

**Stuart Turner**

*Aurubis Belgium Research, Development and Innovation, Olen, Belgium*

## Contents



In this chapter, the characterization of nanodiamond particles by means of transmission electron microscopy is covered. Section 3.1 describes the most commonly used TEM imaging techniques that are applicable in the field of nanodiamond characterization. In Section 3.2, the technique of electron energy-loss spectroscopy is introduced and applied to the study of nitrogen and boron impurities in detonation nanodiamond. Section 3.3 consists of several examples of surface functionalized nanodiamond materials, studied by TEM imaging and electron energy-loss spectroscopy.

# 3.1 IMAGING STRUCTURE, MORPHOLOGY, AND DEFECTS IN NANODIAMONDS BY TEM

## 3.1.1 INTRODUCTION

The focus of this chapter will lie on the characterization of nanodiamond materials by transmission electron microscopy (TEM) and electron energy-loss spectroscopy (EELS) in a transmission electron microscope. It is not meant as a complete reference on (scanning) transmission electron microscopy or EELS; only a short introduction will be given into each technique. The possibilities of these techniques in the characterization of nanodiamond materials will be elaborated through several examples from the literature. The reader is referred to the plethora of excellent books on scanning TEM and electron energy-loss spectroscopy in the literature ([1–4] amongst many others) for a more in-depth discussion of the techniques presented here.

Transmission electron microscopy uses high-energy electrons in combination with electromagnetic lenses to focus an electron beam on a given sample and image a magnified version of the sample on a suitable recording device. Owing to the high acceleration voltages used in TEM (typically several hundred kV), the wavelength of the accelerated electrons is far shorter than the typical wavelength of light, making the resolving power of an electron microscope far greater than that of its optical counterpart. There are two main modes of operation in a transmission electron microscope; in TEM the whole sample is simultaneously illuminated by a wide electron beam, with TEM images typically being acquired on a CCD camera. In scanning TEM (STEM), a fine electron probe is scanned over the sample, and the scattered electrons are collected in each scan point to create an image. The image contrast in STEM depends upon which electrons (which scattering angles) are collected to form the image.

The limiting factor for the (S)TEM image resolution was long not the electron wavelength (approximately 2 pm for a 300 kV acceleration voltage), but rather lens aberrations like spherical and chromatic aberration caused by the use of round electromagnetic lenses in a conventional transmission electron microscope.[5] These aberrations limited the resolution of a typical high resolution TEM microscope to a point resolution at optimal defocus of approximately 1.4–1.7 Å (140–170 pm) at typical acceleration voltage of 300 or 200 kV, far from theoretical resolutions of only a few picometers. However, the successful implementation of spherical aberration-correction in both STEMs and TEMs at the turn of the century means that modern instruments are now able to reach image resolutions below 50 pm (0.5 Å) in both TEM and STEM mode at conventional acceleration voltages of 300 kV. An added benefit of the implementation of aberration correctors is that electron microscopes also be operated at lower voltage of 80, 40, or even 20 kV, while keeping atomic resolution.[6–10] During the past decade low voltage microscopy has been used to great success in imaging beam sensitive materials like nanotubes, graphene, and hexagonal BN,[11,12] and can allow for prolonged and detailed experiments to be performed while minimizing the risk of knock-on damage or surface graphitization in the case of nanodiamond.[13,14]

### 3.1.2 BRIGHT-FIELD TEM IMAGING AND ELECTRON DIFFRACTION OF NANODIAMOND MATERIAL

In bright-field TEM, the sample is imaged using a parallel incoming electron beam. It is the most conventional TEM method, and examples of bright-field TEM images of nanodiamond materials are widespread. In general, only a (small) amount of the electrons that have passed through the specimen are used for imaging in the bright-field imaging mode; an objective aperture can be placed in the back focal plane of the objective lens to cut out all electrons except the transmitted beam or a selected amount of reflections, providing a bright-field image with enhanced contrast. There are two main contrast regimes in bright-field imaging. For non-crystalline materials, *mass-thickness contrast* (also known as absorption contrast) is dominant. In crystalline materials like nanodiamond, this amplitude-type contrast is in direct competition with *diffraction contrast*, which originates from elastic Bragg scattering in the crystal and is dependent upon the orientation of the crystalline specimen with respect to the incoming electron beam.

In an electron microscope diffraction and imaging go hand-in-hand. The back focal plane of the objective lens in a TEM is also the diffraction plane, as the back-focal plane is the Fourier conjugate of the object plane. This means that the microscope operator can switch between the diffraction pattern and the real space image by a simple push of a button (see Fig. 3.1 for detailed ray diagrams). In other words, on top of the information that can be retrieved from bright-field TEM images (like particle size, morphology, and agglomeration), the microscope operator can easily have access to crystallographic information on the e.g. nanodiamonds under study through the electron diffraction pattern.

A typical bright-field TEM image and electron diffraction pattern obtained from an ozone-treated detonation nanodiamond (DND) sample are displayed in Fig. 3.2.[14,15] The DND particles have primary particle sizes ranging between approximately 2 and 20 nm, and are in general agglomerated into larger structures. The agglomeration of the DND particles, which was covered in more detail in Chapter 2, is typical for nanosized material, and is mainly due to van der Waals attractive forces.[16,17]

The crystal structure of the particles can be determined from the inset electron diffraction pattern. As the diffraction pattern is taken from many nanodiamond particles with varying orientations simultaneously, the angular information is lost. This leads to an electron diffraction ring pattern and leaves only the ring interspacings to determine the crystal phase through Braggs law (just like a powder XRD pattern).[18] In this case, only reflections (rings) corresponding to the diamond crystal structure (Fd-3m, Space Group No. 227) are present (111, 220, and 311 reflections are clearly visible). No graphite or n-diamond[19] reflections are visible, an important indication for a high purity of the DND material.

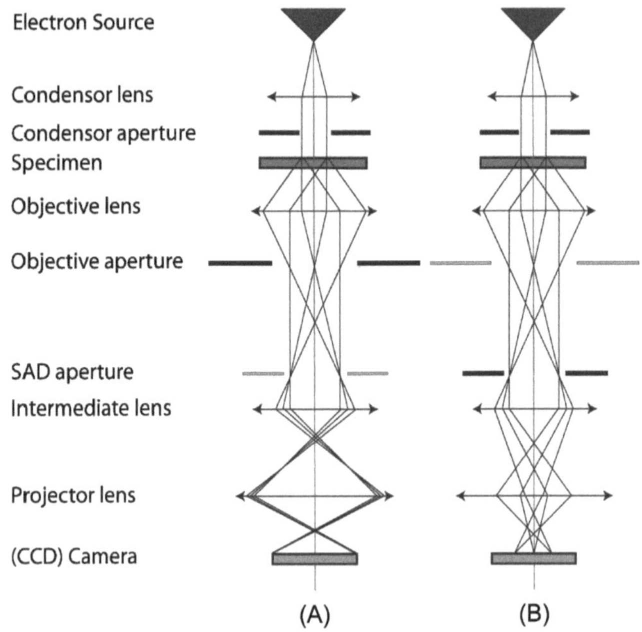


**FIGURE 3.1**

Ray diagrams for two basic modes of operation in a TEM: (A) bright field imaging (B) selected area electron diffraction (SAED)

### 3.1.3 HIGH RESOLUTION TEM IMAGING

Unlike X-ray or neutron diffraction, electron diffraction is highly dynamical, meaning that multiple electron scattering events with relatively large phase changes to the scattered electron beam occur frequently in the materials under study in an electron microscope. In high resolution TEM (HRTEM) a large objective aperture (or no aperture at all) is inserted in the back focal plane of the objective lens, allowing the diffracted beams to interfere both with each other and the transmitted beam, forming an interference (fringe) pattern. As a result, the contrast in such an image is dependent on the phase of the various beams, giving rise to the name phase-contrast imaging for HRTEM imaging. In some cases these interference patterns or high resolution images can be directly interpreted, usually however image simulations need to be carried out in order to correctly interpret the images.

An example of a high resolution, aberration-corrected TEM image of a typical detonation nanodiamond particle sample is presented in Fig. 3.3. To obtain this image, the electron microscope was operated at 80 kV (below the knock-on damage threshold for diamond) with the spherical aberration coefficient ($C_s$) tuned to +15 μm. To extend the information limit of the microscope (which is normally limited by chromatic aberration) the monochromator of the microscope was excited to

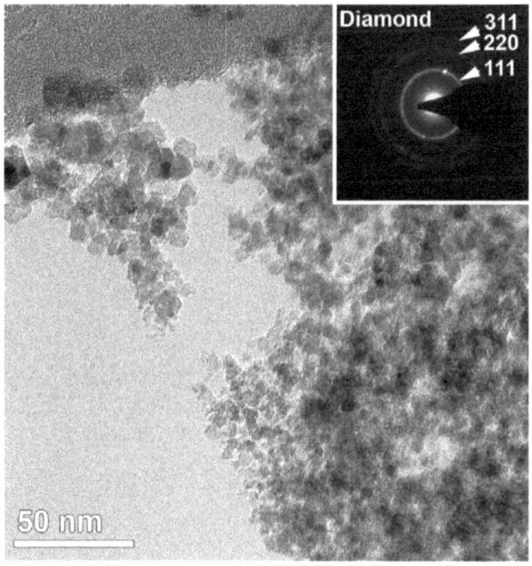


**FIGURE 3.2**

Low magnification image of detonation nanodiamond particles.[15] The particles have sizes ranging from ~2–20 nm. The inset electron diffraction ring pattern evidences the diamond crystal structure

achieve an energy spread of the order of 0.3 eV for the illuminating electron beam. By further extending the information transfer of the microscope, intricate details of the nanodiamond surface become visible (see Fig. 3.3A). The DND particle is imaged along the [011] zone axis orientation (as evidenced by the inset Fourier transform (FT) or "digital diffraction pattern"), allowing {111}-type surface facets and $\Sigma = 3$ {111} twin boundaries to be imaged edge-on. Two coherent $\Sigma = 3$ {111} twin boundaries run through the particle. This type of twin is typical for detonation nanodiamond as it has the lowest defect energy of all diamond defects. It results from the fact that this twin type only demands a small deviation from the ideal lattice, namely in the fourth neighbor lattice positions. The abundance of these defects has been attributed to the increased growth rate of twinned nanodiamonds in comparison to untwinned nanodiamond under detonation synthesis conditions.[20]

The nanoparticle shows mainly {111}-type surface facets with some {100} truncation, a result of the low surface energy of these low-index facets.[18] The {100}-type truncation is imaged in more detail in Fig. 3.3B. Under the imaging conditions used ($C_s = +15$ μm, 4 nm underfocus), the atomic columns are imaged in white, as evidenced by the simulation of a truncated diamond particle in Fig. 3.3C. The atomic potential overlay in Fig. 3.3C also demonstrates that even though the information transfer of the microscope is extended by use of the monochromator, the 89 pm carbon dumbbells are not directly resolved in the image. An elongation of the column contrast in the dumbbell direction is however visible. The combination of the HRTEM image and image simulation allows us to conclude that the surfaces (both the

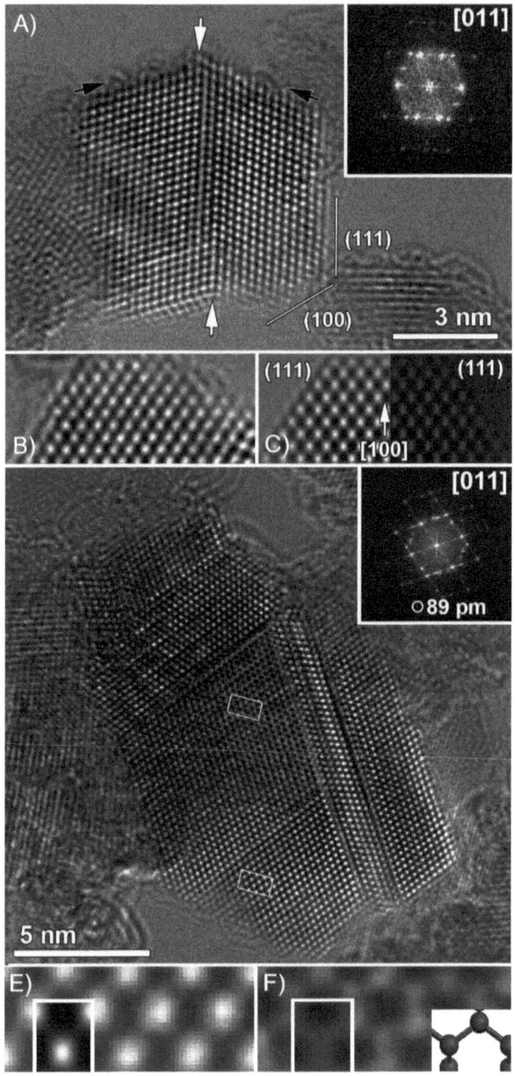


**FIGURE 3.3**

Low voltage, aberration-corrected images of DND particles.[14,22] (A) HRTEM image of a typical detonation nanodiamond particle exhibiting two coherent $\Sigma = 3\ \{111\}$ twin defects. The diamond surfaces show minimal presence of graphitic material. (B) Enlarged image of the (100) truncation right bottom in (A). Both {100} and {111} type surface facets are clean. (C) Image simulation performed for $C_s = 15$ μm, $\Delta f = -4$ nm, the right-hand side has the projected potential overlaid, demonstrating the white image contrast peaks at the atomic positions. (D) HRTEM image of a multiply-twinned DND particle. Inset simulations performed for $C_s = 15$ μm and $\Delta f = -4$ nm at a thickness of 4 and 14 nm

{111} and {100} type surfaces) are extremely clean. In the region cut out in Fig. 3.3B, no graphitic surface material can be made out (although a minimal surface layer is visible on the top two {111} facets in Fig. 3.3A as indicated by black arrows).

A second, multiply-twinned nanoparticle is imaged in Fig. 3.3D. This type of multiply-twinned nanoparticle was previously reported.[21] As was the case for the particle in Fig. 3.3A, this multiply-twinned ND is imaged along the [011] zone axis orientation, with mainly {111} type and some {100}-type surface facets visible. The [011] zone axis orientation is evidenced by the inset FT. Interestingly, this FT (unlike the FT for the smaller ND) shows the presence of a (400) reflection corresponding to the 89 pm dumbbell spacing. In Figs. 3.3E and 3.3F, two enlarged regions of the particle are imaged, together with their corresponding image simulations. In Fig. 3.3E, arising from a thin region of the particle (indicated by a rectangle in Fig. 3.3D), the 89 pm dumbbells are imaged as elongated white contrast (as was the case in Fig. 3.3A). However, in the thicker region of the particle, the dumbbell is clearly resolved, most likely due to dynamical scattering effects. Although the presence of multiple coherent {111} twins was apparent from earlier conventional HRTEM work, several features remained unclear until now. The first is the surface structure, which in this sample exhibits a minimal graphitic presence. Secondly, the details of the multiple twin boundaries are imaged; no indication for an enrichment of amorphous carbon or nitrogen clusters at the twin boundaries can be derived from the HRTEM image. To determine whether amorphous carbon or nitrogen atoms are enriched at the defects and twin boundaries, spatially resolved EELS can be carried out.

### 3.1.4 SCANNING TRANSMISSION ELECTRON MICROSCOPY

In STEM, the electron beam is focused to an angstrom-sized probe and is scanned over the sample by deflection coils. Depending upon the transmitted electrons that are used to form the image, a bright-field STEM image or dark-field STEM image can be formed.[4]

The most-used detector in STEM is by far an annular detector, used to form annular dark-field images. By using the intermediate lenses to change the camera length, the angular range of the electrons falling onto the annular detector can be controlled. In high-angle annular dark-field STEM only the electrons that have been scattered under large angles by the specimen (Rutherford-like scattering) are collected on the annular detector. This type of imaging is also known as Z-contrast imaging, as the Rutherford-like scattering of the electrons scattered to high angles leads to an image intensity that is proportional to atomic number ($Z^{\sim 1.7}$) and to the thickness of the specimen. HAADF-STEM imaging is therefore both chemically sensitive and thickness sensitive. This imaging mode is also known as incoherent imaging, as the image that is formed e.g. in high resolution HAADF-STEM is no longer a phase contrast image like in HRTEM, but is rather a directly interpretable mass-thickness projection of the imaged crystal. An example of the sensitivity of HAADF-STEM imaging is displayed in Fig. 3.5; the images show that typical DND material is covered with individual, heavy Fe atoms from the detonation process, which cannot be seen in

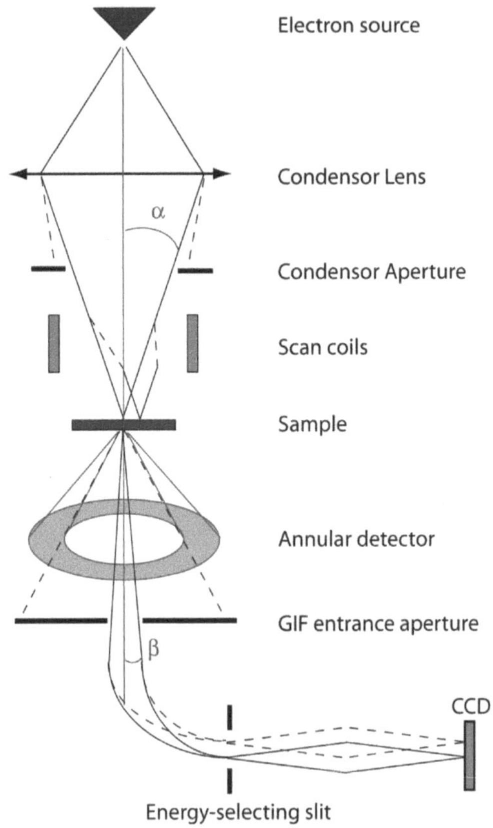


**FIGURE 3.4**

Simplified setup of the electron-optical path for (HA)ADF-STEM-EELS. The fine electron probe is scanned over the specimen by deflection coils and the (high-angle) scattered rays are collected on an annular detector. Simultaneously, the forward scattered electrons can be detected on a bright field or second annular dark field detector, or can be used to collect an EELS spectrum in the post-column energy filter (GIF)

conventional HRTEM images. This type of impurities can also be measured by techniques like ICP-MS, as covered in Chapter 5.

By using an annular detector to collect the electrons scattered by the sample over larger angles, image acquisition and the collection of forward scattered electrons using an energy filter for electron energy-loss spectroscopy can happen simultaneously. (See Fig. 3.4.) This makes (HA)ADF-STEM-EELS ideally suited to the study of 0-D nanomaterials, as imaging can be directly combined with local chemical information down to the angstrom level.

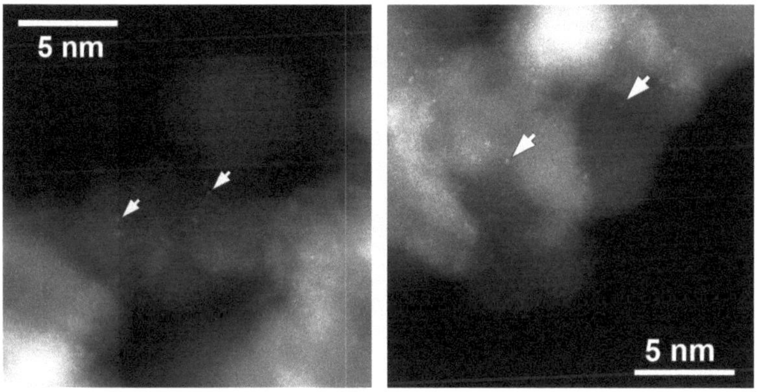


**FIGURE 3.5**

Aberration-corrected HAADF-STEM images of ozone-treated DND particles. Individual heavy Fe atoms from the detonation process decorate the low-contrast diamond nanoparticles and are clearly visible due to the Z-contract regime of HAADF-STEM

## 3.2 PROBING THE ELECTRONIC STRUCTURE OF DIAMOND AND ITS DOPANTS BY ELECTRON ENERGY-LOSS SPECTROSCOPY

### 3.2.1 INTRODUCTION

When electrons are scattered inelastically in a solid, conservation of energy dictates that the energy transferred in a single-electron excitation or in the excitation of a phonon or plasmon is equal to that lost by the scattered fast electron. In electron energy-loss spectroscopy, the processes of electron excitation are studied by measuring the energy-loss of the primary fast electrons. EEL spectroscopy is particularly well-suited to study light elements like nitrogen, oxygen, boron, and carbon. Heavy elements like gold and platinum are generally not well-suited to EELS.

A typical EELS spectrum consists of three major regions, as visible in Fig. 3.6:

**(a)** The zero-loss peak at 0 eV from the electrons that are elastically scattered (the majority of the fast electrons) and from electrons that have transferred less energy than the energy resolution of the energy filter and the incoming electron beam.

**(b)** The low-loss region from 1–100 eV showing characteristic peaks from the bandgap, interband transitions and from the excitation of bulk and surface plasmons.

**(c)** The core-loss region from 100 eV onwards showing the element-specific ionization edges on top of a decaying background, which can usually be approximated by a power-law function ($A \cdot E^{-r}$). The excitation edges correspond to the

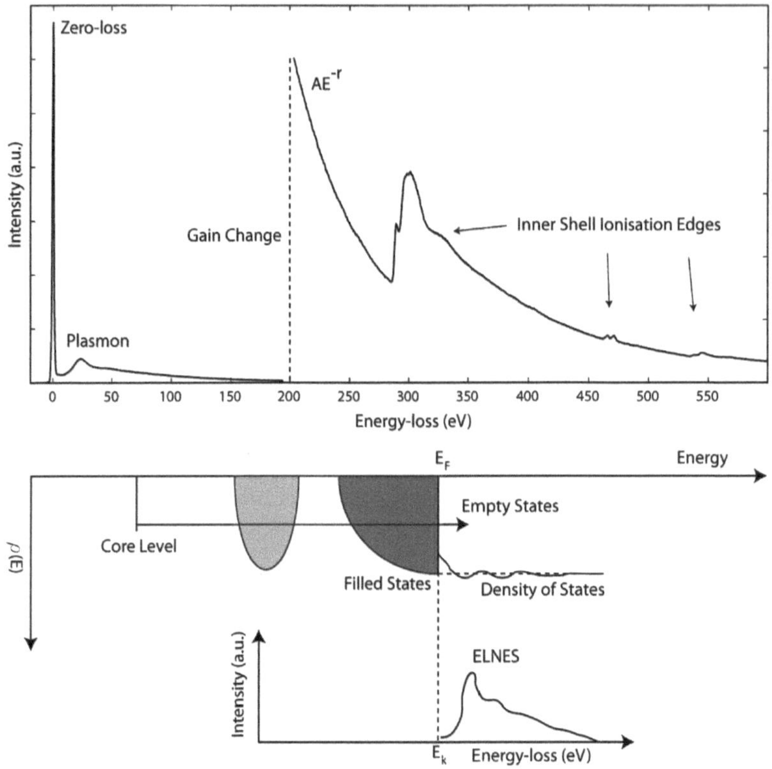


**FIGURE 3.6**

(Top) typical EELS spectrum showing zero-loss, low-loss, and core-loss regions. (Bottom) ELNES signature that reflects the density of unoccupied states[23]

energy-loss of inner-shell electrons that are excited to a higher, unoccupied state above the Fermi level.

## 3.2.2 LOW-LOSS EELS

Reference low-loss spectra for diamond, graphite and amorphous carbon are shown in Fig. 3.7, together with low-loss EELS data from three individual clean nanodiamonds with different sizes. The low-loss EELS spectra in Fig. 3.7B all show a prominent plasmon peak around 34 eV; close to the plasmon energy of bulk diamond (33 eV) confirming that the bulk of DND material is diamond.[1,24,25] A small shoulder is measured around 22 eV, which is attributed to the diamond surface plasmon. The peak does not arise from graphitic or amorphous material, as no $\pi$-plasmon is visible. The smaller the particle, the more prominent the surface plasmon peak becomes, as the

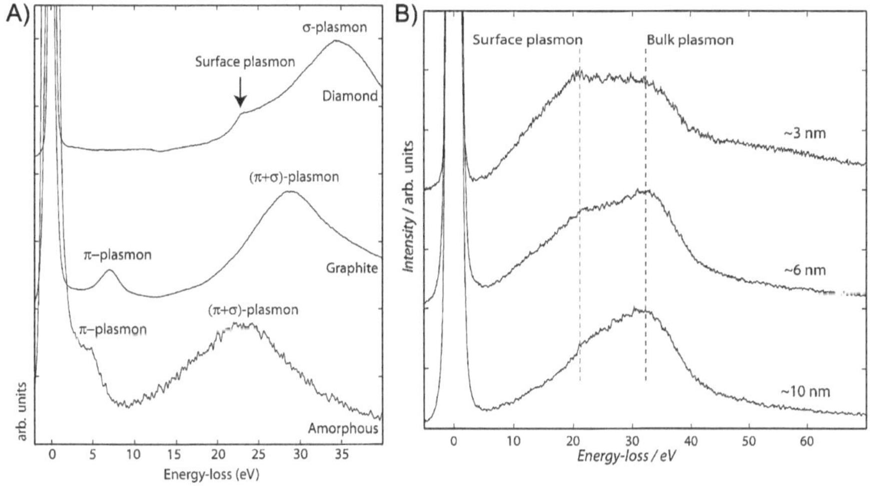


**FIGURE 3.7**

Diamond, graphite and amorphous carbon low-loss spectra (A) together with low loss spectra (B) taken from individual DND particles[26,28]

surface to volume ratio increases.[26] The bandgap around 5–6 eV can also be measured under specific experimental conditions.[27]

## 3.2.3 CORE-LOSS EELS

### *3.2.3.1 Nanodiamond carbon K-edge ELNES*

In the core-loss EELS region, the shape of the excitation edges is linked to this unoccupied density of states and is called the energy loss near edge structure (ELNES). The ELNES therefore contains information about the bonding and electronic state of the probed material and as a result can be used to distinguish between allotropes of the same material and to determine local bonding. A typical example is the difference between the fine structure of the carbon K-edge in the case of diamond ($sp^3$ bonded carbon) and graphite or amorphous carbon (sp, $sp^2$ bonded carbon), as is clear from Fig. 3.8A. Typical EELS spectra taken from cleaned DND samples are displayed in Fig. 3.8B.[21]

All nanodiamond EELS spectra are aligned to the carbon K-edge onset at 284 eV.[1] The spectra show typical diamond-like ($\sigma^*$) energy-loss near edge structures starting at 292 eV (see diamond reference). The spectra also show a small carbon pre-peak, typical for graphitic material at 285 eV ($\pi^*$ contribution – see also graphite reference),[1] arising from surface carbon.[25] In many nanodiamond samples, a clear signal for nitrogen (nitrogen K-edge) can be measured, with a half maximum at 403 eV and peaking at 406 eV. A typical background subtracted nitrogen K-edge is shown in the

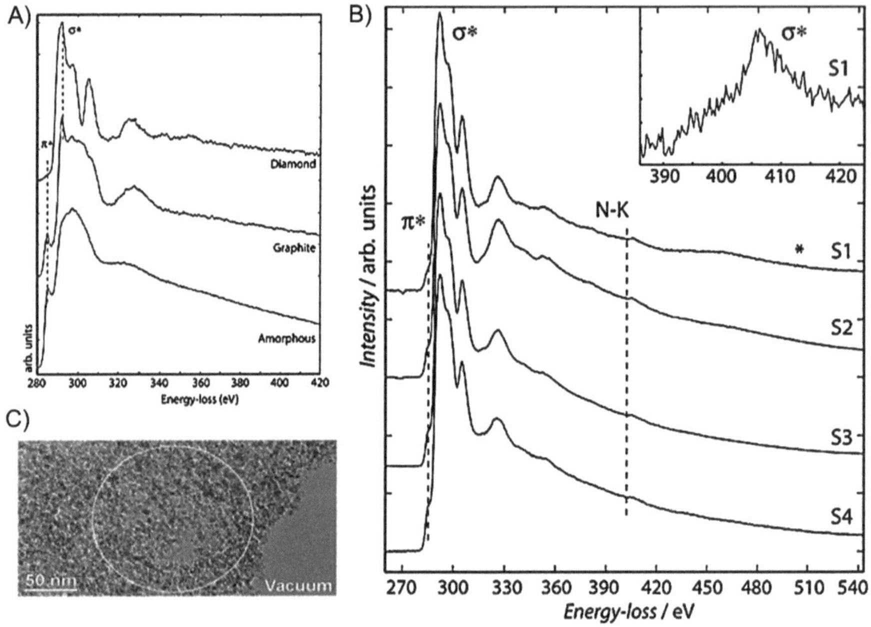


**FIGURE 3.8**

Diamond, graphite, and amorphous carbon K-edge ELNES signatures taken as a reference from a thin diamond film, a thin graphite film, and an amorphous carbon support (top panel) together with background subtracted, normalized core-loss EELS spectra from several cleaned DND samples.[21] Nitrogen is detected in all samples through the K-edge with half maximum at 403 eV. No oxygen is detected at 532 eV (indicated by $*$). (Inset) Background subtracted nitrogen K-edge signal showing the nitrogen $\sigma^*$ peaking at 406 eV. The image in (C) indicates a typical investigated area for TEM-EELS

inset as an example. No oxygen signal (oxygen K-edge at 532 eV) was detected in these investigated samples.

The area under the ionization edges can be used for elemental quantification. In conventional quantification, the decaying background is removed and the integrated intensity under the edges is measured and quantified using a suitable atomic cross section (which is effectively a probability for inelastic scattering). Model-based quantification can be used to obtain the most precise and accurate results possible.[29] Typical DND particles can contain up to several at.% of embedded nitrogen.[21,30]

### 3.2.3.2 Monochromatic EELS data

Modern TEM instruments can be equipped with an electron monochromator. These monochromators allow obtaining a far smaller energy spread of the electron beam than inherently present for the electron gun. This monochromatic electron beam can then be used for electron energy-loss spectroscopy. The enhanced energy resolution

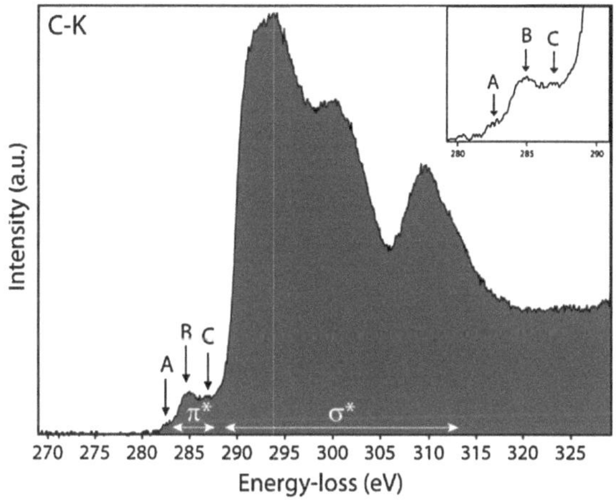


**FIGURE 3.9**

Experimental carbon K-edge fine structure (ELNES) acquired with monochromatic electron illumination. The $\sigma^*$ feature starting at 290 eV is typical for diamond. The $\pi^*$ contribution around 285 eV consists of 3 pre-peaks, at 282.7 eV (A), 285 eV (B), and 286.6 eV (C)

offered by an electron monochromator allows small details in ELNES signatures to be studied. An example of a carbon K-edge EELS spectrum taken from a nanodiamond sample illuminated with a monochromatic electron source ($\Delta E = 250$ meV) is plotted in Fig. 3.9. The spectrum is dominated by the $1s \rightarrow \sigma^*$ feature starting at 290 eV, which is indicative of carbon in an $sp^3$ coordination. As is typical for detonation nanodiamond, a small pre-peak at 285 eV is also present, which is absent for bulk diamond. This $1s \rightarrow \pi^*$ feature is indicative of carbon in a lower or distorted $sp^3$ coordination. As a result of the improved energy resolution provided by the electron monochromator, the data shows a splitting of the $\pi^*$ contribution into three pre-peaks at 282.7 eV (A), 285 eV (B), and 286.6 eV (C). These three peaks have been experimentally observed in UNCD diamond films and in interstellar nanodiamond by EELS,[31,32] in detonation nanodiamond using EXAFS[33] and have been theoretically predicted by tight-binding molecular dynamics simulations[34] and density functional theory.[33] Peak A has been observed in amorphous carbon and diamond surfaces and is associated with mid-gap states localized at dangling bonds, which can be stable under UHV conditions, and distorted $sp^3$ sites. Peaks B and C are attributed to unoccupied $\pi^*$ antibonding states of $sp^2$ coordinated carbon, most likely arising from a fullerene-like surface reconstruction and deviation from the ideal $sp^3$ coordinated carbon symmetry at defect sites.

### 3.2.3.3 Nitrogen detection by core-loss EELS

Besides confirming the presence of impurities like nitrogen in nanodiamond particles through core-loss EELS of many particles simultaneously, spatially resolved EELS

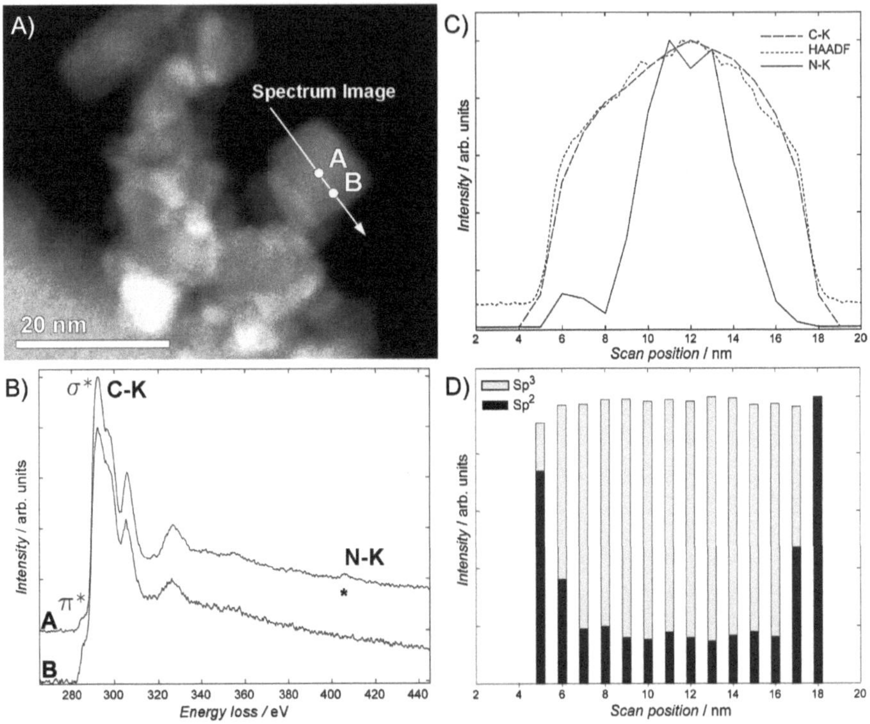


**FIGURE 3.10**

Central nitrogen embedding in a DND nanoparticle[28,37] (A) HAADF-STEM image of a large DND nanoparticle. The 20 point line scan is indicated by an arrow. (B) Background subtracted, normalized EELS spectra from positions A and B from the large DND particle. Position A on the particle shows a clear nitrogen peak at 403 eV; position B closer to the surface does not. (C) Normalized nitrogen peak strength, carbon peak strength, and HAADF-STEM image intensity for the line scan from (A). (D) Normalized $sp^2$ and $sp^3$ content at each scan position modeled by a lorentzian at 285 and 292 eV for the $\pi^*$ and $\sigma^*$ contribution, respectively. The $sp^3$ content at the center is ~95%, at the edge ~86%

in a scanning TEM can be used to locally determine the position of e.g. nitrogen in the diamond lattice.[35,36] Using STEM-EELS, point analysis, 2D line scans and 3D "spectrum images" can be carried out by acquiring an EELS spectrum at each probe position. In modern aberration-corrected instruments, probe sizes well below 1 Å are feasible, allowing routine atomic resolution analysis.[8]

An example of a line scan experiment is displayed in Fig. 3.10. The particle is a single diamond crystal, approximately 13 nm in diameter. The performed line scan is displayed by an arrow in Fig. 3.10A. In total 20 EELS spectra were recorded in the line scan of the 13 nm grain. Two typical spectra – one at position A and one at position B on the particle – are displayed in Fig. 3.10B. Both spectra show a carbon K-edge with $\sigma^*$ contributions typical for diamond at 292 eV and small $\pi^*$ contributions at 285 eV arising from $sp^2$ hybridized surface carbon. Spectrum B shows a larger graphitic $\pi^*$ contribution than spectrum A, which is to be expected as a larger surface contribution is measured at the thinner position B. The spectrum at position A clearly shows a nitrogen signal at 403 eV (indicated by an asterisk), while the spectrum from position B does not. Using model-based quantifications methods the nitrogen content was determined over the whole line scan. It is plotted in Fig. 3.10C as the nitrogen signal strength, together with the carbon signal strength and the HAADF-STEM signal intensity, which may be regarded as a thickness profile. In Fig. 3.10D, the normalized $sp^2$ and $sp^3$ carbon content is plotted. These values were obtained by modeling the $\pi^*$ and $\sigma^*$ contribution using a lorentzian at 285 and 292 eV, respectively. To obtain the plot, the area under the modeled lorentzian peaks was divided by the total intensity of the $\pi^*$ and $\sigma^*$ contribution. In this way the $sp^2$ carbon content was determined to be only 5% at the core positions, rising at the edge of the particle to a maximum of 14% $sp^2$ carbon. It can be seen from the plots in (C) and (D) that there is only a detectable nitrogen signal within the inner 6 nm core of the DND particle. Towards the surface of the particle where the $sp^2$ ratio is higher, no nitrogen is detected. The maximum nitrogen content in the core of the particle was measured to be 3.5 at.%. The lack of an N signal towards the surface is not due to noise, as can be seen when comparing spectrum A and B in Fig. 3.10B. Similar line-scan experiments were carried out by other groups on high nitrogen-doped DND material. These results also indicated the nitrogen was preferentially embedded in the diamond cores.[30]

A second example of STEM-EELS results acquired on a multiply-twinned nanodiamond particle is displayed in Fig. 3.11. The HRTEM image of the multi-twin (taken at 120 kV acceleration voltage) again shows the presence of simple $\Sigma = 3$ coherent twin boundaries and lamellar multiply-twinned regions, probably containing a large degree of other defects. The particle surface appears to be nearly non-$sp^3$ carbon free. For the nanodiamond particle in Fig. 3.11, a $53 * 67$ pixel spectrum image was acquired at 2 Å intervals, recording the carbon K-edge (285 eV) and nitrogen K-edge (403 eV) simultaneously. $Sp^3$ and non-$sp^3$ carbon maps were generated from the acquired spectra by fitting references for amorphous carbon and diamond to the acquired carbon K-edge. The nitrogen map was generated by integrating the intensity under the background-subtracted nitrogen K-edge at 403 eV. The generated maps are displayed in Fig. 3.11C ($sp^3$ C), Fig. 3.11D (non-$sp^3$ C), and Fig. 3.11E (N).

The $sp^3$ carbon map clearly indicates the size of the nanodiamond core, which is almost equal to the particle size from the ADF image. Although only minimal graphitic material was visible in the HRTEM image, the non-$sp^3$ C map clearly follows the surface contours of the nanodiamond particle. The origin of this signal is therefore attributed mainly to remaining fullerene-like $sp^2$ material and dangling bonds, which can be stable under the UHV conditions in the microscope, at the surface of the nanodiamond particle. A significant non-diamond C signal is also measured from both the coherent $\Sigma = 3$ twin boundaries (example indicated by A) as

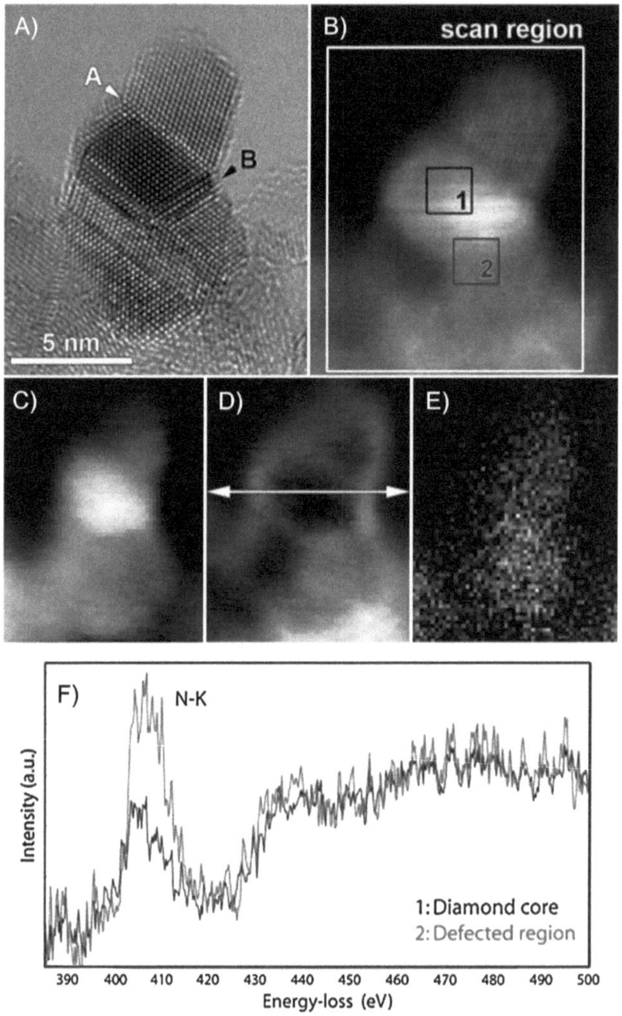


**FIGURE 3.11**

Spatially resolved EELS analysis of a multiply-twinned nanodiamond; (A) HRTEM image of a typical DND particle showing the presence of a single coherent twin indicated by A and multiple parallel twin regions indicated by B. (B) ADF-STEM image of the same particle; the white rectangle depicts the STEM-EELS scanned region ($53 * 67$ pixel spectrum image). (C) $sp^3$ C map; (D) non-$sp^3$ C map; (E) N map; and (F) N K-edge spectra from region 1 and 2 indicated in (B). The N content is $2.7 \pm 0.2$ and $3.8 \pm 0.2$ at.% in regions 1 and 2, respectively. Defected region 2 shows a clear enrichment of nitrogen. (G) Intensity profiles for non-$sp^3$ C (dashed line), $sp^3$ C (black line), and N (bars) taken over the region indicated by the arrow (D) (5 px integration width)

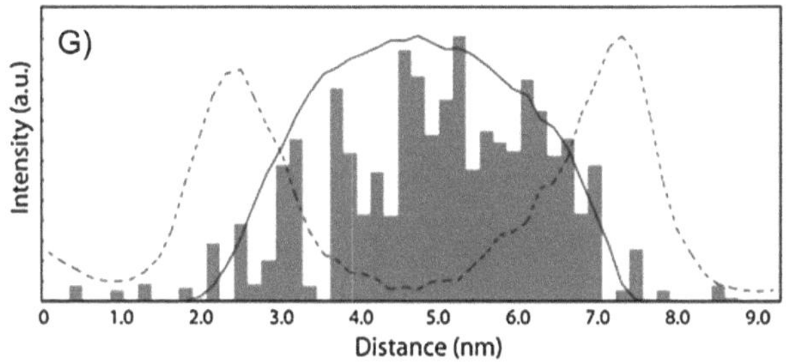


**FIGURE 3.11**

*Continued*

well as the lamellar twinned regions (example indicated by B). This signal again arises from $sp^2$ or distorted $sp^3$ coordinated carbon, indicating that even the coherent twin boundaries can deviate from their ideal symmetry, and contain some non-$sp^3$ coordinated carbon.[38] The nitrogen K-edge signal is mapped in Fig. 3.11E. As the nitrogen content in the nanodiamond particle is low, the elemental N map displays a relatively high noise level. It is, however, clear that the signal is localized in the diamond core of the particle, in accordance with previous measurements.[39] The line profile in Fig. 3.11G, taken over the region indicated in Fig. 3.11D, confirms the presence of nitrogen in the diamond core.

Interestingly, an enrichment of nitrogen at the multiply twinned region is also apparent from the map. The nitrogen K-edge signal summed over two regions, region 1 being the non-defected diamond region indicated in Fig. 3.11B and region 2 being the multiply-twinned region, are plotted in Fig. 3.11F. The spectra have been aligned to their continuum level to allow easy comparison. It is clear that both regions contain significant amounts of nitrogen, however, the multiply twinned area is nitrogen enriched as evidenced by the enhanced signal. The nitrogen content in both regions was quantified using standard EELS quantification methods (see experimental section). The average N content in pristine diamond core region 1 is 2.7 at.%. In defected region 2, the nitrogen content is more than 1 at.% higher (3.8 at.%). On the other hand, no clear nitrogen enrichment was measured at coherent twin boundary A (not displayed).

As mentioned above, the fine structure of EELS edges is known to be related to the local symmetry and environment of the excited atomic species.[1,23] Therefore, the shape of the nitrogen K-edge fine structure should provide clues about the form in which nitrogen is embedded into the nanodiamond particles. In previous experiments, the shape and position of the nitrogen K-edge fine structure lead our and other groups to the conclusion that nitrogen was tetrahedrally embedded in the diamond particles.[21,30]

The averaged experimental nitrogen K-edge fine structure from a diamond core region (both raw data and smoothed) and the experimental carbon K-edge of nanodiamond are plotted in Fig. 3.12 (left panel). The experimental nitrogen K-edge shows

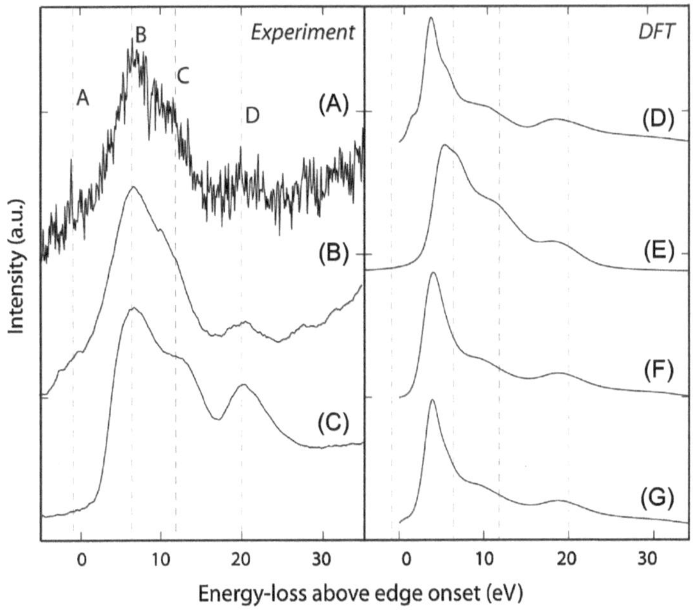


**FIGURE 3.12**

Analysis of the nitrogen K-edge fine structure (ELNES).[14] (Left) Experimental data: (A) experimental N-K ELNES, (B) 1.5 eV smoothed N-K ELNES, (C) experimental nanodiamond C-K ELNES. For nitrogen, peak B corresponds to 406 eV in absolute energy scale; for carbon, peak B corresponds to 292 eV. (Right) DFT calculated data: (D) single substitutional nitrogen $N_s$, (E) A-center (2N), (F) $NV^0$, and (G) $NV^-$

a distinct triangular shape with three pronounced peaks B, C, and D and a less pronounced pre-peak A (the size of which can vary). This nitrogen ELNES shape was found to be similar in other DND samples, even in samples with very high-nitrogen doping.[30] A striking similarity exists between the nitrogen and the carbon (diamond) K-edge fine structures. Peaks B, C, and D are all present in both ELNES spectra, and in the case of the carbon K-edge the pre-peak A at 285 eV is known to be related to sp, $sp^2$ hybridized carbon. The striking similarity between the N and C ELNES points to the fact that nitrogen is embedded into an environment that is highly similar to carbon embedded in diamond; namely a dominant $sp^3$ hybridization. In an attempt to provide more accurate information on type of nitrogen embedding, detailed DFT calculations were performed to calculate the nitrogen K-edge ELNES for four possible embedded states of nitrogen, namely,

- Single substitutional nitrogen; $N_s$ or P1 center
- Two neighboring embedded nitrogen atoms; A-center
- The charge neutral Nitrogen-Vacancy center; $NV^0$

- The negatively charged Nitrogen-Vacancy center; $NV^-$

The DFT calculated spectra, performed in the GGA approximation on $2 \times 2 \times 2$ supercells in the full core hole approximation using the code WIEN2k, are plotted in Fig. 3.12 (right panel). At first glance, all four calculated N-K edge spectra look similar, a result of the similarity of the nitrogen coordination in all four cases. However, small differences can be made out. The spectra for $NV^-$ and $NV^0$ centers are near-identical, and clearly differ from the experimental nitrogen K-edge. Although all calculated spectra have been broadened by 1 eV (to correspond to the experimental broadening), the spectra for $NV^-$ and $NV^0$ are narrower and display fewer features than the experimental spectra. This appears to indicate that the nitrogen present in as-synthesized nanodiamond is not present in large amounts of $NV^-$ or $NV^0$ centers, which is in good agreement with the literature.[39] These optically active structures are only obtained in large amounts after vacancy creation and sintering treatments,[40–42] as described in detail in Chapter 7.

The triangular spectrum for single substitutional nitrogen $N_s$ shows more similarities to the experimental nitrogen K-edge ELNES, reproducing features B, C, and D, as well as a pre-peak to the main feature B. This so-called $\pi^*$ contribution arises from bond length elongation of one of the C–N bonds by approximately 30%, a result of the energy relaxation step in the calculation, and which is in good agreement with the literature.[43] Due to this elongation, the nitrogen atom is present in a more planar coordination with respect to the other 3 bonding carbon atoms, giving rise to the $\pi^*$ contribution. As was the case for the $NV^-$ and $NV^0$ spectra, the spectrum is clearly less broadened than the experimental spectrum. The best correspondence is clearly given by the calculated spectrum for 2 neighboring, substitutional nitrogen atoms (the A-center model). All three peaks at B, C, and D are reproduced in this model, and the peak positions fit well with the experimental positions. The small prepeak A that is present in the experimental ELNES is, however, not reproduced by the A-center model, even though it was reproduced in earlier ELNES simulations using multiple-scattering theory.[44]

#### 3.2.3.4 Boron detection by core-loss EELS

A second technologically relevant dopant in nanodiamonds is boron. For long, the production of boron-doped nanodiamond had remained unsuccessful. However, recent experiments, in which CVD grown B:diamond was milled down to nanosized particles, have proven successful (see Chapter 2). An example of this type of milled boron-doped nanodiamond is displayed in Fig. 3.13. In order to study the morphology of the boron-doped nanodiamond, and investigate the distribution of boron within individual diamond nanoparticles, the boron-doped nanoparticles were studied using a combination of high resolution transmission electron microscopy and spatially resolved electron energy-loss spectroscopy.

Fig. 3.13A is an overview bright-field TEM image of the produced material. The nanodiamond particle size is relatively monodisperse with particles ranging from ap-

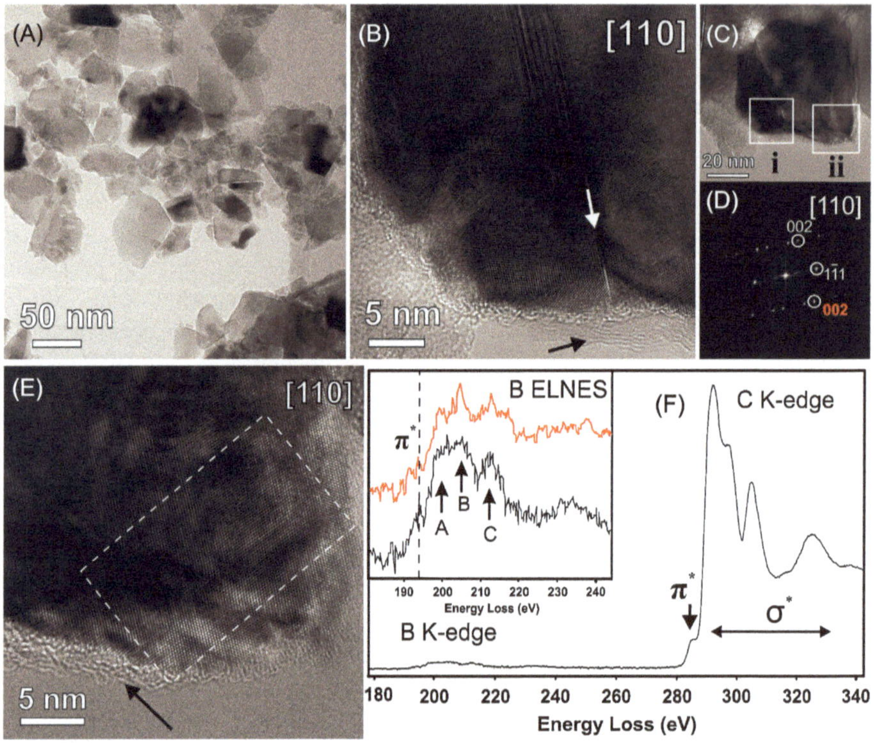


**FIGURE 3.13**

*TEM imaging and EELS spectroscopy of B:NDs.* (A) Low magnification TEM image showing the diamond particles have sizes ranging from ~30 to 70 nm in diameter. (B) and (E) HRTEM images recorded from selected areas (i) and (ii) of a single diamond grain in (C), (D) Fourier transform of the defect indicated by a white arrow in (B), evidencing it is a $\Sigma3$ {111} twin boundary. Both HRTEM images show amorphous and graphitic carbon covering the diamond particle surface (indicated by black arrows). (F) EELS spectrum acquired from a pristine diamond region (indicated by the white dashed rectangle in (E)) demonstrating the presence of boron. (Inset, black spectrum) The B K-edge fine structure. (Inset, red spectrum) B K-edge fine structure reference for B tetrahedrally embedded in CVD diamond

proximately 30–70 nm in diameter. No debris from the milling procedure was found in between the nanoparticles. An overview image of a single 40 nm diameter nanodiamond particle is shown in Fig. 3.13C, with two details of the image presented in (B) and (E). Even though the particles have been milled down to the nanoscale, individual particles can still contain defects, like the $\Sigma = 3$ {111} twin boundary indicated by the white arrow in (B). The nature of the $\Sigma = 3$ twin is evidenced through the Fourier transform of the HRTEM image, displayed in (D). The surface of the nanoparticle is covered by a non-diamond carbon shell, with a thickness of 3–5 nm (see example regions indicated by black arrows in (B) and (E)). As all TEM experiments were carried out at a relatively low acceleration voltage of 120 kV, the observed shell, which was stable during the TEM measurements, is not formed by exposure to the electron beam, but is the result of the high mechanical impact during milling.[45]

The presence of boron incorporated into the individual nanoparticles is evidenced by electron energy-loss spectroscopy. An EELS spectrum, acquired from the region indicated by the white rectangle in Fig. 3.13E, is plotted in Fig. 3.13F. The presence of boron is evidenced through the B K-edge at 190 eV. Using standard EELS quantification techniques, the boron content was determined to be $1.3 \pm 0.2$ at.% ($\sim$13,000 ppm $\approx 2.3 \times 10^{21}$ $cm^{-3}$). This concentration is somewhat below the boron concentration determined for the B-doped microcrystalline diamond film prior the nanoparticle formation. This loss of boron has its origin in the breaking down of grain boundaries and defective regions in the material during milling, where B is known to be enriched.

The fine structure of the carbon K-edge, starting at 285 eV, mainly shows features associated with $sp^3$ hybridized carbon in diamond ($\sigma^*$ contribution). The small $\pi^*$ contribution at 285 eV arises from the graphitic material at the nanoparticle surface. The coordination of B in the nanoparticles can be determined through investigation of the energy-loss near edge structure of the boron K-edge. The ELNES signature is known to be sensitive to the local environment of the probed atomic species. In the inset to Fig. 3.13F, the fine structure of the B K-edge (in black) is plotted together with a spectrum acquired from a boron doped CVD diamond film (in red).[46] Interestingly, both boron ELNES signatures are highly similar to the carbon/diamond fine structure. Apart from a similar small $\pi^*$ pre-peak, three peaks, A, B, and C, which are present in the diamond ELNES, are also present in the boron fine structure. This implies that excited boron atoms are embedded in a similar local environment to the excited carbon atoms. In other words, most boron atoms in the diamond-rich areas are present as substitutional impurities. The small $\pi^*$ contribution in the B fine structure most likely arises from B embedded in a planar configuration in the non-diamond carbon surface shell or at defects in the material, but might also originate from heavily distorted tetrahedral B sites.

In an attempt to provide more accurate information on the type of boron embedding in the diamond grains, DFT calculations were performed to calculate the boron K-edge ELNES in the case of substitutional boron in diamond. One ELNES calculation was performed with full relaxation of the atomic positions surrounding the boron impurity; another was performed without relaxation. The results of the DFT calculations are plotted in Fig. 3.14 (top panel) together with the experimental boron K-edge and carbon K-edge fine structure from B:CVD diamond grain (bottom panel). After relaxation, the boron ELNES signature is found to be highly similar to the diamond ELNES, with peaks A, B, and C being reproduced. Interestingly, a small pre-peak is present in the calculated boron spectrum, due to new states in the diamond band gap. This peak, 8.7 eV before peak A, is clearly present at a different energy-loss position than the $\pi^*$ peak, and has recently been observed experimentally.[46] The ratio of

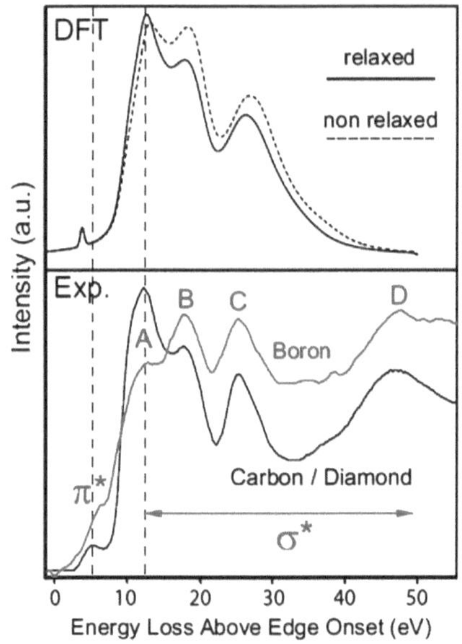


**FIGURE 3.14**

Comparison between the DFT calculated ELNES and experimental data. (Top) DFT calculated B K-edge ELNES for substitutional B in diamond with structure relaxation (full line) and without internal structure relaxation (B at ideal C position; dotted line). (Bottom) A comparison between the 2 eV smoothed boron ELNES and the simultaneously acquired carbon ELNES from a diamond region. The carbon edge has been shifted to align with the boron fine structure

the peaks A, B, and C in the calculated ELNES signature does not, however, fully coincide with the experimental B ELNES, even though the peak ratios can in general be reproduced well for e.g. the carbon K-edge in diamond by DFT.[47] The ratio changes of peaks A, B, and C are probably due to small changes in the local bond length, similar to the peak ratio changes seen in substitutional nitrogen in diamond.[44] This is corroborated by the ELNES signature for the non-relaxed substitutional boron system; in this case the ratio of peaks A and B changes. This could indicate that the environment surrounding the B impurities is not fully relaxed but rather strained or stressed, causing the peak ratios of peaks A, B, and C to change.

## 3.3 IMAGING AND SPECTROSCOPY OF NANODIAMOND SURFACE FUNCTIONALIZATION

Over the past decade, surface functionalization of nanodiamonds has become increasingly important, as further elucidated in Chapter 8. This surface functionalization can open up new applications field for nanodiamonds e.g. enhancing the luminescent properties of nanodiamond, increasing their bio-compatibility, or introducing click-chemistry at the surface of nanodiamonds for flexible additive grafting in the nanodiamond surface.[48]

A first example of the characterization of surface functionalized nanodiamonds is given in Fig. 3.15. Here, carbon dot functionalized nanodiamonds are under study, which demonstrate exceptional luminescent properties.[49] (HR)TEM images of the carbon dot decorated DND material are displayed in Fig. 3.15A. The white arrows points to a round graphitic structure, a 1–2 nm carbon dot, at the nanodiamond surfaces. These structures are clearly present all over the nanodiamond material surfaces and are attached to the nanodiamond surfaces. These nanodiamond surfaces are generally covered by 1–2 monolayers of graphitic material, which is in turn attached to the carbon dots.

The nature of the carbon dots was further investigated using spatially resolved electron energy-loss spectroscopy, performed in the same microscope as the HRTEM imaging but operated in scanning TEM (STEM) mode. To do so, we adopted the so-called spectrum imaging (SI) technique. In this technique, the sample area of interest, here containing dots-decorated nanodiamonds (Fig. 3.15B), is scanned in small steps using a fine (~1.5 Å) electron probe. At each point of the scan, an EELS signal is acquired containing the C-K (285 eV), N-K (400 eV), and O-K (532 eV) edge, allowing the chemical nature of the material to be investigated. Several nanometer-sized structures, presumably the carbon dots, are visible at the surface of the nanoparticles. Averaged EELS spectra from regions marked by 1 and 2 in Fig. 3.15B are plotted in Fig. 3.15F. The top EELS spectrum from region 1, a nanodiamond region, shows the presence of carbon at this position through the C-K edge starting at 285 eV, but also a small amount of nitrogen though the N-K edge around 400 eV. The fine structure of the C-K edge, with a large $\sigma^*$ contribution starting at 292 eV, corresponds largely to the diamond energy-loss near edge (ELNES) signature.[1] The small $\pi^*$ contribution at 285 eV in this spectrum arises from graphitic material at the top and bottom nanodiamond surfaces.

The spectrum from a single carbon dot at the nanodiamond surface, spectrum 2, looks distinctly different from spectrum 1. The fine structure of the C-K edge has a large $\pi^*$ contribution at 285 eV, and a broader $\sigma^*$ contribution. The $\sigma^*$ contribution in spectrum does not show the presence of sharply defined peaks, as is the case for graphite, but is also not without any definition as is the case for amorphous carbon. As was clear from the HRTEM images, the dots are made up of graphitic material that is highly deformed or defected. No significant nitrogen or oxygen signal is detected in the carbon dot structure in position 1. As the signal-to-noise (S/N) ratio in the spectrum is relatively low, due to the small size of the carbon dot structure, an averaged spectrum from several dots is also plotted in Fig. 3.15F. The spectrum confirms that no significant nitrogen of oxygen signal is measured in the carbon dots.

To plot the distribution diamond and non-diamond carbon species in the studied region, we fitted a diamond reference as well as the reference obtained from several carbon dots to all the acquired spectra. Plotting the intensity of the two fitted compo-

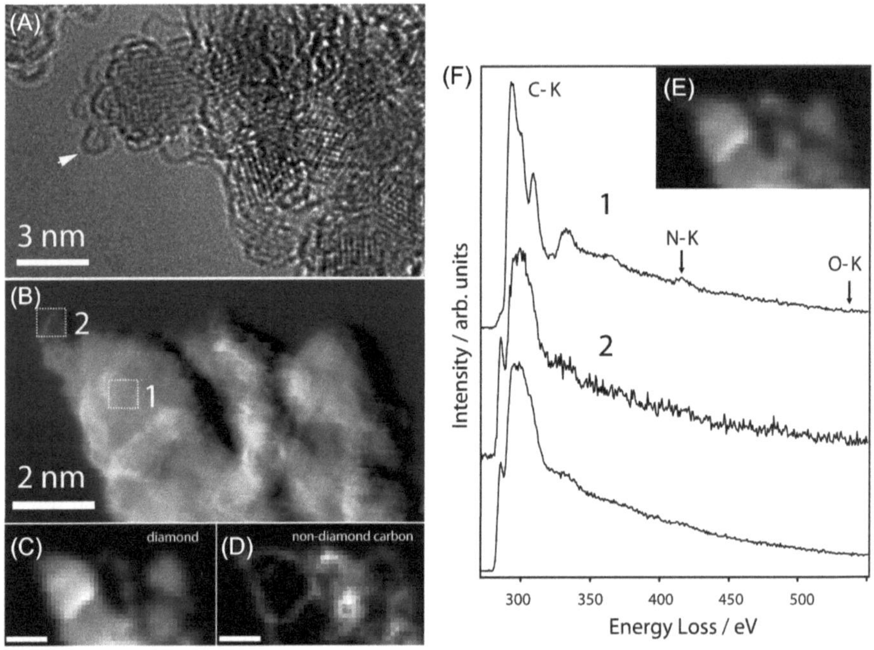


**FIGURE 3.15**

Investigation of CDD-ND material. (A) High resolution TEM image showing carbon dots (example indicated by arrows) decorating the nanodiamond surfaces, which are generally covered by 1–2 monolayers of graphitic material. (B) ADF-STEM overview image of the region investigated by spatially-resolved EELS. (C) Diamond map; (D) Non-diamond carbon map; (E) Color maps showing the non-diamond shell and dots covering the diamond nanoparticle cores. (F) Extracted EELS spectra from region 1 (nanodiamond) and region 2 (carbon dot) in the EELS spectrum image. The bottom spectrum is a summed spectrum from several dots

nents provides the diamond and non-diamond (graphitic material) maps displayed in Figs. 3.15C & D. A color overlay of both is also plotted in Fig. 3.15E.

A final example of nanodiamond surface functionalization, this time for biological applications, is displayed in Fig. 3.16. In this example, rounded nanodiamonds from HPHT origin are coated with PEG and silica, in order to make them biocompatible.[50] In order to confirm the presence of the various coatings on this sample directly, we performed spatially resolved electron energy-loss spectroscopy measurements using the spectrum imaging technique in STEM mode. A 72 $*$ 129 pixel spectrum image with EELS spectra containing the Si $L_{2,3}$-edge and the C K-edge was acquired from the region indicated by a white rectangle in Fig. 3.16.

The acquired EELS data were used to generate an element-specific map for silicon, and element- & bond-specific maps for amorphous carbon and diamond. The

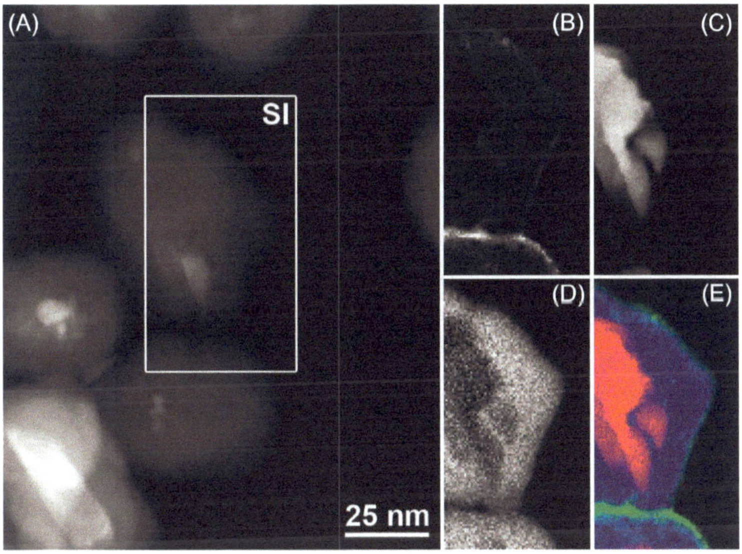


**FIGURE 3.16**

STEM-EELS analysis of PEG coated nanodiamonds. ADF-STEM survey image (A) with the spectrum image region indicated by the white rectangle. (B–E) STEM-EELS maps for amorphous carbon (B), diamond (C), silicon (D), and a color map (E) showing diamond (red), silicon (blue), and amorphous carbon (green)

results are displayed in Fig. 3.16B–E. The diamond map in Fig. 3.16C confirms the size and irregular shape of the diamond core. The Si map shows the ~10–20 nm thick silica shell surrounding the diamond core. Some amorphous carbon is measured at the surface of the diamond core in the amorphous carbon map, which is to be expected in most nanodiamond samples. The predominant amorphous carbon signal, however, arises from the thin layer of PEG at the surface of the silica coating.

In conclusion, this chapter described the characterization of nanodiamond particles by means of transmission electron microscopy. Several examples covering the most commonly used TEM imaging techniques were presented, as well as the application of electron energy-loss spectroscopy to the detection of nitrogen and boron impurities in nanodiamond. Finally, the possibilities of TEM and EELS for the characterization of surface functionalized nanodiamonds were covered.

## REFERENCES

1. Egerton, R. F. *Electron Energy-Loss Spectroscopy in the Electron Microscope;* Springer, 2011.
2. Williams, D. B.; Carter, C. B. *Transmission Electron Microscopy: A Textbook for Materials Science;* Springer, 2009.

3. Erni, R. *Aberration-Corrected Imaging in Transmission Electron Microscopy;* Imperial College Press, 2010.
4. *Nanocharacterisation;* RSC Publishing, 2007.
5. Scherzer, O. Über einige Fehler von Elektronenlinsen. *Z. Phys. Hadrons Nucl.* **1936**, *101*, 593–603.
6. Haider, M.; et al. A Spherical-Aberration-Corrected 200 kV Transmission Electron Microscope. *Ultramicroscopy* **1998**, *75*, 53–60.
7. Krivanek, O. L.; Dellby, N.; Lupini, A. R. Towards Sub-angstrom Electron Beams. *Ultramicroscopy* **1999**, *78*, 1–11.
8. Erni, R.; Rossell, M. D.; Kisielowski, C.; Dahmen, U. Atomic-Resolution Imaging With a Sub-50-pm Electron Probe. *Phys. Rev. Lett.* **2009**, *102*.
9. Kisielowski, C.; Erni, R.; Freitag, B. Object-Defined Resolution Below 0.5 Å in Transmission Electron Microscopy – Recent Advances on the TEAM 0.5 Instrument. *Microsc. Microanal.* **2008**, *14* (supplement 2), Ref Type: Abstract.
10. Bell, D. C.; Russo, C. J.; Kolmykov, D. V. 40 keV Atomic Resolution TEM. *Ultramicroscopy* **2012**, *111*, 31–37.
11. Alem, N.; et al. Atomically Thin Hexagonal Boron Nitride Probed by Ultrahigh-Resolution Transmission Electron Microscopy. *Phys. Rev. B* **2009**, *80*, 155425.
12. Girit, C. O.; et al. Graphene at the Edge: Stability and Dynamics. *Science* **2009**, *323*, 1705–1708.
13. Krasheninnikov, A. V.; Nordlund, K. Ion and Electron Irradiation-Induced Effects in Nanostructured Materials. *J. Appl. Phys.* **2010**, *107*, 071301.
14. Turner, S.; et al. Aberration-Corrected Microscopy and Spectroscopy Analysis of Pristine, Nitrogen Containing Detonation Nanodiamond. *Phys. Status Solidi A* **2013**, *210*, 1976–1984.
15. Shenderova, O.; et al. Surface Chemistry and Properties of Ozone-Purified Detonation Nanodiamonds. *J. Phys. Chem. C* **2011**, *115*, 9827–9837.
16. Min, Y. J.; Akbulut, M.; Kristiansen, K.; Golan, Y.; Israelachvili, J. The Role of Interparticle and External Forces in Nanoparticle Assembly. *Nat. Mater.* **2008**, *7*, 527–538.
17. *Ultrananocrystalline Diamond;* William-Andrew: New York, 2006.
18. *Characterization of Nanophase Materials;* Wiley-VCH Verlag Gmbh: Weinheim, 2000.
19. Synthesis and Crystal Structure of n-Diamond. *Int. Mater. Rev.* **2007**, *52*, 131–151.
20. Daulton, T. L.; Eisenhour, D. D.; Bernatowicz, T. J.; Lewis, R. S.; Buseck, P. R. Genesis of Presolar Diamonds: Comparative High-Resolution Transmission Electron Microscopy Study of Meteoritic and Terrestrial Nano-Diamonds. *Geochim. Cosmochim. Acta* **1996**, *60*, 4853–4872.
21. Turner, S.; et al. Determination of Size, Morphology, and Nitrogen Impurity Location in Treated Detonation Nanodiamond by Transmission Electron Microscopy. *Adv. Funct. Mater.* **2009**, *19*, 2116–2124.
22. Shenderova, O.; et al. Surface Chemistry and Properties of Ozone-Purified Detonation Nanodiamonds. *J. Phys. Chem. C* **2011**, *115*, 9827–9837.
23. Brydson, R. Electron Energy-Loss Spectroscopy and Energy Dispersive X-Ray Analysis. In Kirkland, A. I., Hutchison, J. L., Eds., *Nanocharacterisation*; RSC Publishing: Cambridge, 2007.
24. Peng, J. L.; Bulcock, S.; Belobrov, P. I.; Bursill, L. A. Surface Bonding States of Nano-Crystalline Diamond Balls. *Int. J. Mod. Phys. B* **2001**, *15*, 4071–4085.
25. Belobrov, P. I.; Bursill, L. A.; Maslakov, K. I.; Dementjev, A. P. Electron Spectroscopy of Nanodiamond Surface States. *Appl. Surf. Sci.* **2003**, *215*, 169–177.

26. Zhang, L.; Turner, S.; Brosens, F.; Verbeeck, J. Model-Based Determination of Dielectric Function by STEM Low-Loss EELS. *Phys. Rev. B* **2010**, *81*.
27. Iakoubovskii, K.; Mitsuishi, K.; Furuya, K. High-Resolution Electron Microscopy of Detonation Nanodiamond. *Nanotechnology* **2008**, *19*, 155705.
28. Turner, S. *Transmission Electron Microscopy Characterisation of 0-D Nanomaterials;* University of Antwerp, 3-4-2010, Ref Type: Thesis/Dissertation.
29. Verbeeck, J.; Van Aert, S. Model Based Quantification of EELS Spectra. *Ultramicroscopy* **2004**, *101*, 207–224.
30. Pichot, V.; et al. High Nitrogen Doping of Detonation Nanodiamonds. *J. Phys. Chem. C* **2010**, *114*, 10082–10087.
31. Garvie, L. A. J. Surface Electronic States of Meteoritic Nanodiamonds. *Meteorit. Planet. Sci.* **2006**, *41*, 667–672.
32. Birrell, J.; et al. Bonding Structure in Nitrogen Doped Ultrananocrystalline Diamond. *J. Appl. Phys.* **2003**, *93*, 5606–5612.
33. Raty, J. Y.; Galli, G.; Bostedt, C.; van Buuren, T. W.; Terminello, L. J. Quantum Confinement and Fullerene-Like Surface Reconstructions in Nanodiamonds. *Phys. Rev. Lett.* **2003**, *90*.
34. Cleri, F.; Keblinski, P.; Colombo, L.; Wolf, D.; Phillpot, S. R. On the Electrical Activity of sp(2)-Bonded Grain Boundaries in Nanocrystalline Diamond. *Europhys. Lett.* **1999**, *46*, 671–677.
35. Kvit, A. V.; Zhirnov, V. V.; Tyler, T.; Hren, J. J. Aging Effect and Nitrogen Distribution in Diamond Nanoparticles. *Composites, Part B, Eng.* **2004**, *35*, 163–166.
36. Shenderova, O. A.; et al. Nitrogen Control in Nanodiamond Produced by Detonation Shock-Wave-Assisted Synthesis. *J. Phys. Chem. C* **2011**, *115*, 14014–14024.
37. Vlasov, I. L.; et al. Hybrid Diamond-Graphite Nanowires Produced by Microwave Plasma Chemical Vapor Deposition. *Adv. Mater.* **2007**, *19*, 4058–4062.
38. Erni, R.; et al. Atomic Scale Analysis of Planar Defects in Polycrystalline Diamond. *Microsc. Microanal.* **2006**, *12*, 492–497.
39. Vlasov, I. I.; et al. Nitrogen and Luminescent Nitrogen-Vacancy Defects in Detonation Nanodiamond. *Small* **2010**, *6*, 687–694.
40. Baidakova, M.; Vul', A. New Prospects and Frontiers of Nanodiamond Clusters. *J. Phys. D, Appl. Phys.* **2007**, *40*, 6300–6311.
41. Soltamova, A.; et al. Electron Paramagnetic Resonance Detection of the Giant Concentration of Nitrogen Vacancy Defects in Sintered Detonation Nanodiamonds. *JETP Lett.* **2010**, *92*, 102–106.
42. Baranov, P. G.; et al. Enormously High Concentrations of Fluorescent Nitrogen-Vacancy Centers Fabricated by Sintering of Detonation Nanodiamonds. *Small* **2011**, *7*, 1533–1537.
43. Kajihara, S. A.; Antonelli, A.; Bernholc, J.; Car, R. Nitrogen and Potential n-Type Dopants in Diamond. *Phys. Rev. Lett.* **1991**, *66*, 2010–2013.
44. Brydson, R.; Brown, L. M.; Bruley, J. Characterizing the Local Nitrogen Environment at Platelets in Type IaA/B Diamond. *J. Microsc.* **1998**, *189*, 137–144.
45. Heyer, S.; et al. Toward Deep Blue Nano Hope Diamonds: Heavily Boron-Doped Diamond Nanoparticles. *ACS Nano* **2014**, *8*, 5757–5764.
46. Lu, Y. G.; et al. Direct Visualization of Boron Dopant Distribution and Coordination in Individual Chemical Vapor Deposition Nanocrystalline B-Doped Diamond Grains. *Appl. Phys. Lett.* **2012**, *101*.

47. Dadsetani, M.; Titantah, J.; Lamoen, D. Ab Initio Calculation of the Energy-Loss Near-Edge Structure of Some Carbon Allotropes: Comparison With n-Diamond. *Diam. Relat. Mater.* **2010**, *19*, 73–77.
48. Meinhardt, T.; Lang, D.; Dill, H.; Krueger, A. Pushing the Functionality of Diamond Nanoparticles to New Horizons: Orthogonally Functionalized Nanodiamond Using Click Chemistry. *Adv. Funct. Mater.* **2011**, *21*, 494–500.
49. Shenderova, O.; et al. Carbon-Dot-Decorated Nanodiamonds. *Part. Part. Syst. Charact.* **2014**, *31*, 580–590.
50. Rehor, I.; et al. Fluorescent Nanodiamonds Embedded in Biocompatible Translucent Shells. *Small* **2014**, *10*, 1106–1115.

CHAPTER

# 4 Raman investigations on nanodiamonds

**Michel Mermoux**[*,†]

*Grenoble Alpes University, LEPMI, Grenoble, France* CNRS, LEPMI, Grenoble, France†*

**Contents**



## 4.1 INTRODUCTION AND SCOPE

Nanocrystalline or ultra-dispersed diamond, often simply referred to as Nanodiamond (ND), belongs to a broad family of carbon-based nanomaterials that are promising for a wide range of applications. NDs can be available as thin films produced using chemical vapor deposition techniques and as powders obtained via different production methods. These methods are described in Chapter 2. Both kinds of material have been the focus of several reviews; see, for example, [1–4], and references therein. In this chapter, we will focus on ND powders produced by different techniques, in particular detonation nanodiamond (DND). Details concerning the Raman analysis of nanocrystalline diamond films may be found elsewhere.[2]

There are various sources of NDs.[1–4] A large subgroup of NDs is produced via so-called dynamic synthesis techniques, harnessing the energy of explosives and different carbon-containing precursor materials. Different classes of NDs synthesized by shock-wave-assisted methods are now manufactured on an industrial scale, in particular NDs produced by direct phase transformation of graphite induced by an external detonation shock wave (shock wave ND); or NDs produced from the carbon contained in high-energy explosives themselves (so-called detonation ND or DND, or ultradispersed diamond). When sub-10 nm NDs are required, DND is the typical choice at present. Most often, they have a characteristic, narrow size distribution of 5 ± 2 nm. However, separation of individual DNDs from the detonation soot and powder desegregation are complex processes, see Chapter 2. Another source of NDs arises from mechanical grinding of high-quality HPHT (high-pressure high-temperature) diamond microcrystals.[5] Their main advantage over DNDs is perhaps an expected more uniform crystallographic structure with a lower concentration of lattice defects such as dislocations or twins. This may be particularly important when spintronic or sensing applications are considered, as they may stably host various color centers, see Chapters 7 and 17. Ultrasmall HPHT NDs, down to 1 nm in diameter, recently received growing attention.[5] However, the finest HPHT NDs available nowadays on the marketplace have a median particle size of about 20 nm. This particle size thus approaches that of DNDs, although its size distribution is broader due to the grinding processes they arise from. Up to now, it seems that little attention was paid to material characterization of the separated sub-10 nm NDs. Finally, ND synthesis from laser-assisted techniques was also reported. The mean particle size of such NDs is comparable to that of DNDs. An advantage of this synthesis method is that it allows NDs of high purity to be obtained, which are metal-free in particular, see e.g. Chapter 5 in [3].

In material sciences, nanoparticles have great potential for a large variety of applications, including composite materials, lubricants, polishing compositions, electrode materials, seeds for further CVD diamond growth, etc. In addition, they have recently attracted wide interest in the field of biomedical applications[6,7] because of some remarkable properties such as low toxicity and biocompatibility,[8–10] possibility of sophisticated surface functionalization,[11,12] and their ability to fluoresce when excited with ultraviolet or visible wavelengths.[13] For all these applications, material purity in terms of $sp^2$ carbon content as well as a controlled particle size and a well-defined surface termination is mandatory because surface properties may significantly alter their chemical affinity. Moreover, the surface properties, or more exactly the surface functional groups of ND particles, have a strong influence on the aggregation behavior of the low-concentrated NDs aqueous suspensions.

NDs, DNDs in particular, are characterized by a high surface/volume ratio. Rough calculations show that 5 nm diamond nanoparticles are made of about 11000 carbon atoms. NDs, irrespective of their synthesis method, are also commonly described according to a core–shell model including a crystalline diamond core that contains crystallographic defects like twins or dislocations[5,14,15] surrounded by a more or less amorphous "shell" that may carry various surface functional groups. For DNDs, this disordered shell should contain at least 30% of all carbon atoms. Atomistic modeling,[16–18] Fourier transform infrared spectroscopy (FTIR),[19–21] nuclear magnetic resonance (NMR),[21–28] high resolution transmission electron microscopy

(HRTEM),[14,29] and X-ray photoelectron spectroscopy (XPS)[21,30] studies showed that the surface of diamond nanoparticles may have rather complex structures that may include mixed $sp^2/sp^3$ bonds, an $sp^2$-like carbon surface, different reconstructed diamond surfaces (the so-called fullerene-like reconstructions[16]), radicals, and a large variety of oxygen-containing functional groups. DND containing large amount of $sp^2$ carbons are often labeled "bucky diamond"[16,29] or "onion-like carbon" when they are fully graphitized.[31,32]

Raman spectroscopy is one of the characterization methods that can be used for the analysis of NDs. More generally, Raman spectroscopy is an experimental technique that is now commonly used to characterize all carbon materials and carbon nanostructures from three to zero dimensions (3D–0D), such as 3D graphite or diamond, 2D graphene, 1D carbon nanotubes, and 0D fullerenes.[33,34] Raman spectroscopy has also the ability to distinguish between different forms of amorphous carbon-based materials.[35] Raman spectroscopy of carbons provides not only vibrational and crystallographic information, but also some information about physical properties that are relevant to electrons and phonons. When a Raman process is combined with the optical absorption to (or emission from) an excited state, the Raman intensity may be significantly enhanced by a process called resonance Raman spectroscopy. Due to this resonance effect, one can routinely observe the Raman signal of a single layer of graphene, that is, an atomic layer of the hexagonal lattice of carbon, or an isolated single-wall carbon nanotube. In particular, the analysis of the observed Raman spectra of $sp^2$ carbons and the resonance conditions, when combined with theory, provides information on the electronic states, phonon energy dispersion, and electron–phonon interaction in $sp^2$ carbon nanostructures.[33,34] Furthermore, in some specific cases, Raman spectroscopy can also be used to control surface modifications of carbon materials, in particular when strongly conjugated molecules are grafted.[36,37] In such a case, monolayers can be detected on $sp^2$ graphitic materials, and their chemical or electrochemical reactivity can be studied.

The fundamental requirement for Raman activity, leading to light scattering, is that there must be a net change in bond polarizability during the vibration for the molecule, the crystal or the functional group under study. Another important form of vibrational spectroscopy is infrared spectroscopy, which is complementary to Raman spectroscopy. The selection rules for Raman spectroscopy are different from those for infrared spectroscopy, and in this last case a net change in dipole moment during the vibration must be observed for a transition to be infrared active. In the following, the general physics of Raman spectroscopy is discussed only briefly because this is covered in depth in textbooks and/or recent review articles; see, for example, [48] and [49].

Initially, the use of Raman spectroscopy was mainly limited to simple identification of the diamond phase through the observation of the so-called "diamond line" at about 1332 $cm^{-1}$. However, over the last decade, there is an ongoing effort to understand the different features appearing in the Raman spectrum of DND.[5,26,38–47] In this chapter, we describe the contributions of Raman spectroscopy for the analysis and the understanding of these nanoparticles.

Finally, recent progress in the instrumentation allows high sensitivity image detection both in vitro and in vivo at the cell level. Among various techniques, fluorescence or photoluminescence spectroscopy in particular, Raman spectroscopy was recently effectively used in many bio- and medical studies.[9,50] We will briefly address the use of Raman imaging techniques for such a purpose.

## 4.2 METHODOLOGY AND EXPERIMENTAL ISSUES

Getting Raman spectra of NDs is significantly more difficult than that of CVD diamond films. The first experimental issue is related to the strong photoluminescence (PL) of such materials when illuminated with the usual excitation wavelengths given by visible lasers. Depending on the specimen, the Raman signal is more or less observable superimposed on a huge PL background. In such a case, the Raman signal may be obtained at the expense of intensive baseline corrections. However, such baseline subtractions are often more or less subjective in nature. Because most of the strong PL emission is emitted in the visible frequency range, increasing the laser excitation energy to above ca 3.7 eV effectively allows a clear observation of the characteristic features of the ND Raman spectra. These excitation wavelengths are usually given by He–Cd lasers (325 nm) or frequency-doubled argon ion lasers (257 or 244 nm). The benefit of UV Raman spectroscopy to analyze ND was demonstrated in a number of papers.[38,41–47]

A second difficulty is related to the inherent low thermal conductivity of fine-grained absorbing materials. Actually, the heat conductivity of such powder samples is much lower than that of a crystal or a thin film. Local heating caused by the focused laser light must be taken in account, since it may affect the Raman spectrum even at relatively low laser power levels. Although it is difficult to quantify the thermal effect, overheating is expected to be all the more important when the grain size is small. In addition, overheating depends on sample packaging as well as the excitation wavelength. To limit the overheating induced by the probing laser beam, the incident power density employed in the Raman analysis of NDs powder samples has to be at least an order of magnitude lower than that usually used. In particular, it was observed that UV excitations used to get PL-free spectra induce considerable sample heating compared to visible ones, and may lead to thermal damage and/or changes in sample composition.[42,47] Even if no dramatic changes in the shape of the spectra could be observed with respect to the power at the sample, some damages at some sample surfaces can be visualized using simple optical microscopy. Cooling agents, water in particular, may be used to limit sample heating. However, in most cases, there is some overlapping between the Raman spectra of the solvents and the NDs. Moreover, some possible chemical modifications when using deep UV wavelengths may be anticipated.

Last, a third experimental problem is related to the rather low response such materials, implying long acquisition times.

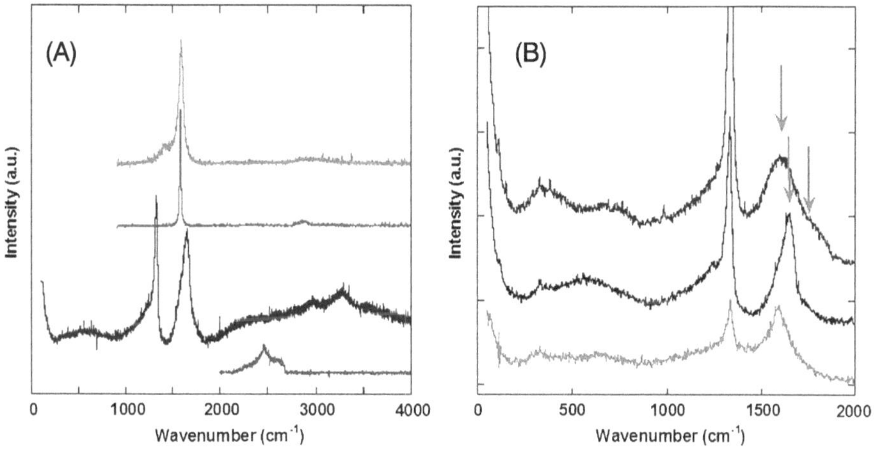


**FIGURE 4.1**

(A) Example of Raman spectrum of a DND sample (Nanocarbon Research Institute). For comparison purposes, from top to bottom, we also give the spectra of a carbon sample annealed at 1600°C, graphite (HOPG sample) and the diamond second-order spectrum. (B) Examples of Raman spectra of NDs samples. From bottom to top: shock wave ND, detonation ND, HPHT ND (mean particle size 30 nm). The arrows indicate the position of different apparent maxima. The excitation wavelength is systematically the 325 nm line of an He–Cd laser

## 4.3 CHARACTERISTIC FEATURES OF THE RAMAN SPECTRA OF NDS – SIMILARITIES AND DIFFERENCES

The UV-Raman spectra of different NDs are shown in Fig. 4.1. Here, the excitation wavelength was the 325 nm line of the He–Cd laser, allowing one to get PL-free backgrounds. They are compared to spectra of three references carbon samples analyzed with similar acquisition conditions: a diamond single crystal (for clarity, the first order line at 1332 $cm^{-1}$ is not shown, only the second-order region is displayed), a graphite (Highly Oriented Pyrolitic Graphite or HOPG) crystal, and a carbon sample heat-treated at 1600°C. These spectra are given in Fig. 4.1A.

For the materials investigated here, the Raman spectra can be divided into first- and second-order regions. In the first-order region (up to ca 1800 $cm^{-1}$), reference materials spectra are dominated by the first-order Raman-allowed graphite and diamond doubly degenerated $E_{2g}$ and triply degenerated $T_{2g}$ modes at about 1580 and 1332 $cm^{-1}$, respectively. Concerning the microcrystalline graphite sample, another weak feature is observed at about 1435 $cm^{-1}$. This feature corresponds to the disorder-induced “D” line which is usually observed at lower frequencies and with a much stronger intensity with conventional visible excitations. Features due to second-order processes are usually observed in the 2100–3400 $cm^{-1}$ region. Again, using

UV excitation wavelengths, the "2D" band intensity observed at about 2875 $cm^{-1}$ undergoes severe quenching. Raman spectroscopy of graphitic materials is extensively described in [33] and [34]. For the sake of completeness, the UV Raman spectra of some amorphous diamond-like materials may be found in [35].

The main feature in the NDs first-order spectrum is a sharp asymmetric peak centered at ca 1330 $cm^{-1}$; see Fig. 4.1B. Obviously, this line corresponds to the first-order Raman signal of diamond. The second strong feature in the Raman spectra of the NDs samples is a broad band rising between 1500 and 1800 $cm^{-1}$. Here, the fine structure of this line depends on the sample origin (HPHT, Shock Wave or detonation). The different band profiles suggest that they might be composed of at least three overlapping peaks at about 1590, 1650, and 1750 $cm^{-1}$. The observation of a line with a maximum at ca 1590 $cm^{-1}$ may be the indication of the presence of some graphitic carbon in a particular ND sample. Besides the diamond peak of NDs samples, broad features are also observed in the 300–1000 $cm^{-1}$ and the 1000–1350 $cm^{-1}$ range. Finally, high frequency lines at about 2900 and 3300 $cm^{-1}$ may be also distinguished, depending on the sample. Most often, spectra obtained for different purified DNDs origins overlap almost perfectly. This suggests that the characteristic features of the DND Raman spectrum do not strongly depend on their origins and are rather independent on the purification treatments, providing their mean sizes are similar. However, as discussed in the following, the line shape of these spectra may depend on the purification degree of the samples.

As we will see below, the line shape of the spectrum can more or less depend on the excitation wavelength.

## 4.4 THE DIAMOND LINE PROFILE – PHONON CONFINEMENT

First, let us focus on the diamond Raman line. Compared to the monocrystalline diamond standard, the diamond Raman line of the different ND samples shows significant line shift and asymmetric broadening towards lower wavenumbers. The line shift and broadening are all the more significant that the mean crystal size is low, see Fig. 4.2A. In particular, for the DND samples, the full width at half maximum (FWHM) may reach values as high as 40 $cm^{-1}$, along with an apparent line shift of about 6–8 $cm^{-1}$. On the contrary, the diamond peak of HPHT NDs exhibits only a minor shift and broadening (FWHM $\approx$ 15 $cm^{-1}$) in comparison with the monocrystalline diamond standard. Finally, comparing the different samples, strong overall signal intensity variations are observed.

Phonon confinement effects are mainly considered as the physical origin for these changes.[45,51–53] In a Raman scattering process with incident photon energy $\hbar\omega_l$, the scattered photon of energy $\hbar\omega_s$ has lost (Stokes scattering) or gained (anti-Stokes scattering) the energy $\hbar\omega$ of an elementary excitation, here a particular phonon,

$$\hbar\omega_s = \hbar\omega_l \pm \hbar\omega. \tag{4.1}$$

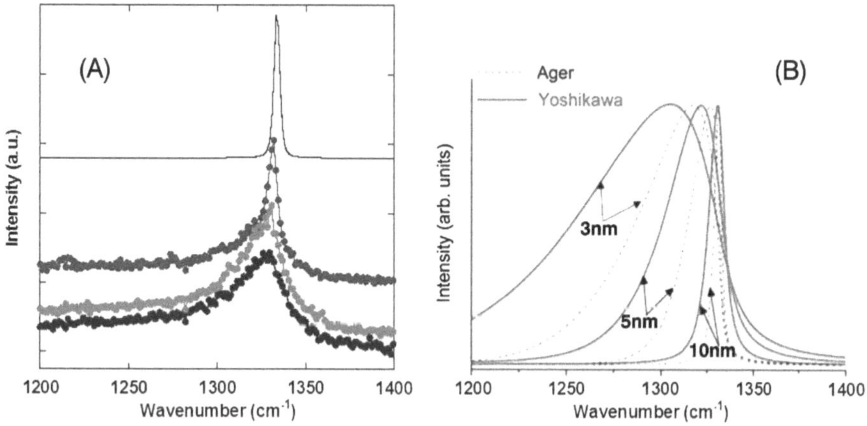


**FIGURE 4.2**

(A) Raman spectra of different DNDs, compared to the first-order Raman signal of a diamond single crystal. The excitation wavelength was the 325 nm line of an He–Cd laser. (B) Examples of Raman scattering patterns simulated for nanocrystallites, adapted with permission from: Osswald, S., et al. Phys. Rev. B 2009, 80, 075419,[45] using both the Ager et al. (reference [29] in [45]) and Yoshikawa et al. (reference [34] in [45]) parameters

The other important selection rule for light scattering processes in crystals is the wavevector conservation

$$\hbar k_s = \hbar k_l \pm \hbar q \tag{4.2}$$

where $k_s(k_l)$ denotes the wavevector of the plane waves of the scattered (incident) light, $q$ is the wavevector transferred to the phonon. The energies of crystalline vibrations probed by Raman spectroscopy are usually lower than 250 meV (about 2000 cm$^{-1}$), and those of the exciting visible laser light are about 1.5–5 eV, corresponding to the infrared or deep UV excitations that can be *a priori* used. Therefore, we get $\omega_s \approx \omega_l$ and the wavevector in a typical backscattering experiment will be on the order of $q \approx 4\pi n/\lambda_l$, much smaller than the wavevector $q = 2\pi/a$ at the boundary of the Brillouin zone, where $a$ is the lattice constant and $n$ denotes the refractive index of the sample. Therefore, this selection rule implies the contributions of phonons at the center of the Brillouin zone ($\Gamma$ point). Thus, for the case of an ideal single crystal of diamond (graphite), i.e. when the translational symmetry is preserved over large distances, only the $T_{2g}$ Raman-allowed 1332 cm$^{-1}$ mode ($E_{2g}$ mode at 1580 cm$^{-1}$) is observed in the Raman spectrum. (A second Raman active mode for graphite is predicted at low frequency (about 42 cm$^{-1}$). It is hardly observed and not useful for sample characterization.)

However, when the size of the crystal is quite limited, the q $\approx$ 0 selection rule is no longer valid. In disordered systems, the lack of translational invariance may induce the activation of forbidden modes. A striking example of such a breakdown of the q ≈ 0 selection rule is the observation of the so-called "D" line peaking at about 1350 $cm^{-1}$ when excited at 514 nm (note that the frequency of the "D" line has a strong dependence on the excitation wavelength[33,34]) in defective or microcrystalline graphitic materials.

To characterize and interpret the observed diamond line profile, a phenomenological phonon confinement model (also known as spatial-correlation model) is commonly used. It is briefly outlined below. The simplest model was originally proposed by Richter, Wang, and Ley (RWL)[51] for the analysis of the shift and the broadening of the Raman signal from microcrystalline silicon films. It was later extended by Campbell and Fauchet,[52] and by others; see, for example [54], and references therein.

In an infinite crystal, the plane wave-like phonon wavefunction of a phonon with wavevector $q_0$ is

$$\phi\,(q_0, r) = u(q_0, r)e^{-q_0 r} \tag{4.3}$$

where $u(q_0, r)$ has the periodicity of the lattice. In a crystallite assumed to be spherical with a diameter $L$, the phonon is restricted to the volume of the crystallite. Thus, the bulk, plane wave-like phonon wavefunctions cannot exist within a crystallite with a diameter lower than ca 10–20 nm. Therefore, introducing a weighting function $W\,(r, L)$ for the phonon amplitude, where $L$ is the size of a nanocrystal, the vibrational wavefunction of the nanocrystal may be approximated by

$$\psi\,(q_0, r) = W\,(r, L)\,\phi\,(q_0, r) = \psi'(q_0, r)u(q_0, r). \tag{4.4}$$

Here, the localization of the phonon in the nanocrystal is imposed by the weighting $W\,(r, L)$ function. The weighting function $W\,(r, L)$ describes the phonon confinement, although it is not a known function *a priori*. Thus, the phonon is no longer described by a plane wave; it is rather a "wave packet" whose spatial dimensions are similar to the crystallite size. Therefore, phonons out of an interval of $q$ wavevectors centered around $q = 0$ will contribute to the Raman spectra.

An uncertainty in wavevector $\Delta q \approx 1/L$ is thus introduced, which will be larger for smaller grain sizes because the "wave packet" becomes more localized in real space. Since the phonon dispersion curves are in general not flat, new frequencies will be introduced in the Raman line when contributions from phonons having significantly different wave vectors start playing a role. Therefore, the confined phonon is composed of a sum of bulk-like phonons with a range of wavevectors around the zone center. $\psi'(q_0, r)$ can be expanded in a Fourier series, the Fourier coefficients being

$$C\,(q_0, q) = \frac{1}{(2\pi)^3}\int d^3r\,\psi'(q_0, r)e^{-iq_0 r}. \tag{4.5}$$

The Raman line shape is constructed by superimposing Lorentzian line shapes centered at $\omega(q)$, weighted by the wavevector uncertainty caused by the confinement.

The dependence of the Raman line intensity on the frequency is given by

$$I(\omega) = \int \frac{|C(q)|^2}{(\omega - \omega(q))^2 - \left(\frac{\Gamma_0}{2}\right)^2} d^3q. \tag{4.6}$$

The integration is carried out over the whole Brillouin zone, and $\Gamma_0$ is the intrinsic Raman line width which is inversely proportional to the intrinsic phonon lifetime. In the absence of phonon confinement (infinite $L$), Eq. (4.6) simply reduces to a simple Lorentzian type function. Obviously, the magnitude of the effect of nanocrystallinity on different materials depends on the optic phonon dispersion relation in the region centered at the $\Gamma$ point.

In many publications, Eq. (4.6) is solved by choosing a weighting function and a dispersion curve fitting the experimental results, although this has no physical significance. For example, as discussed by Campbell and Fauchet, a Gaussian weighting function,

$$W(r, L) = \exp\left(-\frac{8\pi^2 r^2}{L^2}\right), \tag{4.7}$$

may be introduced in the initial RWL model. Some other weighting functions were also applied in other works.[34] The Fourier coefficients for the Gaussian weighting functions are

$$|C(q)|^2 \cong \exp\left(-\frac{q^2 L^2}{4}\right), \tag{4.8}$$

with q in units of $2\pi/a$, where $a$ is the lattice constant, and $L$ and $r$ are in units of $a$.

Approximating the three-dimensional Brillouin zone with a one-dimensional integration in a spherical Brillouin zone, and using an appropriately averaged isotropic dispersion curve, for example, given by

$$\omega(q) = A + B\cos(\pi q), \tag{4.9}$$

the dependence of the Raman line intensity on the frequency becomes:

$$I(\omega) \cong \int \frac{\exp\left(-\frac{q^2L^2}{4}\right) 4\pi q^2}{(\omega - \omega(q))^2 - \left(\frac{\Gamma_0}{2}\right)^2} dq. \tag{4.10}$$

Examples of calculated line profiles are given in Fig. 4.2B. As expected, the model qualitatively reproduces the downshift and the asymmetric broadening of the diamond line: a broadening and low-energy shift are evident as the crystallites become smaller, in agreement with the simple qualitative model of phonon confinement discussed earlier. However, examining the different results, some scatter exists between the different models, in particular when the line shape of the smallest particles has to be reproduced, see Fig. 4.2B. Clearly, this illustrates the effects of the choice of a particular weighting function, and the choice of the representation of the dispersion curve and parameters used. Recently, more sophisticated 3D representations of the dispersion curve were introduced in an attempt of developing a physically sound quantitative approach for the phonon confinement effect.[55]

However, particle defects, micro-strains and even slight heating of the sample may contribute to significantly affect the position and lineshape of the diamond first-order band, making it difficult to extract a true coherent domain size from the Raman measurements. Moreover, the nanocrystal shapes are not often well-defined and the nanocrystals have a size distribution. Finally, the structure of the particles is certainly not perfectly crystalline. This point will be further discussed in the following.

## 4.5 LOW FREQUENCY BROAD SIGNALS – CORE SHELL MODEL

As already mentioned, broad signals are also observed in the 300–1000 and 1000–1350 $cm^{-1}$ range. In particular, a shoulder centered around 1250 $cm^{-1}$ is clearly visible in most diamond nanoparticles spectra. As already discussed in [45] and [55], according to the phonon confinement model, this shoulder may track the presence of the finest grains in the particle size distribution. However, these features may also nicely find their origins in the breakdown of selection rules because of coherent domains of very small sizes, which allow phonons from high-symmetry points at zone boundary to contribute to Raman spectra. Again, as the translational symmetry which characterizes the crystalline structure is lost, in particular by introducing some bond angle disorder,[56,57] the optical vibrations become localized instead of being extended in the whole lattice. If the coherence length is sufficiently small, the Raman spectrum basically reflects the broadened vibrational density of states (VDOS) of the compound. The inset in Fig. 4.3 gives the experimental and calculated VDOS of crystalline diamond according to the calculations of Pavone et al.[58] and the measurements of Bosak et al.[59] It is seen that there is a close match between the diamond VDOS and the broad signals observed in the DND Raman spectra.

Note that similar broad signals were observed for ion-beam-amorphized diamond.[60,61] Such lattice distortions in DNDs were already suggested from NMR[24] and X-ray or neutron diffraction data.[62] In agreement with most of the HRTEM studies, both methods have shown that a major part of DND volume constitutes a diamond lattice with few apparent defects.[14] However, both methods suggest that this diamond core is bordered by a distorted/strained diamond lattice which forms the surface shell. In DND, the thickness of such a partially disordered shell has been estimated at about 0.5–0.7 nm. A 0.7 nm thick disordered shell should contain at least 30% of all carbon atoms of the particles. Thus, these broad bands may reflect some disorder in the outer region of the DND. As shown in Fig. 4.1B, they are observed for all NDs origins, detonation, shock wave or HPHT.

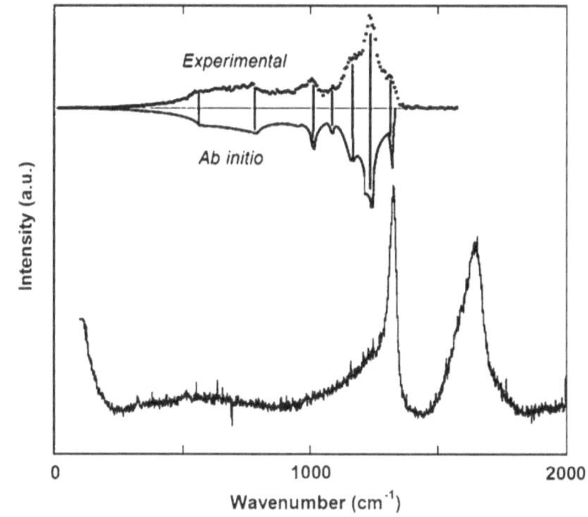


**FIGURE 4.3**

Raman spectra of one as-received DND sample obtained with the 325 nm excitation wavelength. The inset gives the diamond experimental and calculated phonon density of states, adapted with permission from: Bosak, A., Krisch, M. Phys. Rev. B 2005, 72, 224305[59]

## 4.6 RAMAN SPECTROSCOPY OF DETONATION NANODIAMONDS

The second dominant feature in the Raman spectra of DNDs is a broad band rising between 1500 and 1800 $cm^{-1}$. Again, this feature is observed for all NDs origins, detonation, shock wave, HPHT, or from laser-assisted techniques. However, Fig. 4.1B suggests that the line shape of this feature, the "peak" frequency and width in particular, may be sample dependent. This feature cannot be assigned to any pure $sp^3$ form of carbon, being far away from the cut-off frequency of the diamond VDOS ($\approx$1336 $cm^{-1}$). Depending on the sample, this feature may be also different from the graphite "G" band, being upshifted by about 30–50 $cm^{-1}$ from its expected position, see Fig. 4.1. The comparison of the UV spectra obtained for different samples suggests that it might be composed of at least three overlapping peaks: the so-called G-band at 1590 $cm^{-1}$ from in plane vibrations of graphitic carbon, a peak centered around 1640–1650 $cm^{-1}$ that has been ascribed either to O–H bending vibrations from adsorbed water,[38] to small $sp^2$fragments, mixed $sp^2$ and $sp^3$-coordinated carbon fragments,[41,48,53] localized interstitial C=C pairs, or similar defects within the diamond lattice [63–65] and a shoulder at 1760 $cm^{-1}$ related to the stretching mode of the surface C=O functional groups.[38,46]

The assignment of this broad signal is still questionable. To obtain additional information about the origin of this band, it seems necessary to find some means to modify the line shape of the characteristic DND spectrum. For such a purpose, different approaches may be considered: (i) probing the modifications induced by a change of the excitation wavelength, as it was done earlier for the analysis of CVD nanocrystalline films[2] or amorphous carbon films,[35] (ii) modifying the surface chemistry or the surface structure of the particles, (iii) performing analyses at low and/or high temperatures under controlled atmospheres. These different alternatives will be discussed in the following.

### 4.6.1 WAVELENGTH DEPENDENT EFFECTS: UV EXCITATIONS

Because Raman scattering from $sp^2$ carbon materials is most often a resonant process, it is always useful to compare spectra obtained for different excitation wavelengths. In general, when the exciting photon frequency does not correspond to an interband transition, the Raman effect is a second-order virtual process. In this process, the incident photon excites the system to a virtual electronic state, and on de-excitation, the emitted photon can be upshifted by phonon absorption or downshifted by phonon emission, thereby constituting the Raman lines in the emission spectrum. On the other hand, the intensities of the Raman lines are greatly increased when the laser frequency approaches or matches the energy of an interband transition. This effect is called resonant enhancement. Single incoming or outgoing resonances are widely known in Raman spectroscopy and frequently used to study the electronic and vibrational properties of crystals or molecules. They occur if the energy of the incoming or the scattered photon matches the transition energy of an allowed electronic transition. Closely related is the idea of double resonant Raman scattering, where, in addition to the incoming or outgoing resonances, the elementary excitation makes a real transition. Double resonances are much stronger than single resonances. In particular, such a double resonance process explains why the defect modes (the so-called "D" mode and its second-order peak) are so strong compared to the graphite Γ point vibration.[33,34] The "D" mode is related to the finite crystallite size and disappears for perfect crystals.

In the case of a multicomponent material, because the Raman cross-section for each Raman line is frequency dependent, the relative intensities of the various spectral features depend on the laser frequency. Indeed, it is well-known that the Raman spectra of carbon-based materials may strongly depend on the excitation wavelength.[2,33–35,66] Most often, $sp^2$ phases are probed with near infrared and/or visible excitations, whereas the diamond Raman signal will be clearly visible for higher excitation energies. A striking example is that of nanocrystalline diamond layers obtained by microwave plasma CVD processes, see Fig. 4.4B. Although such layers may contain more than 95% of $sp^3$ carbon, the diamond Raman signal is usually obscured by that of $sp^2$ phases when conventional visible excitation wavelengths are used, while a clear diamond signature is obtained for excitation wavelengths in the deep UV range.[2,66] Note that the indirect bandgap of diamond is 5.4 eV. This corresponds to a wavelength of about 230 nm. This means that for the wavelengths used in this study, resonant excitation of the diamond line is not really expected.

As an example, the Raman spectra of the as received DNDs excited at wavelengths ranging from 244 nm to 514 nm are shown in Fig. 4.4A. Here, the spectra

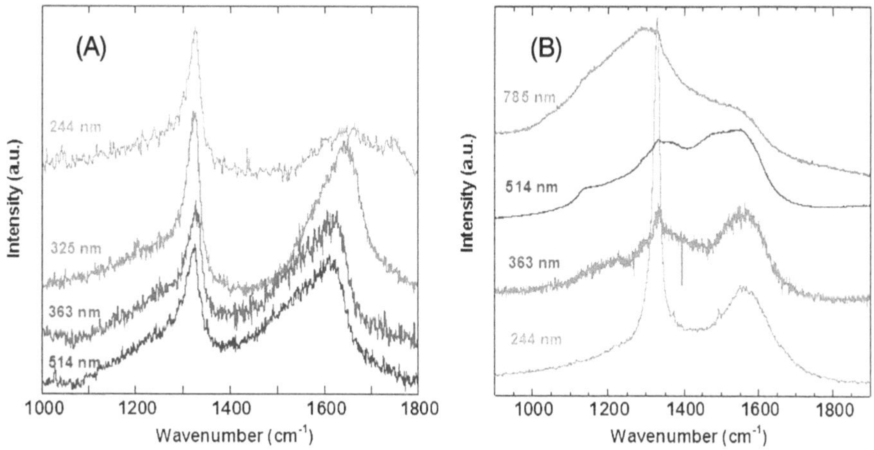


**FIGURE 4.4**

(A) Raman spectra of the as-received DND sample obtained with different excitation wavelength, from 514.5 nm (lower trace) to 244 nm (upper trace). (B) Raman spectra of nanocrystalline CVD diamond film obtained with different excitation wavelength, from 244 nm (lower trace) to 785 nm (upper trace)

were normalized to the diamond line intensity. One immediately sees that the shape of the spectra only slightly changes with the excitation wavelength. In contrast to nanocrystalline CVD layers, the intensity ratio of the diamond line to the 1650 $cm^{-1}$ band is more or less constant. As expected, the profile of the diamond line and the broad line at about 1250 $cm^{-1}$ are not affected by the wavelength change. Here, even for the low power levels used, the slight frequency shifts of the diamond line are immediately ascribed to uncontrolled heating effects. The fact that the low frequency part (1000–1450 $cm^{-1}$) of the spectra is unaffected by the wavelength change also precludes the presence of a strong graphitic "D" band in this frequency range that should be easily observable with visible excitations. Regarding the 1650 $cm^{-1}$ band, two small evolutions can be seen with the change of excitation wavelength. First, there is a noticeable apparent shift of the maximum of this line, from $\approx$1625 $cm^{-1}$ at 514 nm to $\approx$1660 $cm^{-1}$ at 244 nm. Second, a clear broadening of this line is also observed. Part of this broadening is explained by the development of a high frequency shoulder peaking at about 1750 $cm^{-1}$. As reported earlier,[38,46] this line may track the presence of C=O bonds located at the DND particle surface. The corresponding frequency is usually found in most of the FTIR spectra of DND particles. Finally, note that a third apparent maximum can be clearly isolated at 244 nm, at about 1590 $cm^{-1}$.

Such frequency dependent measurements suggest the observation of three different maxima in this wavenumber range. They further show that the graphitic carbon content is necessarily weak, in agreement with most of the published NMR measurements. Finally, if the shift of the component centered around 1650 $cm^{-1}$ is real, it

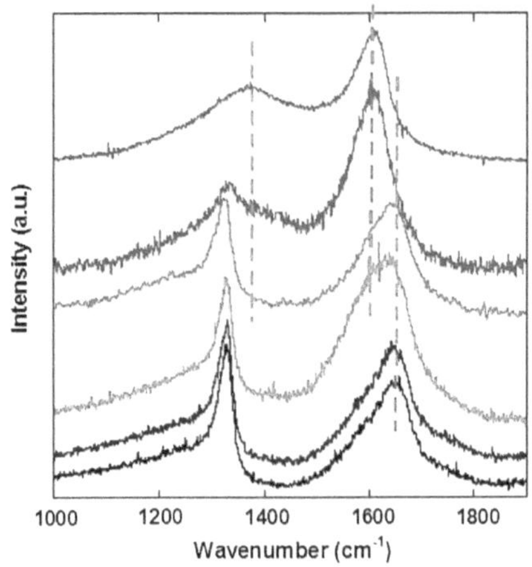


**FIGURE 4.5**

Raman spectra of modified DND samples. From the bottom to the top traces: as-received, air-annealed at 400°C, plasma hydrogenated, fluorinated, vacuum-annealed at 700°C, vacuum-annealed at 1100°C. Excitation wavelength was the 325 nm line of an He–Cd laser. The dashed lines indicate apparent maxima

precludes the presence of a simple zone-center mode, for which such dispersion is not expected.

### 4.6.2 DNDS WITH DIFFERENT SURFACE CHEMISTRIES

Annealing under different atmospheres may be used to modify the surface properties of DND. These treatments confer different surface properties, in particular in terms of surface terminations and surface charge. The modifications induced by different gases can be divided into three categories: in reductive atmospheres, e.g. hydrogen; in an oxidative atmosphere, air, fluorine; and in inert atmospheres, argon or *in vacuum*. The effects of those different treatments have already been discussed starting from NMR,[21–28] X-ray photoelectron spectroscopy (XPS),[21] or infrared measurements.[19–21] Note that Raman spectroscopy has been used only scarcely for such a purpose.[41,43,46,47] Here, we examine the effects of air, vacuum annealing, plasma, or high temperature hydrogenation and fluorination. As an example, characteristic spectra that concern surface-modified DNDs are given in Fig. 4.5, excitation wavelength was at 325 nm.

Examining this figure, we do not show major changes in the line shape of the spectra. They are still dominated by the diamond signal and the broad feature peaking at about 1600–1650 cm$^{-1}$. On the contrary to infrared spectroscopy results,[19–21] we do not see the appearance or the modification of some lines peaking in the 1000–1600 cm$^{-1}$ range that could be characteristics of the different surface functional groups induced by the annealing under different atmospheres. Moreover, because of their low intensity, it is difficult to comment on the observed signals at about 2800–3000 $cm^{-1}$ and 3250 $cm^{-1}$, even if they are observed in a frequency range where are expected the C–H and hydroxyl stretching modes. With the current acquisition conditions used, the overall line shape of this spectrum was nearly independent of the surface chemistry, oxygen, hydrogen, or fluorine terminated surfaces. Even if the surface functional groups may involve up to 12–14% of all the carbon atoms of a DND particle, Raman spectroscopy is actually rather insensitive to the surface chemistry of DNDs and it may be concluded that the whole 1500–1800 $cm^{-1}$ region in Raman spectrum comes from non-diamond carbon. Indeed, the scattering cross-section of such chemical bonds with low polarizability is known to be low.

For this particular example, the observed spectral changes mostly concern the line profile of the broadband centered in the 1600–1650 $cm^{-1}$ wavenumber range. Using the 325 nm excitation wavelength, the spectra of the as-received, air-annealed, and fluorinated samples are very similar in shape. The main observable difference between both samples is perhaps a slight increase of the signal near 1750 $cm^{-1}$ for the air-annealed sample, the position at which the C=O bonds are detected. A slight increase of the 1650 $cm^{-1}$ line for the as received sample may be also mentioned. By contrast, there is a clear increase in intensity and broadening of the 1650 $cm^{-1}$ component for the hydrogenated sample. A simple subtraction of the air-annealed and hydrogenated samples spectra immediately reveals two broad lines peaking at about 1600 and 1430 $cm^{-1}$, the usual positions of the graphitic so-called "D" and "G" lines when excited at 325 nm. The spectrum of the sample vacuum-annealed during one hour at 700°C is rather similar in shape to the hydrogenated one. Again, the spectrum subtraction with the air-annealed sample clearly evidences a clear contribution of the "D" and "G" graphitic lines. For the vacuum-annealing at the highest temperature, i.e. 1100°C, the Raman spectrum now only exhibits the "D" and "G" lines while the signal characteristic of DND is nearly lost. Thus, the asymmetry of this broad component may be described with two contributions at about 1590 and 1650 $cm^{-1}$. A similar trend is observed using an excitation at 514 nm. In this case, the presence of other apparent maxima may also be discussed.[47]

The similarity of the spectra of this specific DND sample treated *in vacuum* at 700°C to those treated in hydrogen plasma also indicates that the temperature may play an important role in the formation or the destruction of the surface functional groups. In particular, it may be thought that the generally acknowledged hydrogenation reaction is not the only reactions occurring in the process of treating DND in hydrogen. However, the answer given by Raman spectroscopy is incomplete in so far, at present, this method does not clearly detect most of the different surface functional groups.

Moreover, the issue of the stability of the different surface functional groups under UV irradiation arises. Using the 325 and 244 nm excitations, signals at about 1750 $cm^{-1}$ have been almost systematically detected and have been previously assigned to C=O bond stretching vibrations. If this assignment is correct, these chemical moieties are present on every DND surface, irrespective of the surface treatment and the surface terminations. In the particular case of the hydrogenated sample, carboxylic groups could not be detected by FTIR,[47] although this method is highly sensitive to these functional groups. Thus, at present, we cannot ascertain that this method is not destructive, especially from the stability of surface chemical bonds viewpoint. In our opinion, this is still an issue one has to work on.

### 4.6.3 ANNEALING DND IN DIFFERENT ATMOSPHERES: IN SITU RAMAN SPECTROSCOPY STUDIES

Owing to modern instrumentation, Raman spectroscopy is a powerful tool for the in situ studies of molecules, materials, and/or surfaces under working conditions. Raman characterizations can be carried out in different environments, for example, in liquids or in controlled atmospheres at low or high temperatures. In situ Raman spectroscopy was used to control the reactivity of DNDs at high temperatures under different atmospheres.[31,47,67] Using in situ Raman spectroscopy, one is able to monitor the related structural changes of carbon nanostructures in "real time", in order to identify optimum process conditions. In a typical in situ Raman study, samples are heated in a heating stage, operated in vacuum or in controlled environment. Most often, in situ Raman studies are performed using heating/cooling stages placed under the Raman microscope.

From such studies, Osswald et al. first proposed a simple and environmentally-friendly route involving oxidation in air to selectively remove the amorphous $sp^2$-bonded carbon soot from raw DND at a moderate temperature.[67] In situ Raman spectroscopy data revealed a narrow temperature window, 400–430°C, in which the oxidation of $sp^2$-bonded carbon efficiently occurred with no or minimal loss of diamond. A similar study from the same group nicely showed that annealing of DND powders in an inert environment first led to the reconstruction of the mixed $sp^2/sp^3$ bonding of carbon atoms located in the outer shell into simple $sp^2$ bonding which exists in nanocrystalline graphite.[31] This reconstruction begins with the smallest DND crystallites for temperatures as low as 600–700°C. Increasing the annealing temperature, the broad 1650 $cm^{-1}$ band became narrower and shifted to lower wavenumbers down to the frequency of the graphite "G" line. At the same time, the diamond peak decreased in intensity. At around 900°C, the diamond peak disappeared while the disorder induced graphitic "D band" emerged at about 1400 $cm^{-1}$. At higher temperatures (>1100°C), the majority of the DND crystals were partially or fully converted to carbon onions. However, this work based on the comparison of the Raman data and transmission electron microscopy (TEM) shows that it is difficult to extract quantitative data ($sp^3/sp^2$ ratio, or the diamond to graphite conversion rate) based on Raman data only. A similar conclusion has been reached during the analysis of CVD diamond films.

A similar annealing study was conducted in flowing hydrogen.[47] Again, the reading of the spectra suggested the progressive development of graphitic signals for temperatures as low as 175°C, as evidenced by the development of a shoulder on the low frequency side of the 1650 $cm^{-1}$ band. This signal clearly grew in inten-

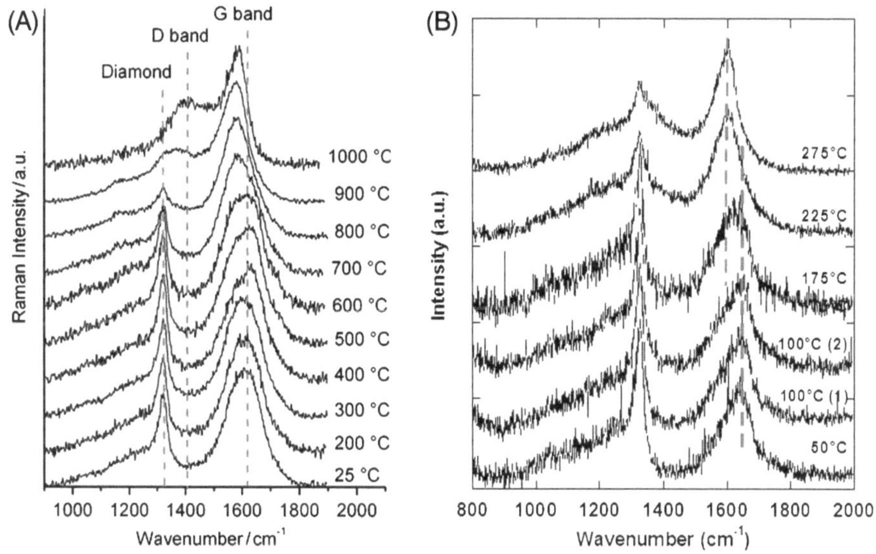


**FIGURE 4.0**

In situ UV (325 nm) Raman spectra recorded during annealing of DND powder flowing argon (A, adapted with permission from: Cebik, J., et al. Nanotechnology 2013, 24, 205703[31]) and in flowing hydrogen (B, adapted with permission from: Mermoux, M., et al. J. Phys. Chem. C 2014, 118, 23415–23425[47]). The dashed lines indicate apparent maxima

sity up to 275°C. It was also mentioned that an increase of the incident laser power contributed to modify the observed signals. Again, UV in situ Raman spectroscopy effectively senses the initial stage of the transformation, revealing the reconstruction of the mixed $sp^2/sp^3$ bonding of carbon atoms located in the outer shell into simple $sp^2$ graphitic bonding.

From a methodological viewpoint, it is difficult to completely control these experiments. In particular, in most cases, the temperature is a simple thermocouple reading that does not take into account some additional heating induced by the focused laser beam. This may explain why the spectral changes are observed at such low temperature in the last series of experiments. This may also explain some of the reproducibility issues mentioned in the work described above.

## 4.7 BEYOND SIMPLE MATERIAL CHARACTERIZATION: IN SITU RAMAN SPECTROSCOPY AND RAMAN IMAGING

The use of excitation lasers with wavelengths in the UV, visible, and near-infrared regions permits efficient coupling of Raman spectrometers with optical microscopes. Such micro-Raman spectrometers allow mapping of the properties or the characteristics of a particular sample with near-diffraction-limited spatial resolution (300–500 nm). The most common method for obtaining Raman spectral images is by raster-scanning the sample through the laser spot or scanning the laser spot across the sample. However, wide-field techniques are also reported.[9]

ND has been proven to be biocompatible and proposed for various biomedical applications. Recently, nanometer-sized diamonds have been demonstrated as an effective Raman/fluorescence probe for bio-labeling, as well as for drug delivery. Biomedical applications making use of luminescent NDs usually employ the high-efficiency luminescence of the N-V centers and/or the intense broadband PL, which presumably originates from structural defects at the ND surface. However, Raman signals may also serve as markers in various biomedical applications. Moreover, different studies showed that such diamonds are non-toxic in human cells. NDs luminescence properties, their applications in biology and nanoscale medicine, as well as toxicity issues, are described in Chapters 7, 12, and 16.

Recent development in spectroscopic techniques allows high sensitive image production both in vitro and in vivo at the cell level. The penetration of NDs into the cells (NDs internalization) was effectively analyzed using confocal fluorescence imaging and Raman imaging methods.[9,50] In particular, with a confocal configuration, one can achieve high spectral (well below than 1 $cm^{-1}$) and spatial (roughly half of the excitation wavelength) resolution, meaning that spectral overlapping can be largely avoided. When HPHT NDs are concerned, the "diamond peak" is rather intense and simple enough to be discriminated from the signal of the cells and used for the location of the diamond nanoparticles.

Cell mapping is usually performed by raster scanning the sample with point spacing of about 1 μm or lower, see Fig. 4.7A.[50] However, such scans may be rather time-consuming. To reduce the acquisition times, global or wide-field Raman imaging can also be used. Moreover, because the diamond Raman spectrum is superimposed on a high luminescence background due to N-V centers, a much higher contrast may be obtained making the analysis in a wavelength range where the photoluminescence background is intense. In Fig. 4.7B, each bright spot indicates the location of a diamond nanoparticle, as clearly confirmed by the Raman analysis which exhibited the expected diamond signal at each bright spot. Thus, the Raman and fluorescence mappings may provide quick localization of NDs in cells.

## 4.8 SUMMARY

Raman spectroscopy is an experimental technique that is now widely used to study all carbon materials and carbon nanostructures and as such is one of the characterization methods that can be used for the analysis of NDs. However, on the contrary to other carbon based materials, getting Raman spectra of NDs is more difficult. Experimental issues are related to the strong photoluminescence of such materials when illuminated with the usual excitation wavelengths given by visible lasers and to the inherent low thermal conductivity of fine grained absorbing materials. Thus, experimental condi-

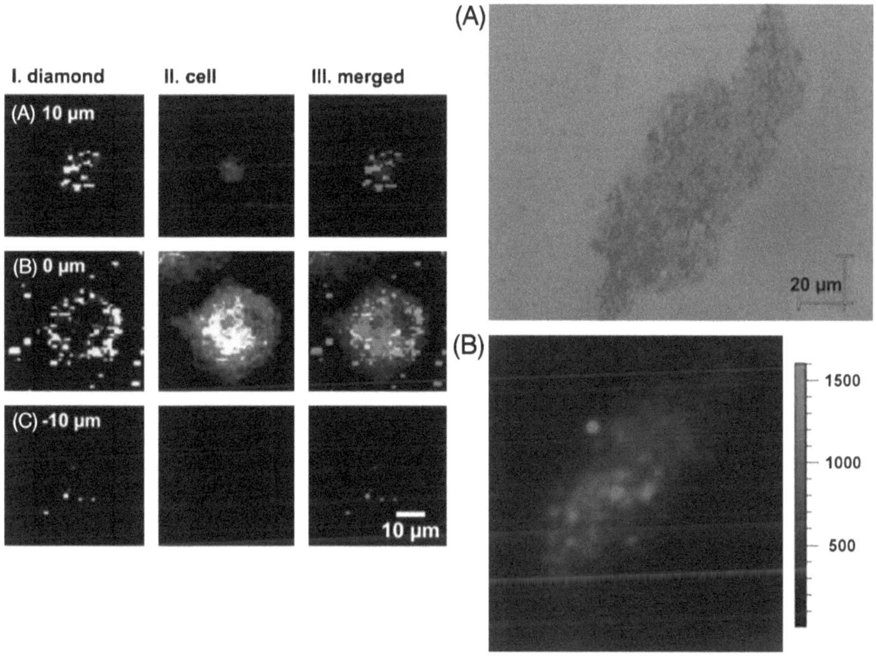


**FIGURE 4.7**

(Left) Confocal Raman mapping image of A549 cell and HPHT ND (mean particle size 100 nm). The images shown are at different z position scans: (A) at $z = 10$ μm with diamond collected in 1320–1340 $cm^{-1}$ and cell collected in 1432–1472 $cm^{-1}$, (B) at $z = 0$ μm position, and (C) at $z = -10$ μm position. Reproduced with permission from: Cheng, C. Y., et al. Appl. Phys. Lett. 2007, 90, 163903.[50] (Right) Example of wide-field Raman and PL microscopy of a HepG2 cell exposed to ND-20 nm. Optical image (A) and corresponding photoluminescence (PL) image recorded at about 3500 $cm^{-1}$ from the 514 nm excitation wavelength of a HepG2 cell (B). Each bright spot in the PL image indicates the location of a diamond nanoparticle. The HPHT ND mean particle size was 20 nm. Adapted with permission from: Paget, V., et al. Nanotoxicology 2014, 8, 46–56[9]

tions are to be adapted to minimize laser induced effects on NDs. Moreover, a better description of those compounds is reached using different excitation wavelengths.

Analyzing NDs after different surface treatments, Raman spectroscopy was shown to be poorly sensitive, or not at all, to the different surface terminations. At present, a much better sensitivity to these functional groups is reached with NMR or FTIR, see Chapters 6 and 9. On the contrary, graphitic reconstructions are clearly recognized from the spectra with a high sensitivity. Even if it has to be related to $sp^2$ bonded carbon, the assignment of the 1650 $cm^{-1}$ line systematically observed in the spectra of DND of various origins is still in no way unanimous. Similarly, it remains to establish whether or not Raman spectroscopy may allow monitoring the grafting of specific molecules on DND surfaces. At present, this is still a challenging task.

Finally, the possibilities offered by the method (in situ analysis and imaging) can be used to extend the application field of this method.

## REFERENCES


1. Shenderova, O. A.; Zhirnov, V. V.; Brenner, D. W. Carbon Nanostructures. *Crit. Rev. Solid State* **2002**, *27*, 227–356.
2. Shenderova, O. A., Gruen, D. M., Eds. *Ultrananocrystalline Diamond: Synthesis, Properties, and Applications*; William Andrew Publishing: Norwich, USA, 2006.
3. Williams, O. A. Nanodiamond; RSC Nanoscience & Nanotechnology No. 31; The Royal Society of Chemistry, 2014.
4. Vul', A.; Shenderova, O. Detonation Nanodiamonds: Science and Applications; Pan Stanford Publishing, Taylor & Francis Group, 2013.
5. Stehlik, S.; Varga, M.; Ledinsky, M.; Jirasek, V.; Artemenko, A.; Kozak, H.; Ondic, L.; Skakalova, V.; Argentero, G.; Pennycook, T.; Meyer, J. C.; Fejfar, A.; Kromka, A.; Rezek, B. Size and Purity Control of HPHT Nanodiamonds Down to 1 nm. *J. Phys. Chem. C* **2015**, *119*, 27708–27720.
6. Ho, D., Ed. *Nanodiamonds: Applications in Biology and Nanoscale Medicine*; Springer: New York, USA, 2010.
7. Schrand, A. M.; Ciftan Hens, S. A.; Shenderova, O. A. Nanodiamond Particles: Properties and Perspectives for Bioapplications. *Crit. Rev. Solid State* **2009**, *34*, 18–74.
8. Schrand, A. M.; Huang, H.; Carlson, C.; Schlager, J. J.; Osawa, E.; Hussain, S. M.; Dai, L. Are Diamond Nanoparticles Cytotoxic? *J. Phys. Chem. B* **2007**, *111*, 2–7.
9. Paget, V.; Sergent, J. A.; Grall, R.; Altmeyer-Morel, S.; Girard, H. A.; Petit, T.; Gesset, C.; Mermoux, M.; Bergonzo, P.; Arnault, J. C.; et al. Carboxylated Nanodiamonds Are Neither Cytotoxic Nor Genotoxic on Liver, Kidney, Intestine and Lung Human Cell Lines. *Nanotoxicology* **2014**, *8*, 46–56.
10. Arnault, J. C. Surface Modifications of Nanodiamonds and Current Issues for Their Biomedical Applications. In Yang, N., Ed., *Novel Aspects of Diamond: From Growth to Applications*; Springer: 2015.
11. Krueger, A.; Lang, A. Functionality Is Key: Recent Progress in the Surface Modification of Nanodiamond. *Adv. Funct. Mater.* **2012**, *22*, 890–906.
12. Krueger, A.; Liang, Y.; Jarre, G.; Stegk, J. Surface Functionalisation of Detonation Diamond Suitable for Biological Applications. *J. Mater. Chem.* **2006**, *16*, 2322–2328.
13. Chang, B. M.; Lin, H. H.; Su, L. J.; Lin, W. D.; Lin, R. J.; Tzeng, Y. K.; Lee, R. T.; Lee, Y. C.; Yu, A. L.; Chang, H. C. Highly Fluorescent Nanodiamonds Protein-Functionalized for Cell Labeling and Targeting. *Adv. Funct. Mater.* **2013**, *23*, 5737–5745.
14. Turner, S.; Lebedev, O. I.; Shenderova, O. A.; Vlasov, I. I.; Verbeeck, J.; Van Tendeloo, G. Determination of Size, Morphology, and Nitrogen Impurity Location in Treated Detonation Nanodiamond by Transmission Electron Microscopy. *Adv. Funct. Mater.* **2009**, *19*, 2116–2124.
15. Shames, A.; Mogilyansky, D.; Panich, A. M.; Sergeev, N. A.; Olszewski, M.; Boudou, J. P.; Osipov, V. Y. XRD, NMR, and EPR Study of Polycrystalline Micro- and Nanodiamonds Prepared by a Shock Wave Compression Method. *Phys. Status Solidi A* **2015**, *212*, 2400–2409.

16. Raty, J. Y.; Galli, G.; Bostedt, C.; van Buuren, T. W.; Terminello, L. J. Quantum Confinement and Fullerene-Like Surface Reconstructions in Nanodiamonds. *Phys. Rev. Lett.* **2003**, *90*, 037401.
17. Lai, L.; Barnard, A. S. Modeling the Thermostability of Surface Functionalisation by Oxygen, Hydroxyl, and Water on Nanodiamonds. *Nanoscale* **2011**, *3*, 2566–2575.
18. Paci, J. T.; Man, H. B.; Saha, B.; Ho, D.; Schatz, G. C. Understanding the Surfaces of Nanodiamonds. *J. Phys. Chem. C* **2013**, *117*, 17256–17267.
19. Jiang, T. L.; Xu, K.; Ji, S. FTIR Studies on the Spectral Changes of the Surface Functional Groups of Ultradispersed Diamond Powder Synthesized by Explosive Detonation After Treatment in Hydrogen, Nitrogen, Methane and Air at Different Temperatures. *J. Chem. Soc. Faraday Trans.* **1996**, *92*, 3401–3406.
20. Ji, S.; Jiang, T.; Xu, K.; Li, S. FTIR Study of the Adsorption of Water on Ultradispersed Diamond Powder Surface. *Appl. Surf. Sci.* **1998**, *133*, 231–238.
21. Shenderova, O. A.; Panich, A. M.; Moseenkov, S.; Hens, S. C.; Kuznetsov, V.; Vieth, H. M. Hydroxylated Detonation Nanodiamond: FTIR, XPS, and NMR Studies. *J. Phys. Chem. C* **2011**, *115*, 19005–19011.
22. Shames, A. L.; Panich, A. M.; Kempiński, W.; Alexenskii, A. E.; Baidakova, M. V.; Dideikin, A. T.; Osipov, V. Y.; Siklitski, V. I.; Osawa, E.; Ozawa, M.; et al. Defects and Impurities in Nanodiamonds: EPR, NMR and TEM Study. *J. Phys. Chem. Solids* **2002**, *63*, 1993–2001.
23. Panich, A. M.; Shames, A. I.; Vieth, H.-M.; Osawa, E.; Takahashi, M.; Vul', A. Y. Nuclear Magnetic Resonance Study of Ultrananocrystalline Diamonds. *Eur. Phys. J. B* **2006**, *52*, 397–402.
24. Fang, X.; Mao, J.; Levin, E. M.; Schmidt-Rohr, K. Nonaromatic Core-Shell Structure of Nanodiamond From Solid-State NMR Spectroscopy. *J. Am. Chem. Soc.* **2009**, *131*, 1426–1435.
25. Panich, A. M. Nuclear Magnetic Resonance Studies of Nanodiamonds. *Crit. Rev. Solid State Mater. Sci.* **2012**, *37*, 276–303.
26. Dubois, M.; Guerin, K.; Batisse, N.; Petit, E.; Hamwi, A.; Hayat Kharbache, N.; Pirotte, P.; Masin, F. Solid State NMR Study of Nanodiamond Surface Chemistry. *Solid State Nucl. Magn.* **2011**, *40*, 144–154.
27. Dubois, M.; Guérin, K.; Petit, E.; Batisse, N.; Hamwi, A.; Komatsu, N.; Giraudet, J.; Pirotte, P.; Masin, F. NMR Study of Nanodiamonds Produced by the Detonation Technique. *J. Phys. Chem. C* **2009**, *113*, 10371–10378.
28. Cui, J. F.; Fang, X. W.; Schmidt-Rohr, K. Quantification of C=C and C=O Surface Carbons in Detonation Nanodiamond by NMR. *J. Phys. Chem. C* **2004**, *118*, 9621–9627.
29. Iakoubovskii, K.; Mitsuishi, K.; Furuya, K. High-Resolution Electron Microscopy of Detonation Nanodiamond. *Nanotechnology* **2008**, *19*, 155705.
30. Zeppilli, S.; Arnault, J. C.; Gesset, C.; Bergonzo, P.; Polini, R. Thermal Stability and Surface Modifications of Detonation Diamond Nanoparticles Studied With X-Ray Photoelectron Spectroscopy. *Diam. Relat. Mater.* **2010**, *19*, 1117–1123.
31. Cebik, J.; McDonough, J. K.; Peerally, F.; Medrano, R.; Neitzel, I.; Gogotsi, Y.; Osswald, S. Raman Spectroscopy Study of the Nanodiamond-to-Carbon Onion Transformation. *Nanotechnology* **2013**, *24*, 205703.
32. Xu, Q.; Zhao, X. Bucky-Diamond Versus Onion-Like Carbon: End of Graphitization. *Phys. Rev. B* **2012**, *86*, 155417.
33. Dresselhaus, M. S.; Jorio, A.; Saito, R. Characterizing Graphene, Graphite, and Carbon Nanotubes by Raman Spectroscopy. *Annu. Rev. Condens. Matter Phys.* **2010**, *1*, 89–108.

34. Dresselhaus, M. S.; Dresselhaus, G.; Eklund, P. C. *Science of Fullerenes and Carbon Nanotubes: Their Properties and Applications;* Academic Press, 1996.
35. Ferrari, A. C.; Robertson, J. Raman Spectroscopy of Amorphous, Nanostructured, Diamond-Like Carbon and Nanodiamond. *Philos. Trans. R. Soc. Lond. A* **2004**, *362*, 2477–2512.
36. Itoh, T.; McCreery, R. L. In Situ Raman Spectroelectrochemistry of Azobenzene Monolayers on Glassy Carbon. *Anal. Bioanal. Chem.* **2007**, *388*, 131–134.
37. Actis, P.; Caulliez, G.; Shul, G.; Opallo, M.; Mermoux, M.; Marcus, B.; Boukherroub, R.; Szunerits, S. Functionalization Of Glassy Carbon With Diazonium Salts in Ionic Liquids. *Langmuir* **2008**, *24*, 6327–6333.
38. Mochalin, V.; Osswald, S.; Gogotsi, Y. Contribution of Functional Groups to the Raman Spectrum of Nanodiamond Powders. *Chem. Mater.* **2009**, *21*, 273–279.
39. Chen, P.; Huang, F.; Yun, S. Characterization of the Condensed Carbon in Detonation Soot. *Carbon* **2003**, *41*, 2093–2099.
40. Kozak, H.; Remes, Z.; Houdkova, J.; Stehlik, S.; Kromka, A.; Rezek, B. Chemical Modifications and Stability of Diamond Nanoparticles Resolved by Infrared Spectroscopy and Kelvin Force Microscopy. *J. Nanopart. Res.* **1568**, *2013*, 15.
41. Aleksenskii, A. E.; Baidakova, M. V.; Vul, A. Y.; Davydov, V. Y.; Pevtsova, Y. A. Diamond-Graphite Phase Transition in Ultradisperse-Diamond Clusters. *Phys. Solid State* **1997**, *39*, 1007–1015.
42. Yushin, G. N.; Osswald, S.; Padalko, V. I.; Bogatyreva, G. P.; Gogotsi, Y. Effect of Sintering on Structure of Nanodiamond. *Diam. Relat. Mater.* **2005**, *14*, 1721–1729.
43. Osswald, S.; Havel, M.; Mochalin, V.; Yushin, G.; Gogotsi, Y. Increase of Nanodiamond Crystal Size by Selective Oxidation. *Diam. Relat. Mater.* **2008**, *17*, 1122–1126.
44. Slocombe, D.; Porch, A.; Bustarret, E.; Williams, O. A. Microwave Properties of Nanodiamond Particles. *Appl. Phys. Lett.* **2013**, *102*, 244102.
45. Osswald, S.; Mochalin, V. N.; Havel Yushin, G.; Gogotsi, Y. Phonon Confinement Effects in the Raman Spectrum of Nanodiamond. *Phys. Rev. B* **2009**, *80*, 075419.
46. Mykhaylyk, O. O.; Solonin, Y. M.; Batchelder, D. N.; Brydson, R. Transformation of Nanodiamond Into Carbon Onions: A Comparative Study by High-Resolution Transmission Electron Microscopy, Electron Energy-Loss Spectroscopy, X-Ray Diffraction, Small-Angle X-Ray Scattering, and Ultraviolet Raman Spectroscopy. *J. Appl. Phys.* **2005**, *97*, 074302.
47. Mermoux, M.; Crisci, A.; Petit, T.; Girard, H. A.; Arnault, J. C. Surface Modifications of Detonation Nanodiamonds Probed by Multiwavelength Raman Spectroscopy. *J. Phys. Chem. C* **2014**, *118*, 23415–23425.
48. Weber, W. H.; Merlin, R. *Raman Scattering in Materials Science;* Springer-Verlag: Berlin, Germany, 2000.
49. Long, D. A. *The Raman Effect;* John Wiley & Sons Ltd: Chichester, UK, 2002.
50. Cheng, C. Y.; Perevedentseva, E.; Tu, J. S.; Chung, P. H.; Cheng, C. L.; Liu, K. K.; Chao, J. I.; Chen, P. H.; Chang, C. C. Direct and In Vitro Observation of Growth Hormone Receptor Molecules in A549 Human Lung Epithelial Cells by Nanodiamond Labeling. *Appl. Phys. Lett.* **2007**, *90*, 163903.
51. Richter, H.; Wang, Z. P.; Ley, L. The One Phonon Raman Spectrum in Microcrystalline Silicon. *Solid State Commun.* **1981**, *39*, 625–629.
52. Campbell, I. H.; Fauchet, P. M. The Effects of Microcrystal Size and Shape on the One Phonon Raman Spectra of Crystalline Semiconductors. *Solid State Commun.* **1986**, *58*, 739–741.

53. Yoshikawa, M.; Mori, Y.; Obata, H.; Maegawa, M.; Katagiri, G.; Ishida, H.; Ishitani, A. Raman Scattering From Nanometer-Sized Diamond. *Appl. Phys. Lett.* **1995**, *67*, 694–696.
54. Arora, A. K.; Rajalakshmi, M.; Ravindran, T. R.; Sivasubramanian, V. Raman Spectroscopy of Optical Phonon Confinement in Nanostructured Materials. *J. Raman Spectrosc.* **2007**, *38*, 604–617.
55. Korepanov, V. I.; Witek, H.; Okajima, H.; Osawa, E.; Hamaguchi, H. O. Three-Dimensional Model for Phonon Confinement in Small Particles: Quantitative Bandshape Analysis of Size-Dependent Raman Spectra of Nanodiamonds. *J. Chem. Phys.* **2014**, *140*, 041107.
56. Beeman, D.; Tsu, R.; Thorpe, M. F. Structural Information From the Raman Spectrum of Amorphous Silicon. *Phys. Rev. B* **1985**, *32*, 874–878.
57. Shuker, R.; Gammon, R. W. Raman-Scattering Selection-Rule Breaking and the Density of States in Amorphous Materials. *Phys. Rev. Lett.* **1970**, *25*, 222–225.
58. Pavone, P.; Karch, K.; Schütt, O.; Windl, W.; Strauch, D.; Giannozzi, P.; Baroni, S. Ab Initio Lattice Dynamics of Diamond. *Phys. Rev. B* **1993**, *48*, 3156–3163.
59. Bosak, A.; Krisch, M. Phonon Density of States Probed by Inelastic X-Ray Scattering. *Phys. Rev. B* **2005**, *72*, 224305.
60. Prawer, S.; Nugent, K. W.; Jamieson, D. N.; Orwa, J. O.; Bursill, L. A.; Peng, J. L. The Raman Spectrum of Nanocrystalline Diamond. *Chem. Phys. Lett.* **2000**, *332*, 93–97.
61. Kalish, R.; Reznik, A.; Prawer, S.; Saada, D.; Adler, J. Ion-Implantation-Induced Defects in Diamond and Their Annealing: Experiment and Simulation. *Phys. Status Solidi A* **1999**, *174*, 83–99.
62. Palosz, B.; Pantea, C.; Grzanka, E.; Stelmakh, S.; Proffen, T.; Zerda, T. W.; Palosz, W. Investigation of Relaxation of Nanodiamond Surface in Real and Reciprocal Spaces. *Diam. Relat. Mater.* **2006**, *15*, 1813–1817.
63. Prawer, S.; Nugent, K. W.; Jamieson, D. N.; Orwa, J. O.; Bursill, L. A.; Peng, J. L. The Raman Spectrum of Nanocrystalline Diamond. *Chem. Phys. Lett.* **2000**, *332*, 93–97.
64. Kalish, R.; Reznik, A.; Prawer, S.; Saada, D.; Adler, J. Ion-Implantation-Induced Defects in Diamond and Their Annealing: Experiment and Simulation. *Phys. Status Solidi A* **1999**, *174*, 83–99.
65. Hyde-Volpe, D.; Slepetz, B.; Kertesz, M. The [V–C=C–V] Divacancy and the Interstitial Defect in Diamond: Vibrational Properties. *J. Phys. Chem. C* **2010**, *114*, 9563–9567.
66. Wagner, J.; Wild, C.; Koidl, P. Resonance Effects in Raman Scattering From Polycrystalline Diamond Films. *Appl. Phys. Lett.* **1991**, *59*, 779–781.
67. Osswald, S.; Yushin, G.; Mochalin, V.; Kucheyev, S. O.; Gogotsi, Y. Control of $sp^2/sp^3$ Carbon Ratio and Surface Chemistry of Nanodiamond Powders by Selective Oxidation in Air. *J. Am. Chem. Soc.* **2006**, *128*, 11635–11642.

## ACKNOWLEDGEMENTS

The author would like to thank Jean Charles Arnault and Hugues Girard (CEA/LIST, Saclay, France), Tristan Petit (Helmholtz Zentrum Berlin, Germany), Marc Dubois (Institut de Chimie de Clermont-Ferrand, France) for providing some of the nanodiamond samples analyzed in this work, and for fruitful discussions around this topic. Likewise, the author would like to thank Alexandre Crisci (SIMAP, Grenoble, France) for technical support and special attention for this work.

CHAPTER

# 5 Elemental analysis of nanodiamonds by inductively coupled plasma hyphenated methods

**Pavel N. Nesterenko**[*,†], **Dimitar Mitev**[*], **Brett Paull**[*,†]

*Australian Center for Research on Separation Science (ACROSS), School of Physical Sciences, University of Tasmania, Hobart, TAS, Australia[*] ARC Centre of Excellence for Electromaterials Science (ACES), School of Physical Sciences, University of Tasmania, Hobart, TAS, Australia[†]*

## Contents



## 5.1 SIGNIFICANCE OF THE PROBLEM

Undoubtedly, the purity of nanodiamond (ND), in particular of detonation nanodiamond (DND) synthesized under heterogeneous conditions, has an extremely important impact in various applications in sensitive areas such as bioimaging, nanomedicine, adsorption and separation related technologies. The majority of elemental impurities are located upon the developed surface of diamond nanoparticles, in either physically or chemically adsorbed form, so even a trace presence of $10^{-6}$–$10^{-8}$ wt.% of certain elements could be responsible for dramatic changes in observed properties. Beside the general physical effects such as decreasing zeta-potential promoting aggregation of ND in suspensions,[1] elemental impurities can be responsible for more serious secondary impacts, such as unwanted side effects in drug delivery using nanodiamond. For example, the excessive delivery of sodium ions into the cells[2] or toxic effects connected with intracellular delivery of heavy metals, such as $Cu^{2+}$, $Cd^{2+}$, $Cr^{3+}$, and $Ni^{2+}$, by DND particles, also known as the "Trojan horse" effect[3] have been reported.

For the last years, a strong interest in the quantitation of individual elements present in diamond at extremely low concentration levels has developed. As discussed in Chapter 2, the purification and disaggregation of DND agglutinates is not a trivial task, due to the specific and complex nature of the surface chemistry of DND. However, for a considerably long time the progress in the purification of DND has been limited by the absence of accurate and precise methods for the quantitative determination of trace elemental impurities and possible sources of contamination.

From an opposite point of view, there are many extraordinary changes in the physical properties of diamond which can be achieved by implanting various elements within the crystal structure or by their immobilization on the surface. For example, doping of diamond with boron dramatically improves its electrical conductivity and allows applying this material as stable and robust electrodes[4] in electrochemical systems. Another example is the doping of diamond with silicon for the formation of optically active silicon-vacancy centers (SiV) (see Chapter 7), displaying bright luminescence at 738 nm,[5] while the concentration of dopant in this material could be as low as $1.4 \cdot 10^{-5}$ wt.%.[6] The presence of nickel related defects in combination with nitrogen vacancy (NV) centers can be used for extremely sensitive analytical techniques based on optically detected electron resonance.[7] The use of $Gd^{3+}$-nanodiamond conjugates for magnetic resonance imaging (MRI) contrast enhancement in medical diagnostics has also been reported.[8] Finally, metal coated nanodiamonds are efficient catalysts (see Chapter 18) in heterogeneous catalysis.[9] Thus, once again it can be seen that there is a strong demand for the accurate and sensitive determination of minor elements present within nanodiamond.

## 5.2 TYPES AND SOURCES OF ELEMENTAL IMPURITIES IN DIAMOND

The major elemental constituents in DND include carbon (81.8–93.1 wt.%), oxygen (1.4–15.8 wt.%), nitrogen (1.9–2.6 wt.%), and hydrogen (0.5–3.0 wt.%), with all other elements considered as minor impurities.[10,11] The major elements (carbon, hydrogen, nitrogen, and CHN) in DND are determined by classical elemental analysis based on measurement of gaseous products of combustion of diamond at high temperatures of 850–1200°C.[10] Because of hygroscopic properties of DND, which in "dried in air form" may contain up to 5.0–7.0 wt.% of water,[12] the accuracy of elemental analysis depends strongly on sample preparation. The complete removal of moisture from the samples by heating in vacuum is often impractical, as the dried samples quickly regain original mass by adsorption of water from air. For this reason, drying of DND in air at 105–150°C until constant mass is observed has been accepted as a standard pre-treatment procedure for elemental analysis.[10]

**Table 5.1** Sources of elemental impurities in nanodiamonds

| Process | Source | Elemental impurities |
|---|---|---|
| Detonation synthesis | Explosives | Various |
| | Construction material of reaction chamber | Fe, Ni, Cr, Mo, W, Si |
| | Coolant (water, ice) | Ca, Mg, Na |
| | Detonator/initiator | Pb, Ba, Sb, Al, Cu, Zn, Hg, As |
| Isolation and purification | Residues of oxidants and washing reagents after wet chemical oxidation of amorphous carbon | Cr, Mn, S from oxidants and various impurities from washing reagents |
| | Catalysts of oxidation | Ag, Ce |
| | Inhibitors of oxidation | B |
| Disaggregation | Sonication with sonotrode | Ti, V, Al, Mo, W |
| | Milling with microbeads | Zr, Y, Hf, Si, Al, Mg |
| | Disaggregation media (solid salts, electrolytes in suspensions, etc.) | Na, K, Mg, Ca, Al, Cl, P, S |
| Processing and storage | Adsorption of impurities from chemical reagents | Various |
| | Laboratory glassware | Na, K, B, Si |
| Targeted doping | Various methods | B, N, Si, Ni, and others |

X-ray photoelectron spectroscopy (XPS) has also been reported for the determination of major elements using C 1s, O 1s, N 1s signals,[13] together with some other elements (Si 2p, W 4f, Cl 2p, and Na 1s signals) found in nanodiamond at concentrations higher than 0.1 at.%. However, it should be noted that XPS has an average depth of analysis around 5 nm, so XPS data mainly reflects the composition of the surface layer in a diamond sample, which is normally different from the composition of bulk material in aggregates of nanodiamond particles. Thus, the XPS data obtained for five nanodiamond samples in published works[13,14] reports the presence of 82.3–91.3 at.% of carbon, 8.5–15.8 at.% of oxygen, and 0.4–1.9 at.% of nitrogen.

In many applications, a small variation of major elements (C, O, N, and H) content in DND is not too significant, but the presence of tiny minor elemental impurities could be responsible for dramatic differences in the optical and magnetic properties of the material. Traditionally, the total content of minor elemental impurities in DND is estimated by the weight of incombustible residue (ICR), which typically varies across the range 0.35–2.0 wt.%, although for some non-purified DND containing products (e.g. detonation soot) ICR could be as high as 7.5 wt.%.[15] The major reasons for such a high contamination levels of DND include the heterogeneous conditions of detonation synthesis and the low purity of explosives and cooling reagents used, as shown in Table 5.1. Interestingly, substantial further contamination of nanodiamonds can also occur during their purification and isolation from the soot via wet oxidation of the

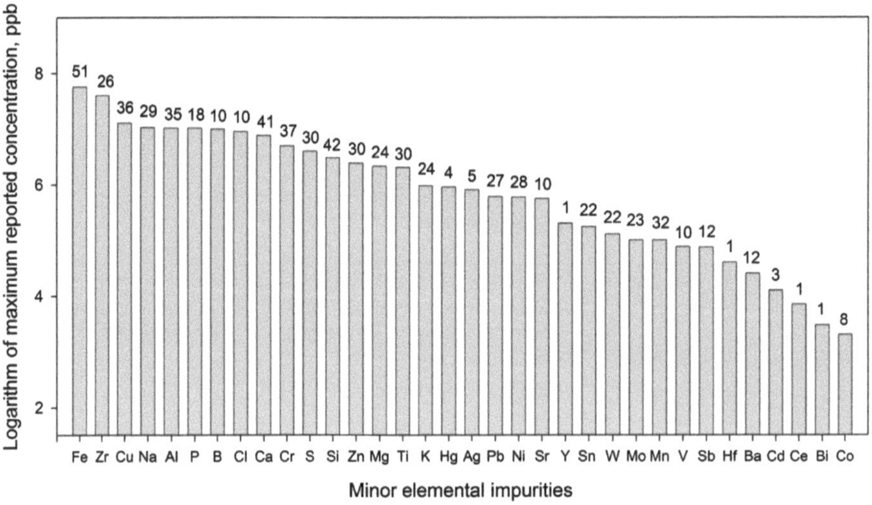


**FIGURE 5.1**

Bar diagram showing maximal concentrations (as log C) reported in the literature for 34 minor impurities in DND. The number at the top of the bar indicates the number of the samples for which the quantitative determination of the selected element was reported. The number of minor elemental impurities identified in DND samples depends on their purity and the sensitivity of analytical instrumentation used. Adapted from: Mitev, D., et al. J. Mater. Sci. 2014, 49(10), 3573–3591,[17] and updated

amorphous graphite, as well as during routine laboratory work. For example, disaggregation of DND by milling with zirconia microbeads is the suspected reason for the elevated concentration of zirconium in the NanoAmando diamond product (NanoCarbon Research Institute Co., Japan), where zirconium is found at up to 0.5 wt.%.[16]

Unfortunately, the information on the type and concentration of minor elemental impurities, such as individual metals, can rarely be found in quality certificates of commercially produced DND. At the same time, recent investigations using the ultra-sensitive method of inductively coupled plasma – mass spectrometry (ICP-MS) show that the total number of possible inorganic impurities which can be found in DND is close to 50–60 elements, present at concentrations above the instrumental limits of detection (LOD), which is around $10^{-7}$–$10^{-8}$ wt.%.[17] A summary of literature data on maximal concentrations (as log C) and abundancy of 34 minor elemental impurities found within DND is presented in Fig. 5.1.

Apart from during synthesis, where some amounts of elemental impurities can be entrapped within the crystal structure of diamond, more substantial contamination of DND can occur due to post-synthesis adsorption of inorganic ions. DND has a well-developed surface area of 190–330 $m^2\,g^{-1}$, covered by various functional groups. Many of these groups are capable of interacting with both positively and negatively charged inorganic ions.[18] Indirectly, the zwitterionic nature of the ion-exchange prop-

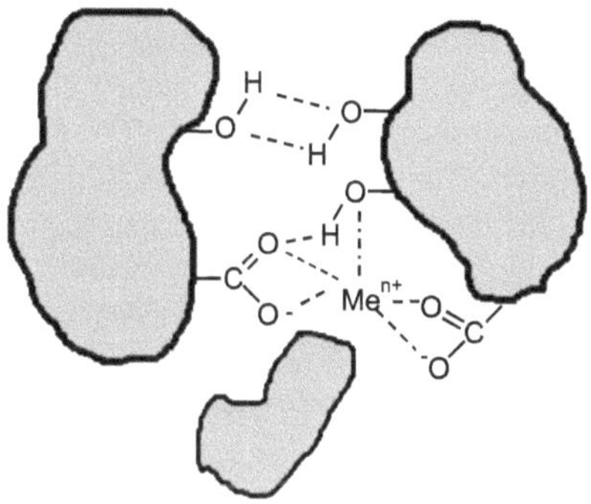


**FIGURE 5.2**

Ion-exchange, donor–acceptor coordination of metal ions, and hydrogen bonding as possible reasons for aggregation and contamination of diamond nanoparticles

erties of DND has been confirmed by the switching of surface charge (zeta potential), from positive to negative, with increase in pH of the DND suspension. Maximum values reported for zeta potential of DND vary across a broad range from −80 to +55 mV, depending on pH of the suspension, degree of oxidation, and applied dis-aggregation procedures.[17] The large negative charge of the surface is associated with a high concentration of carboxylic groups, which could be as high as 2.1 groups per $nm^2$. By accepting the surface area of DND sample as 330 $m^2\,g^{-1}$, one can calculate a total cation-exchange capacity of 1.05 mequiv $g^{-1}$.[19] It should be noted that this value of ion-exchange capacity is comparable with ion-exchange capacity values for common commercial ion-exchange resins.[20]

Multiple donor-acceptor, ion-exchange, and hydrogen bonding interactions between carboxylic and hydroxyl groups upon the surface of nanodiamond particles and multivalent metal cations and inorganic anions are responsible for the contamination of the surface by impurities. As multivalent metal cations can be coordinated by two or even three functional groups from the same or different nanoparticles, the presence of such metal impurities could be one of the reasons for the formation of extremely stable DND aggregates,[1] known as agglutinates (Fig. 5.2). However, due to steric hindrance, the coordination of three nanoparticles around a single metal cation is impossible, so the aggregation can be described as follows:

$$2[DND^-H^+] + Me^{3+} = [(DND^-)_2Me]^+\downarrow + 2H^+. \qquad (5.1)$$

Alternatively, the treatment of DND with excess amounts of monovalent metal salts (chlorides, carbonates) is often used for the preparation of stable suspensions of dis-aggregated nanodiamonds.[15] Technically, this treatment reflects the conversion of the hydrogen form of ion-exchange groups from DND into sodium form according to the following equation:

$$[DND^-H^+] + Na^+ = [DND^-Na^+] + H^+. \qquad (5.2)$$

The disaggregation occurs due to lack of coordination of sodium ions by two functional groups from different nanoparticles, and due to the suppression of hydrogen bonding between protic functional groups when in sodium form.[15] In the case of complete conversion of this ion-exchanger from $H^+$ form into sodium form, the maximum mass of sodium electrostatically retained by DND can be evaluated as the product of ion-exchange capacity of the DND (up to 0.9–1.05 mmol[19]) and sodium ion mass. Interestingly, the obtained value of 2.4 wt.% is close to the typical 0.5–2.2 wt.% content of minor elemental impurities reported for DND. Probably, the producers of DND should specify the ionic form of DND products (e.g. $H^+$ form, $Na^+$ form, etc.) as an absence of such information produces uncertainty for the definition of contamination, and the interpretation and comparison of results obtained by the different research groups. Otherwise, the amount of electrostatically retained metals, e.g. sodium contribute a significant part of the incombustible residue (ICR) and, hence, be counted as an impurity of the DND.[2]

Recent investigations[19] have demonstrated the ability of DND to adsorb simultaneously inorganic anions and metal cations from solutions due to the presence of both negatively charged (carboxyl, hydroxyls) and positively charged (protonated carbonyls) functional groups upon the surface. It was found that the adsorption of multiple charged inorganic anions (e.g. sulphate, phosphate) also promotes the increased adsorption of metals cations. Additionally, the high concentration of hydroxyl-groups at the surface of diamond is deemed responsible for the adsorption of borate anions,[19] and therefore substantial contamination of DND is possible during routine laboratory operations in borosilicate glassware. Again, the simultaneous adsorption of both metal cations and inorganic anions has a very complex impact upon the surface charge of nanoparticles and, therefore, on their aggregation and the stability of colloidal suspensions.[21]

## 5.3 INSTRUMENTAL METHODS APPLICABLE TO THE ANALYSIS OF IMPURITIES IN DIAMOND

Depending on sample preparation, the methods used for analysis of minor elemental impurities in diamond can be divided into two groups, namely destructive and non-destructive methods. Early work in the area of analysis of metal content in DND included ashing of the sample, dissolution of residues and final quantification using a suitable spectroscopic technique.[22] As a rule, this procedure requires a relatively large (up to 10 g) DND sample and prolonged (for 8 hours) heating at temperatures of 700°C,[23] 900°C,[24] 950°C,[25] or even 1200°C,[26] followed by eventual dissolution of the ICR (up to 2.2 wt.%) and metal determination using atomic emission spectroscopy (AES),[27] inductively coupled plasma-atomic emission spectroscopy (ICP-AES),[23] or atomic absorption spectroscopy (AAS).[21]

According to previous reports,[24,26] DND combustion ash contains a mixture of Ca and Al sulphates, chromates, Si, Fe, Al, Cr oxides, and various metal silicates and carbides. It should be noted that using ICR for analysis includes a number of systematic errors affecting the accuracy and precision of the determination of individual elements due to many separate reactions between species taking place during the combustion or ashing of the DND. For example, the use of ICR does not consider the possibility of the elimination of some elemental impurities in the form of volatile products (e.g. sulphur oxides), as well as the formation of new insoluble products such as metals oxides, sulphates, silicates, and other compounds,[28,29] all of which contribute to extra uncertainty in this type of analysis. As a result, incorrect quantitative data is often obtained for Ca, Si, Pb, Ni, Mn, and some other elements, especially, if only the acid extraction of ashes is used for the subsequent quantitative analysis. To overcome these issues, the application of additional time and labor consuming fusion techniques using lithium metaborate ($LiBO_2$) for the decomposition of silicates and oxides is required.[22] However, there is also here a probability of cross-contamination from fusion reagents.

The obvious solution to the above mentioned problems is the use of non-destructive methods of analysis, where minimal or no sample preparation is required for the determination of trace elements. The characteristics of analytical methods suitable for the non-destructive determination of the minor elemental composition of diamond are included within Table 5.2.

For many years, the rationale behind the analysis of impurities in diamond was associated with the evaluation of quality and identity, as well as an understanding of the reasons for the occurrence of specific optical properties, and the geological origin of the diamond. The first publication in this area dates back to 1943, when Raal reported the application of a spectrographic method of analysis for the determination of impurities (Si, Mg, Al, Ca, Fe, Cu, Ti, and Cr) in natural diamond at concentrations ranging from $10^{-4}$ to $10^{-2}$ wt.% and used this information for identification of elements responsible for the color of these precious stones.[30] The spectrographic analysis technique was designed for the analysis of powder samples, which should be weighted and accurately transferred into hollow graphite cathode cuvette followed by ignition via an electrical arc and measurement of intensity of specific lines in emission spectra during an exposition time of 120–180 s. At that time, the intensity of the signal was measured as density (absorbance) of lines developed in photographic plates. It is worth noting that the required process of crushing the natural diamonds definitely introduced contamination into the samples and caused some errors in the determination of silicon and aluminum, but this problem is not relevant to the analysis of DND powders. This technique is obviously rather labor consuming and has lost popularity in recent years.

Neutron Activation Analysis (NAA) was the first non-destructive analytical method applied to the analysis of diamonds, which was found to be extremely useful for the testing of authenticity and identification of the origin of precious natural stones and diamond containing jewelry, by fingerprinting the trace elemental impurities.[31] NAA method is based on neutron induced nuclear reactions when the sample is exposed to a flux of neutrons. As a result of these reactions, a neutron enriched radioactive nucleus forms, which has a characteristic short half-life period connected with the emission of a beta- and, more importantly, gamma-radiation. The intensity

**Table 5.2** Comparison of analytical methods suitable for direct (without special sample pre-treatment) analysis of minor elemental impurities in nanodiamond and corresponding references

| Method | Sample requirements | Elements detected at concentrations above LOD | Limits of detection, wt.%[a] | Refs |
|---|---|---|---|---|
| High Resolution ICP-MS | 0.05–0.10 mg $mL^{-1}$ suspension in water | Na, S, Cl, Au | $10^{-5}$–$10^{-7}$ | 3,11,13,17,28,43,46 |
| | | B, Al, Si, P, K, Ca | $10^{-8}$–$10^{-9}$ | |
| | | Mg, Ti, Mn, Fe, Ni, Cu, Zn, As, Se, Hg | $10^{-9}$–$10^{-8}$ | |
| | | Li, Be, V, Cr, Co, Ga, Rb, Sr, Ge, Zr, Mo, Rh, Ag, Cd, Sn, Ba, W, Re, Os, Ir, Pb, Ce | $10^{-10}$–$10^{-9}$ | |
| | | Sc, Y, Ru, Pd, In, Sb, Te, Cs, La, Hf, Tl, Bi, Pr, Nd, Sm, Gd, Dy, Ho, Er, Tm, U | $10^{-11}$–$10^{-10}$ | |
| | | Nb, Ta, Pt, Eu, Tb, Yb, Lu, Th | $<10^{-11}$ | |
| ICP-OES/ICP-AES | 10 mg $mL^{-1}$ suspension in water | Cs, Rb | $10^{-2}$ | 12 |
| | | Na, As, Ge, Se, Si, Tl | $10^{-3}$ | |
| | | Ag, Al, As, Au, B, Ba, Be, Bi, Ca, Cd, Ce, Co, Cr, Cu, Dy, Er, Eu, Fe, Ga, Gd, Ge, Ho, Hf, Hg, In, Ir, K, Li, Lu, Mg, Mn, Mo, Nb, Nd, Ni, Os, P, Pb, Pd, Pr, Pt, Re, Rh, Ru, S, Sb, Sc, Sm, Sn, Sr, Ta, Tb, Th, Ti, Tm, U, V, W, Y, Yb, Zn, Zr | $10^{-4}$ | |

**Table 5.2** (*Continued*)

| Method | Sample requirements | Elements detected at concentrations above LOD | Limits of detection, wt.%[a] | Refs |
|---|---|---|---|---|
| Neutron activation analysis (NAA) | Solid sample | F, Si, P, Ca, Nb | $10^{-4}$ | 32 |
| | | Ne, Mg, Ti, Fe, Ni, Ge, Sr, Y, Zr, Rh, Nd, Os | $10^{-5}$ | |
| | | Al, Cl, K, Cr, Cu, Zn, Rb, Mo, Ru, Pd, Cd, Sn, Ba, Ce, Gd, Er, Pt, Hg | $10^{-6}$ | |
| | | Na, V, Co, Ga, Kr, Ag, Xe, Cs, Pr, Eu, Tb, Ho, Tm, Ta, Re, Th | $10^{-7}$ | |
| | | Ar, Sc, Mn, Se, Br, Sb, Te, I, La, Sm, Dy, Yb, Hf, W, Ir, U | $10^{-8}$ | |
| | | As, In, Lu, Au | $10^{-9}$ | |
| Gamma activation analysis (GAA) | Solid sample | Cl, Ti, Cr, Fe, Ni, Zr, Mo, Sb, Sr, Mn, U, Eu | $1 \cdot 10^{-4}$–$3 \cdot 10^{-2}$ | 14 |
| Pyrolysis flameless atomic absorption spectroscopy (PFAAS) | Burning solid sample | Hg | $10^{-6}$ | 40 |

(continued on next page)

**Table 5.2** (*Continued*)

| Method | Sample requirements | Elements detected at concentrations above LOD | Limits of detection, wt.%[a] | Refs |
|---|---|---|---|---|
| Particle Induced X-ray Emission (PIXE) | Pressed tablets | Na, Mg | $10^{-2}$–$10^{-1}$ | 34 |
| | | Al, Si, P, S, Cl, K, Ru, Rh, Pd, Ag, Cd, In, Sb | $10^{-3}$–$10^{-2}$ | |
| | | Ca, Sc, Ti, V, Br, Rb, Sr, Y, Zr, Nb, Mo, Tc, Sn, Te, I, Cs, Ba, La, Ce, Pr, Nd, Pm, Sm, Eu, Gd, Tb, Dy, Hf, Ta, Re, Os, Ir, Pt, Au, Hg, Tl, Pb, Bi, Th, U | $10^{-4}$–$10^{-3}$ | |
| | | Cr, Mn, Fe, Co, Ni, Cu, Zn, Ga, Ge, As, Se, Ho, Er, Tm, Yb, Lu, W | $10^{-5}$–$10^{-4}$ | |
| X-ray fluorescence | Solid sample | Fe, Cu, Cr, Zr | $10^{-2}$–$10^{-1}$ | 40,49 |
| X-ray photoelectron spectroscopy (XPS) | Solid sample | Si, W, Cl, Na | $10^{-1}$ | 13 |
| Spectrographic analysis | Powder | Ag, Al, B, Ca, Cu, Fe, Mg, Mn, Ni, Si | $10^{-4}$–$10^{-1}$ | 30,50 |

[a] *In some case the concentration range of detected elements is used*

of the latter can be accurately measured using high-resolution spectrometers after defined periods of time and used for both qualitative and quantitative analysis. The detection of 71 elements at competitively low concentration levels of $10^{-4}$–$10^{-9}$ wt.% can be achieved with instrumental mode of NAA (INAA), although this method can be more sensitive for determination of certain metals, e.g. Au and As, than any other method listed in Table 5.2. The advantage of NAA is its relatively high precision, but this method has also some big disadvantages, such as high cost of analysis connected with access to an expensive neutron source for the activation of samples, handling of radioactive samples, and long analysis times (a few weeks) required for radioactive cooling and measurements. Until this work (see results shown in Table 5.3) no data on the application of INAA to the analysis of impurities in DND have been published. However, Merchel et al. successfully used INAA for the determination of 35 metals in nanodiamonds isolated from meteorites.[32]

Similar principles are applied in gamma activation analysis (GAA), where gamma radiation is used for the activation of atomic nuclei. The sensitivity of GAA is usually less ($10^{-2}$–$10^{-5}$ wt.%) than INAA, but some modes of GAA can be used for the determination of light elements, such as carbon and oxygen. Recently, Yakovlev et al.[14] reported data on the quantitative determination of 12 minor impurities in DND at concentration levels of 0.1–3.0 wt.% using GAA.

One more type of activation analysis suitable for multi-element analysis of diamond is Particle Induced X-ray Emission spectroscopy (PIXE). In this case, samples have to be irradiated with a beam of protons, causing activation of electrons, followed by emission of a characteristic X-ray spectrum. The intensity of emitted radiation according to element specific energies can be used for the determination of trace elements from Al to U in nanodiamond powders. The sensitivity of PIXE ranges from $10^{-4}$ to $10^{-2}$ wt.% depending upon the element. The nanodiamond samples used for analysis should be pressed into tablets of size suitable for the representative irradiation by the beam of protons, which typically covers an area with a diameter of 1 to 10 mm. There are communications reporting the use of PIXE for the analysis of impurities (0.12% Cr, 0.55% Fe) in Cu/ND composites after milling[33] and for the determination of impurities (Fe, Cr, Ni, Al, Si, Ca, Cd, Ni, etc.) in DND[34,35] at 0.1–1.0 wt.% concentrations.

Another family of non-destructive analytical methods includes Energy Dispersive X-ray spectroscopy (EDX, EDAX, EDS)[29,36–38] and related X-ray Fluorescence (XRF) spectroscopy and X-ray photoelectron spectroscopy (XPS). These methods are frequently used for the screening of major impurities in DND, which provide complementary information together with scanning electron microscopy (SEM) or primary information on surface chemistry. Typically, LODs values for detection of impurities using these techniques are rather limited, with the majority of them ranging between 0.01 and 0.1 wt.%. However, this is sufficient for the identification of main minor elemental impurities in DND such as Fe, Cu, Cr, Zr, Si, W, and others. Usually, EDX is less sensitive in the determination of lighter elements. Also, the accuracy and precision of EDX related methods in the analysis of powders like DND heavily depends on the quality of sample preparation and the availability of calibra-

**Table 5.3** Comparison of data obtained with different analytical methods for the determination of 59 minor elemental impurities in NanoAmando (NanoCarbon Research Institute Co., Ltd., Japan) and SDND (PlasmaChem GmbH, Germany) detonation nanodiamond

| Element | NAA[a], μg g$^{-1}$ | ICP-AES, μg g$^{-1}$ | | | ICP-MS, μg g$^{-1}$ | |
|---|---|---|---|---|---|---|
| | NanoAmando | | | PL-SDND-5p | | |
| | 1 | 2[c] | 3[d] | 4[e] | 5[e] | 6[f] |
| Ag | <0.3[b] | <1 | <1 | 660 ± 100 | 475.4 | 0.7 |
| Al | 165 ± 6 | 700 ± 100 | 1450 ± 200 | 530 ± 80 | 483.40 | 713.74 |
| As | <0.3 | <10 | <10 | <10 | 0.39 | <0.01 |
| Au | 0.005 ± 0.001 | <1 | <1 | <1 | no data | no data |
| B | no data | <1 | <1 | 540 ± 80 | 407.37 | 2122.36 |
| Ba | <5.6 | 4 ± 1 | 6 ± 1 | 25 ± 4 | 30.89 | 5.66 |
| Bi | no data | <1 | <1 | 30 ± 5 | 35.39 | 0.21 |
| Br | 2.65 ± 0.1 | no data | no data | no data | no data | no data |
| Ca | 24 ± 6 | 12 ± 2 | 9 ± 2 | 280 ± 40 | 299.49 | 1088.53 |
| Cd | <5.7 | <1 | <1 | <1 | 0.14 | 0.16 |
| Ce | <0.95 | <1 | <1 | 7 ± 1 | 7.62 | 48.01 |
| Cl | 85 ± 4 | no data | no data | no data | no data | no data |
| Co | 0.23 ± 0.01 | <1 | 4 ± 1 | 1.9 ± 0.3 | 0.39 | 3.04 |
| Cr | 480 ± 20 | 7 ± 1 | 10 ± 2 | 11 ± 2 | 8.66 | 3.64 |
| Cs | 0.46 ± 0.03 | <100 | <100 | <100 | <0.01 | 0.16 |
| Cu | 13 ± 1 | 4.0 ± 1 | 30 ± 5 | 20 ± 3 | 22.10 | 76.40 |
| Dy | <0.02 | <1 | <1 | <1 | 0.01 | 0.01 |
| Er | <3 | <1 | <1 | <1 | 0.01 | <0.01 |
| Eu | <0.009 | <1 | <1 | <1 | 0.01 | <0.01 |
| Fe | 800 ± 30 | 564 ± 85 | 480 ± 70 | 1780 ± 270 | 1701.01 | 3472.69 |
| Gd | <1.9 | <1 | <1 | <1 | 0.06 | 0.23 |
| Hf | 430 ± 20 | 40 ± 6 | 80 ± 12 | <1 | 0.27 | 0.73 |
| Hg | 82 ± 3 | <1 | <1 | <1 | no data | no data |
| Ho | <0.4 | <1 | <1 | <1 | <0.01 | 0.01 |
| K | <104 | <1 | 20 ± 3 | 20 ± 3 | 36.85 | 214.57 |
| La | <0.01 | no data | no data | no data | 0.07 | 0.43 |
| Lu | <0.9 | <1 | <1 | <1 | <0.01 | 0.02 |
| Mg | <18 | <1 | <1 | 40 ± 6 | 62.63 | 78.49 |
| Mn | 2.5 ± 0.1 | 6 ± 1 | 12 ± 2 | 21 ± 3 | 17.78 | 158.95 |
| Mo | 58 ± 3 | 7 ± 1 | 5 ± 1 | 13 ± 2 | 3.67 | 0.93 |
| Na | 29 ± 1 | 41 ± 6 | 7 ± 1 | 10800 ± 1600 | 9576.74 | 4164.40 |
| Nb | <65 | <1 | <1 | <1 | 1.09 | 0.27 |
| Nd | <0.02 | <1 | <1 | <1 | 0.04 | 0.21 |
| Ni | <175 | 9 ± 1 | 20 ± 3 | 15 ± 2 | 13.79 | 11.68 |
| P | no data | <1 | <1 | 11 ± 2 | 8.72 | 57.14 |

**Table 5.3** *(Continued)*

| Element | NAA[a], μg g$^{-1}$ | ICP-AES, μg g$^{-1}$ | | | ICP-MS, μg g$^{-1}$ | |
|---|---|---|---|---|---|---|
| | NanoAmando | | | PL-SDND-5p | | |
| | 1 | 2[c] | 3[d] | 4[e] | 5[e] | 6[f] |
| Pb | no data | <1 | <1 | 5 ± 1 | 7.43 | 33.29 |
| Pr | <1.7 | <1 | <1 | <1 | 0.01 | 0.06 |
| Rb | <1.2 | <100 | <100 | <100 | 0.12 | 0.36 |
| S | <192 | 180 ± 30 | 200 ± 30 | 38 ± 6 | 57.99 | <0.01 |
| Sb | 0.57 ± 0.03 | <1 | <1 | 10 ± 2 | 14.77 | 1.35 |
| Sc | 0.0048 ± 0.0005 | <1 | <1 | <1 | 0.02 | 0.12 |
| Si | <210400 | 300 ± 45 | 660 ± 100 | 1470 ± 220 | 1142.03 | 954.50 |
| Sm | <0.02 | <1 | <1 | <1 | 0.07 | 0.10 |
| Sn | <33 | <1 | 5 ± 1 | 170 ± 25 | 163.51 | 17.94 |
| Sr | <10 | <1 | <1 | 5 ± 1 | 5.66 | 1.12 |
| Ta | 1.05 ± 0.04 | <1 | <1 | <1 | 0.89 | <0.01 |
| Tb | <0.02 | <1 | <1 | <1 | <0.01 | 0.01 |
| Te | <4.3 | no data | no data | no data | 4.23 | 0.88 |
| Th | <0.2 | <1 | <1 | <1 | 0.17 | 0.11 |
| Ti | 81 ± 4 | 14 ± 2 | 15 ± 2 | 285 ± 40 | 230.23 | 460.63 |
| Tl | no data | <10 | <10 | <10 | 0.01 | 0.02 |
| Tm | <0.4 | <1 | <1 | <1 | <0.01 | 0.01 |
| U | <0.26 | <1 | <1 | <1 | 0.02 | 0.02 |
| V | 9.9 ± 0.4 | <1 | <1 | <1 | 1.21 | 0.32 |
| W | 99 ± 4 | 85 ± 13 | 200 ± 30 | 60 ± 9 | 37.99 | 9.40 |
| Y | 1140 ± 50 | 200 ± 30 | 480 ± 70 | <1 | 0.09 | 0.27 |
| Yb | 1.6 ± 0.2 | <1 | <1 | <1 | <0.01 | 0.02 |
| Zn | <4400 | 4 ± 1 | 30 ± 5 | 30 ± 5 | <0.01 | 77.12 |
| Zr | 17860 ± 630 | 3200 ± 48 | 3200 ± 48 | 17 ± 3 | 13.60 | 37.10 |
| **Total** | **21365** | **5350 ± 800** | **10500 ± 1600** | **16800 ± 2500** | **14398.84** | **13817.59** |

[a] *Data are kindly provided by Dr T. Plakhotnik (University of Queensland, Australia)*
[b] *Less than (<X.X) means that the content of this impurity is less than LOD for the method*
[c,d] *Samples obtained from NanoCarbon Research Institute in 2009 and 2012, respectively*
[e,f] *Samples obtained from PlasmaChem GmbH in 2008 and 2012, respectively*

tion or reference samples. The literature data on the use of XRF for determination of impurities in some commercially available DNDs are included in Table 5.2. The application of one more related method, namely Synchrotron micro-X-ray fluorescence ($\mu$SXRF) spectroscopy,[39] for the analysis of inclusions of micron size (>10 microns) natural diamonds has also been reported. However, its suitability for this analysis is limited due to many factors, such as the cost of analysis, absence of reference samples for calibration of instruments, availability of synchrotron sources for routine analysis, and difficulties in analysis of powder samples.

The extremely low LOD value of $10^{-6}$ wt.% for the determination of mercury in DND was achieved using pyrolysis flameless atomic absorption spectroscopy

(PFAAS).[40] This method has a higher sensitivity than ICP-AES and was successfully applied for analysis of 20 commercial DND samples. Surprisingly, the concentration of mercury in these samples varied significantly from 21 ng $g^{-1}$ to 717 μg $g^{-1}$.

The applicability of secondary ion mass spectrometry (SIMS) to analysis of DND has been also been considered,[37] but no quantitative results have been provided.

## 5.4 HYPHENATED METHODS BASED UPON INDUCTIVELY COUPLED PLASMA

Inductively coupled plasma or ICP (an ionized gas containing the same number of positive ions and electrons) is an extremely powerful source of sample atomization and ionization. Argon is normally used to generate the ICP because of its inertness and limited ability to form stable compounds in the plasma with other elements. At the same time, this monoatomic gas has a relatively simple emission spectrum and the plasma has sufficient power to excite and ionize most of the elements in the periodic table. Ionization of argon takes place in a torch consisting of three concentric quartz tubes, under an intensive high-frequency field (radio frequency, RF), which is applied through an induction coil located next to the torch. A stable, high temperature ICP of 6000–10000 K can be generated in analytical instruments, which is higher than the temperature produced by the other excitation sources used for atomic emission analysis of solid materials, such as electric arc (3000–4000 K) or spark (4000–5000 K). For this reason, ICP related analytical methods (ICP-MS, LA-ICP-MS, ICP-AES, and ICP-OES) can be applied to the direct (by spraying of diamond nanoparticles suspension in plasma) quantitative analysis of impurities in DND. The laser ablation (LA) produces a hole of diameter 50–200 μm with depth of 15–60 μm in the diamond material during LA-ICP-MS analysis,[41,42] thus the suitability of this analytical method is rather related to the analysis of reasonably big crystals and limited by the necessity of forming tablets through applied pressure to DND powders, together with fillers or by mixing with epoxide resin and curing.

The ICP-atomic emission spectrometry (ICP-AES), also known as an ICP-Optical Emission Spectroscopy (ICP-OES), is a type of emission spectroscopy, where the emission spectrum emitted by the atoms and ions excited within the ICP is used for the sensitive determination of traces of elements. The spectrum is divided into discrete lines corresponding to the specific emission wavelength. Each element is characterized by its own distinctive set of lines in the emission spectrum, while the intensity of these lines provides the required information on the concentration of the element in the sample. ICP-AES is a multi-element method of analysis allowing quantitative determination up to 68 elements[12] present in concentrations above the corresponding LOD values as specified in Table 5.2.

High resolution ICP-MS provides excellent detection sensitivity for 71 elemental impurities, which normally can be quantified at low pg $mL^{-1}$ concentration levels in aqueous solutions. ICP-MS method is approximately two to three orders of magnitude more sensitive over a similar range of elements as compared with ICP-AES.[17,28]

For this reason, more elemental impurities can be detected and quantified in DND using this technique. An important feature of ICP-MS is the possibility to obtain isotopic information on the samples, which is valuable in analysis of natural (meteorite) nanodiamonds.[32]

## 5.5 COMPARISON OF ANALYTICAL METHODS FOR DETERMINATION OF ELEMENTAL IMPURITIES IN NANODIAMOND

The results from three analytical methods, namely NAA, IC-OES, and ICP-MS, each of which has been used for the multi-elemental analysis of minor impurities in two popular types of detonation nanodiamond NanoAmando and single digit nanodiamond (SDND), are presented in Table 5.3. Obviously, NAA is designed for the analysis of solid samples and normally 100–200 mg of DND sample pressurized with an appropriate filler into tablets is sufficient for detection of minor impurities. For ICP-AES and ICP-MS analysis of detonation nanodiamond, the analytical matrix is an aqueous suspension. As ICP-MS is a much more sensitive analytical technique, as compared with ICP-AES, the use of very dilute 50–100 $\mu g\,mL^{-1}$ aqueous suspensions of nanodiamonds allowed the simultaneous determination up to 37–47 minor elemental impurities by ICP-MS.[17,28] Only 15–28 impurities could be detected by ICP-AES in significantly more concentrated (10 $mg\,mL^{-1}$) DND suspensions.[12] It should be noted that the high concentration of nanodiamond in suspension must be avoided to maintain signal stability, which influences precision and accuracy, and reliable sample delivery in these techniques.

A comparison of the results obtained with same brand of commercial DNDs shows that these methods differ in sensitivity, precision, and accuracy of measurements. The LOD values of ICP-MS are significantly smaller ($5 \cdot 10^{-6}$ wt.%) for the analysis of impurities in DND, being two orders of magnitude smaller than those obtained with ICP-AES. Hence, the detection of certain impurities present at very low concentration levels in detonation nanodiamond is possible only by using ICP-MS. For example, the concentrations of alkali metals (Rb, Cs), heavy metals (V, As, Cd, Te, Hf, Ta, Tl, Th, U) and rare earth elements (Sc, La, Pr, Nd, Sm, Eu, Gd, Tb, Dy, Ho, Er, Tm, Yb, Lu) in the sample PL-SDND-5p (results presented in columns 4–6 of Table 5.2) are below of limits of detection (LOD) of ICP-AES and normally cannot be detected at low contents.

A significant question arises when trying to correlate results between different techniques. Table 5.3 contains results on impurities detected in three samples of Nanoamando, where two of them, produced in 2009 and 2012, are analyzed by ICP-AES, and one (production date is not specified) by INAA. Unfortunately, there is a big difference between contents of elemental impurities detected in different batches of Nanoamando (samples 1–3 in Table 5.3), which could be explained by the rather low reproducibility of the production technique, rather than by low precision of the compared analytical methods. However, the presence of characteristic impurities such as

Fe, Zr, Y, Hf, and Al can be observed in all samples. Three of these contaminants (Zr, Y, and Hf) clearly originated from ball milling disaggregation, and are very distinctive for Nanoamando type of DND,[17] which can be considered as a sort of correlation between results obtained by NAA and ICP-AES in two different research institutes (ANSTO and Moscow State University).

A similar comparison of results obtained with ICP-AES and ICP-MS is carried out for three samples of another popular brand of DND, PL-SDND-5p, which is also known as Single Digit Nanodiamond produced by PlasmaChem GmbH in 2008 (Sample 4 and 5 in Table 5.3) and 2012 (Sample 6 in Table 5.3). Again, there is some discrepancy in profiles of impurities in DND obtained for different batches (Sample 5 and 6) using the same analytical technique ICP-MS. Obviously, this is again related to the reproducibility of production of this type of nanodiamond. However, very strong correlation ($r^2 = 0.9992$, $n = 27$, see Fig. 5.3) is obtained between Samples 4 and 5 belonging to the same batch of DND analyzed by ICP-AES and ICP-MS. Interestingly, the unusually high content of silver (1000-fold as compared with common DND samples) was recorded for both of these samples, which confirms the accuracy of both methods. It should be noted that ICP-AES allowed the detection of 27 elemental impurities present in PL-SDND-5p at concentrations above LODs, but the more sensitive ICP-MS method allowed the determination of 48 elements (see Table 5.3). Surprisingly, the slope of correlation plot shown in the top of Fig. 5.3 is equal to 1.128 that means a systematic error resulting in higher values of detected impurity concentrations. ICP-AES also displays a bigger uncertainty range for the obtained results.

Thus, both ICP hyphenated methods are suitable for the detection of many minor elemental impurities, but ICP-MS is the preferential choice if ultra-low concentrations and small changes in composition of DND are to be monitored.

## 5.6 APPLICATIONS OF ICP-HYPHENATED TECHNIQUES

The primary reasons for using ICP-MS in the analysis of DND include the identification of key impurities and sources of contamination, monitoring of purification processes, and the improvement of corresponding technologies and final quality of produced DND. The strong retention of minor elemental impurities within DND and their difficult removal is connected with the specifics of the occurrence of the minor elemental impurities in separate nanoparticles and in DND aggregates.[1,19] Because of a highly developed surface area, the majority of elemental impurities present in DND are in the form of adsorbed species, while inorganic species are retained due to ion-exchange or complexation interactions with various functional groups.[19] ICP-MS and ICP-AES have been used for the monitoring of minor elemental impurities in various types of commercial DND products.[12,17,28,43]

Obviously, the content of metal impurities can be decreased by intensive washings with concentrated mineral acids[11,17,44] or complexing reagents such as EDTA,[11,45]

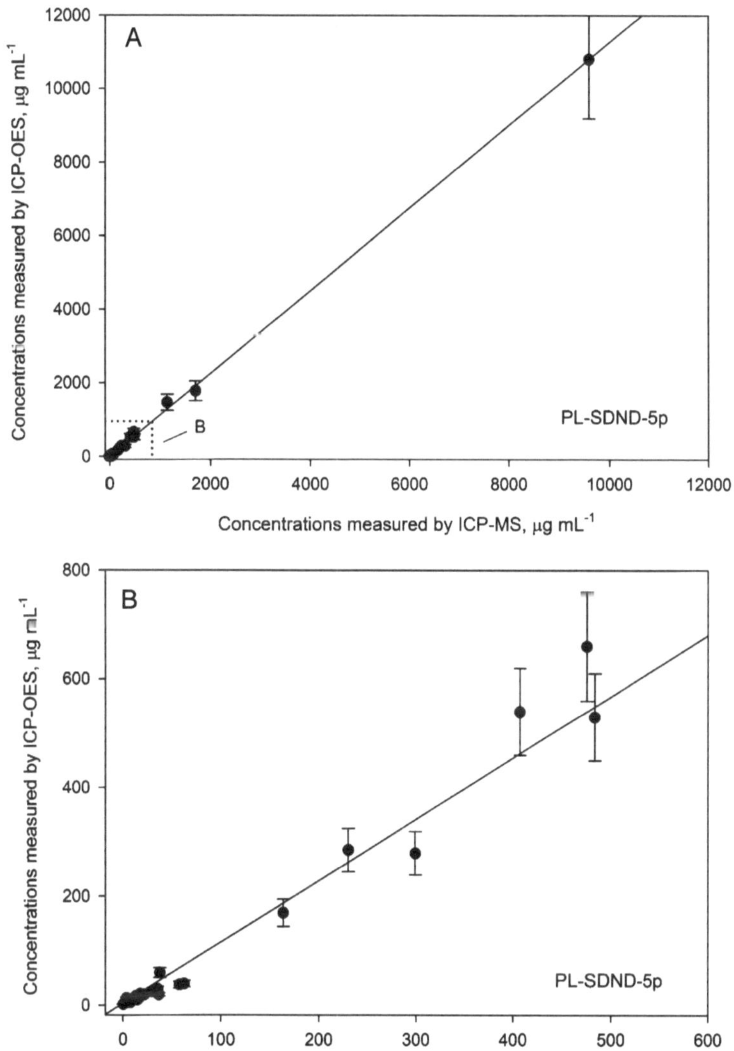


**FIGURE 5.3**

Comparison of the results (see columns 4 and 5 in Table 5.3) obtained by ICP-OES and ICP-MS on measurement of 27 elemental impurities detected in single digit nanodiamond sample (PL-SDND-5p product of Plasmachem GmbH) at concentrations above LOD values of both analytical methods

unithiol,[45] or dipicolinic acid.[11] The complete removal of ionic impurities is not a trivial task, as the zwitterionic nature of DND allows adsorption of both cations and anions.[19] For example, the washing of DND with mineral acids, e.g. sulphuric

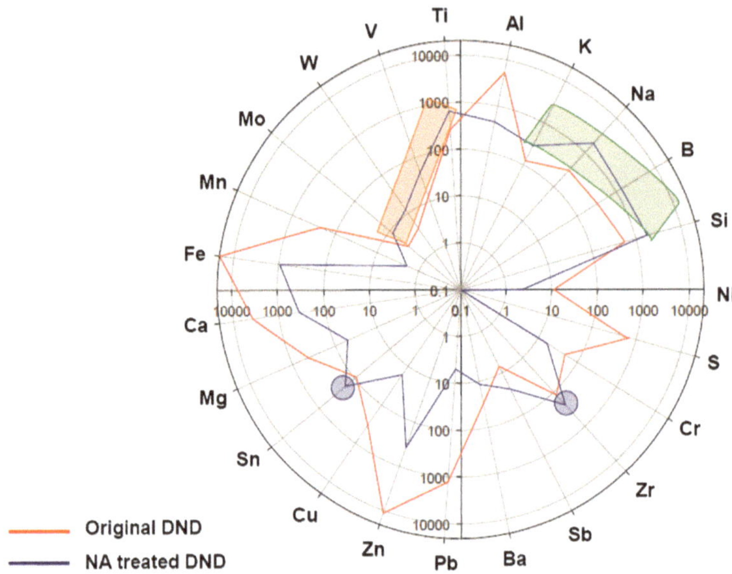


**FIGURE 5.4**

Radian diagram of impurities contents in DND before and after nitric acid treatment using borosilicate glassware and ultrasonication with sonotrode. The concentrations are shown in ng mL$^{-1}$

acid, removes adsorbed metal cations by replacing them from negatively charged carboxylic groups with $H^+$, but at the same time converts positively charged lactone groups into sulphate associated form. Therefore, elevated concentrations of sulphur and oxygen should be expected in the purified sample of DND, which can be considered as processing derived "contamination". Practically, it means that every step in any purification technology should be carefully controlled.

ICP-MS has been used for purity control of DND samples prepared by microwave assisted purification with different wet oxidizing mixtures as described in a recent study.[11] It was noted that the application of this extremely sensitive and precise analytical technique allowed identification of new sources of impurities. It was found that the use of sonotrode electrodes and borosilicate glassware re-contaminated the purified DND with Mo, W, V, Ti and Na, K, B, Si, respectively. The corresponding profiles of 23 most abundant minor (not including oxygen, nitrogen, and hydrogen) elemental impurities in the DND sample before and after nitric acid treatment using ultra sonication with a sonotrode in borosilicate glassware beaker are presented in Fig. 5.4.

The concentrations of some elements in a treated sample can be 2–5 times higher as compared with those for untreated sample. A similar type of contamination is also possible during the prolonged storage of DND in glass bottles and vials, and some producers (PlasmaChem GmbH, Sigma-Aldrich) are continuing to supply purified DND products to the customers in small glass bottles and vials.

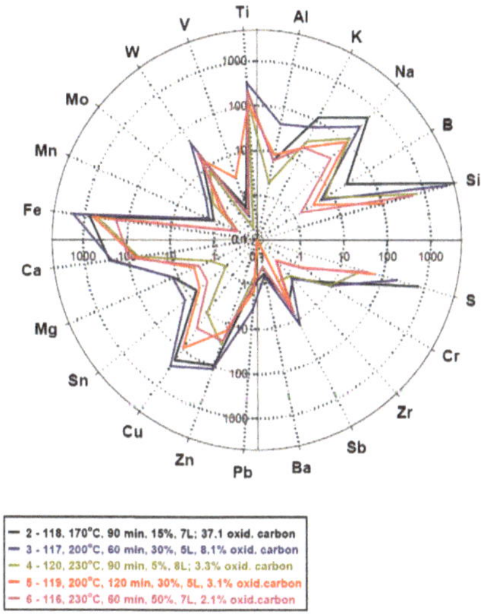


| Sample | Conditions ($t$ °C, time, concentration and volume of $HNO_3$) used for treatment of 100 g of detonation soot | Oxidisable carbon, wt % |
|---|---|---|
| Soot | | 65.9 |
| 118 | 170°, 90 min, 15 %, 7 L | 37.1 |
| 117 | 200°, 60 min, 30 %, 5 L | 8.1 |
| 120 | 230°, 90 min, 5 %, 8 L | 3.3 |
| 119 | 200°, 120 min, 30 %, 5 L | 3.1 |
| 116 | 230°, 60 min, 50 %, 7 L | 2.8 |

**FIGURE 5.5**

Variation of impurities in samples prepared conditions applied for oxidation of graphitic carbon and isolation of nanodiamonds from detonation soot. Samples prepared as described[44] by FSUE Technolog (St. Petersburg, Russia)

DNDs prepared by microwave-assisted purification[11] have been recently used for the evaluation of their cytotoxicity using two cell models: a human osteosarcoma cell line, MG-63, and primary rat mesenchymal stem cells (rMSCs).[46] A clear correlation between toxicity and purity of DNDs was found. The most toxic sample obtained after treatment with mixture of nitric, sulphuric, and hydrochloric acids (NASH sample, total content of minor elemental impurities 1.17 wt.% measured here and later by ICP-MS) contained non-diamond carbon and caused suppression of cell proliferation and cell detachment in 24 hours after addition to the cells. YTM (YTM ARGE A.S, Istanbul, Turkey) and DND-30 (Beijing Grish Hitech Co., Ltd, China) samples containing 0.95 and 1.34 wt.%, respectively, also affected cell viability to a different degree. The lowest cytotoxicity was demonstrated using NSFPA nanoparticles isolated from the detonation soot after treatment with a mixture of nitric, sulphuric, hydrofluoric, and perchloric acids, and containing less than 0.05 wt.% of minor elemental impurities.[11] This study confirmed that cytotoxicity of DND particles depends both on purity, size, and shape of DND aggregates. However, knowledge of exact impurity profiles can be used for the identification of the most toxic elements for specific types of cells.

Another example of using ICP-MS for monitoring of impurities in detonation nanodiamonds is shown in Fig. 5.5. In this case samples were prepared by wet oxidation of the graphitic component in the detonation soot with nitric acid at different temperatures and times. Obviously, the key target of such treatment was the optimization of conditions for the preparation of the products containing the maximum content of diamond phase; however, the concentration and profile of residual impurities is also of great importance. The obtained results show that the content of minor elemental impurities depends mainly on temperature of the treatment and to a lesser degree on the concentration and volume of nitric acid.

Direct ICP-MS analysis of suspensions of nanodiamonds isolated from meteorites was also demonstrated[47] with a focus on isotopic and elemental abundance patterns in rare earth elements (REE), which provides important information on the origin of the meteorites and physico-chemical processes within the cosmos.

Finally, the possibility of using ICP-MS for characterization of surface modification was exploited in a recent work,[29] where DND treatment with various silanes was used for hydrophobization of the surface and disaggregation of nanoparticles. The accurate measurements of silicon concentrations using ICP-MS before and after treatment of DND samples provided information on the coating degree and surface concentration of the bonded groups. The Si-content increased dramatically to 24,700–29,300 $\mu g\,g^{-1}$ in samples, realizing an increase of over 570–610 times as compared to the original DND. These values correspond to the surface coating of 0.88–0.95 $mmol\,g^{-1}$ as calculated from the increase in Si-content after silylation, which is 2–3 times higher as compared with 0.3–0.4 $mmol\,g^{-1}$ reported for DND surface groups esterification with alkanoic acid chlorides.[48]

## CONCLUSIONS AND ACKNOWLEDGMENTS

Undoubtedly, the application of inductively coupled plasma hyphenated analytical techniques provides significant advancement in our knowledge of the purity of nanodiamonds, the source of possible contaminations, identification of their toxicity, and applicability in nanomedicine. ICP-related methods are simple, sensitive, and provide accurate and precise quantification of 50–55 minor elemental impurities. The use of ICP-MS for monitoring impurities can provide drastic improvements in technology of purification of nanodiamonds.

This work was supported by the Australian Research Council (DP110102046, DP150101518 and CE140100012). The authors would also like to acknowledge the Central Science Laboratory (University of Tasmania) and Australian Nuclear Science and Technology Organisation (ANSTO) for substantial instrumental support and service. The authors are thankful to A. Townsend (University of Tasmania), T. Plakhotnik (University of Queensland, Australia), and K. Korolev (FSUE Technolog, St. Petersburg, Russia) for the provided data.

## REFERENCES

1. Aleksenskiy, A. E.; Eydelman, E. D.; Vul', A. Y. *Nanosci. Nanotechnol. Lett.* **2011**, *3* (1), 68–74.

2. Zhu, Y.; Li, W.; Zhang, Y.; Li, J.; Liang, L.; Zhang, X.; Chen, N.; Sun, Y.; Chen, W.; Tai, R.; Fan, C.; Huang, Q. *Small* **2012**, *8* (11), 1771–1779.
3. Zhu, Y.; Zhang, Y.; Shi, G.; Yang, J.; Zhang, J.; Li, W.; Li, A.; Tai, R.; Fang, H.; Fan, C.; Huang, Q. *Part. Fibre Toxicol.* **2015**, *12* (2), 1–11.
4. Luong, J. H. T.; Male, K. B.; Glennon, J. D. *Analyst* **2009**, *134* (10), 1965–1979.
5. Vlasov, I. I.; Barnard, A. S.; Ralchenko, V. G.; Lebedev, O. I.; Kanzyuba, M. V.; Saveliev, A. V.; Konov, V. I.; Goovaerts, E. *Adv. Mater.* **2009**, *21* (7), 808–812.
6. Barjon, J.; Rzepka, E.; Jomard, F.; Laroche, J. M.; Ballutaud, D.; Kociniewski, T.; Chevallier, J. *Phys. Status Solidi A – Appl. Mater. Sci.* **2005**, *202* (11), 2177–2181.
7. Jelezko, F.; Wrachtrup, J. *Phys. Status Solidi A – Appl. Mater. Sci.* **2006**, *203* (13), 3207–3225.
8. Manus, L. M.; Mastarone, D. J.; Waters, E. A.; Zhang, X. Q.; Schultz-Sikma, E. A.; MacRenaris, K. W.; Ho, D.; Meade, T. J. *Nano Lett.* **2010**, *10* (2), 484–489.
9. Chiganova, G. A.; Chul'myakova, D. A.; Mordvinova, L. E.; Petrova, T. I. *Russ. J. Appl. Chem.* **2012**, *85* (2), 177–181.
10. Dolmatov, V. Y. *J. Superhard Mater.* **2009**, *31* (3), 158–164.
11. Mitev, D. P.; Townsend, A. T.; Paull, B.; Nesterenko, P. N. *Diam. Relat. Mater.* **2014**, *48*, 37–46.
12. Volkov, D. S.; Proskurnin, M. A.; Korobov, M. V. *Carbon* **2014**, *74*, 1–13.
13. Singh, B.; Smith, S. J.; Jensen, D. S.; Jones, H. F.; Dadson, A. E.; Farnsworth, P. B.; Vanfleet, R.; Farrer, J. K.; Linford, M. R. *Anal. Bioanal. Chem.* **2016**, *408* (4), 1107–1124.
14. Yakovlev, R. Y.; Dogadkin, N. N.; Kulakova, I. I.; Lisichkin, G. V.; Leonidov, N. B.; Kolotov, V. P. *Diam. Relat. Mater.* **2015**, *55*, 77–86.
15. Bondar', V. S.; Puzyr', A. P. *Phys. Solid State* **2004**, *46* (4), 716–719.
16. Osawa, E. *NCRI Tech. Bull.* **2009**, *3*, 1–8.
17. Mitev, D.; Townsend, A.; Paull, B.; Nesterenko, P. *J. Mater. Sci.* **2014**, *49* (10), 3573–3591.
18. Peristyy, A. A.; Fedyanina, O. N.; Paull, B.; Nesterenko, P. N. *J. Chromatogr. A* **2014**, *1357*, 68–86.
19. Peristyy, A.; Paull, B.; Nesterenko, P. N. *Adsorption* **2016**, *22* (7), 371–383.
20. Haddad, P. R.; Nesterenko, P. N.; Buchberger, W. *J. Chromatogr. A* **2008**, *1184* (1–2), 456–473.
21. Chiganova, G. A.; Bondar, V. A.; Chiganov, A. S. *Colloid J. Russ. Acad. Sci.* **1993**, *55* (5), 774–775.
22. Singh, N.; Rashmi Sarkar, A. K. *J. Mater. Sci.* **2004**, *39* (5), 1665–1669.
23. Sakurai, H.; Ebihara, N.; Osawa, E.; Takahashi, M.; Fujinami, M.; Oguma, K. *Anal. Sci.* **2006**, *22* (3), 357–362.
24. Bogatyreva, G. P.; Voloshin, M. N.; Padalko, V. I. *Diam. Relat. Mater.* **2008**, *17* (3), 213–216.
25. Shenderova, O.; Petrov, I.; Walsh, J.; Grichko, V.; Grishko, V.; Tyler, T.; Cunningham, G. *Diam. Relat. Mater.* **2006**, *15* (11–12), 1799–1803.
26. Dolmatov, V. Y. *Russ. J. Appl. Chem.* **2008**, *81* (10), 1747–1753.
27. Dolmatov, V. Y. *Russ. Chem. Rev.* **2001**, *70* (7), 607–626.
28. Mitev, D. P.; Townsend, A. T.; Paull, B.; Nesterenko, P. N. *Carbon* **2013**, *60*, 326–334.
29. Xu, N. S.; Chen, J.; Deng, S. Z. *Diam. Relat. Mater.* **2002**, *11* (2), 249–256.
30. Raal, F. A. *Am. Mineral.* **1957**, *42* (5–6), 354–361.
31. Fesq, H. W.; Bibby, D. M.; Sellscho, J. P.; Watterso, J. I. *J. Radioanal. Chem.* **1973**, *17* (1–2), 195–216.

32. Merchel, S.; Ott, U.; Herrmann, S.; Spettel, B.; Faestermann, T.; Knie, K.; Korschinek, G.; Rugel, G.; Wallner, A. *Geochim. Cosmochim. Acta* **2003**, *67* (24), 4949–4960.
33. Nunes, D.; Correia, J. B.; Carvalho, P. A.; Shohoji, N.; Fernandes, H.; Silva, C.; Alves, L. C.; Hanada, K.; Osawa, E. *Fusion Eng. Des.* **2011**, *86* (9–11), 2589–2592.
34. Wang, Y.; Jaiswal, M.; Lin, M.; Saha, S.; Ozyilmaz, B.; Loh, K. P. *ACS Nano* **2012**, *6* (2), 1018–1025.
35. Rakha, S. A.; Raza, R.; Munir, A. *Polym. Compos.* **2013**, *34* (6), 811–818.
36. Spitsyn, B. V.; Davidson, J. L.; Gradoboev, M. N.; Galushko, T. B.; Serebryakova, N. V.; Karpukhina, T. A.; Kulakova, I. I.; Melnik, N. N. *Diam. Relat. Mater.* **2006**, *15* (2–3), 296–299.
37. Koshcheev, A. P. *Russ. J. Gen. Chem.* **2009**, *79* (9), 2033–2044.
38. Puzyr, A. P.; Burov, A. E.; Bondar, V. S.; Purtov, K. V.; Hwang, K. C.; Rhee, C. K. *Patent Russ. Fed.* **2012**, 2458858.
39. Sitepu, H.; Kopylova, M. G.; Quirt, D. H.; Cutler, J. N.; Kotzer, T. G. *Am. Mineral.* **2005**, *90* (11–12), 1740–1747.
40. Volkov, D. S.; Proskurnin, M. A.; Korobov, M. V. *Diam. Relat. Mater.* **2014**, *50*, 60–65.
41. Rege, S.; Jackson, S.; Griffin, W. L.; Davies, R. M.; Pearson, N. J.; O'Reilly, S. Y. *J. Anal. At. Spectrom.* **2005**, *20* (7), 601–611.
42. Frick, D. A.; Gunther, D. *J. Anal. At. Spectrom.* **2012**, *27* (8), 1294–1303.
43. Duffy, E.; Mitev, D. P.; Thickett, S. C.; Townsend, A. T.; Paull, B.; Nesterenko, P. N. *Appl. Surf. Sci.* **2015**, *357*, 397–406.
44. Korolev, K. M.; Makarov, I. A.; Shorikov, D.; Sushchev, V. G. Controlled Oxidation of Detonation Blend. In *11th Internat. Conf. on Advanced Carbon Nanostructures;* ACN-2013, 1–5 July 2013, St Petersburg, Russia, 2013, pp. 4–14.
45. Dolmatov, V. Y.; Vehanen, A.; Myllymaki, V.; Rudometkin, K. A.; Panova, A. N.; Korolev, K. M.; Shpadkovskaya, T. A. *Russ. J. Appl. Chem.* **2013**, *86* (7), 1036–1045.
46. Keremidarska, M.; Ganeva, A.; Mitev, D.; Hikov, T.; Presker, R.; Pramatarova, L.; Krasteva, N. *Biotechnol. Biotechnol. Equip.* **2014**, *28* (4), 733–739.
47. Yin, Q. Z.; Lee, C. T. A.; Ott, U. *Astrophys. J.* **2006**, *647* (1), 676–684.
48. Krueger, A.; Boedeker, T. *Diam. Relat. Mater.* **2008**, *17* (7–10), 1367–1370.
49. Golubina, E. V.; Erokhin, A. V.; Lokteva, E. S.; Kharlanov, A. N.; Petrova, N. I. Effect of Nanodiamond Surface Composition on Adsorption of Metal Ions and Catalytic Properties of Supported Metal Particles. In *11th Internat. Conf. on Advanced Carbon Nanostructures;* ACN-2013, 1–5 July 2013, St. Petersburg, Russia, 2013, p. 179.
50. Angus, J. C.; Will, H. A.; Stanko, W. S. *J. Appl. Phys.* **1968**, *39* (6), 2915–2922.

CHAPTER

# Paramagnetic defects in nanodiamonds

# 6

**Alexander I. Shames, Alexander M. Panich**

*Department of Physics, Ben-Gurion University of the Negev, Be'er-Sheva, Israel*

**Contents**



## 6.1 INTRODUCTION TO MAGNETIC RESONANCE AND NUCLEAR RELAXATION

The phenomenon of electron paramagnetic resonance, EPR (also called electron spin resonance, ESR), is a resonant absorption of electromagnetic radiation by unpaired electron spins subjected to an external static magnetic field. For theoretical background and application to studies of a large variety of systems, the reader is referred to remarkable books by Weil/Bolton[1] and Abragam/Bleaney.[2] This non-destructive method provides scientists with unique possibilities in identification of the electron-nuclear structure of various substances, from gases to complex biological molecules.

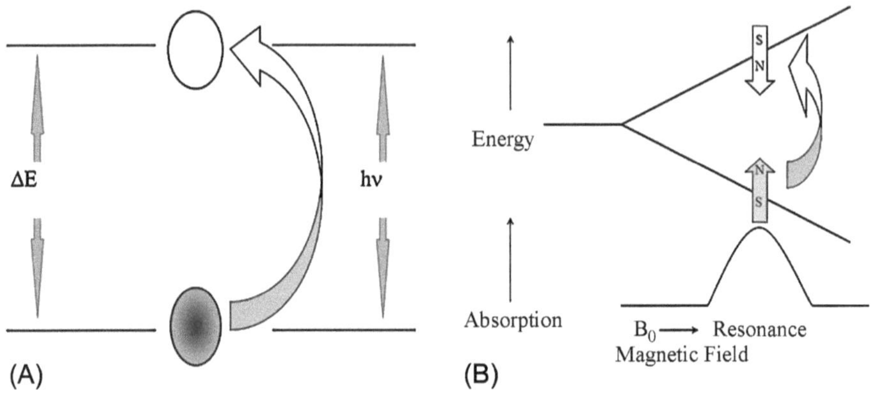


**FIGURE 6.1**

(A) Transitions between discrete energy levels of a quantum system initiated by absorption of electromagnetic radiation energy by the system; (B) spin state energy plotted vs. magnetic field

Unlike many other spectroscopic and analytical methods, EPR offers important information on the structure and dynamics of a system, including the chemical and physical processes occurring there.

By spectroscopy one commonly understands measurement of the differences between the energies of discrete states of atoms or molecules, as well as their interpretation. Knowledge of these differences allows one to look into a quantum system and determine the characteristic features, structure, and dynamics of the system under study.

By Planck's law, an electromagnetic wave may be absorbed by an atom or a molecule only if the relation

$$\Delta E = hv \tag{6.1}$$

is satisfied, where $h$ is the Planck constant ($6.620693(11) \times 10^{-34}$ J · s) and $v$ is frequency of electromagnetic radiation (Fig. 6.1A).

The dependence of the intensity of radiation absorption on the frequency $v_{res}$ of electromagnetic radiation defines the EPR spectrum. The frequency range employed in EPR spectroscopy lies in the gigahertz waveband (1–100 GHz). The energy difference, which is the object of the study by EPR spectroscopy, is the result of interaction of unpaired electrons in the material subjected to an external magnetic field, which results in a splitting of the energy levels (Zeeman effect).

Electron has an intrinsic magnetic moment, spin, and in the external magnetic field $B_0$ can occupy one of the two states with the spin vector directed along or opposite to the vector $B_0$. These two states can be specified in accordance with the projections of the electron spin vector $M_s$ on the direction of the external magnetic field. The absolute value of electron spin is $S = 1/2$, the direction along the field is identified with the spin projection $M_s = -1/2$, and that opposite to the magnetic field with $M_s = +1/2$.

The main equations describing the EPR are

$$E = g\mu_B B_0 M_s = \pm\frac{1}{2} g\mu_B B_0 \tag{6.2}$$

and

$$\Delta E = hv = g\mu_B B_0, \tag{6.3}$$

where $g$ is the $g$-factor, or the spectroscopic splitting factor, a proportionality constant, $g \cong 2$ for most samples, which varies only weakly with electron configuration of a radical or ion [for free electrons $g_e = 2.0023193043718$ (75)], and $\mu_B$ is the Bohr magneton, a unit of the electron magnetic moment [$\mu_B = 9.27400949$ (80) $\times 10^{-24}$ $J \cdot T^{-1}$]. As evident from Eqs. (6.2) and (6.3) and Fig. 6.1B, the energy gap between two spin states increases linearly with magnetic field. In conventional EPR spectroscopy the signal is registered by keeping a constant frequency of electromagnetic radiation and scanning the magnetic field (Fig. 6.1B).

The absorption of electromagnetic radiation appears to be maximal when the energy difference between the energy levels is exactly equal to the radiation energy. Because a spectrum may be measured at different frequencies and, accordingly, in different magnetic field ranges, the resonant field is not a unique characteristic of the material under study (its "fingerprint", as it were). On the other hand, the $g$-factor in Eq. (6.3) is independent of microwave frequency and is much better suited for identification of a material. Significantly, in EPR spectra obtained at a fixed frequency, larger values of the $g$-factor appear in weaker magnetic fields, and vice versa. Thus, the most common commercial EPR spectrometers work at the X-band, $v = 9$–$10$ GHz, $B_{res}(g = 2.00) = 0.322$–$0.357$ T; Q-band, $v = 34$–$35$ GHz, $B_{res}(g = 2.00) = 1.22$–$1.25$ T; and W-band, $v = 94$ GHz, $B_{res}(g = 2.00) = 3.4$ T.

While measurement of $g$-factors provides very useful information on the material under study, the knowledge of the $g$-factor alone is far from being adequate for forming a comprehensive idea of the molecular structure of a material. Fortunately, the unpaired electron involved in the EPR is very sensitive to its local environment (other electrons and atomic nuclei). Indeed, nuclear spins possess quite frequently a nonzero magnetic moment generating a local magnetic field, which acts upon the electron. The interaction between an electron and a nucleus causes a change in the EPR spectrum and formation of a hyperfine structure in this spectrum, which provides important information on the sample under study. It is the hyperfine structure that makes possible the identification of the atoms composing a molecule or a cluster and their location with respect to the unpaired electron.

The nature of hyperfine interaction can be readily unraveled if we assume the nucleus of an atom to act as a small magnet generating a magnetic field $B_I$ in which the electron resides. This field $B_I$ is capable of increasing or lowering the external magnetic field in which a maximum in EPR is observed. In an EPR spectrum obtained for a substance containing an unpaired electron interacting with a nucleus with spin $I = 1/2$ (such nuclear spins have, for instance, 100% of fluorine nuclei, 99.99% of hydrogen nuclei, 1.07% of carbon nuclei or 2.12% of iron nuclei), the only absorption signal (a singlet) will split into two signals (a doublet), each shifted by $B_I$ relative to the original singlet signal.

If a molecule contains one more nucleus with spin 1/2, each of the signals of the doublet will be split once more into a pair of signals to form a hyperfine 4-line (quadruplet) hyperfine structure. As the number of magnetic nuclei increases, the number of signals in the hyperfine structure grows sharply by complex rules which have to account for the fact that even nuclei with the same spin can be nonequivalent, i.e., located differently with respect to the electron (for instance, at different distances).

Significantly, the magnitude of the EPR signal contains important information concerning the number of EPR-active centers (radicals, transition metal ions, defects or unpaired charge carriers) in the object under study. In the language of spectroscopy, the "magnitude of the signal" is defined as the integrated intensity, i.e., the area under the absorption curve.

The integrated intensity of the EPR signal is proportional to the concentration of paramagnetic centers or, termed otherwise, to the magnetic susceptibility of a sample at the corresponding microwave frequency. It should be borne in mind, however, that the intensities of observed signals depend not only on the concentration of EPR-active centers but on the power of the microwave radiation as well. If one does not operate at very high levels of microwave power, the intensity of the observed signal grows proportional to the square root of power. At high levels of microwave power one observes broadening and drop of the EPR peak intensity. To ensure correct measurement of the line width and shape, as well as of weakly separated hyperfine lines, one should avoid this saturation effect.

A characteristic feature of most of modern EPR spectrometers is the specific technique employed in recording EPR signals; indeed, in contrast to the optical and NMR spectrometers, EPR signals are presented in the form of the first derivative of the absorption signal. Typical X-band (9.464 GHz) EPR spectra are shown in Fig. 6.2; see the next section.

EPR and related techniques (like Electron Nuclear Double Resonance – ENDOR, etc.) being applied to crystalline diamond samples provide useful information on anisotropy of electronic $g$ tensor; zero field couplings in defects with $S > 1/2$; hyperfine interaction with principle nuclei of surrounding, etc.[2,3] Traditionally, EPR and related phenomena are interpreted within the framework of the spin-Hamiltonian concept, where spin-Hamiltonian is an energy operator describing all existing interactions (electric, magnetic, exchange) of the electron spins and nuclei with their surrounding and external fields applied.[1,2,4] Useful data obtained from EPR spectra are presented as corresponding spin-Hamiltonian parameters ($g$-values, hyperfine splittings $A$, zero-field splittings $D$, $E$, etc.). Well-developed theoretical tools in combination with state-of-the-art experimental techniques allow obtaining tiny details of defects nature, surrounding, and interactions. However, EPR study of ND has its specificity dictated by the form of ND samples. Conventional EPR (excluding fast developing single-electron/single-molecule EPR techniques) deals with bulky samples. In contrast to well-ordered single crystalline diamond and crystalline diamond films, the most of bulky ND samples are in polycrystalline form. Moreover, a typical ND sample is an ensemble of nanoparticles characterized by a distribution of individual sizes and properties of the ND particles. Thus, EPR spectra of ND samples are spatially averaged, which causes certain decrease in spectral resolution, reduction in sensitivity (due to additional line broadening), and complication in interpretation of the EPR data obtained. On studying defects in ND by means of magnetic resonance all the aforementioned aspects must be taken into account.

NMR is a phenomenon of a similar nature and represents absorption of electromagnetic radiation of radio frequency $f$ by nuclear spins subjected to magnetic field. As EPR detects paramagnetic defects directly, NMR obtains information about them via electron-nuclear interactions.

$^{13}C$ Magic Angle Spinning (MAS) spectrum reveals $sp^3$ carbons of perfect diamond core and partially disordered shell, as well as some $sp^2$ carbons fractions located on the ND surface.[5–7] The latter are represented by C—C, C—O, and C—O–OH groups, and can be efficiently removed by additional deep purification.[5–7] Static $^1H$ spectrum reveals broad and narrow components attributed to (i) rigid hydrocarbon and hydroxyl groups and (ii) to moisture adsorbed on the DND surface, respectively.[5,8–10] Since NMR spectroscopy probes local magnetic effects, solid-state NMR spectra are not sensitive to a small amount of paramagnetic defects typically found in diamonds, except for a slight line broadening. Herewith paramagnetic defects strongly affect nuclear spin-lattice relaxation.

From the point of view of NMR, a system of spin 1/2 nuclei placed in an external magnetic field, which is strong enough to consider the magnetic dipolar coupling as a small perturbation of the Zeeman energy, relaxes due to the contact between the nuclei and the lattice that acts as a heat reservoir. Nuclear spin-lattice relaxation in diamagnetic insulators is realized due to the lattice vibrations that cause energy exchange between the nuclear spin system and lattice phonons. Since nuclear transition frequencies are small compared to the frequencies of lattice vibrations, the number of phonons at the NMR frequencies is too small to be effective in the nuclear spin-lattice relaxation. For example, $^{13}C$ spin-lattice relaxation in natural diamonds lasts from several hours to days. However, the spin-lattice relaxation time $T_1$ in ND is much shorter and varies in the range of around 0.5–100 s. Such an effect is provided by the interaction of nuclear spins with fluctuating electron spins of localized paramagnetic defects,[5,11–14] which are coupled to the various modes of lattice vibrations. This interaction opens new, fast relaxation channel and significantly accelerates nuclear spin-lattice relaxation (Scheme 6.1). For a magnetically diluted system, in which the density of paramagnetic defects is of several orders of magnitude smaller than that of nuclei, the nuclear spin-lattice relaxation rate $R_1 = \frac{1}{T_1}$ is proportional to the defects' density.[15] Spin-lattice relaxation rate of a nuclear spin that interacts with

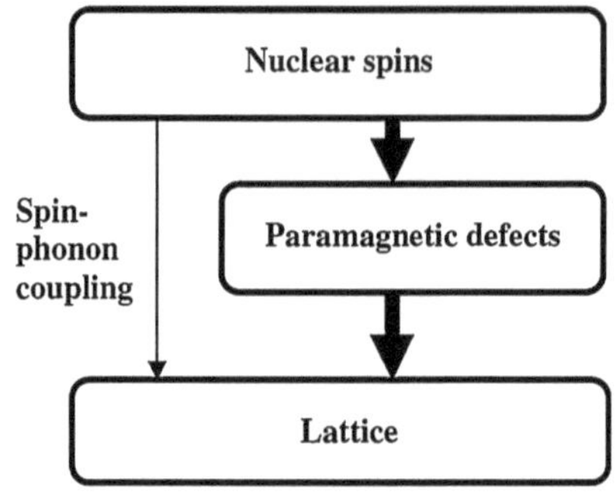


**SCHEME 6.1**

Nuclear spin-lattice relaxation passway via paramagnetic defects

the unpaired electron spin of a paramagnetic defect is given by the expression[15]

$$\frac{1}{T_1(r_{ik})} = \frac{2}{5}\gamma_S^2\gamma_I^2\hbar^2 S(S+1)\frac{\tau_S}{1+\omega_0^2\tau_S^2}\left(\frac{1}{r_{ik}^6}\right) \tag{6.4}$$

where $r_{ik}$ is the distance from the $i$th nucleus to the $k$th paramagnetic defect, $\gamma_S$ and $\gamma_I$ are the electron and nuclear gyromagnetic ratios, $\omega_0$ is the nuclear Larmor frequency, and $\tau_S$ is the electron correlation time which is actually determined by the spin-lattice relaxation time of the unpaired electron of the paramagnetic defect. The relaxation of each nucleus is considered to be exponential. Herewith, different nuclear spins are fixed at different distances from the paramagnetic defect. This circumstance results in a multi-exponential relaxation. Since it is not possible to distinguish between many $T_1$'s in a multi-exponential product, an effective mathematical approach, in which nuclear magnetization recovery is depicted by a stretched exponential $M(t) = M_\infty\{1-\exp[-(\frac{t}{T_1})^\alpha]\}$ is usually used to describe this process.[5,11–14] Here $M_\infty$ is the equilibrium magnetization, and parameter $\alpha$ varies in the range of $0.5 < \alpha < 1$. In diamond nanoparticles, additional reasons of non-exponential relaxation may originate from a distribution in particle size and in the concentration of paramagnetic defects often observed in nano-objects.

In solid insulators (i) the dipolar interaction between electron and nuclear spins and (ii) spin diffusion among nuclear spins are the dominant nuclear spin-lattice relaxation mechanisms. The latter allows spin magnetization of the distant nuclei to be spatially transferred to paramagnetic defects.[16] Low natural abundance of $^{13}C$ isotope, which accounts for 1.07%, makes carbon materials to be nuclear spin diluted, which, in turn, makes the $^{13}C$ nuclear spin diffusion in diamonds to be quite slow. Details on spin diffusion and its dependence on characteristic distances between a paramagnetic center and surrounding nuclei in carbon systems may be found in [17,18].

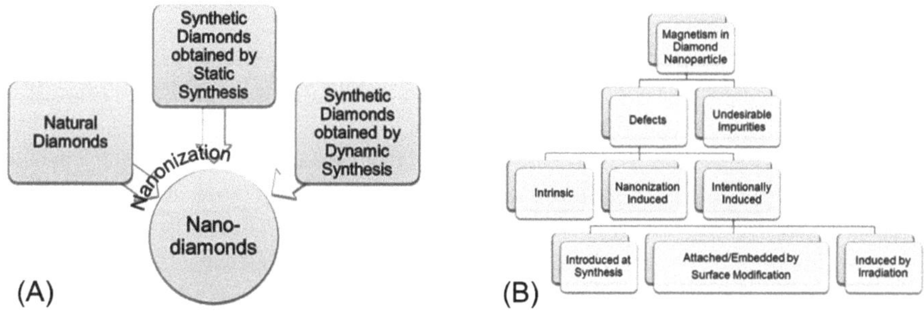


**SCHEME 6.2**

(A) Sources of NDs by origin and synthesis techniques; (B) classification of sources of magnetism in NDs by origin

## 6.2 MAGNETISM IN ND

Real ND samples demonstrate a large variety of magnetic properties descended mainly from origin of these samples. Scheme 6.1 shows sources of ND: first, nano-sized diamond particles may be obtained by nanonization (sub-micronization by various milling/grinding/pulverizing techniques) of bulk diamonds (from millimeter to micron sizes), both natural and synthetic ones. All these bulk diamonds were synthesized or naturally created in static (natural and high pressure high temperature "HPHT" synthetic diamonds) or quasi-static (CVD grown synthetic diamonds) conditions. In contrast, dynamic techniques of the diamond synthesis (detonation of explosives, "DuPont" shock wave synthesis method and laser ablation) directly provide diamonds of nanometric sizes.

Following Scheme 6.2A, magnetic properties of the ND samples are strongly dependent on both the origin and synthesis techniques (see Chapter 2). NDs obtained by the nanonization of bulk diamonds inherit, first of all, magnetic features of their ancestors. It means almost entire set of paramagnetic species (transition ions and structural defects carrying non-coupled electron spins) and potential ferromagnetic inclusions, contained in the bulk samples before nanonization, will be transferred into these nanodiamonds. Moreover, the nanonization process itself creates a huge amount of new carbon inherited paramagnetic defects within surface layers of ND particles and can be a source of technologically originated para- and ferromagnetic species.

Scheme 6.2B provides simplified classification of the sources causing magnetic properties of ND. First of all, there should be a clear discrimination between defects, either inherent or induced in the ND core or on the ND surface by its origin and synthesis/nanonization/post-processing techniques, and between impurities arising due to the same aforementioned techniques. In both research and practical applications of ND, uncontrolled impurities are undesirable species which may (and have to) be removed by special purification techniques; see, for instance, [6,12,19,20]. In this respect, EPR acts as a powerful tool for control of initial magnetic impurities content and the efficacy of purification procedures, complementary to ICP-MS analysis (see Chapter 5). Besides this aspect, the magnetism in ND due to undesirable impurities is beyond the scope of the present chapter.

Non-impurity magnetism in ND originates from three main sources: (i) intrinsic paramagnetic defects inherited from source materials where these defects were created in the process of raw diamond synthesis; (ii) defects induced by nanonization; and (iii) intentionally induced defects aiming to modify properties of the initial diamond material. The latter type of defects may be sub-classified into defects embedded into the diamond structure during a synthesis; defects attached to the diamond surface and even subsurface interface layers by the modification of the surface of ND particles which is chemically active and has great capacity for carrying guest metal ions and molecular groups; and defects induced by ionizing irradiation/ion implantation of raw diamond materials or ready ND particles. Further, each type of these defects found in ND is described in more details.

## 6.2.1 ND ORIGINATING FROM NATURAL DIAMONDS AND DIAMONDS PRODUCED BY STATIC SYNTHESIS

### 6.2.1.1 *Intrinsic defects in starting materials*

The main distinctive feature of these NDs is the fact they are manufactured from natural or synthetic diamonds of larger sizes, most often from the commercial micron-sized HPHT diamonds of various grades.[20,21] Thus, almost the entire set of defects appearing in the bulk of starting diamond material is transferred into their nanosized derivatives. Here it is worth mentioning that nanonization processes (crushing, grinding, nano-milling, purification in boiling acids, etc.) may cause quenching of some initial paramagnetic defects which may not be observed in nanosized derivatives.

Large single crystalline diamonds can be produced with low impurity concentrations and high structural perfection. However, unless precautions are taken, nitrogen is readily incorporated into HPHT and CVD diamonds in relatively high concentrations (up to a few hundred ppms, see Chapters 2 and 3). Nitrogen is also the major impurity in about 98% of natural diamond and its content may reach 3000 ppm in some stones. Hydrogen may be a significant source of defects in CVD diamond. Some of metal solvents and catalysts (like nickel, cobalt, etc.) may be also incorporated in HPHT diamonds on growth. A variety of dopants such as boron, hydrogen, phosphorus and silicon can be intentionally introduced into synthetic diamond. Extensive information on various types of defects in bulk diamonds may be found in the classical review article of Loubster and van Wyk[22] and the modern review of Newton.[3] Reference [23] is focused especially on numerous types of nitrogen related defects in diamonds.

Fig. 6.2 represents EPR spectra of commercial micron-sized type Ib synthetic HPHT diamonds (SYP series, Van Moppes and Sons SA) recorded at $\upsilon = 9.464$ GHz and several temperatures. Just diamonds of this type (from various commercial man-

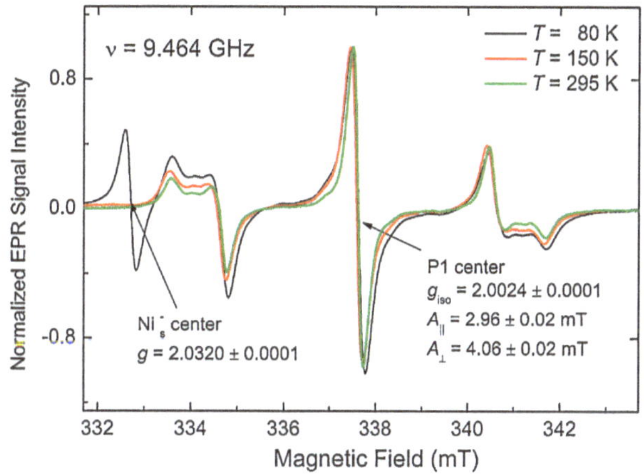


**FIGURE 6.2**

X-band EPR spectra of 105–125 μm SYP diamonds recorded at $\upsilon = 9.464$ GHz and various temperatures: $T = 80$ K (black trace), $T = 150$ K (red trace), and $T = 295$ K (green trace; room temperature, RT). Spectra intensities are normalized to the intensity of the central line

ufacturers incidentally different in fabrication techniques and compositions of the impurity elements) are main sources for ND production by nanonization. Central triplet-like pattern observed within the broad temperature range represents typical polycrystalline spectrum of dispersed substitutional nitrogen in the diamond lattice – so-called P1 (or $N_s$, or $N^0$) centers. The uncoupled electron in this defect occupies an N–C antibonding orbital. P1 center is neutral and has $C_{3v}$ symmetry. The spin-Hamiltonian parameters for this singlet ($S = 1/2$) paramagnetic center are $g_{iso} = 2.0024 \pm 0.0001$; axially anisotropic hyperfine splittings $A_{||} = 4.06 \pm 0.02$ mT and $A_{\perp} = 2.96 \pm 0.02$ mT. Defects of this type are the most abundant in bulk diamonds, both natural and synthetic. The density of P1 centers in this sample is $6.7 \times 10^{18}$ spin/g, which corresponds well with the nitrogen content declared by the manufacturer ($\sim$100 ppm of nitrogen). It indicates that the most of nitrogen contained in this diamond is in the form of paramagnetic substitutional nitrogen which is more or less homogeneously distributed over the diamond bulk. The singlet narrow line with $g = 2.0320 \pm 0.0001$ observed in the EPR spectrum at $T = 80$ K is attributed to substitutional nickel center designated as $Ni_s^-$. The density of $Ni_s^-$ centers is $\sim 10^{17}$ spin/g. These paramagnetic defects appear during the growth of diamond crystallites from Ni-contained metal solvents and supposed to be secondly (after nitrogen) abandoned in HPHT grown diamonds.[24,25] At temperatures above 150 K, the EPR line of $Ni_s^-$ drastically broadens and becomes unobservable by conventional EPR.

It seems to be obvious supposing that well-defined EPR signatures of the above defects will migrate from starting micron-sized diamond crystallites into their nanosized derivatives. Indeed, these defects are inherent for diamond bulk and only part of them will be lost (quenched) by nanonization. However, ND is not as easy as it seems. On the long way to nanosized particles, many amazing things happen with paramagnetic defects.

#### 6.2.1.2 What does really happen in nanonization?

Synthetic NDs of various particle sizes are produced from initial micron-sized HPHT crystallites by nanonization that includes intensive milling (in some cases milling is a multistage process) and subsequent separation into fractions of different particle sizes with precision grading.[20,21] The commercial samples underwent purification to remove metallic and metallic oxide impurities and non-diamond carbons. Fig. 6.3A demonstrates changes occurring in RT EPR spectra of resulting ND of different nanoparticle sizes.[20] It is clearly seen that decreasing of the average particle size leads to a consecutive decrease of the P1 signal contribution and the appearance of a new singlet line, whose contribution grows with decreasing particle size; see spectral decomposition in Fig. 6.3B. Remarkably, when decreasing ND size, the P1 spectrum remains the same, showing no line broadening just reduction of its intensity. For ND sizes below 86 nm, no traces of the P1 signal are observed. EPR spectra of smallest SYP ND (particle sizes below 86 nm) provide an intensive singlet signal with $g = 2.0028 \pm 0.0001$ having complicated line shape. Simulation of this signal by a superposition of two Lorentzian lines (broad and narrow), having practically indistinguishable $g$-factors (even in W-band),[21] is shown in Fig. 6.3C. The total density of the paramagnetic centers sequentially increases up on decrease of average particle size: from $6.7 \times 10^{18}$ spin/g in starting micron-sized particles to $3.3 \times 10^{19}$ spin/g in smallest ND particles of this series.[20] Thus, the first consequence of nanonization is the appearance in ND of new paramagnetic defect with $g = 2.0028$ and line width $\Delta H_{pp} = 0.93$ mT responsible for the growth of total spin density. In fact, intensive milling of the initial diamond particles results in countless amounts of ND particles characterized by very broad size distribution – from relatively large particles of submicron sizes to very small NDs having sizes of a few tens of nm. Then, after separation of these particles into groups characterized by specific average particle size, one can find that each smaller ND particle carries more paramagnetic defects induced by the milling process. This fact allows supposing that the origin of new defects is mechanical deformation induced carbon-inherited defects (mainly uncoupled electron spins of broken bonds) located on the ND surface or within the distorted diamond subsurface layers. On the other hand, nanonization leads to a decrease of the characteristic P1 spectrum contribution in the spectra of smaller NDs. Does it really mean that in small ND fractions the density of P1 centers drops down with the particle size? The authors of [21] suppose that the lack of P1 EPR signals in ND samples of small particle size (5–25 nm) could be explained by either conversion of the neutrally charged nitrogen $N^0$ to paramagnetic-silent charge state or due to the nitrogen flushing outside of the diamond. However, there are well-founded doubts about such an interpretation. First of all, it was found that, when diminishing the average

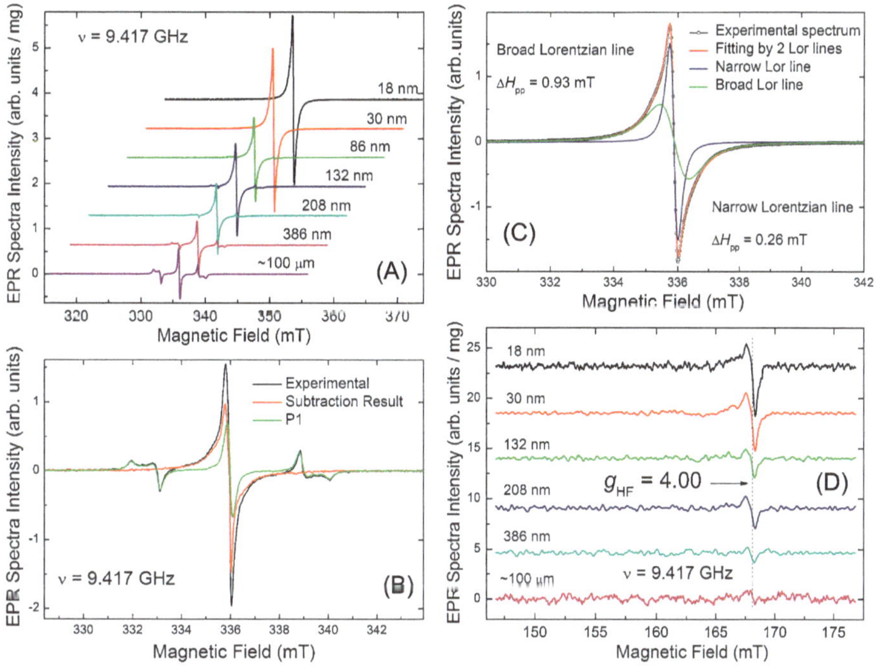


**FIGURE 6.3**

(A) $g = 2.00$ region of X-band RT EPR spectra of SYP ND as a function of the average particle size (as designated in the figure), $\upsilon = 9.417$ GHz. Spectra intensities are normalized per unit mass. Spectra are shifted in both vertical and horizontal directions for better presentation; (B) decomposition of the EPR spectrum of 386 nm SYP ND into spectrum of P1 centers and a singlet line; (C) simulation (red trace) of the experimental EPR spectrum of 18 nm SYP ND (black open circles) by superposition of two Lorentzian lines, broad (green trace) and narrow (blue trace); (D) half-field $g = 4.00$ region of X-band RT EPR spectra of SYP diamonds as a function of average particle size (as designated in the figure), $\upsilon = 9.417$ GHz

particle size, the P1 signal really drops down in its intensity but remains of the same shape, and even line width which is very sensitive to interactions with neighboring paramagnetic centers remains unchanged. It indicates that P1 defects observed in ND samples are quite far from increasing mechanically induced surface paramagnetic defects and do not feel interaction with the latter. Here it is worth mentioning that any ND sample characterized by a certain average particle size is an ensemble of ND particles having fixed particle size distribution (well known for each sample). In that ensemble one can find a certain amount of larger and smaller particles. Coarser grade ND samples (132–386 nm) still contain enough larger size particles of 240–400 nm size to provide the same P1 signals as observed in initial micron-sized samples. Naturally, the smaller the average particle size in the ND sample of a certain grade, the fewer larger particles. The finest ND grades have a very narrow distribution of sizes around the average value and practically do not contain larger particles at all. This consideration may describe the size-dependent reduction of the conventional P1 signal quite reasonably. However, the question "what does really happen with P1 defects in small ND particles" still remains. Careful analysis of the line shape of the EPR signal in fine ND fractions can help in answering the question. Indeed, as it is clearly seen in Fig. 6.3C that this signal consists of two Lorentzian signals with the same (within the experimental error) $g$-factor. One of them, broad with $H_{pp} = 0.93$ mT, is attributed to nanonization induced carbon inherited defects. Another one, narrow with $H_{pp} = 0.26$ mT, appears in all ND samples, and its intensity grows with decreasing particle size. In the finest ND sample (18 nm), the contribution of this signal to the total EPR signal intensity is about 20%, which corresponds to spin density of $\sim 6.6 \times 10^{18}$ spin/g. This value is in a perfect agreement with the density of P1 defects determined in the starting HPHT micron-sized crystallites ($6.7 \times 10^{18}$ spin/g)! One can suppose that, in spite of the lack of conventional P1 hyperfine pattern in EPR spectra, fine ND fractions contain practically the same amount of P1 centers, as in the starting material, but the P1 spectrum itself underwent drastic transformation from the characteristic extended hyperfine pattern to the narrow singlet Lorentzian line. Such a transformation of spectrum occurs due to dipole–dipole and exchange interactions of induced spins located on the surface or within the subsurface layers with spins of P1 defects located in the diamond core of ND particles. Up on decreasing ND particle size, both the growing density of induced spins and reducing inter-spin distances intensify these interactions. The proposed model exhaustively describes size-dependent behavior of EPR spectra shown in Fig. 6.3A. Thus, the main paramagnetic defects found in ND obtained by nanonization are paramagnetic substitutional nitrogen (P1 centers) and mechanically induced carbon-inherited defects (non-saturated dangling bonds carrying unpaired electron spin).

In addition to the above mentioned defects, EPR reveals another type of defect induced by nanonization. Spectra recorded in the range of so-called "half-fields" (HF, around $g \sim 4.0$) show appearance of a new EPR signal which may be characterized by effective $g$-factor $g_{HF} = 4.00 \pm 0.01$; see Fig. 6.3D. This signal is practically not observed in initial micron-sized crystallites, but grows up on nanonization, reaching its maximal intensity in small ND fractions. The intensity of this signal is a couple of orders of magnitude weaker than that for the EPR signals of main defects, observed within the $g = 2.00$ region. Recently these narrow $g_{HF} = 4.00$ signals were attributed to the so-called 'forbidden' $\Delta M_S = 2$ transitions in polycrystalline EPR spectra of a certain triplet ($S = 1$) center[26,27] with zero field splitting $D \sim 0.03$ cm$^{-1}$. It was proposed[26] that these centers do not relate to atomic nitrogen substitutions in the diamond lattice and are, therefore, caused solely by carbon-inherited defects, namely intrinsic multivacancy defects, formed by exchange coupling of singlet ($S = 1/2$) defects. Progressive strengthening of the $g = 4.00$ signals just in the EPR spectra of the fine fractions ND samples supports this hypothesis. Indeed, the fine fraction ND samples are characterized by the enormous density of the nanonization induced singlet paramagnetic defects, which, in turn, increases the probability of these defects coupling.

#### *6.2.1.3* ***Irradiation induced defects: fabrication of fluorescent ND***

Fluorescent nanodiamonds (FND) containing $NV^-$ defects are novel objects that have a number of innovative applications in magnetic sensing, cell microbiology, and quantum information processing; see Chapters 7 and 17. The large variety of these applications is based on the unique properties of the negatively charged paramagnetic $NV^-$ color centers with triplet spin properties [Chapters 7 and 17]. $NV^-$ (as well as $NV^0$) centers can be created in bulk synthetic HPHT diamonds naturally containing substitutional nitrogen. Irradiation of diamond micron-sized crystals by a beam of high-energy ($\leq$20 MeV) particles creates a large number of vacancies. Subsequent annealing of the irradiated diamond at $T \geq 850°C$ causes thermal diffusion of the vacancies and the appearance of $NV^-$ complexes consisting of vacancies stabilized in structural positions adjacent to the substitutional nitrogen. $NV^-$ has a characteristic luminescence with a zero-phonon line (ZPL) at a wavelength of 638 nm and a broad (100 nm width) emission sideband due to strong phonon coupling.

One of techniques to obtain FND is nanonization of fluorescent micron-size (average crystallite sizes of diamond powders) to fine fractions that are characterized by an average particle size below 100 nm.[28,29] This method was found to be more efficient and technologically justified in comparison with the methods that apply the aforementioned irradiation/annealing treatment to ND particles obtained by nanonization of non-irradiated bulk diamonds. However, using the first, more efficient technique, one has to pay a certain price. A detailed study of the impact of nanonization to $NV^-$ defects induced in bulky synthetic diamonds by irradiation and annealing is reported in [19]. This section is devoted to specific features of $NV^-$ defects revealed in FND by EPR.

X-band EPR spectrum of coarser FND fraction (particles sizes ~100 nm) shows intensive EPR signals of intrinsic P1 centers as well as nanonization induced dangling bond defects with the total spin density $\sim 8 \times 10^{19}$ spin/g. In addition to these signals, less intensive group of signals is observed in low- and high-field regions of the spectrum; see Fig. 6.4, black trace. The EPR lines in the half-field (HF) region ($g = 4$) are attributed to 'forbidden' transitions in the EPR spectra of triplet paramagnetic centers, both $NV^-$ ($g_{HF1} = 4.26 \pm 0.01$) and the multi-vacancies ($g_{HF2} = 4.00 \pm 0.01$). The satellite lines observed aside of intense $g = 2.00$ lines and separated by ~100 and ~200 mT are attributed to 'allowed' transitions in EPR spectra of triplet $NV^-$ defects.[27]

The spin density of $NV^-$ defects in the coarse FND is 30% reduced in comparison with the same density in irradiated/annealed micron-sized samples and found to be $3.6 \times 10^{17}$ spin/g. The spectrum of fine (<20 nm) FND fraction (Fig. 6.4, red trace) reveals changes occurred with defects due to mechanical reduction of particle size. The main signals at $g = 2.00$ undergo the same modification described in the sec-

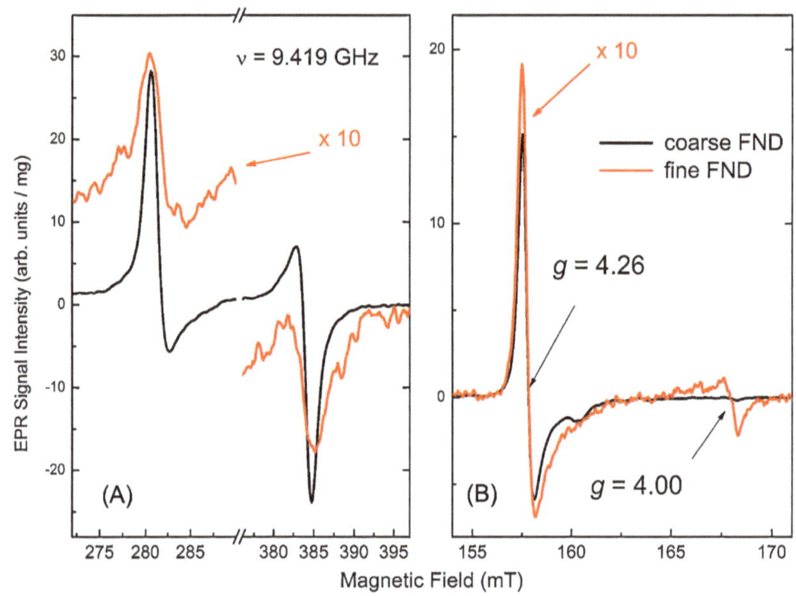


**FIGURE 6.4**

X-band RT EPR spectra of triplet $NV^-$ defects in FND, $\upsilon = 9.419$ GHz: (Black traces) coarse FND fraction, (red traces) fine FND fraction. Spectra intensities are normalized per unit mass. Intensity of the fine FND fraction spectrum is plotted multiplied by a factor of 10. (A) Spectra of $x$ and $y$ components of low- and high-field 'allowed' lines; intense signal at $g = 2.00$ is not presented; (B) spectra of half-field 'forbidden' lines

tion above: an increase of the nanonization induced defects up to $\sim 1 \times 10^{19}$ spins/g and transformation of the hyperfine P1 pattern into a narrow singlet line. At the same time, triplet spectrum signals report on significant perturbations in the system of $NV^-$ defects. First of all, the density of these defects drops down by about an order of magnitude to $\sim 5 \times 10^{17}$ spins/g relative to the same density in initial irradiated/annealed fluorescent diamond. The reduction of the total amount of $NV^-$ centers in fine FND may occur due to surface induced charge state conversion of single color NV defects hosted in ND, which solely comes from surface electron traps. As a result of such a conversion, EPR observable $NV^-$ centers turn to practically EPR silent $NV^0$ centers.[30]

Secondly, such characteristic features of the triplet spectrum as 'allowed' lines become practically unobservable, which significantly complicates correct EPR-attribution of the remaining spectral features to $NV^-$ defects (Fig. 6.4A). Fortunately, the half-field 'forbidden' line with $g_{HF1} = 4.26$, in spite of its dropping intensity, remains well observable and becomes the main EPR signature of the presence of $NV^-$ centers in FND (Fig. 6.4B). It should be noted that changes in integral intensity of the 'forbidden' line agree quite well with the changes in the $NV^-$ content obtained by the correct procedure of integration of the entire $NV^-$-related EPR spectrum. Thus, the EPR spectrum of 'forbidden' line in the HF region may be used as a reliable tool for comparative estimations of the $NV^-$ content in FND even in those cases when weak 'allowed' lines are not (or hardly) observed.

## 6.2.2 ND OBTAINED BY DYNAMIC SYNTHESIS

Dynamic techniques of diamond synthesis directly provide diamond particles of nanometric sizes (Chapter 2). In most cases, the size distribution of dynamically obtained ND (DyND) is quite narrow with the average particle sizes below 25 nm: 4–6 nm for detonation ND and typical laser ablation ND,[9,31] and ~20–25 nm for ND synthesized by "DuPont" method.[7] In fact, real DyND samples rarely consist of well-separated single nanoparticles, but are rather quite large (hundreds of nm or even micron-sized) polycrystalline aggregates of single NDs. Various techniques are used for disaggregation of these entities into solitary ND particles, but all of them are much "softer" than usual nanonization processes applied to bulk crystalline diamonds. It has been demonstrated that disaggregation does not create additional mechanically induced paramagnetic defects.[6,31] Thus, all paramagnetic defects in initial (non-modified) DyND are intrinsic defects, emerged in DyND particles at their generation.

### 6.2.2.1 *Main singlet ($S = 1/2$) defects in DyND*

Early EPR studies of detonation DyND already revealed well-pronounced paramagnetism of those DyNDs. DyND samples synthesized by different techniques and obtained from different manufacturers exhibited similar EPR characteristics, more specifically, the absence of P1 type of EPR signals and presence of a strong Lorentzian-like singlet signal with $g = 2.003$ and line width $\Delta H_{pp} \sim 0.7–0.9$ mT; see black trace in Fig. 6.5A. The results of measurements done on various detonation DyND powders are summarized in [12]. Precise measurements of the $g$-factor for the main signal gave $g = 2.00282 \pm 0.00003$ within the temperature range of 4.2–300 K. The observed shift of this $g$-factor from that of the free electron ($g_e = 2.0023$) may be interpreted in terms of the model of unpaired spin localized at a mesoscopic object (to which NDs also belong). This interpretation allows estimating the size of a DyND particle as 20–60 Å, which correlates well with the DyND dimensions derived from XRD and HRTEM. A careful SQUID and EPR study of well-purified detonation DyND revealed that the density of singlet spins in DyND is $6.3 \times 10^{19}$ spin/g [32] which corresponds to 13–15 singlet defects per each DyND particle. These numbers vary with the differences not exceeding 30% for DyND sample of different origins. Further EPR study revealed that the singlet Lorentzian-like line observed consists of two main Lorentzian components, broad and narrow, similar to the main EPR line observed in fine ND fractions of milled bulk diamonds – see Fig. 6.3C.[20,31,33] The only dissimilarities found are the line widths of both components, which in DyND

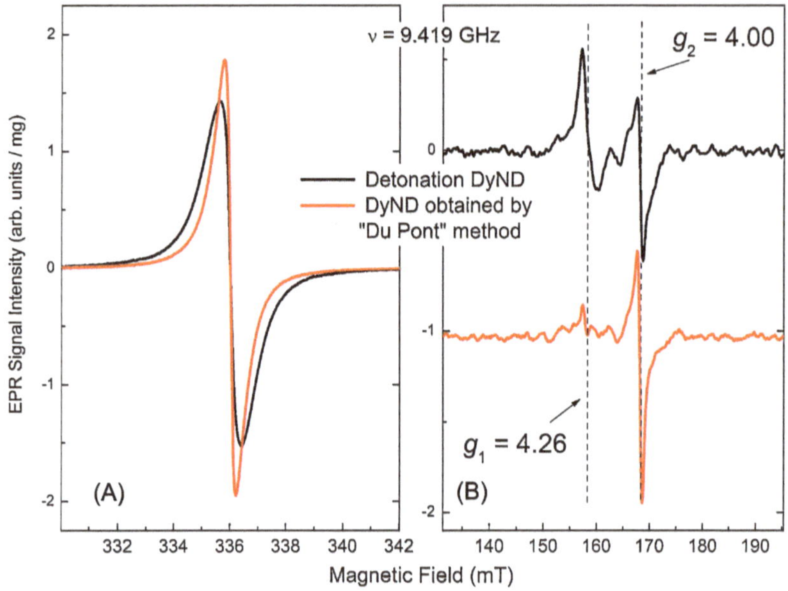


**FIGURE 6.5**

X-band RT EPR spectra of well-purified samples of detonation DyND (black traces) and commercial SSX 0.05 DyND obtained by "DuPont" method (red traces), $\upsilon = 9.419$ GHz. (A) Spectra of the main intense signal at $g = 2.00$; (B) spectra of half-field 'forbidden' lines

are usually larger than those in fine ND fractions, and relative contributions of these components to the total signal. However, the latter varies also for DyND of different origin[31] – compare black and red traces in Fig. 6.5A. It was suggested that broad and narrow signals are due to two types of singlet paramagnetic defects somehow differing by their origin and location. Modification of DyND surface by intentionally introduced transition metal ions, itself being a promising and useful method for the design of ND with controlled characteristics, helped in attribution of broad and narrow lines.

### 6.2.2.2 Modification of DyND surface by transition metal ions

Both the modification method and comprehensive study of superclean DyND particles, whose surface was modified by divalent ions of copper, cobalt and trivalent ions of gadolinium by ion exchange in water suspensions, were reported in [33,34]. Direct observation of $Cu^{2+}$ EPR signals allowed acceptable estimation of the number of $Cu^{2+}$ ions attached to the DyND surface. These ions may be used as paramagnetic probes which interact (by dipole–dipole and exchange channels) with the intrinsic singlet paramagnetic defects. Varying modification conditions, one can control the strength of the mentioned interaction which, in turn, allowed estimating distances between paramagnetic probes and different types of intrinsic paramagnetic defects.

It was found that surface located $Cu^{2+}$ ions cause a broadening of the singlet defects EPR lines, but this broadening is more pronounced for the broad line and less for the narrow line.[35] Such a behavior indicates that singlet defects responsible for the broad lines are located closer to the DyND surface than those responsible for the narrow line. The estimated depths of occurrence for two types of intrinsic singlet defect were found to be ~0.8 and ~1.5 nm from the DyND surface for the shallow and deeper defects, respectively.

#### 6.2.2.3 *Location and origin of singlet defects in DyND*

Summarizing the results reported in the above two sections, it may be concluded that the intrinsic singlet defects in DyND belong to at least two main groups differing first of all by their location inside the DyND particle. Thus, there are singlet defects located at average distances of less than 1 nm from the surface. These defects are responsible for the broad Lorentzian component of the main EPR signal. Defects of another group, associated with the narrow Lorentzian component, are located deeper from the surface – at average distances ~1.5 nm. May defects of the first group be attributed to surface defects or are they "hidden" within the subsurface interface layers? EPR provides univocal answer to this question. References [11,12] report on the experiments with evacuated DyND samples. Air evacuation increases integrated intensity of the singlet EPR signal for about 10% with no changes in line shape, widths, and positions of the signal. When the sample was open to air again, the intensity of the signal returned to its initial value. Such an effect may be understood it terms of exchange broadening of singlet spins due to encounters with molecules of triplet oxygen ($O_2$), which is a well-known magnetic agent contained in air. Thus, there are some defects in the first group which are located on the surface or close enough to the surface to interact with molecular oxygen. Following a 10% reduction, there may be about 1 singlet defect per DyND particle. Other defects in this group are well-isolated from encounters with oxygen molecules because of their location inside subsurface interface layers. Following direct analogy with carbon-inherited defects, induced in diamond particles by nanonization, this group of singlet defects may be attributed to dangling bonds located within the distorted diamond structure of interface layers.

The second group of inherent singlet defects in DyND, which is located at least twice deeper inside the ND particle, is responsible for the narrow Lorentzian component of the EPR signal. The width of this line is usually broader in detonation DyND samples and narrower in shock wave synthesized samples (Fig. 6.5A). The same narrow line is observed in EPR spectra of fine ND fractions obtained by nanonization (see Section 6.2.1.2). There it was supposed that such a line originates from nitrogen substitution defects (P1 centers) strongly interacting with large amount of mechanically induced dangling bonds. It is logical applying the same model to DyND. Indeed, typical P1 hyperfine split patterns have never been reliably observed in EPR spectra of detonation DyND (only a few authors reported on observation of weak P1-like signals on vaguely characterized or heat treated samples). On the other hand, high abundance of nitrogen in detonation DyND is well known. Thus, the absence of characteristic P1 signals sounds quite puzzling. However, the hypothesis proposed in Section 6.2.1.2

may successfully solve the puzzle: namely, narrow Lorentzian component in EPR spectra of DyND is due to nitrogen substitution defects (P1) which are located in the crystalline core of a DyND particle and strongly interact with uncoupled electron spins of multiple dangling bonds located in the interface layers. In such a configuration (and there is no other configuration in small DyND particles!), the conventional hyperfine split pattern just cannot be observed.

#### 6.2.2.4 Inherent triplet centers in DyND

X-band ($\upsilon = 9.4$ GHz) EPR study at $4 \leq T \leq 300$ K of well-purified detonation ND[32] revealed the presence in the half-field region of two additional (to the main intense singlet signal) weak and narrow signals with $g_1 = 4.26(1)$, $\Delta H_{pp} = 2.9(1)$ mT and $g_2 = 4.00(1)$, $\Delta H_{pp} = 1.4(1)$ mT at $T = 4$ K), separated by the distance of 10.4 mT (data at $T = 4$ K). The integral intensity of these signals was found to be several orders of magnitude lower than that for the main signal. It was supposed that these doublet signals appear due to 'forbidden' fine structure transitions between Zeeman states of some triplet centers. Further, the same doublet signals were found in the X- and Q-band (9.4 and 34 GHz, respectively) of RT EPR spectra of a large variety of intact, purified, and modified DyND obtained from various manufacturers and synthesized by different techniques (detonation, laser, "DuPont").[26] Careful analysis of EPR spectra collected from multiple samples showed that these signals, in spite of having frequency dependent $g$-factors and splitting, as well as sample dependent line widths and relative intensities ratios, are always present in EPR spectra of DyND and may be attributed to two different types of triplet paramagnetic centers: the X-band $g_1 = 4.26$ signal, designated as TR1, is characteristic for $NV^-$ defects, whereas the X-band $g_2 = 4.00$ signal, designated as TR2, is characteristic for exchange coupled $S = 1/2$ defects (multivacancies); see Fig. 6.5B. Here it is worth mentioning that the 'allowed' lines, always accompanying 'forbidden' lines in EPR spectra of $NV^-$ triplets recorded on bulky fluorescent diamonds and coarse FND fractions (see Section 6.2.1.3), have never been observed in DyND. An assumed nature of this phenomenon observed in fine FND fractions is discussed in [19,26].

These two universal types of triplet center are minor rare defects in all DyND independently of the type of dynamic synthesis and source material used. The ultimate confirmation of the correctness of the above attribution of TR1 and TR2 was done by comparison of half-field EPR spectra of DyND with similar spectra of electron irradiated FND samples obtained by nanonization.[27] Thus, it may be concluded that each DyND sample contains a certain amount of inherent (i.e. created by the synthesis itself, not induced by intentional high energy irradiation and/or nanonization) triplet $NV^-$ and multivacancy defects. The actual content of each of these defects depends on the synthesis technique, starting materials, and conditions, as well as post-processing. Thus, the highest abundance of inherent $NV^-$ defects, well compared with the same abundance in fine FND samples, was found in DyND obtained by the detonation technique, whereas the lowest abundance of inherent $NV^-$ defects was found in DyND of SSX series fabricated by "DuPont" method;[6,27] see Fig. 6.5B. This finding is quite obvious taking into account that detonation technique uses high

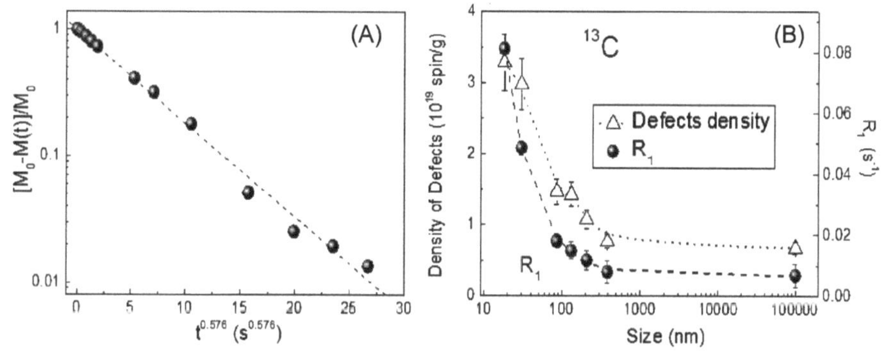


**FIGURE 6.6**

(A) $^{13}C$ magnetization recovery (by $T_1$ measurements) in the SYP 0–0.05 ND sample on a semi-logarithmic scale; (B) size dependence of density of paramagnetic defects (triangles) and $^{13}C$ spin-lattice relaxation rate $R_1$ (filled circles) in SYP ND. Dashed line is simulation using Eq. (6.5)

nitrogen content explosives (TNT, hexogen) as a starting material in contrast to the DuPont method which uses pure graphite and where nitrogen in those samples may come only from the atmosphere.

## 6.3 NUCLEAR SPIN-LATTICE RELAXATION VIA PARAMAGNETIC DEFECTS IN ND

### 6.3.1 SIZE DEPENDENT $^{13}C$ NUCLEAR SPIN-LATTICE RELAXATION IN ND PRODUCED BY STATIC SYNTHESIS

Samples of SYP ND show that (i) $^{13}C$ nuclear spin-lattice relaxation is described by a stretched exponential and (ii) the spin-lattice relaxation time $T_1$ is short and varies in the range of ~0.5–100 s (Fig. 6.6). Such an effect is evidently provided by the interaction of nuclear spins with fluctuating electron spins of localized paramagnetic defects as explained in the Introduction. Moreover, the $T_1$ is found to be particle size dependent, revealing noticeable relaxation acceleration with decreasing the ND particle size (Fig. 6.6).[20] We showed, both experimentally and theoretically, that this effect is caused by the contribution to relaxation coming from the interaction of nuclear spins with the surface paramagnetic centers induced by the sample milling,[20] which is proportional to the density of the surface paramagnetic centers $n_{surf}$ and inversely proportional to the particle size $Z$, namely

$$\frac{1}{T_1} = R_1 = \frac{1}{T_{1core}} + \frac{n_{surf}}{aZ}. \tag{6.5}$$

Here the first term comes from the interaction of nuclear spins with the paramagnetic defects located in the diamond core, and $a$ is a constant.

The aforementioned relaxation measurements have been carried out using non-spinning samples to prevent the hindrance of the spin diffusion among $^{13}C$ nuclei caused by magic angle spinning (MAS) of the sample. Herewith the aforementioned SYP ND samples produced by HPHT synthesis and subsequent crushing show noticeable slowdown of both $^{13}C$ spin-lattice ($R_1$) and spin-spin ($R_2$) relaxation rates and reduction in nuclear spin diffusion coefficient $D$ with increasing MAS rate.[36,37] This effect was attributed to the average of nuclear spin–spin interactions and suppression of nuclear spin diffusion by MAS. Actually one can neglect spin diffusion among $^{13}C$ nuclei under MAS rate faster than ~3–4 kHz.[17,36] We developed a theoretical approach that describes the MAS rate dependence of $R_1$, $R_2$, and $D$, allows quantitative analysis of the data, and shows good compliance with the experiment. The effect is not observed in ND particles with size less than 12 nm and with paramagnetic defects density above $4 \times 10^{19}$ spin/g,[37] in which spin diffusion does not take place. This finding is in accordance with the data on bulk diamonds,[17,38–40] showing that at low density of paramagnetic defects in the material, spin diffusion contributes to the $^{13}C$ spin-lattice relaxation, while high concentration of paramagnetic defects ($>4.2 \times 10^{19}$ $cm^{-3}$) makes spin diffusion among $^{13}C$ spins very weak and unable to transfer the polarization to the distant $^{13}C$ nuclei.

## 6.3.2 NUCLEAR SPIN-LATTICE RELAXATION IN DyND

The aforementioned size dependence of $T_1$ was found by us also in nanodiamonds prepared by the laser-assisted technique.[9] DyND of SSX series prepared by "DuPont" method did not show size dependence of the relaxation rate.[6] The reason is that these samples are of a polycrystalline origin, each particle of which initially consists of numerous nanocrystallites, approximately 20–25 nm in size, with adjoining surfaces. Crushing of such polycrystalline aggregates only diminishes the size of the aggregates, leaving the constituting nanocrystallites and their intrinsically facet surfaces mainly untreated, and therefore does not produce new paramagnetic defects.

NMR measurements of detonation nanodiamonds of the size of ~4.5 nm show that spin-lattice relaxation time $T_1$ of the carbon spins belonging to the diamond core varies in the range from 270 to 615 ms.[5,8,9,14] Herewith $T_1$'s of the shell carbons are shorter than those of the core,[5,8] revealing somewhat larger concentration of the paramagnetic defects in the partially disordered shell compared with that of the core. Surface dangling bonds in detonation DyND are mainly terminated by hydrogen atoms that form hydrocarbon, hydroxyl, and carboxyl groups. Parameter $\alpha$ of stretched exponential varies in the range of 0.5–0.66.[5,8,9,13,14] Significant reduction in $T_1$ compared with that in natural diamonds and stretched exponential behavior of the relaxation unambiguously reflect the nuclear relaxation via paramagnetic defects.

### 6.3.3 NUCLEAR SPIN-LATTICE RELAXATION IN MODIFIED DETONATION DyND

Functionalization of the DyND surface with targeted species allows fabrication of DyND with specified chemical, physical, and electronic properties. For example, fluorinated, chlorinated, and hydroxylated detonation DyNDs have a number of chemical and technological applications, while DyND particles grafted by magnetic ions are suggested to be used in spintronics, magnetic recording, and magnetic resonance imaging (MRI).

Fluorinated DyND shows significant number of paramagnetic defects ($\sim 10^{20}$ spin/g) resulting in faster $^{13}C$ nuclear spin-lattice relaxation comparing with the initial non-fluorinated sample.[41] A similar effect was obtained in chlorinated DyND.[42] but was not observed in hydroxylated OH-DyND.[43] Such findings indicate a specific role of the chemically active fluorine and chlorine atoms in breaking the C–C covalent bonds on the diamond surface and creating new stable dangling bonds carrying uncoupled electron spins.

Modification of highly purified DyND particles by copper, cobalt and gadolinium[10,33,34,44] leads to acceleration of $^{13}C$ and $^{1}H$ spin-lattice relaxation proportionally to the transition metal ion content. This finding unambiguously shows that Cu, Co and Gd ions are bound to the detonation DyND surface but not present in the material as a separate phase, since in the latter case the effect of paramagnetic species on relaxation would be negligible. It opens an opportunity for biomedical applications of ND, for example, for MRI contrast enhancement.[10] Annealing of Cu- and Co-grafted samples at 550 and 900°C in a stream of hydrogen yields a reduction of the relaxation rate, indicating that upon annealing the surface paramagnetic complexes are destroyed.[10,33,44]

## 6.4 CONCLUDING TIPS

- The origin of paramagnetic defects in ND depends on starting diamond material, nanonization, and synthesis techniques;
- Nanonization induces additional defects, namely solitary dangling bonds and multivacancies;
- In small nanosized ND particles, nanonization induced defects interact with defects inherited from starting diamond material and modify EPR signature of inherited defects;
- Nanonization significantly reduces number of fluorescent $NV^-$ centers induced by irradiation of starting diamonds;
- Defects in dynamically synthesized ND (DyND) are created in the process of synthesis. Their content depends on the starting materials (graphite, explosives) and technique;

- The primary paramagnetic defects in both nanonized ND and DyND are P1 centers and dangling bonds, whereas the secondary paramagnetic defects are $NV^-$ centers and multivacancies;
- $NV^-$ centers in DyND are inborn, their abundance depends on nitrogen content in the starting materials;
- Increasing the density of paramagnetic defects in ND is the main factor responsible for the significant acceleration of nuclear spin-lattice relaxation.

## REFERENCES


1. Weil, J. A.; Bolton, J. R. *Electron Spin Resonance. Elementary Theory and Practical Applications*. 2nd ed.; Wiley–Interscience, 2007.
2. Abragam, A.; Bleaney, B. *Electron Paramagnetic Resonance of Transition Ions*. Oxford Classic Texts in the Physical Sciences, reprinted ed.; Oxford University Press, 2012.
3. Newton, M. E. EPR, ENDOR and EPR Imaging of Defects in Diamond. *Electron Paramagn. Reson.* **2007**, *20*, 131–156.
4. Rudowicz, C.; Misra, S. K. Spin-Hamiltonian Formalisms in Electron Magnetic Resonance (EMR) and Related Spectroscopies. *Appl. Spectrosc. Rev.* **2001**, *36* (1), 11–63.
5. Panich, A. M. Nuclear Magnetic Resonance Studies of Nanodiamonds. *Crit. Rev. Solid State Mater. Sci.* **2012**, *37*, 276–303.
6. Shames, A. I.; Mogilyansky, D.; Panich, A. M.; Sergeev, N. A.; Olszewski, M.; Boudou, J.-P.; Osipov, V. Yu. XRD, NMR, and EPR Study of Polycrystalline Micro- and Nano-Diamonds Prepared by a Shock Wave Compression Method. *Phys. Status Solidi A* **2015**, *212* (11), 2400–2409.
7. Panich, A. M.; Shames, A. I. Comment on "Quantification of C=C and C=O Surface Carbons in Detonation Nanodiamond by NMR". *J. Phys. Chem. C* **2015**, *119*, 21286–21287.
8. Panich, A. M.; Shames, A. I.; Vieth, H.-M.; Osawa, E.; Takahashi, M.; Vul', A. Ya. Nuclear Magnetic Resonance Study of Ultrananocrystalline Diamonds. *Eur. Phys. J. B* **2006**, *52*, 397–402.
9. Panich, A. M.; Shames, A. I.; Zousman, B.; Levinson, O. Magnetic Resonance Study of Nanodiamonds Prepared by Laser-Assisted Technique. *Diam. Relat. Mater.* **2012**, *23*, 150–153.
10. Panich, A. M.; Altman, A.; Shames, A. I.; Osipov, V. Yu.; Alexenskiy, A. E.; Vul', A. Ya. Proton Magnetic Resonance Study of Diamond Nanoparticles Decorated by Transition Metal Ions. *J. Phys. D, Appl. Phys.* **2011**, *44* (1–5), 125303.
11. Shames, A. I.; Panich, A. M.; Kempiński, W.; Baidakova, M. V.; Osipov, V. Yu.; Enoki, T.; Vul', A. Ya. Magnetic Resonance Study of Nanodiamonds. In Gruen, D. M., Shenderova, O. A., Vul', A. Ya., Eds., *Synthesis, Properties and Applications of Ultrananocrystalline Diamond;* NATO Science Series II, Vol. 192; Springer: 2005, pp. 271–282.
12. Shames, A. I.; Panich, A. M.; Kempinski, W.; Alexenskii, A. E.; Baidakova, M. V.; Dideikin, A. T.; et al. Defects and Impurities in Nanodiamonds: EPR, NMR and TEM Study. *J. Phys. Chem. Solids* **2002**, *63*, 1993–2001.
13. Panich, A. M.; Furman, G. B. Nuclear Spin-Lattice Relaxation and Paramagnetic Defects in Carbon Nanomaterials. *Diam. Relat. Mater.* **2012**, *23*, 157–161.

14. Dubois, M.; Guerin, K.; Petit, E.; Batisse, N.; Hamwi, A.; Komatsu, N.; Giraudet, J.; Pirotte, P.; Masin, F. Solid-State NMR Study of Nanodiamonds Produced by the Detonation Technique. *J. Phys. Chem. C* **2009**, *113*, 10371–10378.
15. Abragam, A. *The Principles of Nuclear Magnetism;* Clarendon: Oxford, 1961.
16. Bloembergen, N. On the Interaction of Nuclear Spins in a Crystalline Lattice. *Physica* **1949**, *15*, 386–426.
17. Reynhardt, E. C.; High, G. L. Nuclear Magnetic Resonance Studies of Diamond. *Prog. Nucl. Magn. Reson. Spectrosc.* **2001**, *38*, 37–81.
18. Hoch, M. J. R.; Reynhardt, E. C. Nuclear Spin-Lattice Relaxation of Dilute Spins in Semiconducting Diamond. *Phys. Rev. B* **1988**, *37*, 9222–9226.
19. Shames, A. I.; Osipov, V. Yu.; Boudou, J.-P.; Panich, A. M.; von Bardeleben, H.-J.; Treussart, F.; Vul', A. Ya. Magnetic Resonance Tracking of Fluorescent Nanodiamond Fabrication. *J. Phys. D, Appl. Phys.* **2015**, *48* (1–13), 155302.
20. Panich, A. M.; Sergeev, N. A.; Shames, A. I.; Osipov, V. Yu.; Boudou, J.-P.; Goren, S. D. Size Dependence of $^{13}C$ Nuclear Spin-Lattice Relaxation in Micro- and Nanodiamonds. *J. Phys. Condens. Matter* **2015**, *27* (1–7), 072203.
21. Yavkin, B. V.; Mamin, G. V.; Gafurov, M. R.; Orlinskii, S. B. Size-Dependent Concentration of N0 Paramagnetic Centres in HPHT Nanodiamonds. *Magn. Reson. Solids* **2015**, *17* (1–7), 15101.
22. Loubster, J. H. N.; van Wyk, J. A. Electron Spin Resonance in the Study of Diamond. *Rep. Prog. Phys.* **1978**, *41*, 1201–1248.
23. Baker, J. M.; Newton, M. E. Nitrogen in Diamond Studied by Magnetic Resonance. *Appl. Magn. Reson.* **1995**, *8*, 207–228.
24. Samoilovich, M. I.; Bezrukov, G. N.; Butuzov, V. P. Electron Paramagnetic Resonance of Nickel in Synthetic Diamond. *Pis'ma Zh. Eksp. Teor. Fiz.* **1971**, *14*, 551, *JETP Lett.* **1971**, *14*, 379.
25. Isoya, J.; Kanda, H.; Uchida, Y. EPR Studies of Interstitial Ni Centers in Synthetic Diamond Crystals. *Phys. Rev. B* **1990**, *42*, 9843–9852.
26. Shames, A. I.; Osipov, V. Yu.; von Bardeleben, H.-J.; Vul', A. Ya. Spin S = 1 Centers: A Universal Type of Paramagnetic Defects in Nanodiamonds of Dynamic Synthesis. *J. Phys. Condens. Matter* **2012**, *24* (1–8), 225302.
27. Shames, A. I.; Osipov, V. Yu.; von Bardeleben, H.-J.; Boudou, J.-P.; Treussart, F.; Vul', A. Ya. Native and Induced Triplet Nitrogen-Vacancy Centers in Nano- and Micro-Diamonds: Half-Field Electron Paramagnetic Resonance Fingerprint. *Appl. Phys. Lett.* **2014**, *104* (1–5), 063107.
28. Boudou, J.-P.; Curmi, P. A.; Jelezko, F.; Wrachtrup, J.; Aubert, P.; Sennour, M.; et al. High Yield Fabrication of Fluorescent Nanodiamonds. *Nanotechnology* **2009**, *20* (1–11), 235602.
29. Boudou, J.-P.; Tisler, J.; Reuter, R.; Thorel, A.; Curmi, P. A.; Jelezko, F.; Wrachtrup, J. Fluorescent Nanodiamonds Derived From HPHT With a Size of Less Than 10 nm. *Diam. Relat. Mater.* **2013**, *37*, 80–86.
30. Felton, S.; Edmonds, A.; Newton, M.; Martineau, P.; Fisher, D.; Twitchen, D *Phys. Rev. B* **2008**, *77* (1–4), 081201.
31. Casabianca, L. B.; Shames, A. I.; Panich, A. M.; Shenderova, O.; Frydman, L. Factors Affecting DNP NMR in Polycrystalline Diamond Samples. *J. Phys. Chem. C* **2011**, *115*, 19041–19048.

32. Osipov, V. Yu.; Shames, A. I.; Enoki, T.; Takai, K.; Baidakova, M. V.; Vul', A. Ya. Paramagnetic Defects and Exchange Coupled Spins in Pristine Ultrananocrystalline Diamonds. *Diam. Relat. Mater.* **2007**, *16*, 2035–2038.
33. Shames, A. I.; Panich, A. M.; Osipov, V. Yu.; Aleksenskiy, A. E.; Vul', A. Ya.; Enoki, T.; Takai, K. Structure and Magnetic Properties of Detonation Nanodiamond Chemically Modified by Copper. *J. Appl. Phys.* **2010**, *107* (1–10), 014318.
34. Panich, A. M.; Shames, A. I.; Sergeev, N. A.; Osipov, V. Yu.; Alexenskiy, A. E.; Vul', A. Ya. Magnetic Resonance Study of Gadolinium-Grafted Nanodiamonds. *J. Phys. Chem. C* **2016**, *120*, 19804–19811.
35. Shames, A. I.; Osipov, V. Yu.; Aleksenskiy, A. E.; Ōsawa, E.; Vul', A. Ya. Locating Inherent Unpaired Orbital Spins in Detonation Nanodiamonds Through the Targeted Surface Decoration by Paramagnetic Probes. *Diam. Relat. Mater.* **2011**, *20*, 318–321.
36. Panich, A. M.; Sergeev, N. A.; Olszewski, M. Effect of Magic Angle Spinning on $^{13}C$ Spin-Lattice and Spin-Spin Relaxation in Nanodiamonds. *J. Phys. Condens. Matter* **2015**, *27* (1–4), 365302.
37. Sergeev, N. A.; Panich, A. M.; Olszewski, M.; Shenderova, O.; Goren, S. D. $^{13}C$ Spin-Lattice Relaxation in Nanodiamonds in Static and Magic Angle Spinning Regimes. *Solid State Nucl. Magn. Reson.* **2015**, *66–67*, 51–55.
38. Henrichs, P. M.; Cofield, M. L.; Young, R. H.; Hewitt, J. M. Nuclear Spin-Lattice Relaxation via Paramagnetic Centers in Solids. $^{13}C$ NMR in Diamonds. *J. Magn. Reson.* **1984**, *58*, 85–94.
39. Duijvestijn, M. J.; van der Lugt, C.; Smidt, J.; Wind, R. A.; Zilm, K. W.; Staplin, D. C. $^{13}C$ NMR Spectroscopy in Diamonds Using Dynamic Nuclear Polarization. *Chem. Phys. Lett.* **1983**, *102*, 25–28.
40. Pruski, M.; Lang, D. P.; Hwang, S.-J.; Jia, H.; Shinar, J. Structure of Thin Diamond Films: A $^{1}H$ and $^{13}C$ Nuclear-Magnetic-Resonance Study. *Phys. Rev. B* **1994**, *49*, 10635–10642.
41. Panich, A. M.; Vieth, H. M.; Shames, A. I.; Froumin, N.; Ōsawa, E.; Yao, A. Structure and Bonding in Fluorinated Nanodiamond. *J. Phys. Chem. C* **2010**, *114*, 774–782.
42. Panich, A. M.; Sergeev, N. A. M.; Froumin, N.; Dideykin, A. T.; Sokolov, V. V.; Vul', A. Ya. Structure and Bonding in Chlorine-Functionalized Nanodiamond – Nuclear Magnetic Resonance and X-Ray Photoelectron Spectroscopy Study. *J. Nanosci. Nanotechnol.* **2015**, *15*, 1030–1036.
43. Shenderova, O.; Panich, A. M.; Moseenkov, S.; Hens, S. C.; Kuznetsov, V. L.; Vieth, H. M. Hydroxylated Detonation Nanodiamond: FTIR, XPS and NMR Studies. *J. Phys. Chem. C* **2011**, *115*, 19005–19011.
44. Panich, A. M.; Shames, A. I.; Medvedev, O.; Osipov, V. Yu.; Alexenskii, A. E.; Vul', A. Ya. Magnetic Resonance Study of Detonation Nanodiamonds With Surface Chemically Modified by Transition Metal Ions. *Appl. Magn. Reson.* **2009**, *36*, 317–329.

CHAPTER

# 7 Photoluminescence of color centers in nanodiamonds

**François Treussart*, Igor I. Vlasov†**
*Laboratoire Aimé Cotton, Orsay, France* General Physics Institute RAS, Moscow, Russia†*

**Contents**



## 7.1 THE NITROGEN-VACANCY COLOR CENTER

In high-pressure high temperature (HPHT) synthesis of diamond, as well as in detonation production method (see Chapter 2), nitrogen (N) happens to be the most abundant impurity, sitting in place of carbon atoms. In type Ib diamond, nitrogen atoms are present as isolated impurities in the diamond lattice.[1] Under blue or green light excitation of such a material, photoluminescence (PL) is present in the red and near infrared spectral domains (Fig. 7.1D), with characteristic narrow lines at 637 nm (Fig. 7.2B) and 575 nm (Fig. 7.3B), followed at higher wavelengths with a broad band. These spectral signatures are more pronounced in diamond that has been exposed to a high energy particle beam (Fig. 7.1E), which is capable of displacing

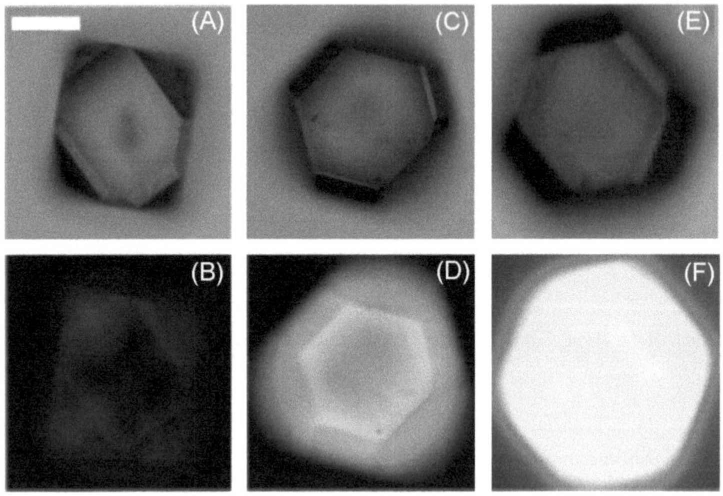


**FIGURE 7.1**

**Creation of NV color center in high pressure high temperature synthetic diamond microcrystals.** (A),(C), and (E) White light illumination optical microscopy image of a diamond microcrystal at different stages of NV center creation process: as-received Micron+ MDA 0.1 (Element6 Ltd, England) microcrystals (A), microdiamond irradiated with a 8 MeV energy electron beam, at fluence of $2.10^{18}$ electrons/$cm^2$ but not annealed, and (C) irradiated and high temperature annealed (800°C, 2 hours, under vacuum) microcrystal. (D) (resp., (E) and (F)) Same microcrystal as in (A) (resp., (B) and (C)) observed in epifluorescence microscopy. Epifluorescence images were acquired in identical conditions: excitation band wavelength 500–550 nm, detection with a longpass filter at wavelengths above 600 nm, and CCD array detector integration time of 1.5 s. The annealing induces the migration of the vacancies, leading to the formation of a high concentration of NV center, the photoluminescence of which saturates the CCD array. Scale bar, 50 μm

carbon atoms and leaving a vacancy (V) behind, and even more intense after high temperature (≈800°C) annealing (Fig. 7.1F).

These defects were therefore suspected since 1965 by Du Preez[2] to involve both nitrogen impurities and vacancies. The trigonal symmetry of the 637 nm emitting color center was then inferred from photoluminescence and uniaxial stress studies confirming that the defect must originate from an NV pair with the nitrogen-vacancy axis direction along the [111] diamond crystalline axis.[3,4] One year after, using electron paramagnetic resonance (EPR, see Chapter 6 for more information on this technique) under light excitation, Loubser and van Wyk observed in an irradiated and annealed diamond sample an EPR triplet signature that they attributed to the negative charge state ($NV^-$).[5,6] In photoluminescence, $NV^-$ is the defect responsible for the 637 nm narrow line (1.945 eV in energy),[4] which is the zero-phonon line (ZPL) corresponding to the deexcitation of the excited $NV^-$ center from the lowest energy

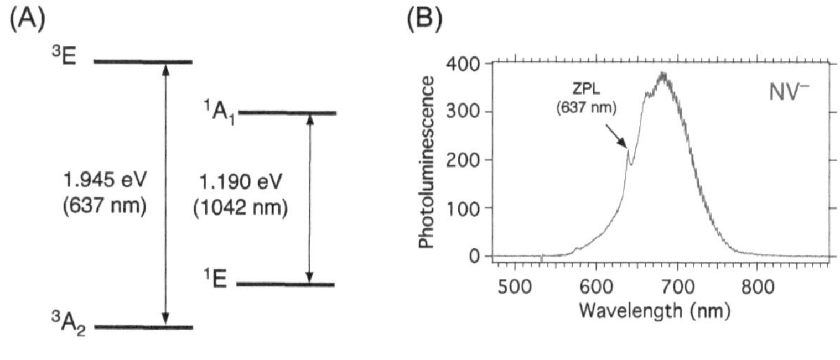


FIGURE 7.2

**Simplified electronic structure of negatively charged NV$^-$ and associated photoluminescence.** (A) NV$^-$ simplified electronic structure showing the triplet $^3A_2$–$^3E$ system, the singlet $^1E$–$^1A_1$ metastable states and the energy associated to the zero-phonon lines of each system of levels. (B) Photoluminescence spectrum of a single NV$^-$ center at room temperature, under laser excitation at wavelength of 532 nm. The ZPL appears as a narrow line at wavelength 637 nm (energy 1.945 eV), and the phonon replica as a broad band at higher wavelengths

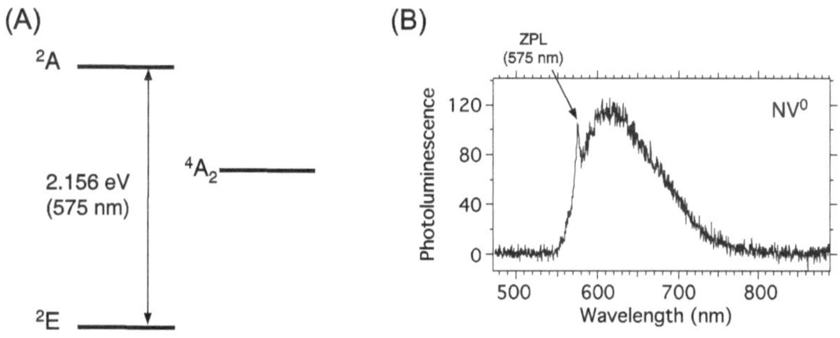


FIGURE 7.3

**Simplified electronic structure of neutral NV$^0$ and associated photoluminescence.** (A) NV$^0$ simplified electronic structure showing the doublet $^2E$–$^2A$ system, the $^4A_2$ metastable state and the energy associated to $^2E$–$^2A$ zero-phonon line. (B) Photoluminescence spectrum of a single NV$^0$ center at room temperature, under laser excitation at wavelength of 532 nm. The ZPL appears as a narrow line at wavelength 575 nm (energy 2.156 eV), and the phonon replica as a broad band at higher wavelengths

vibrational level of the excited state to the lowest vibrational level of the ground state. The defect with the ZPL centered on 575 nm (2.156 eV) was attributed to the neutral state NV$^0$ by Mita[7] in 1996. Both NV$^-$ and NV$^0$ have a PL that extends from the ZPL to higher wavelength due to vibronic phonon sidebands (Fig. 7.2 and Fig. 7.3). The sharp ZPLs of the nitrogen vacancy indicate that they are deep-level defects in the diamond energy bandgap.

### 7.1.1 NV CENTER ELECTRONIC STRUCTURE AT ROOM TEMPERATURE (SIMPLIFIED PRESENTATION)

Researches on NV center electronic structure are still very active because contradictory results have been obtained by different approaches and a full understanding of $NV^-$ most remarkable properties, i.e. the spin polarization by light, is still lacking. An in-depth recent review of the NV center properties is presented in Doherty et al.[8] A molecular model of $NV^-$ providing a complete electronic solution is also presented in Doherty et al.[9] Briefly, for $NV^-$, the calculation is based on (i) the NV center $C_{3v}$ point group symmetry, and (ii) a model involving six electrons,[5] five of them being electrons unpaired between the vacancy, the three nearest carbon atoms and the nitrogen atom, plus an additional electron trapped at the defect center. This even number of electrons is associated to an integer spin value. The center electronic state can then be described by configurations of molecular orbitals constructed according to group theory rules,[10] as linear combinations of $sp^3$ orbitals of the vacancy nearest carbon and nitrogen atoms (see a full development of $NV^-$ molecular model in M. Doherty et al.[9]). These combinations include two degenerate orbitals that transform as the $E$ irreducible representation (e-orbitals) and two separate orbitals of $A_1$ symmetry ($a_1$-orbitals). One $a_1$-orbital is always occupied by two electrons, leaving three possible electronic configurations: $a_1{}^2e^2$, $a_1e^3$, and $e^4$. Taking into account the electron–electron Coulomb interaction leads to the separation of the multi-electron states into triplet (spin $S = 1$) $^3A_2$ and $^3E$, and singlet (spin $S = 1/2$) $^1E$ and $^1A_1$ levels (Fig. 7.2A). It was deduced from hole burning spectra and temperature dependence of magnetic circular dichroism signal[11] that the ground state is the triplet $^3A_2$, and that the PLs originate from the excited triplet state $^3E \rightarrow {}^3A_2$ transition. The $^3E$ excited state is an orbital doublet. Its degeneracy is lifted by nonaxial strain into two orbital branches $E_x$ and $E_y$, each of them being formed by three spin states $S_x$, $S_y$, and $S_z$, resulting in six different levels that can be observed at cryogenic temperature.[12] For the ground $^3A_2$ triplet level, the spin–spin interaction is responsible for a zero field splitting, i.e. a separation between $m_S = 0$ and $m_S = \pm 1$ of $D = 2.88$ GHz (i.e. in the microwave domain). The energy order of the spin projections (along N–V crystallographic direction) $m_S = 0, \pm 1$ of the $^3A_2$ triplet, leading to $m_S = 0$ as the lowest energy state, was assessed by Raman heterodyne-detected-electron–nuclear double resonance.[13]

Importantly, at room temperature the $^3E$ excited state fine structure is impacted by phonon mediated averaging of the two low-temperature fine structure branches,[14] resulting in only three detectable levels of spin projection $m_S = 0$ and $m_S = \pm 1$ separated by a zero field splitting $D' = 1.42$ GHz due to the spin–orbit coupling in this case (Fig. 7.4).

The relative energy positions of the intermediate $^1E$ and $^1A$ singlet states have long been debated. A near-infrared (NIR) emission with a ZPL at 1042 nm has been recently reported,[15] and based on uniaxial stress measurements it was attributed to the $^1E \leftrightarrow {}^1A$ transition (or $^1E' \leftrightarrow {}^1A$ in [16]). The NIR emission intensity is about four orders of magnitude weaker than for the far red $^3E \rightarrow {}^3A_2$ transition, and has

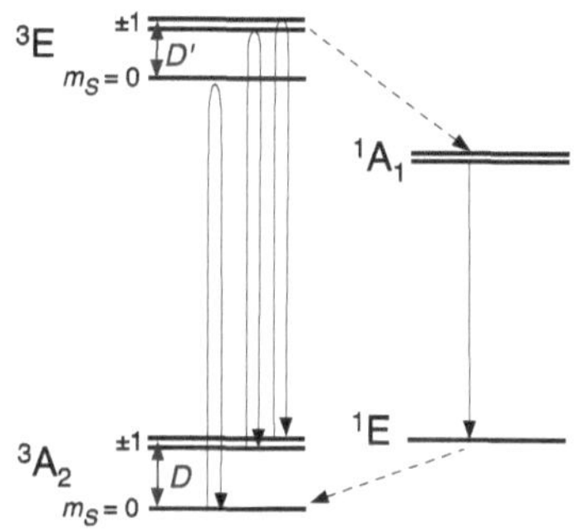


**FIGURE 7.4**

**Fine structure of electronic levels of NV⁻ center at room temperature, without strain.** This structure is the result of phonon mediated averaging of the two low-temperature fine structure branches (total of 6 levels).[14] Values $D = 2.88$ GHz and $D' = 1.42$ GHz are the zero-field splitting of the ground and excited states, respectively. Plain arrows display spin projection $m_S$ preserving transitions, while dashed arrows indicate intersystem crossings (non-radiative). The polarization of $NV^-$ in $m_S = 0$ ground state spin projection by optical pumping is made possible by these intersystem crossings that take place mainly from $m_S = \pm 1$ excited state

a lifetime of about 1 ns,[17] both observations indicating a decay dominated by non-radiative pathways. Very recently, further uniaxial stress measurements (shifts and splitting) at cryogenic temperature interpreted with the electronic model of $NV^-$ center, provided evidence for $^1$E to be the lowest energy state of the intermediate states system (as displayed on Fig. 7.4), and also indicated that the large splitting of the 1042 nm ZPL observed under stress is most likely due to electron–electron interaction.[18]

The neutral $NV^0$ has attracted less attention than $NV^-$ because it was hard to detect its EPR signature. $NV^0$ has five electrons associated to always occupied $a_1$-orbitals and, in the ground state, to the $a_1{}^2e$ configuration yielding the multiple electron ground state $^2$E. When an electron is promoted from the singlet to the excited doublet, the configuration becomes $(a_1'^2)a_1{}^1e^2$ with associated multiple electron states $^2$A, $^4A_2$, and $^2$E (Fig. 7.3A). The observation by EPR of spin $S = 3/2$ $^4A_2$ level populated by optical excitation was realized only recently,[19] and the $S = 1/2$ has $^2$E still never been detected by the same approach, probably due to molecular orbital distortions.

## 7.1.2 OPTICALLY DETECTED MAGNETIC RESONANCE AND SPIN POLARIZATION

The strong interest of a broad community (point defects in solid, nanomagnetism, quantum information processing, quantum optics, etc.) for the negatively charged $NV^-$ center stems from its optically detectable spin $S = 1$ ground states. Optical Detection of Magnetic Resonance (ODMR) is a double resonance technique combining optical detection with electron spin resonance spectroscopy. It was initially realized using the phosphorescence (triplet state transitions) of aromatic molecules embedded in crystals, first in ensemble[20,21] and later, owing to the improvement of optical detection techniques, at the level of single molecule.[22] ODMR signal is present only under the right experimental conditions (i.e. with appropriate transition rates) for which the magnetic resonance transitions between the spin sublevels of a triplet state produce changes in the system luminescence. The $NV^-$ center meets these conditions, and like for molecules, ODMR was first observed in ensemble,[23] and then at the single emitter level,[24] paving the way to the huge variety of the afore mentioned applications.

The polarization selection rules inferred from group theory[25] indicate that $^3A_2 \leftrightarrow {}^3E$ transitions are orbitally allowed. Moreover, the electrical dipolar operator cannot change the spin projection $m_S$. Therefore, $^3A_2 \leftrightarrow {}^3E$ transitions can only happen with $\Delta m_S = 0$ (see curved arrows on Fig. 7.4). The NV center is usually excited off-resonance with a laser of higher energy than the one of the ZPL transition, leading to the population of the excited state vibronic levels first and then, by non-radiative decay, towards one of the three possible $^3E$ sublevels $m_S = 0$ or $m_S = \pm1$. Intravibronic transition does not affect the spin projection, so that $m_S$ is preserved throughout the excitation–emission cycle, between $m_S = 0$, $m_S = +1$, or $m_S = -1$. If the systems were closed, the transition intensities would be similar. However, the addition of a microwave (MW) field with a frequency tuned at the zero field splitting value of 2.88 GHz leads to a decrease of $NV^-$ luminescence (Fig. 7.4). The MW field induces a transfer of population from $m_S = 0$ to $m_S = \pm1$. The photoluminescence intensity decrease at the MW resonant frequency (signature of ODMR) indicates that the transitions in the $m_S = +1$ or $m_S = -1$ subsystems are dimmer than the one in the $m_S = 0$ subsystem, which is why the ground $m_S = 0$ substate is often called the "bright" state.

The origin of $NV^-$ center ODMR stems from the intersystem (triplet → singlet) crossing $^3E \rightarrow {}^1A$ made possible by spin–orbit coupling which mixes triplet and singlet.[26] This mixing allows relaxation between states that transforms with the same symmetry, and calculations indicates that the transfer is predominantly from $m_S = \pm1$. Therefore $^3E$, $m_S = \pm1$ levels have an additional decay pathway through the single levels, which explains why $^3E(m_S = \pm1) \rightarrow {}^3A_2(m_S = \pm1)$ transitions are associated to a dimmer emission compared to $^3E(m_S = 0) \rightarrow {}^3A_2(m_S = 0)$ one.

Moreover, it has been known for a long time that strong EPR signal from $NV^-$ center ground state comes from its polarization by light excitation (optical pumping)[5] with a preferred orientation in the $^3A_2$, $m_S = 0$ sublevel.[27] This property is also at the heart of quantum measurements based on $NV^-$ center. Like in the case of ODMR, the interpretation of this spin polarization by light involves intersystem crossing $^3E \rightarrow {}^1A$, but a full understanding of the spin polarization mechanism is still lacking. Indeed, after $^3E \rightarrow {}^1A$ intersystem crossing $^1A$ decays to $^1E$ by emitting light at 1042 nm wavelength. Then, $^1E$ decays non radiatively towards $^3A_2$. Symmetry rules indicate a decay preferentially into E symmetry states (*i.e.* states with

$m_S = \pm 1$ spin projection),[9,28] but this is in contradiction with the observed polarization in $m_S = 0$ state.[27] The spin polarization (change of $m_S = \pm 1$ to $m_S = 0$) therefore results from a forbidden singlet-triplet transition that can be explained by first order terms of spin–orbit interaction (indirectly accessible experimentally).[9]

### 7.1.3 NV CENTER PRODUCTION IN NANODIAMOND: STATE-OF-THE-ART AND CHALLENGES

The production of NV centers in nanodiamonds involves a similar treatment as the one in bulk diamond slab, i.e. the creation of vacancies by high energy particle irradiation in synthetic diamond, followed by high temperature (800–900°C) annealing[29] required for the vacancies to migrate and eventually stabilize in lattice sites adjacent to nitrogen atoms. The vast majority of synthetic diamond used is produced by HPHT method naturally containing nitrogen impurities in place of carbon atoms. However, nanodiamonds synthesized by detonation process (leading to particle primary size ≈5 nm, as described in Chapter 2) have also been used, but NV center photoluminescence is less stable in these ultra small nanocrystallites[29] because the centers are in close proximity to the surface, allowing energy transfer to defects that offer many non-radiative decay pathways. The high energy (a few keV to a few MeV) particles employed for irradiation are either electrons, protons, or helium $He^+$ ions (and occasionally neutrons[7]). The particles displace the carbon atoms at interstitial positions, leaving a vacancy behind.

To selectively produce $NV^-$ center and not the neutral form $NV^0$, it has been demonstrated that a concentration of a few 100 ppm is required[30] (type Ib HPHT diamond), and that based on the following two-steps conversion mechanism $V^0 + 2N^0 \rightarrow NV^0 + N^0 \rightarrow NV^- + N^+$, the negative charge is gained from one nitrogen atom a few lattices apart from the NV center considered. Therefore, considering a 100 ppm nitrogen impurity concentration, the maximum reachable $NV^-$ concentration should be 50 ppm,[31] i.e. 50% conversion rate. Moreover, an initial 10 ppm nitrogen concentration would not allow the production of $NV^-$.

For application to biolabeling, the largest concentration in NV centers is sought to obtain the brightest possible nanoparticles. A mass production method is also necessary. To this aim, two approaches are used (Fig. 7.5): (i) direct irradiation and annealing of commercially available nanocrystals, or (ii) irradiation and annealing of HPHT microcrystals (size ≈150–190 μm) followed by their milling into nanocrystals. After either one of these treatments, some graphitic layers and $sp^2$ carbon phases appear at the surface that need to be removed because they provide additional non-radiative decay pathways and also contain traps of the electronic charge[32] of $NV^-$.

The first mass production technique based on irradiation of nanocrystals was developed by Chang et al.[29] using a 40 keV energy $He^+$ ion beam for irradiation (Fig. 7.5A). In this technique, nanodiamonds of size <50 nm are deposited in a monolayer (because of small penetration depth of 40 keV $He^+$) on a few meter long copper ribbon that is slowly moved in front of the ion beam. The advantage of such an approach is the use of a compact ion source that can be installed in any laboratory,

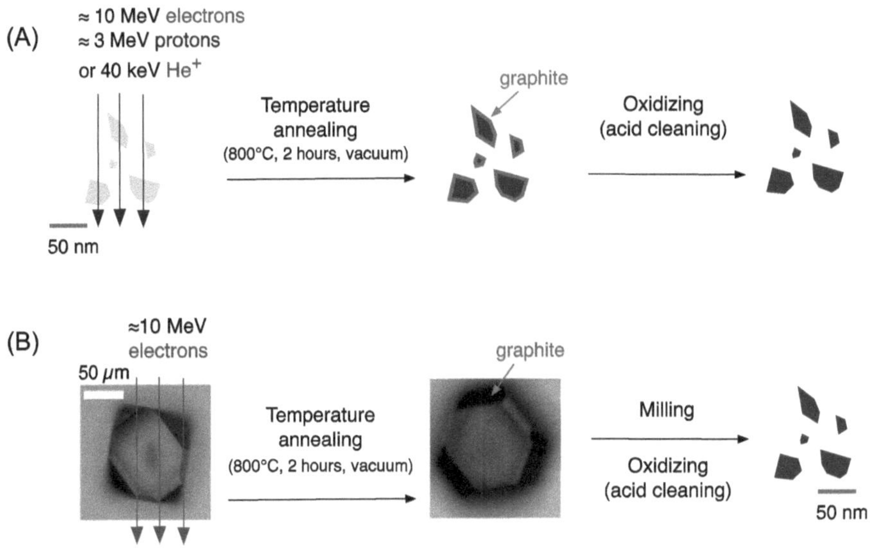


**FIGURE 7.5**

**Process of fluorescent nanodiamond synthesis.** (A) Direct irradiation of synthetic nanodiamond (obtained by milling faceted HPHT microdiamonds, containing ≈100 ppm nitrogen impurities) with high energy particles (protons, electrons, or helium ions, depending on the amount of material to treat) followed by 800°C vacuum annealing, and oxidizing (450°C in air, or/and strong acid cleaning) to remove amorphous $sp^2$ graphitic layer formed during irradiation and the annealing. (B) Irradiation of the HPHT size ≈150 μm microcrystals, followed by high temperature annealing, milling down to size <50 nm nanoparticles, and oxidation to remove amorphous carbon

while proton or electron accelerators are larger equipments subject to constraining safety regulations. Moreover, one 40 keV $He^+$ produces ≈40 vacancies when it penetrates into diamond, while 2 MeV electron and 3 MeV proton beams create 0.1 and 13 vacancies, respectively.[29] However, larger fluences can be achieved with these two last types of beam, compensating for this difference in damage production: up to $10^{20}$ electrons/cm$^2$ with 10 MeV electrons[33] and $10^{17}$ protons/cm$^2$ with 2.4 MeV protons,[34] compared to $6 \times 10^{13}$ $He^+/cm^2$. In addition, the large penetration depth of high energy electrons (1–2 cm for 10 MeV electron, the exact value depending on the design of the reactor in which the diamond powder is placed) allows to treat a large amount (up to a few grams) at once.[35–37]

The second mass production approach consisting in irradiating microcrystals before their milling into nanoparticles (Fig. 7.5B) was expected to yield higher concentration of NV centers. NV complex formation requires the migration of the vacancies during the annealing process, and applying this treatment to microcrystals instead of nanocrystals should reduce the annihilation of the vacancy at the surface.[35] However, this top-down production approach did not yield higher $NV^-$ concentration compared to direct irradiation of ND, as revealed by photoluminescence and EPR measurements.[38] The smallest reported fluorescent ND (FND) containing at least one $NV^-$ center (charge state assessed by PL and ODMR measurements) has size in the 5–10 nm range.[33,39]

Currently, the highest concentrated FNDs of size ≈30 nm contain in average about 15 NV centers, corresponding to about 19 ppm, which would be associated to about 10% conversion efficiency of nitrogen into NV centers in a 200 ppm nitrogen concentrated sample. These bright FNDs are produced by 3 MeV proton beam irradiation[31] at fluence of $2 \times 10^{16}$ protons/cm$^2$. Like in the case of bulk diamond,[34] further increase of the fluence did not result in an increase of NV concentration. Moreover, the use of starting materials containing increasing nitrogen impurities from 100 to 200 ppm resulted in a concomitant increase of NV concentration as expected, but ND with larger $N^0$ concentration of ≈400 ppm yielded an NV concentration similar to the one with 100 ppm $N^0$. This disappointing observation may be due to the presence of structural defects in this highly N-doped diamond (see Chapter 3, for example) that impact NV center formation.[31]

The main applications of FND are in the fields of bioimaging (Chapter 16), physico-chemical sensing,[40,41] and scanning probe magnetometry[42] (Chapter 17). In most of these fields, decreasing the FND size while increasing the NV concentration would enhance the possibilities of applications. In particular, for biolabeling, FND of size <10 nm would give access to region of interest hardly accessible due to steric hindrance, like the synaptic cleft or the dendritic spine neck of neurons. Current challenges in the field of NV center production in nanodiamonds are thus the following: (i) increase the production yield of size <10 nm FND; (ii) increase FND brightness, by increasing the nitrogen → NV conversion efficiency; (iii) improve FND size and brightness homogeneity. Regarding the latter point, irradiation of ND in the liquid phase showed a reduced dispersion of FND brightness,[43] attributed to more homogeneous irradiation dose per particle as a consequence of thermal agitation.

### 7.1.4 DIFFERENT EXCITATION MODALITIES AND PHOTOPHYSICAL PARAMETERS

Thanks to their remarkable photostability, NV centers in nanodiamond can be excited and then observed on a very long time scale, using either a laser beam (PL) as previously discussed or an electron beam (Cathodoluminescence,[44,45] CL) in a scanning transmission electron microscope (STEM). In the latter case, it is possible to get simultaneously an EM contrast (for example, the bright field or high angle angular dark field,[45] explained in Chapter 3) with nanometer resolution and a CL image perfectly correlated (Fig. 7.6). Such integrated correlative microscopy may become an invaluable method for cellular imaging,[46] allowing to observe simultaneously the cell ultrastructure and the FND label, provided the FND has been functionalized to recognize the target of interest.

Regarding the radiative lifetime, measurements gave values different for $m_S = 0 \rightarrow m_S = 0$ or $m_S = \pm 1 \rightarrow m_S = \pm 1$ transitions (see Table 10 of [8]):

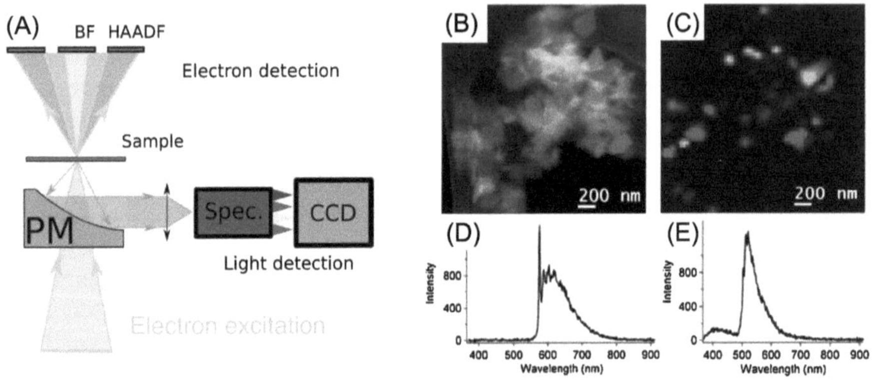


**FIGURE 7.6**

**Integrated spectrally resolved cathodoluminescence (CL) and scanning transmission electron microscopy (STEM).** (A) CL-STEM setup. The electron beam (energy ≈80 keV) comes from the bottom, crosses a parabolic optical mirror (PM), then hits the sample and yields a bright field (BF) or high angled annular dark field (HAADF) TEM image on the electron detectors. The PM directs the CL light towards an imaging spectrograph made of a monochromator (Spec.) and a CCD array detector. (B) HAADF image of a mixture of natural and HPHT nanodiamonds displaying either a green emission (from N-V-N defect centers, also known as H3) or red emission (from $NV^0$) centers, with the corresponding spectra ((D) for $NV^0$ and (E) for H3). The sample holder temperature is about 100 K, which explains that the ZPL of $NV^0$ is narrower than in Fig. 7.3B room temperature PL spectra. (B)–(E) are adapted from: Tizei, L. H. G., et al. Phys. Status Solidi 2013, 210, 2060–2065,[77] with permission, © 2013 Wiley

11.6 to 13.7 ns for $NV^-$ in bulk diamond at room temperature for the transition $^3E\ m_S = 0 \rightarrow {}^3A_2\ m_S = 0$, and 7.3–7.8 ns for $^3E\ m_S = \pm 1 \rightarrow {}^3A_2\ m_S = \pm 1$. However, the diamond nanoparticle host, being smaller than NV emission wavelength, modifies the local density of states of the electromagnetic modes, and therefore the radiative lifetime, which can be strongly increased in such an electrostatic limit.[47,48] Indeed, in practice, due to the presence of numerous non-radiative channels,[49] radiative lifetime of $NV^-$ in nanodiamond is broadly distributed in the 10–40 ns range, with the most frequent value being 33 ns for 100 nm sized nanoparticle and 28 ns for 25 nm sized one. Despite a smaller radiative lifetime increase than expected from electromagnetism theory, values around 30 ns are about one order of magnitude larger than radiative lifetime of autofluorescence which is in the 1–4 ns range (depending on the endogenous pigment). Therefore, using Fluorescence Lifetime Imaging Microscopy (FLIM) or simply time delayed detection (in time window starting from ≈10 ns), one can filter out photons from $NV^-$ emission on a nearly dark background and strongly improve the signal-to-background ratio of fluorescent nanodiamond imaging.[50] Such FLIM or time-delayed detection has been used to investigate the tissue distribution of nanodiamonds after injection in *Caenorhabditis elegans* (*C. elegans*) worm[51] (Fig. 7.7) or in mouse lung sections.[52]

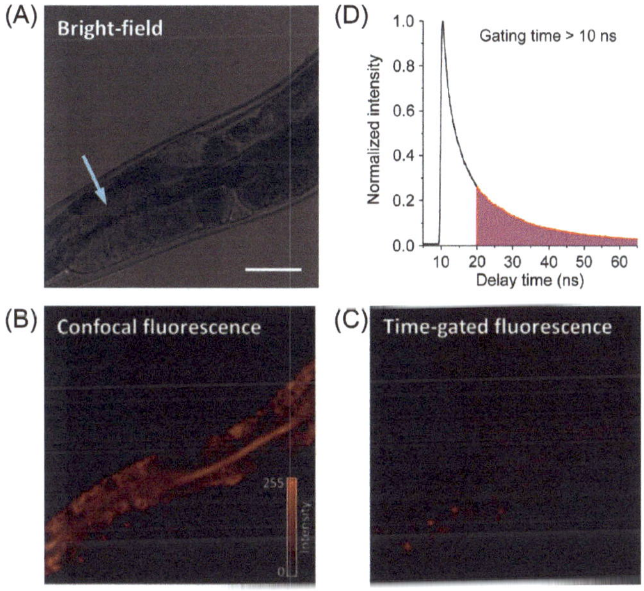


**FIGURE 7.7**

**Background-free imaging of fluorescent nanodiamond in living C. elegans worm, based on time delayed detection.** (A) White light (bright-field) imaging of a portion of a living *C. elegans* worm. The arrow indicates the location of FNDs injection. Scale bar, 50 μm. (B) Fluorescence confocal raster scan of the same region than in (A), showing the FNDs and the worm autofluorescence. (C) Fluorescence decay time trace of size ≈100 nm FNDs suspended in water. The area shaded represents the fluorescence signal collected at the gating times longer than 10 ns. (D) Time-delayed fluorescence raster scan displaying only FNDs fluorescence on a dark autofluorescence background, which allows non-ambiguous identification of FNDs within the worm body. Adapted from: Kuo, Y., et al. Biomaterials, 2013, 34, 8352–8360,[51] with permission, © 2013 Elsevier

Finally, since bright FNDs are intended to be used as fluorescent bioimaging labels, it is of interest to compare their brightness to the one of standard fluorescent nanoparticles like semiconductors nanocrystals (quantum dots, QDs). Such a comparison was done in scanning confocal microscopy on single emitters of each type (single NV center in a nanodiamond, and single QD).[53] It revealed that the excitation saturation intensity is 75 times larger for one NV center in an ND, compared to a single QD, and that the maximum detection rate was about 3 times larger for the QD than for the NV center. These data indicate that an FND containing more than 3 NV centers will be brighter than a single QD, but one also needs to use a higher excitation power. This latter requirement is the consequence of two phenomena: (i) a larger scattering of incident light by the diamond crystal than by the QD, and (ii) about one order of magnitude smaller absorption cross-section of NV center compared to QD.

### 7.1.5 HIGH-RESOLUTION OPTICAL MICROSCOPY WITH FND IN PHOTO- AND CATHODOLUMINESCENCE (SEE CHAPTER 17, TOO)

Owing to its very high stability under photon or electron beam irradiation, fluorescent nanodiamonds have been used in super-resolution microscopy. The record resolution ever achieved with STimulated Emission Depletion (STED) microscopy (which is only limited by the stimulated laser beam power that can be sustained by the sample) was obtained for single NV center in bulk diamond, down to 2.4 nm.[54] In 80 nm sized nanodiamond, STED also allowed imaging individual NV centers with a resolution of $\approx$10 nm,[55] that was limited by thermal drift due to residual absorption of the STED laser beam which maximum mean power was about 1 W, focused onto the FND.

By integrating a pierced parabolic mirror into a Scanning Transmission Electron Microscopy (STEM), with its focal point located at the sample position, Kociak et al. obtained a cathodoluminescence image of a single NV center in a nanodiamond with a sub-wavelength optical resolution of about 100 nm.[56] In such an experiment, the NV center is promoted in its excited state by the recombination of excitons travelling inside the diamond lattice. These excitons result from the de-excitation of a plasmon generated by the interaction of the fast electrons with the diamond material. The optical resolution limit in CL may correspond to the exciton mean free path before their recombination and subsequent NV center excitation.

## 7.2 THE SILICON-VACANCY CENTERS IN NANODIAMONDS

Silicon-vacancy (SiV) color centers possess a number of advantages with respect to their optical properties over NV centers in nanodiamond. Zero-phonon emission (ZPL) from SiV defects is observed at wavelength of 738 nm which is away from the characteristic broadband (450–650 nm) photoluminescence of ND. Narrow-band (a few nanometers) zero-phonon emission is a few percent for the NV centers, whereas it is 70–80% for SiV centers. Both centers demonstrate photo stability and high quantum yield of luminescence even at room temperature. Listed properties of the SiV centers make NDs, containing these centers, promising for application as photoluminescent biomarkers and as single-photon emitters.

### 7.2.1 PHOTOLUMINESCENT SPECTRUM AND ELECTRONIC STRUCTURE OF THE SILICON-VACANCY COLOR CENTER

A characteristic photoluminescence spectrum of diamond containing silicon-vacancy centers is shown in Fig. 7.8. Luminescence of SiV is mostly localized in a zero-phonon line (ZPL) of the center at 738.5 nm, and a weak phonon side-band is placed in the range of 740–820 nm. The ZPL contains 70–80% of the SiV emission, and this feature is favorable for single photon source applications.[57] SiV ZPL splits into four lines at liquid helium temperature.[58] The origin of the splitting is attributed to split ground and excited states (Fig. 7.9).

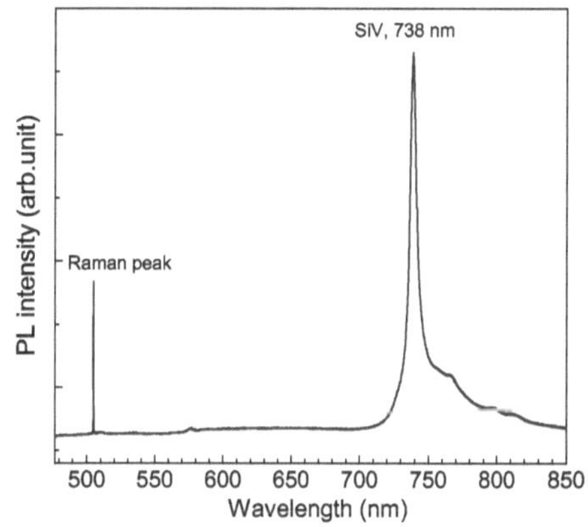


**FIGURE 7.8**

**Characteristic photoluminescence spectrum of diamond containing silicon-vacancy centers emitting at 738 nm**

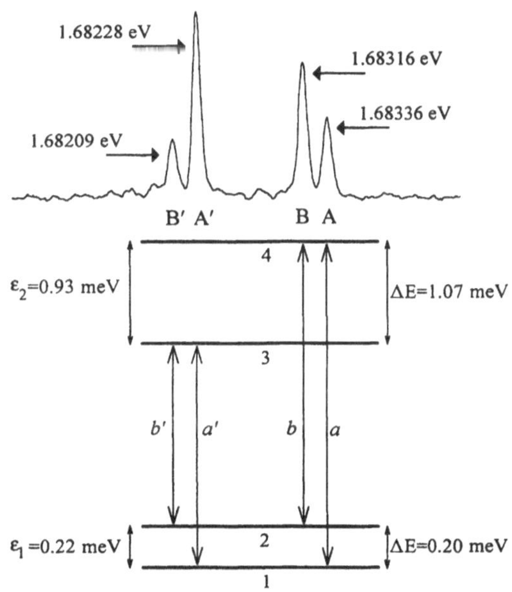


**FIGURE 7.9**

**Absorption spectrum of diamond with SiV centers** measured at 6K (top). Proposed energy-level diagram of SiV center (bottom).[58] From Clark et al., Silicon Defects in Diamond. Phys. Rev. B 1995, 51, 16681–16688

The molecular structure of the SiV center was calculated using spin-polarized local density functional cluster theory. Silicon was predicted to be unstable when placed

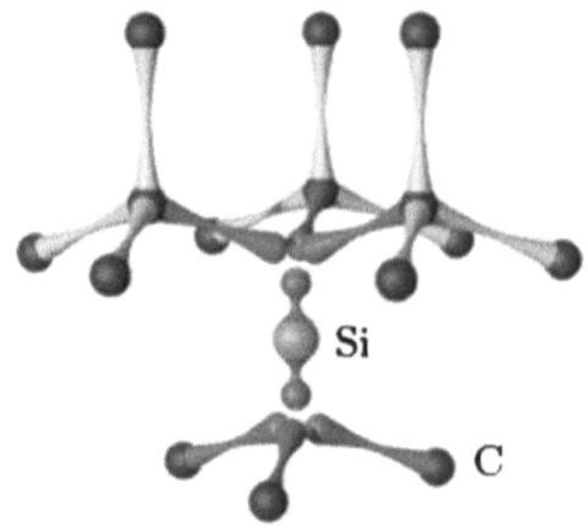


**FIGURE 7.10**

**The molecular structure of the SiV center.** From Rogers et al., All-Optical Initialization, Readout, and Coherent Preparation of Single Silicon-Vacancy Spins in Diamond. Phys. Rev. Lett. 113, 263602 (2014)

in a lattice site next to a vacancy, spontaneously relaxing into a split-vacancy position. Si atom is located at the center of a nearest-neighbor divacancy (Fig. 7.10), forming the color center of $D_{3d}$ symmetry.[59] Recently the $D_{3d}$ symmetry of the SiV centers was confirmed experimentally.[60,61] The results on comparative study of Si-related centers by electron paramagnetic resonance and photoluminescence spectroscopy suggest a negative charge state for the SiV center emitting at 738 nm.[62] Due to this connection, it is currently designed as $SiV^-$.

### 7.2.2 PRODUCTION OF NANODIAMONDS WITH $SiV^-$ CENTERS

Nanotechnology applications of $SiV^-$-based luminescent nanodiamonds, in particular their use as luminescent biomarkers, is constrained by the lack of methods for mass production of such a nanomaterial. Below, we discuss the possibility of the main methods of diamond synthesis and the recent results of their use in production of the luminescent NDs.

#### CVD technique

Chemical vapor deposition (CVD) is a modern technique used for production of polycrystalline and single crystal diamond at typical pressures of ~100 Torr (~0.13 bar) and temperatures in the range 700–1000°C (see Chapter 1).[63] In all versions of the CVD synthesis, the gas phase comprises a volatile carbonaceous material (methane, alcohols, acetone) and hydrogen. The deposition is carried out from activated gas phase. The main methods of activation – thermal (hot filament) and electric – combine different forms of discharge (arc, glow) with the use of both DC and AC (microwave pulse mode). Activation of the gas phase creates a sufficiently high concentration of active carbonaceous particles, atomic carbon, and hydrogen. The active carbon is the key to the subsequent gas-phase and surface chemical reactions necessary to sustain the growth of the diamond. Created atomic hydrogen etches various forms of non-diamond carbon appearing in the synthesis zone. Carbide forming metals (W, Mo, Ta, etc.), silicon, or diamond are used as substrates for CVD diamond growth. For

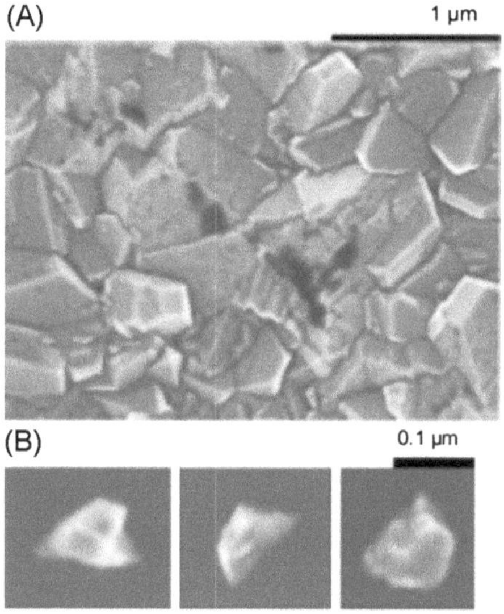


**FIGURE 7.11**

**SEM images** of (A) the polycrystalline starting material and (B) individual NDs after spin coating onto an Si substrate.[65] From Neu et al. Narrowband Fluorescent Nanodiamonds Produced From Chemical Vapor Deposition Films. Appl. Phys. Lett. 2011, 98, 12–15

deposition on non-diamond substrate micro- and nanocrystals of diamond are used as seeds. They are applied to a substrate surface, for example, by sonication.

The CVD production of isolated diamonds nanoparticles doped with Si is described by H. Zhang et al.[64] $SiV^-$ luminescent nanodiamonds were synthesized by standard CVD technique on silicon substrate, followed by sonication to generate free particles. Typical nanodiamond sizes were found to be in the range of 50–250 nm. This approach for production of luminescent nanodiamonds is extremely ineffective. In one working cycle of CVD diamond deposition (typical time is about 1 hour), only ~0.01 mg of 100 nm diamond particles can be produced. One more way for using of CVD for luminescent ND synthesis was suggested by E. Neu et al.[65] There the production of fluorescent NDs containing *in situ* produced $SiV^-$ centers from polycrystalline CVD films via the bead assisted sonic disintegration (BASD) method is reported (see Chapter 2). Produced NDs combine the advantageous $SiV^-$ fluorescence properties and their feasible production during the CVD process with the extended applicability of NDs dispersed in solution. The starting material for ND production was a polycrystalline diamond film grown by hot filament CVD on ND seeded Si substrate. Scanning electron microscope (SEM) images indicate a film thickness of 2 μm and high crystalline quality (grain size 0.5–1 μm; see Fig. 7.11A). The characterized diamond film is used for ND production after removing the substrate by chemical etching. The BASD process, disclosed recently,[66] enables the deagglomeration of strongly agglomerated nanoparticles by charging ceramic microbeads to a sonicated

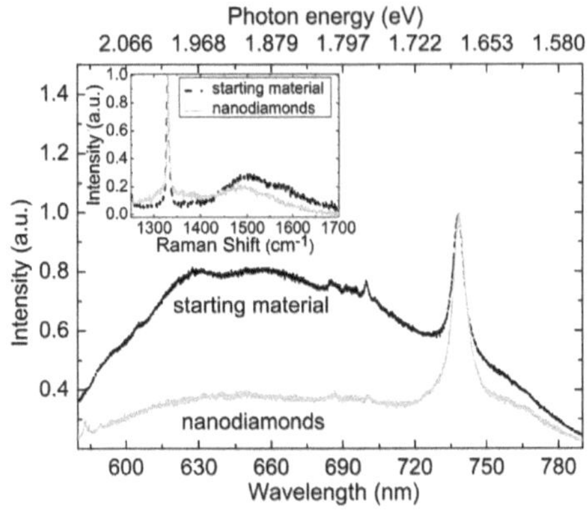


**FIGURE 7.12**

**Photoluminescence spectra recorded under 532 nm laser excitation from the polycrystalline starting material and an ensemble of the produced NDs.** The inset shows the Raman spectrum of the starting material and the NDs recorded with 488 nm laser light. From Neu et al. Narrowband Fluorescent Nanodiamonds Produced From Chemical Vapor Deposition Films. Appl. Phys. Lett. 2011, 98, 12–15

suspension. The microjets or shock waves induced by ultrasonic cavitations propel the beads ($ZrO_2$, 50 μm sized) leading to the disintegration of large agglomerates and the eventual formation of colloidal solutions of primary nanoparticles. So far, the method has only been applied for the deagglomeration of nanoparticles bound by interparticle forces. There, the BASD technique was applied as a top down approach for the production of nanoparticles from a CVD diamond film by separating the grains and crushing the micrometer-sized crystals. The BASD process is followed by several purification steps, delivering a colloidal suspension of NDs in de-ionized water. The average particle size, determined by dynamic light scattering directly in the resulting solution, shows a distribution maximum at 70–80 nm. SEM observation verified the nanometric size of the nanoparticles (Fig. 7.11B). The faceted shape of the individual NDs indicates the crushing of the film's crystallites along lattice planes. No contamination from the BASD process was found in the final colloidal solution. Fig. 7.12 displays a typical fluorescence spectrum of a ND ensemble. Significant reduction in broadband fluorescence was achieved, while the $SiV^-$ luminescence remained basically unchanged (ZPL 738.6 nm, width 6.8 nm). The method of ND production by milling looks more time- and labor-intensive than the direct CVD synthesis of the luminescent NDs. It seems that the size of produced diamond nanoparticles is also limited to 70–80 nm in the milling approach.

## HPHT technique

Currently, there are large-scale industrial productions of synthetic diamonds under high temperature and high pressure (HPHT technique, as explained in Chapter 2). HPHT synthesis occurs usually in the metal (solvent) – carbon (graphite)

**FIGURE 7.13**

**Scanning electron microscopy images of diamond materials obtained from binary mixture $C_{10}H_8$–$CF_{1.1}$.** From Davydov et al., Production of Nano- and Microdiamonds With Si-V and N-V Luminescent Centers at High Pressures in Systems Based on Mixtures of Hydrocarbon and Fluorocarbon Compounds. JETP Lett. 2014, 99, 585–589

system. High-pressure (10–20 GPa) is created in the hard-alloy chambers by powerful presses. Diamonds are crystallized during cooling from the melt (a supersaturated solution of carbon in the metal) under pressure. Synthesized diamonds are separated from the sinter by dissolving the matrix in a mixture of acids. Diamond powders for industrial purposes, as well as single crystals of gem quality, are produced by HPHT technique. Without a metal solvent, the rate of conversion of graphite into diamond is significantly reduced.

Until recently, HPHT technique has not been used for the synthesis of luminescent NDs. In 2014, the preparation of heterogeneous size diamond particles (including nanoscaled) doped with silicon and nitrogen during their synthesis from mixtures of hydrocarbon and fluorocarbon compounds was reported.[67] NV and $SiV^-$ luminescence was observed from nano- and micro-sized fractions of produced diamonds (Fig. 7.13). This result demonstrates a high potential of HPHT technique for a large-scale production of the $SiV^-$-based luminescent nanodiamonds.

## Detonation technique

Synthesis of nanodiamonds by detonation loading of explosives with negative oxygen balance is the cheapest and most efficient way to direct production of nanodiamonds with a size of less than 10 nm. Shenderova et al.[68] have carried out a systematic study of main representative classes of NDs produced by detonation shock wave conversion of different carbon precursor materials (see Chapter 2), namely, graphite, graphite/explosive mixture, and different combinations of pure explosives using different cooling methods (wet or dry cooling). $SiV^-$ luminescence was not observed in any of the tested detonation nanodiamonds (Si was expected as background impurity in those materials). A large number of samples of detonation nanodiamonds (DND)

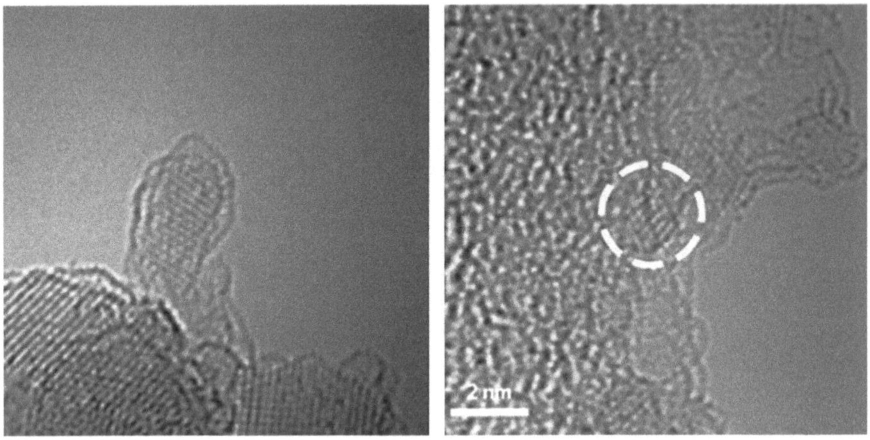


**FIGURE 7.14**

**High-resolution TEM images of meteorite nanodiamonds**: (A) free standing grain (2-nm size), (B) grain (1-nm size) in agglomerate.[73] From Vlasov et al., Molecular-Sized Fluorescent Nanodiamonds. Nat. Nanotechnol. 2013, 9, 54–58

were produced in the presence of various dopants, including silicon.[69] A PL study of these samples in our lab did not reveal any $SiV^-$ luminescence in them. Thus, for today, the detonation technique remains unpromising for production of the $SiV^-$-based luminescent nanodiamonds.

### 7.2.3 SIZE DEPENDED LUMINESCENCE OF $SiV^-$ CENTERS

In developing $SiV^-$-based luminescent nanodiamonds, two main questions have to be answered: (i) how thermodynamic stability of the $SiV^-$ formation depends on a size of diamond nanoparticle, and (ii) at which distance of $SiV^-$ defect from the surface of nanodiamond, a defect–surface interaction becomes essential.

Calculations have predicted nitrogen to be thermodynamically unstable in nanodiamonds with less than 2 nm in size.[70] In contrast, silicon-vacancy defects in hydrogen-passivated truncated octahedral nanodiamonds have been theoretically predicted to be stable in particles of ≈2 nm in diameter.[71,72] Later simulations indicated that the $SiV^-$ centers are thermodynamically stable even in nanodiamonds with sizes ranging from 1.1 to 1.8 nm.[73] Smallest ND shown $SiV^-$ luminescence was found under study of ND extracted from a stony meteorite.[73] Some fraction of meteorite ND has a size below 2 nm (Fig. 7.14). Namely, individual particles of this fraction predominantly contained one or even a few $SiV^-$ centers. The number of the centers in a diamond nanoparticle is determined by measuring the autocorrelation function $g^{(2)}$ of the $SiV^-$ luminescence intensity for an individual particle (see Fig. 7.15 and Chapter 17). The $SiV^-$ emission showed blinking but was stable (no bleaching) over a few hours. We relate the blinking of the $SiV^-$ emission to the interaction of the centers with the defects of diamond located closely to the surface.

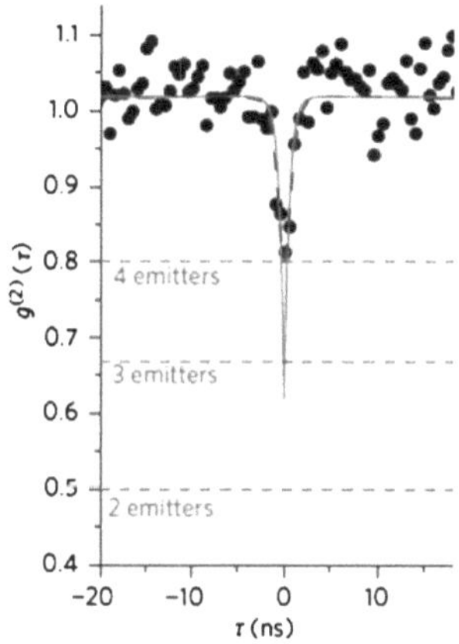


**FIGURE 7.15**

**An autocorrelation function showing the background corrected raw data fitted with a model including the instrument response.** The number of emitters within the analyzed spot was found to be 3. From Vlasov et al., Molecular-Sized Fluorescent Nanodiamonds. Nat. Nanotechnol. 2013, 9, 54–58

In general, a decreasing size of ND containing color centers increases the possibility of their interaction with the diamond surface that usually suppresses the ND luminescence. This happens at the changing charge state of the centers by surface acceptors or/and the emission interruption (the luminescence blinking) until electrons are released back from the surface defects trapping the electrons.[74] A minimum distance of $SiV^-$ and $NV^-$ centers from nanodiamond surface at which these centers cease to "feel" the diamond surface was evaluated by V.A. Shershulin et al.[75] The studied diamond nanoparticles presented a composite structure of "core–outer layer" type. The core was a 20-nm HPHT diamond particle containing $NV^-$ centers, whereas the outer diamond layer containing $SiV^-$ centers was formed by CVD synthesis (Fig. 7.16). The blinking of $NV^-$ emission was found for all samples except for the sample with 44 nm particles (12 nm CVD layer). The duration of the "off" states of $NV^-$ centers was reduced with CVD layer thickness, and the blinking completely disappeared for 44-nm diamond particles. Thus, the 12 nm thickness of CVD layer can be considered as a minimum distance between $NV^-$ centers and ND surface at which their interaction no longer takes place. The 738 nm emission (ZPL) of $SiV^-$ centers was not revealed until the particle size reached 28 nm (4 nm CVD layer). The dramatic increase in $SiV^-$ emission happened when the CVD layer increased by 4 nm, from 8 nm to 12 nm (Fig. 7.17). This effect was interpreted as the termination of $SiV^-$ luminescence suppression by surface defects. Thus, the value of 4 nm may be considered as the largest distance from $SiV^-$ to the diamond surface, beyond which the interaction between $SiV^-$ and diamond surface becomes insignificant. This finding determines a minimum size of nanodiamonds, in which stable luminescent centers could be effectively formed, namely, 8 nm for $SiV^-$ emitters and 24 nm for $NV^-$ emitters.

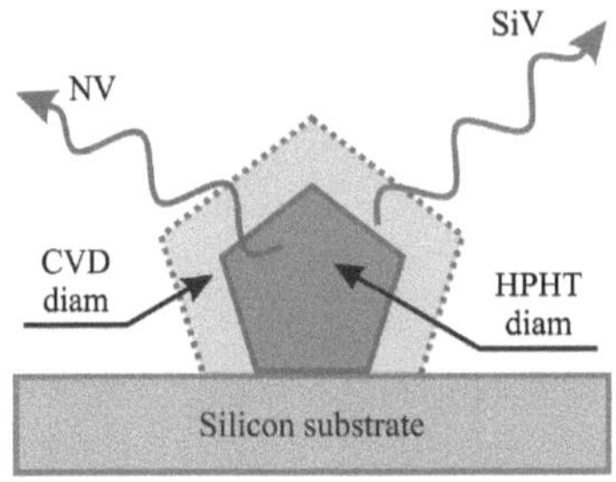


**FIGURE 7.16**

**Schematic representation of the studied diamond nanoparticles presenting composite structure of "core–outer layer" type.** The core is 20 nm HPHT diamond containing NV centers, whereas the outer diamond layer containing $SiV^-$ centers is formed by CVD synthesis. From Shershulin et al., Size-Dependent Luminescence of Color Centers in Composite Nanodiamonds. Phys. Status Solidi 2015, 212, 2600–2605

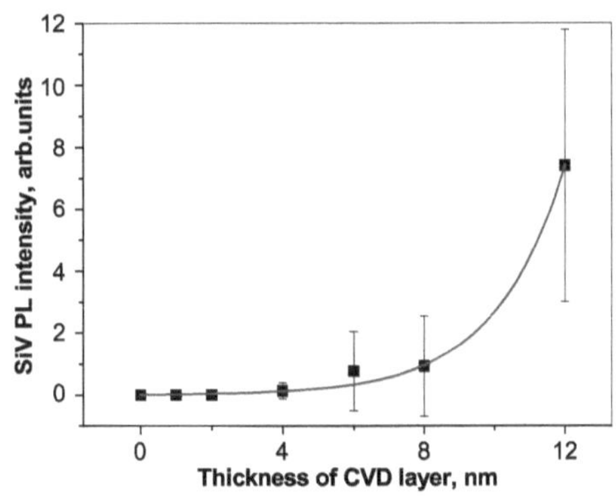


**FIGURE 7.17**

**Dependence of $SiV^-$ luminescence intensity on the thickness of CVD diamond layer.** From Shershulin et al., Size-Dependent Luminescence of Color Centers in Composite Nanodiamonds. Phys. Status Solidi 2015, 212, 2600–2605

## 7.2.4 APPLICATION OF $SIV^-$-CONTAINING NDS AS LUMINESCENT BIOMARKERS

The use of NDs hosting $SiV^-$ defects as effective biomarkers of primary neural precursor cells (NPCs) isolated from the adult mouse brain has been reported by T.D. Merson et al.[76] $SiV^-$ centers were produced in 40 nm high-pressure high-temperature NDs by implantation of Si with a dose of $10^{17}$Si ions/cm$^2$ into the ND powder at room temperature. After implantation, NDs were annealed for 1 h in forming gas to create the defects and remove residual graphite. To examine the spectral

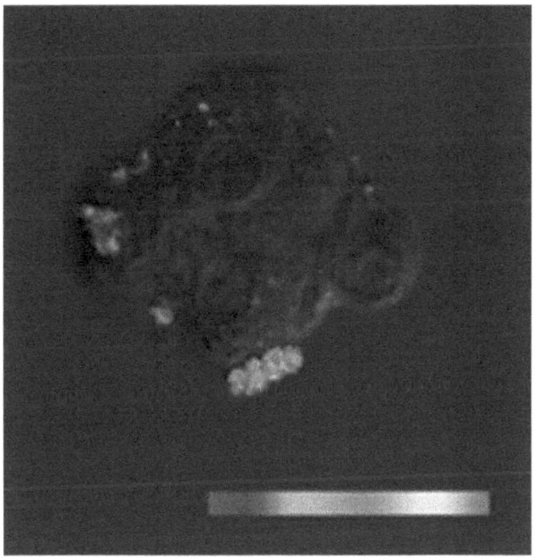

**FIGURE 7.18**

**A confocal image of NPCs grown on NDs merged with a bright field image** of the neurosphere in grayscale collected by Nomarski optics, revealing the exceptional contrast of $SiV^-$-containing NDs over cell background. From Merson, T. D.; Castelletto, S.; Aharonovich, I.; Turbic, A.; Kilpatrick, T. J.; Turnley, A. M. Nanodiamonds with Silicon Vacancy Defects for Nontoxic Photostable Fluorescent Labeling of Neural Precursor Cells. Opt. Lett. 2013, 38, 4170, Reproduced with permission © 2017 Optical Society of America

properties of $SiV^-$ containing NDs for cellular labeling, NPCs were cultured for 4 days as suspension cultures with 40 nm NDs implanted with Si. It was found that NDs containing $SiV^-$ defects labeled cells with great efficiency. Fig. 7.18 shows a confocal image collected using an Olympus FV-1000 microscope of NPCs cultured with 40 nm $SiV^-$-implanted NDs. The pseudo-colored luminescent signal illustrates the range of luminescent intensities detected among labeled cells. Extremely small use of $SiV^-$ -based luminescent nanodiamonds as luminescent biomarker is explained by the lack of methods for mass production of such a nanomaterial.

## 7.3 CONCLUSION

Nanodiamond can have various embedded light emitters, among which the most prominent ones are the NV and SiV color centers presented in this chapter, and that emit in the far red and near infrared spectral regions. Both emitters embedded in nanodiamonds can be observed in photoluminescence or cathodoluminescence and have a very stable emission with a high quantum yield. Moreover, each of them has additional remarkable properties.

The negatively charged $NV^-$ center has a spin triplet ground state that can be optically polarized. Moreover, it has an electron spin resonance that can be detected optically so that $NV^-$'s in nanodiamond have been used as non-invasive (optical) nanoscale magnetic sensors, or sensors of paramagnetic ions with high sensitivity. Finally, $NV^-$ center containing nanodiamond can be produced in mass (grams), which facilitates their application in bioimaging in particular.

However, one drawback of the NV center is its broad emission spectrum of ≈100 nm width, limiting the possibility of multiplexing this fluorophore with only two other colors (green emission like that of fluorescein and blue emission like that of DAPI dye). In comparison, the negatively charged $SiV^-$ center has an emission spectrum mostly contained in the zero-phonon line centered at 738.5 nm and only ≈2 nm broad.

There are various production methods of SiV-containing nanodiamonds either by CVD process of HPHT approaches but none of them has achieved the same production yield than NV-embedded ND, which is why applications of SiV-ND, in particular in bioimaging, have been very limited until now. This may change in a near future because the production of SiV-doped micro- and nanodiamonds by the industrial large scale HPHT process was recently achieved, starting from a mixture of hydrocarbon and fluorocarbon compounds.

In this chapter, we have shown why NV- or SiV-containing nanodiamonds are unique imaging nanolabels and nanosensors of a broad range of applications. Their production is becoming a mature technology. However, there are still some challenges to address. In particular, the smallest fluorescent NDs used in applications up to now have a size of ≈30 nm, but other demanding applications require much smaller size, eventually limited by non-radiative decay towards surface defects which would correspond to a size of a few nanometers. Moreover, a well controlled and versatile bio-functionalization is still lacking. Therefore, research efforts now focus on (i) increasing the production yield of size $<10$ nm fluorescent nanodiamonds and (ii) grafting functional bio-molecule (e.g. antibody) on the nanoparticle, to eventually reduce the gap (in terms of size) between photostable nanodiamonds and bleaching organic fluorophores.

## REFERENCES

1. Dyer, H. B.; Raal, F. A.; Du Preez, L.; Loubser, J. H. N. Optical Absorption Features Associated With Paramagnetic Nitrogen in Diamond. *Philos. Mag.* **1965**, *11*, 763–774.
2. Du Preez, L. *Electron Paramagnetic Resonance and Optical Investigations of Defect Centres in Diamond;* University of the Witwatersrand: South Africa, 1965.
3. Clark, C. D.; Norris, C. A. Photoluminescence Associated With the 1.673, 1.944 and 2.498 eV Centres in Diamond. *J. Phys. C Solid State Phys.* **1971**, *4*, 2223–2229.
4. Davies, G.; Hamer, M. F. Optical Studies of the 1.945 eV Vibronic Band in Diamond. *Proc. R. Soc. A Math. Phys. Eng. Sci.* **1976**, *348*, 285–298.
5. Loubser, J. H. N.; Van Wyk, J. A. Optical Spin-Polarisation in a Triplet State in Irradiated and Annealed Type 1b Diamonds. *Diam. Res.* **1977**, *9*, 11–14.
6. Loubser, J. H. J.; van Wyk, J. A.; van Wyk, J. Electron Spin Resonance in the Study of Diamond. *Rep. Prog. Phys.* **1978**, *41*, 1201.

7. Mita, Y. Change of Absorption Spectra in Type-Ib Diamond With Heavy Neutron Irradiation. *Phys. Rev. B* **1996**, *53*, 11360.
8. Doherty, M. W.; Manson, N. B.; Delaney, P.; Jelezko, F.; Wrachtrup, J.; Hollenberg, L. C. L. The Nitrogen-Vacancy Colour Centre in Diamond. *Phys. Rep.* **2013**, *528*, 1–45.
9. Doherty, M. W.; Manson, N. B.; Delaney, P.; Hollenberg, L. C. L. The Negatively Charged Nitrogen-Vacancy Centre in Diamond: The Electronic Solution. *New J. Phys.* **2011**, *13*, 025019.
10. Coulson, C. A.; Kearsley, M. J. Colour Centres in Irradiated Diamonds. I. *Proc. R. Soc. A Math. Phys. Eng. Sci.* **1957**, *241*, 433–454.
11. Reddy, N. R. S.; Manson, N. B.; Krausz, E. R. Two-Laser Spectral Hole Burning in a Colour Centre in Diamond. *J. Lumin.* **1987**, *38*, 46–47.
12. Batalov, A.; Jacques, V.; Kaiser, F.; Siyushev, P.; Neumann, P.; Rogers, L. J.; McMurtrie, R. L.; Manson, N. B.; Jelezko, F.; Wrachtrup, J. Low Temperature Studies of the Excited-State Structure of Negatively Charged Nitrogen-Vacancy Color Centers in Diamond. *Phys. Rev. Lett.* **2009**, *102*, 1–4.
13. Manson, N. B.; He, X. F.; Fisk, P. T. Raman Heterodyne Detected Electron-Nuclear-Double-Resonance Measurements of the Nitrogen-Vacancy Center in Diamond. *Opt. Lett.* **1990**, *15*, 1094–1096.
14. Rogers, L. J.; McMurtrie, R. L.; Sellars, M. J.; Manson, N. B. Time-Averaging Within the Excited State of the Nitrogen-Vacancy Centre in Diamond. *New J. Phys.* **2009**, *11*, 063007.
15. Rogers, L. J.; Armstrong, S.; Sellars, M. J.; Manson, N. B. Infrared Emission of the NV Centre in Diamond: Zeeman and Uniaxial Stress Studies. *New J. Phys.* **2008**, *10*, 103024.
16. Ma, Y.; Rohlfing, M.; Gali, A. Excited States of the Negatively Charged Nitrogen-Vacancy Color Center in Diamond. *Phys. Rev. B* **2010**, *81*, 1–4.
17. Acosta, V. M.; Jarmola, A.; Bauch, E.; Budker, D. Optical Properties of the Nitrogen-Vacancy Singlet Levels in Diamond. *Phys. Rev. B* **2010**, *82*, 201202.
18. Rogers, L. J.; Doherty, M. W.; Barson, M. S. J.; Onoda, S.; Ohshima, T.; Manson, N. B. Singlet Levels of the $NV^-$ Centre in Diamond. *New J. Phys.* **2015**, *17*, 013048.
19. Felton, S.; Edmonds, A.; Newton, M.; Martineau, P.; Fisher, D.; Twitchen, D. Electron Paramagnetic Resonance Studies of the Neutral Nitrogen Vacancy in Diamond. *Phys. Rev. B* **2008**, *77*, 1–4.
20. Kwiram, A. L. Optical Detection of Paramagnetic Resonance in Phosphorescent Triplet States. *Chem. Phys. Lett.* **1967**, *1*, 272–275.
21. Schmidt, J.; Hesselmann, I. A. M.; De Groot, M. S.; Van der Waals, J. H. Optical Detection of Electron Resonance Transitions in Phosphorescent Quinoxaline. *Chem. Phys. Lett.* **1967**, *1*, 434–436.
22. Wrachtrup, J.; von Borczyskowski, C.; Bernard, J.; Orritt, M.; Brown, R. Optical Detection of Magnetic Resonance in a Single Molecule. *Nature* **1993**, *363*, 244–245.
23. Van Oort, E.; Manson, N. B.; Glasbeek, M. Optically Detected Spin Coherence of the Diamond N-V Centre in Its Triplet Ground State. *J. Phys. C: Solid State Phys.* **1988**, *21*, 4385–4391.
24. Gruber, A.; Dräbenstedt, A.; Tietz, C.; Fleury, L.; Wrachtrup, J.; von Borczyskowski, C. Scanning Confocal Optical Microscopy and Magnetic Resonance on Single Defect Centers. *Science* **1997**, *276* (80), 2012–2014.
25. Maze, J. R.; Gali, A.; Togan, E.; Chu, Y.; Trifonov, A.; Kaxiras, E.; Lukin, M. D. Properties of Nitrogen-Vacancy Centers in Diamond: The Group Theoretic Approach. *New J. Phys.* **2011**, *13*, 025025.

26. Manson, N. B.; Harrison, J.; Sellars, M. Nitrogen-Vacancy Center in Diamond: Model of the Electronic Structure and Associated Dynamics. *Phys. Rev. B* **2006**, *74*, 104303.
27. Harrison, J.; Sellars, M.; Manson, N. Optical Spin Polarisation of the N-V Centre in Diamond. *J. Lumin.* **2004**, *107*, 245–248.
28. Delaney, P.; Greer, J. C.; Larsson, J. A. Spin-Polarization Mechanisms of the Nitrogen-Vacancy Center in Diamond. *Nano Lett.* **2010**, *10*, 610–614.
29. Chang, Y.-R.; Lee, H.-Y.; Chen, K.; Chang, C.-C.; Tsai, D.-S.; Fu, C.-C.; Lim, T.-S.; Tzeng, Y.-K.; Fang, C.-Y.; Han, C.-C.; et al. Mass Production and Dynamic Imaging of Fluorescent Nanodiamonds. *Nat. Nanotechnol.* **2008**, *3*, 284–288.
30. Collins, A. T. The Fermi Level in Diamond. *J. Phys.: Condens. Matter* **2002**, *14*, 3743–3750.
31. Su, L.-J.; Fang, C.-Y.; Chang, Y.-T.; Chen, K.-M.; Yu, Y.-C.; Hsu, J.-H.; Chang, H.-C. Creation of High Density Ensembles of Nitrogen-Vacancy Centers in Nitrogen-Rich Type Ib Nanodiamonds. *Nanotechnology* **2013**, *24*, 315702.
32. Rondin, L.; Dantelle, G.; Slablab, A.; Grosshans, F.; Treussart, F.; Bergonzo, P.; Perruchas, S.; Gacoin, T.; Chaigneau, M.; Chang, H.-C.; et al. Surface-Induced Charge State Conversion of Nitrogen-Vacancy Defects in Nanodiamonds. *Phys. Rev. B* **2010**, *82*, 1.
33. Tisler, J.; Balasubramanian, G.; Naydenov, B.; Kolesov, R.; Grotz, B.; Reuter, R.; Boudou, J.-P.; Curmi, P. A.; Sennour, M.; Thorel, A.; et al. Fluorescence and Spin Properties of Defects in Single Digit Nanodiamonds. *ACS Nano* **2009**, *3*, 1959–1965.
34. Botsoa, J.; Sauvage, T.; Adam, M.-P.; Desgardin, P.; Leoni, E.; Courtois, B.; Treussart, F.; Barthe, M.-F. Optimal Conditions for NV-Center Formation in Type-1b Diamond Studied Using Photoluminescence and Positron Annihilation Spectroscopies. *Phys. Rev. B* **2011**, *84*, 125209.
35. Boudou, J.-P.; Curmi, P. A.; Jelezko, F.; Wrachtrup, J.; Aubert, P.; Sennour, M.; Balasubramanian, G.; Reuter, R.; Thorel, A.; Gaffet, E. High Yield Fabrication of Fluorescent Nanodiamonds. *Nanotechnology* **2009**, *20*, 235602.
36. Dantelle, G.; Slablab, A.; Rondin, L.; Lainé, F.; Carrel, F.; Bergonzo, P.; Perruchas, S.; Gacoin, T.; Treussart, F.; Roch, J.-F. Efficient Production of NV Colour Centres in Nanodiamonds Using High-Energy Electron Irradiation. *J. Lumin.* **2010**, *130*, 1655–1658.
37. Remes, Z.; Micova, J.; Krist, P.; Chvatil, D.; Effenberg, R.; Nesladek, M. N-V-Related Fluorescence of the Monoenergetic High-Energy Electron-Irradiated Diamond Nanoparticles. *Phys. Status Solidi (a)* **2015**, *212*, 2519–2524.
38. Shames, A. I.; Osipov, V. Y.; Boudou, J. P.; Panich, A. M.; von Bardeleben, H. J.; Treussart, F.; Vul', A. Y. Magnetic Resonance Tracking of Fluorescent Nanodiamond Fabrication. *J. Phys. D: Appl. Phys.* **2015**, *48*, 155302.
39. Smith, B. R.; Inglis, D. W.; Sandnes, B.; Rabeau, J. R.; Zvyagin, A. V.; Gruber, D.; Noble, C. J.; Vogel, R.; Osawa, E.; Plakhotnik, T. Five-Nanometer Diamond with Luminescent Nitrogen-Vacancy Defect Centers. *Small* **2009**, *5*, 1649–1653.
40. Schirhagl, R.; Chang, K.; Loretz, M.; Degen, C. L. Nitrogen-Vacancy Centers in Diamond: Nanoscale Sensors for Physics and Biology. *Annu. Rev. Phys. Chem.* **2014**, *65*, 83–105.
41. Hui, Y. Y.; Chang, H.-C. Recent Developments and Applications of Nanodiamonds as Versatile Bioimaging Agents. *J. Chin. Chem. Soc.* **2014**, *61*, 67–76.
42. Rondin, L.; Tetienne, J.-P.; Hingant, T.; Roch, J.-F.; Maletinsky, P.; Jacques, V. Magnetometry With Nitrogen-Vacancy Defects in Diamond. *Rep. Prog. Phys.* **2014**, *77*, 056503.
43. Stursa, J.; Havlik, J.; Petrakova, V.; Gulka, M.; Ralis, J.; Zach, V.; Pulec, Z.; Stepan, V.; Zargaleh, S. A.; Ledvina, M.; et al. Mass Production of Fluorescent Nanodiamonds With a Narrow Emission Intensity Distribution. *Carbon* **2016**, *96*, 812–818.

44. Glenn, D. R.; Zhang, H.; Kasthuri, N.; Schalek, R.; Lo, P. K.; Trifonov, A. S.; Park, H.; Lichtman, J. W.; Walsworth, R. L. Correlative Light and Electron Microscopy Using Cathodoluminescence from Nanoparticles With Distinguishable Colours. *Sci. Rep.* **2012**, *2*, 865.
45. Tizei, L. H. G.; Kociak, M. Spectrally and Spatially Resolved Cathodoluminescence of Nanodiamonds: Local Variations of the NV 0 Emission Properties. *Nanotechnology* **2012**, *23*, 175702.
46. Kimura, E.; Sekiguchi, T.; Oikawa, H.; Niitsuma, J.; Nakayama, Y.; Suzuki, H.; Kimura, M.; Fujii, K.; Ushiki, T. Cathodoluminescence Imaging for Identifying Uptaken Fluorescence Materials in Kupffer Cells Using Scanning Electron Microscopy. *Arch. Histol. Cytol.* **2004**, *67*, 263–270.
47. Beveratos, A.; Brouri, R.; Gacoin, T.; Poizat, J.-P.; Grangier, P. Nonclassical Radiation From Diamond Nanocrystals. *Phys. Rev. A* **2001**, *64*, 061802.
48. Greffet, J.-J.; Hugonin, J.-P.; Besbes, M.; Lai, N. D.; Treussart, F.; Roch, J.-F. Diamond Particles as Nanoantennas for Nitrogen-Vacancy Color Centers. arXiv:1107.0502.
49. Mohtashami, A.; Femius Koenderink, A. Suitability of Nanodiamond Nitrogen-Vacancy Centers for Spontaneous Emission Control Experiments. *New J. Phys.* **2013**, *15*, 043017.
50. Faklaris, O.; Garrot, D.; Joshi, V.; Druon, F.; Boudou, J.; Sauvage, T.; Georges, P.; Curmi, P. A.; Treussart, F. Detection of Single Photoluminescent Diamond Nanoparticles in Cells and Study of the Internalization Pathway. *Small* **2008**, *4*, 2236–2239.
51. Kuo, Y.; Hsu, T.; Wu, Y.; Chang, H.-C. Fluorescent Nanodiamond as a Probe for the Intercellular Transport of Proteins In Vivo. *Biomaterials* **2013**, *34*, 8352–8360.
52. Wu, T.; Tzeng, Y.-K.; Chang, W.-W.; Cheng, C.-A.; Kuo, Y.; Chien, C.; Chang, H.-C.; Yu, J. Tracking the Engraftment and Regenerative Capabilities of Transplanted Lung Stem Cells Using Fluorescent Nanodiamonds. *Nat. Nanotechnol.* **2013**, *8*, 682–689.
53. Faklaris, O.; Garrot, D.; Joshi, V.; Boudou, J.; Sauvage, T.; Curmi, P. A.; Treussart, F. Comparison of the Photoluminescence Properties of Semiconductor Quantum Dots and Non-blinking Diamond Nanoparticles. Observation of the Diffusion of Diamond Nanoparticles in Living Cells. *J. Eur. Opt. Soc. Rapid Publ.* **2009**, *4*, 09035.
54. Wildanger, D.; Patton, B. R.; Schill, H.; Marseglia, L.; Hadden, J. P.; Knauer, S.; Schönle, A.; Rarity, J. G.; O'Brien, J. L.; Hell, S. W.; et al. Solid Immersion Facilitates Fluorescence Microscopy With Nanometer Resolution and Sub-Ångström Emitter Localization. *Adv. Mater.* **2012**, *24*, OP309–OP313.
55. Arroyo-Camejo, S.; Adam, M.-P.; Besbes, M.; Hugonin, J.-P.; Jacques, V.; Greffet, J.-J.; Roch, J.-F.; Hell, S. W.; Treussart, F. Stimulated Emission Depletion Microscopy Resolves Individual Nitrogen Vacancy Centers in Diamond Nanocrystals. *ACS Nano* **2013**, *7*, 10912–10919.
56. Tizei, L. H. G.; Kociak, M. Spatially Resolved Quantum Nano-optics of Single Photons Using an Electron Microscope. *Phys. Rev. Lett.* **2013**, *110*, 153604.
57. Zaitsev, A. M. *Optical Properties of Diamond*. 1st edition; Springer, 2001.
58. Clark, C. D.; Kanda, H.; Kiflawi, I.; Sittas, G. Silicon Defects in Diamond. *Phys. Rev. B* **1995**, *51*, 16681–16688.
59. Goss, J. P.; Jones, R.; Breuer, S. J.; Briddon, P. R.; Öberg, S. The Twelve-Line 1.682 eV Luminescence Center in Diamond and the Vacancy-Silicon Complex. *Phys. Rev. Lett.* **1996**, *77*, 3041–3044.
60. Hepp, C.; Müller, T.; Waselowski, V.; Becker, J. N.; Pingault, B.; Sternschulte, H.; Steinmüller-Nethl, D.; Gali, A.; Maze, J. R.; Atatüre, M.; et al. Electronic Structure of the Silicon Vacancy Color Center in Diamond. *Phys. Rev. Lett.* **2014**, *112*, 036405.

61. Rogers, L. J.; Jahnke, K. D.; Doherty, M. W.; Dietrich, A.; McGuinness, L. P.; Müller, C.; Teraji, T.; Sumiya, H.; Isoya, J.; Manson, N. B.; et al. Electronic Structure of the Negatively Charged Silicon-Vacancy Center in Diamond. *Phys. Rev. B* **2014**, *89*, 235101.
62. Edmonds, A. M.; Newton, M. E.; Martineau, P. M.; Twitchen, D. J.; Williams, S. D. Electron Paramagnetic Resonance Studies of Silicon-Related Defects in Diamond. *Phys. Rev. B* **2008**, *77*, 245205.
63. Varnin, V. P.; Laptev, V. A.; Ralchenko, V. G. The State-of-the-Art in the Growth of Diamond Crystals and Films. *Inorg. Mater.* **2006**, *42*, S1–S18.
64. Zhang, H.; Aharonovich, I.; Glenn, D. R.; Schalek, R.; Magyar, A. P.; Lichtman, J. W.; Hu, E. L.; Walsworth, R. L. Silicon-Vacancy Color Centers in Nanodiamonds: Cathodoluminescence Imaging Markers in the Near Infrared. *Small* **2014**, *10*, 1908–1913.
65. Neu, E.; Arend, C.; Gross, E.; Guldner, F.; Hepp, C.; Steinmetz, D.; Zscherpel, E.; Ghodbane, S.; Sternschulte, H.; Steinmüller-Nethl, D.; et al. Narrowband Fluorescent Nanodiamonds Produced From Chemical Vapor Deposition Films. *Appl. Phys. Lett.* **2011**, *98*, 12–15.
66. Liang, Y.; Ozawa, M.; Krueger, A. A General Procedure to Functionalize Agglomerating Nanoparticles Demonstrated on Nanodiamond. *ACS Nano* **2009**, *3*, 2288–2296.
67. Davydov, V. A.; Rakhmanina, A. V.; Lyapin, S. G.; Ilichev, I. D.; Boldyrev, K. N.; Shiryaev, A. A.; Agafonov, V. N. Production of Nano- and Microdiamonds With Si-V and N-V Luminescent Centers at High Pressures in Systems Based on Mixtures of Hydrocarbon and Fluorocarbon Compounds. *JETP Lett.* **2014**, *99*, 585–589.
68. Shenderova, O. A.; Vlasov, I. I.; Turner, S.; Van Tendeloo, G.; Orlinskii, S. B.; Shiryaev, A. A.; Khomich, A. A.; Sulyanov, S. N.; Jelezko, F.; Wrachtrup, J. Nitrogen Control in Nanodiamond Produced by Detonation Shock-Wave-Assisted Synthesis. *J. Phys. Chem. C* **2011**, *115*, 14014–14024.
69. Dolmatov, V. Y.; Kulakova, I. I.; Myllymäki, V.; Vehanen, A.; Panova, A. N.; Voznyakovskii, A. A. IR Spectra of Detonation Nanodiamonds Modified During the Synthesis. *J. Superhard Mater.* **2014**, *36*, 344–357.
70. Barnard, A. S.; Sternberg, M. Substitutional Nitrogen in Nanodiamond and Bucky-Diamond Particles. *J. Phys. Chem. B* **2005**, *109*, 17107–17112.
71. Vlasov, I. I.; Barnard, A. S.; Ralchenko, V. G.; Lebedev, O. I.; Kanzyuba, M. V.; Saveliev, A. V.; Konov, V. I.; Goovaerts, E. Nanodiamond Photoemitters Based on Strong Narrow-Band Luminescence From Silicon-Vacancy Defects. *Adv. Mater.* **2009**, *21*, 808–812.
72. Barnard, A. S.; Vlasov, I. I.; Ralchenko, V. G. Predicting the Distribution and Stability of Photoactive Defect Centers in Nanodiamond Biomarkers. *J. Mater. Chem.* **2009**, *19*, 360–365.
73. Vlasov, I. I.; Shiryaev, A. A.; Rendler, T.; Steinert, S.; Lee, S.; Antonov, D.; Vörös, M.; Jelezko, F.; Fisenko, A. V.; Semjonova, L. F.; et al. Molecular-Sized Fluorescent Nanodiamonds. *Nat. Nanotechnol.* **2013**, *9*, 54–58.
74. Bradac, C.; Gaebel, T.; Pakes, C. I.; Say, J. M.; Zvyagin, A. V.; Rabeau, J. R. Effect of the Nanodiamond Host on a Nitrogen-Vacancy Color-Centre Emission State. *Small* **2013**, *9*, 132–139.
75. Shershulin, V. A.; Sedov, V. S.; Ermakova, A.; Jantzen, U.; Rogers, L.; Huhlina, A. A.; Teverovskaya, E. G.; Ralchenko, V. G.; Jelezko, F.; Vlasov, I. I. Size-Dependent Luminescence of Color Centers in Composite Nanodiamonds. *Phys. Status Solidi* **2015**, *212*, 2600–2605.

76. Merson, T. D.; Castelletto, S.; Aharonovich, I.; Turbic, A.; Kilpatrick, T. J.; Turnley, A. M. Nanodiamonds With Silicon Vacancy Defects for Nontoxic Photostable Fluorescent Labeling of Neural Precursor Cells. *Opt. Lett.* **2013**, *38*, 4170.
77. Tizei, L. H. G.; Meuret, S.; Nagarajan, S.; Treussart, F.; Fang, C.-Y.; Chang, H.-C.; Kociak, M. Spatially and Spectrally Resolved Cathodoluminescence With Fast Electrons: A Tool for Background Subtraction in Luminescence Intensity Second-Order Correlation Measurements Applied to Subwavelength Inhomogeneous Diamond Nanocrystals. *Phys. Status Solidi* **2013**, *210*, 2060–2065.

## ACKNOWLEDGEMENTS

This work was supported by Russian Science Foundation (grant No. 14-12-01329), by a public grant overseen by the French National Research Agency (ANR) as part of the "Investissements d'Avenir" program (Labex NanoSaclay, reference: ANR-10-LABX-0035), and by ANR grant No 2010-INTB-1002.

CHAPTER

# Current issues and challenges in surface chemistry of nanodiamonds

# 8

**Anke Krueger**

*Institute for Organic Chemistry, Julius-Maximilians-Universität Würzburg, Würzburg, Germany*

## Contents

## 8.1 INTRODUCTION

The surface of a material has a major influence on its actual properties. Carbon, and obviously also diamond, is no exception here. The lattice of diamond needs to be terminated in some way; otherwise, dangling bonds would reach out to the environment. As this is energetically not favorable at all, these sites will be saturated with what is available in the moment of the formation of the diamond. For CVD diamond this is most often hydrogen, and hence CVD diamond films without any further treatment will be hydrogen-terminated in most cases (see Chapter 1). However, other terminations are possible, too. When oxidizing reagents such as concentrated mineral acids or oxygen, ozone, etc., are used, the surface of diamond can be oxygen-terminated (see below for details). However, there are many more options to modify the surface of diamond as it is in the end nothing but an extremely carbon-rich hydrocarbon with complete $sp^3$ hybridization of its atoms. The smallest unit that can be considered the stem system of diamond is adamantane (Fig. 8.1), $C_{10}H_{16}$.

Higher analogues are formed by attaching further $CH_2$ and CH units to form diamantane and the higher diamondoids. Recently, these molecular analogues of diamond have received considerable interest due to their interesting properties.[1] When regarding adamantane and its higher homologues, it becomes clear that not all carbon atoms at the surface are equal. There are those atoms located at the apices of the adamantane cage carrying only one hydrogen and those in the edges carrying two hydrogens (or bonding sites for functional groups). The same is true for nanodiamond. Depending on the orientation of the lattice and the resulting orientation of the surface, the type of surface atoms differs as can be seen in Fig. 8.2. It has been calculated that indeed different reactivities are expected for different surface orientations, shapes, and sizes of nanodiamond. This starts with properties such as electron transfer properties and electrostatic potential and continues to acidity of surface protons, inhomogeneous distribution of surface functional groups, and, e.g., the structure of surface terminations with oxygen (bridged or carbonyl type).[2] In general, it can be said that smaller size and higher curvature of surfaces results in higher reactivity. This can be, e.g., seen in the oxidation of nanodiamond. Typically, the smallest particles disappear first, and larger ones loose sharp tips and edges which are etched away in the oxidation process.

Furthermore, in many cases the diamond nanoparticles are strongly agglomerated (see Chapters 2, 3). When using such materials for chemical transformation it is highly probable that only the outer surface of the agglomerates will be modified with the functional moieties. For many applications this does not cause any problems; it should, however, be considered as a source of inhomogeneity of the products and for unexpected behavior of the material in later process steps (e.g., when agglomerates are redispersed in a later step and thereby exposing unmodified surfaces from the inside of those agglomerates).

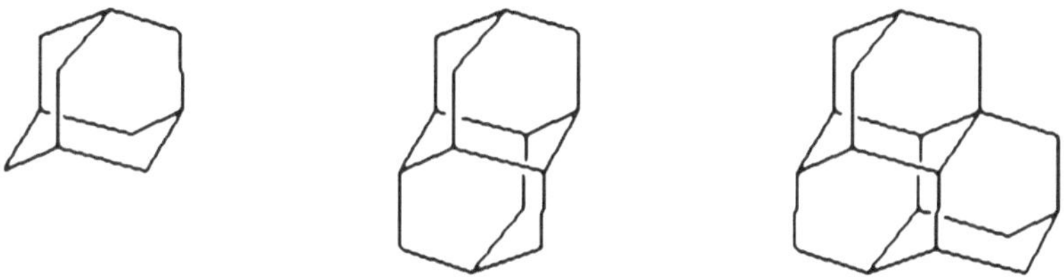

FIGURE 8.1

Adamantane and its higher homologues diamantane and triamantane

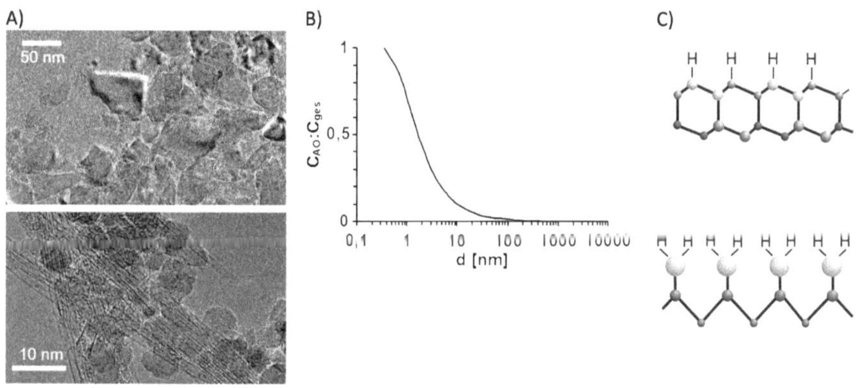


FIGURE 8.2

Size, shape, and surface orientation of the nanodiamond surface have a strong influence on the properties of the material: (A) larger, facetted particles (top) require harsher conditions than more spherical, small detonation nanodiamonds (bottom, in the image adsorbed on single walled carbon nanotubes); (B) the size of the diamond particles determines the number of available surface atoms. In the shown calculation for perfectly octahedral particles, the exponential decrease of surface is obvious; (C) [111] surfaces consist of tertiary carbon atoms (top), [100] surfaces consist of secondary carbon atoms. The reactivity should be different due to the nature of the surface atoms

An additional aspect for the surface modification of all kinds of nanomaterials is the choice between covalent attachment of functional moieties or the non-covalent immobilization by adsorption. Both strategies possess virtues in their own right such as the ease of application and high loadings for non-covalent bonding and well-defined structures and tunable bonding strength for covalent attachment.

We published a detailed account on the surface chemistry of nanodiamond in 2012,[3] and this chapter will mainly focus on recent developments and improvements in the field of surface modifications of nanoscale diamond.

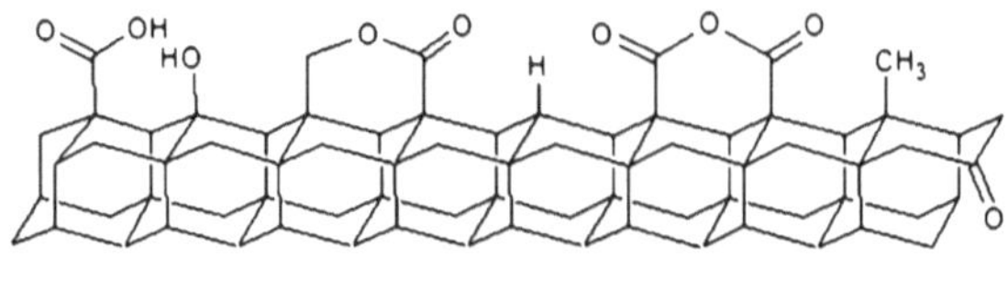


**FIGURE 8.3**

A variety of functional groups is present on nanodiamond surfaces after oxidative cleaning

## 8.2 INITIAL SURFACE TERMINATION

### 8.2.1 GENERAL CONSIDERATIONS

The initial surface termination of a material determines its reactivity in further transformations of the surface and also has a major influence on the properties of the material. Surface hydrophobicity, and hence wettability, strongly depends on the ability of the surface to form non-covalent interactions such as hydrogen bonding or electrostatic interactions. This is clearly seen, e.g., in the hydrophobic character of hydrogen- and fluorine terminated surfaces of diamond, whereas oxygen-terminated diamond shows hydrophilic behavior.[4] Other properties that depend on the surface termination go much deeper: Even the electronic band structure of diamond varies with different surface termination. Hydrogen-terminated diamond's conduction band edge lies above the vacuum level inducing a negative electron affinity, oxygen terminated diamond's conduction band is situated at a significantly lower reductive potential.[5]

How can one influence such a highly important feature of the material as the initial surface termination? For nanoscale diamond, this starts with the method of production (see Chapter 2). In many of the applied methods a highly aggressive cleaning procedure is required after the actual formation of the diamond nanoparticles in order to remove non-diamond carbon and other impurities. Therefore, the most common initial surface termination of nanodiamond is a mixed, oxygenated termination including groups such as carboxylic acids, lactones, anhydrides, different types of OH groups, and even CH bonds (formed during the addition of water to $\pi$-bonds at the diamond surface) (Fig. 8.3).

It is a very challenging task to identify and quantify these groups on the different diamond materials. Identification can be done using advanced IR spectroscopy and, in some cases, NMR measurements. Besides quantitative XPS analysis with suitable peak fitting procedures the wet-chemical quantitative assessment of different acidic surface groups by the Boehm titration is a very useful, yet experimentally rather difficult to handle technique. Reliable results are obtained only with extremely careful titration using fresh reagents and fully degassed solutions.[6]

Such well-characterized materials can be then reliably applied for the production of homogeneously terminated nanodiamond. In the following sections, methods for the production of different surface terminations will be discussed. They then serve as the basis for more complex functionalization strategies and therefore need to be highly reliable.

## 8.2.2 HYDROGEN TERMINATION

For diamond films the most common surface termination is hydrogen, which is a natural consequence from the chemical vapor deposition process that is used for the production of these films. Only the presence of abundant amounts of hydrogen in the feed gas (typically a methane/hydrogen mixture) ensures the formation of diamond not contaminated with large amounts of $sp^2$ carbon.[7] Depending on the orientation of the crystal facet, the surface is either terminated by apices of adamantane units carrying one C–H moiety or by $CH_2$ groups representing the "sidewalls" of the adamantane units (see also Fig. 8.2 for bonding schemes on different crystal facets). The same structures should be present also on hydrogen terminated diamond nanoparticles, typically in mixtures of these types due to the presence of different types of facets on the nanoparticles.

However, for diamond nanoparticles the homogeneous hydrogen termination remains a challenging task. The reason for this are the different production techniques that lead to nanoscale particles made of diamond. Regardless of the used initial fabrication method (detonation, milling of HTHP, or CVD diamond) thorough purification is necessary to remove not only foreign elements (e.g., metal impurities), but also non-diamond carbon. As this is usually done using different types of oxidative protocols (e.g., air oxidation followed by acid washing, oxidative etching using mixtures of mineral acids such as $H_2SO_4$ and $HNO_3$), the surface of the resulting purified particles carries groups containing different amounts of oxygen. The typical termination of surface oxidized diamond consists of carboxyl groups, carbonyl groups, hydroxyl functions, and ether moieties.[8] It is hence necessary to remove these groups and to replace them by hydrogen. Speaking from a chemist's point of view, this means a strong reduction to CH and $CH_2$ moieties starting from C=O groups and tertiary alcohols, not exactly an easy task. Carboxyl groups would incidentally transform into methyl groups if complete reduction can be achieved. For carbonyl groups a variety of reduction methods have been developed by organic chemists such as the Wolff–Kishner, Huang–Minlon, and Clemmenson reduction.[9] However, their value for the surface hydrogenation of oxidized diamond remains unclear (although we saw a reduction in carbonyl groups, we were not able to completely remove the oxygen from the surface, most likely due to the presence of tertiary alcohols that do not undergo reduction under these conditions). Alcohols can be deoxygenated using methods such as the Barton–McCombie or the Markó–Lam reaction,[10] but these methods require inconvenient reagents for the application in nanoparticle suspensions.

Several accounts for thermal hydrogenation of nanodiamond can be found in the literature (Fig. 8.4). Spitsyn et al. reported on the treatment of detonation nanodiamond with hydrogen gas at 1123 K for 1 h and found a significant modification of the surface termination with a high amount of CH and OH formed upon this reductive treatment.[11] This is not unexpected taking the discussion on the nature of reductive

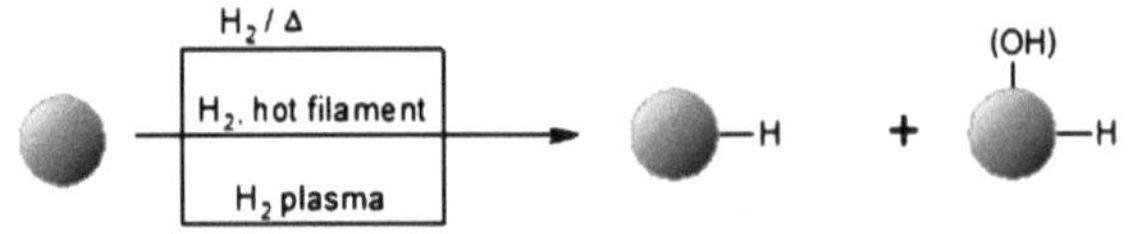


**FIGURE 8.4**

Methods for the hydrogen termination of nanodiamond. The starting material can be not only pristine nanodiamond with mixed O-termination, but also heavily oxidized material

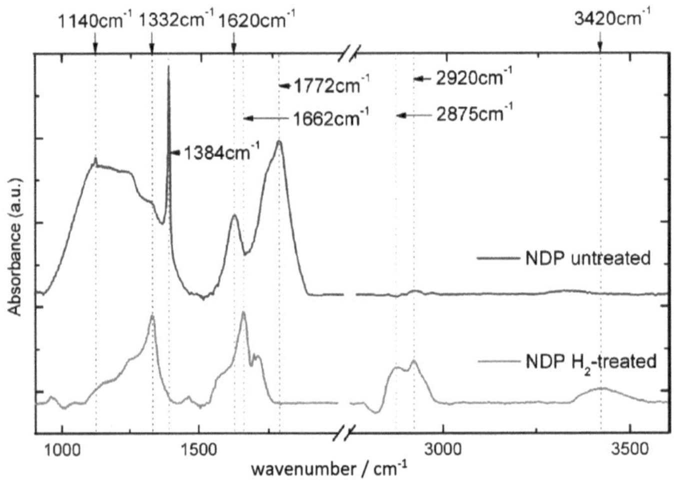


**FIGURE 8.5**

IR spectra of thermally hydrogenated nanodiamond. The band at 1772 $cm^{-1}$ belongs to the stretching mode of C=O groups and disappears after the hydrogenation. Instead, stretching modes of $CH_2$ and $CH_3$ appear at 2875–2920 $cm^{-1}$. Another important change is the appearance of the C–C stretching vibration of the diamond lattice with nitrogen impurities at 1332 $cm^{-1}$ after hydrogen treatment (with permission from ACS; Williams, O. A., et al. ACS Nano 2010, 4, 4824–4830)[12]

transformations into account (see above). Another report showed the possibility to achieve a reaction of hydrogen with detonation diamond at much lower temperatures of about 500°C. In this work hydrogen treatment was continued for 5 h at a pressure of 10 mbar.[12] The authors propose to having achieved a fully hydrogenated surface. The IR spectra show, however, at least signs of oxygen containing groups (Fig. 8.5), and XPS data were unfortunately not available. The observed positive zeta potential could also be due to a mixture of surface CH and OH groups, so this property cannot be used to unambiguously distinguish between OH and H termination. For the moment it remains somewhat unclear if this process leads indeed to a full hydrogen termination or generates a surface with abundant amounts of CH moieties but intermixed with at least a few OH groups. The generation of C–H bonds from car-

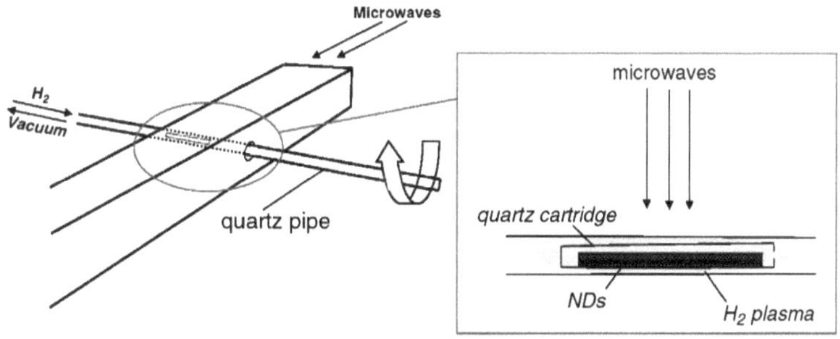


**FIGURE 8.6**

Rotating plasma reactor for the hydrogenation of nanodiamond powders (with permission from Elsevier; Girard, H. A., et al. Diam. Relat. Mater. 2010, 19, 1117–1123)[16]

bonyl and alcohol species could be also catalyzed by the presence of certain transition metal impurities. Many detonation diamond samples contain at least ppm amount of elements such as iron, nickel, cobalt, etc.,[13] which are active hydrogenation catalysts (and would promote the formation of highly reactive chemisorbed H atoms that act then in the hydrogenation). But without question, such thermally hydrogen treated nanodiamond samples are very valuable starting materials for further functionalization as coupling reactions are possible on the generated C–H bonds and the method can be easily used to hydrogenate large amounts of the starting material.

Decomposing the hydrogen molecules by sufficiently high temperature treatment or in a plasma environment is another approach to hydrogenated nanodiamond. Already in 2007 it was reported that hydrogenation using atomic hydrogen generated at a hot tungsten filament leads to the formation of C–H bonds on the nanodiamond surface and efficiently removed (or transformed) $sp^2$ carbon as evidenced by photoelectron spectroscopy (PES).[14] The substrate was heated in situ to ~500 K by the radiation from the hot filament. The actual process for the disappearance of graphitic material could be either the formation of methane or the addition of hydrogen to the $\pi$-bonds on the diamond surface.

Hydrogenation of nanodiamond using a plasma reactor was reported, e.g., by Loh and coworkers.[15] They used an 800 W microwave hydrogen plasma to treat strongly oxidized nanodiamond. In order to complete the transformation and achieve a satisfactory homogeneity of the surface termination, the process was carried out twice. The resulting samples show a strong reduction of carbonyl related groups and an increase of C–H and OH groups. Using a similar process, combined with a rotating quartz reactor, Girard et al. obtained highly hydrogenated nanodiamond that preserved the almost oxygen-free surface when stored under completely inert conditions (Fig. 8.6).[16] Exposure to air and humidity (especially under UV radiation) led to immediate ageing and grafting of OH groups as well as adsorption of water molecules. The same authors reported also on the necessity to maintain sufficiently high substrate temperatures during the hydrogenation process. Only above 700°C a

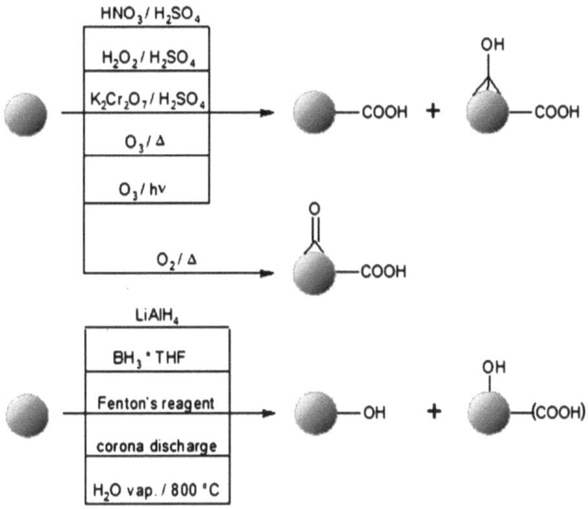


**FIGURE 8.7**

Methods for oxygen termination of nanodiamond. (A) Methods for the generation of carbonyl and carboxyl functions; (B) methods for the grafting of OH groups

complete removal of oxygen containing groups was achieved. Below this temperature the energy is not sufficient to break the C–O bonds.[17] The actual mechanism of this reduction process remains unclear so far. Future work could help to elucidate participating reactive species and to develop less demanding hence still efficient methods for the deoxygenation of O-terminated nanodiamond surfaces.

### 8.2.3 OXYGEN TERMINATION

Oxygen termination of the diamond surface can be done in several different ways including the termination with COOH groups, keto, ether, or hydroxyl functions (Fig. 8.7). The actual structure of the surface not only depends on the orientation of the surface facets, but also on the termination before the oxidative treatment and the respective environment.[18] As has been reported already in the past, wet-chemical treatment of nanodiamond with mineral acids such as mixtures of nitric and sulfuric acid generate a large amount of acidic groups on the diamond surface.[19] Such groups are most likely carboxylic groups; however, certain alcohol groups (tertiary and neopentyl), as well as some carbonyl functions with neighboring hydrogens, can contribute to the overall acidity as well. Typical mixtures for the wet-chemical oxidation of the nanodiamond surface include $HNO_3/H_2SO_4$, $H_2SO_4/H_2O_2$ and mixtures with chromates and dichromates.[20] However, the latter two impose much higher safety and environmental concerns without significant advantages over the conventional acid treatments.

Besides the acid treatment, ozone is another powerful oxidizing agent. First reports on the ozone treatment were published by Shenderova and coworkers.[21] And similar results were reported recently,[22] where ozone treatment under UV irradiation led to a significantly increased content of C=O groups on the surface of the investigated nanodiamond. It remains unclear, however, whether these groups are acidic or not.

Air oxidation is another useful method for the surface oxidation and purification of nanodiamond.[23] It not only removes non-diamond carbon by selective etching, but also terminates the diamond surface with mainly different types of carbonyl and ether functions which are related as they represent two forms of the direct O-termination, with the ether representing the "bridged" bonding scheme, and the C=O groups the "single atom" bonding scheme. Air oxidation needs to be carried out carefully as the diamond nanoparticles are prone to oxidation as well! Their combustion temperature is lower with smaller particle size. Rabeau and coworkers used this to reduce the size of nanodiamond during air oxidation.[24] Typically, the air oxidation is carried out at a temperature around 425°C for detonation diamond and slightly higher temperatures for larger and more faceted crystallites.[25]

Hydroxylation of the diamond surface can be achieved by several methods. First successful attempts include the reduction of surface carbonyl groups by borane or lithium aluminum hydride.[26] Other options are the treatment with Fentons reagent or underwater corona discharge.[27] Both techniques rely on the formation of highly reactive OH radicals that can react with the diamond surface. Such treatment should be most effective either on hydrogenated or thermally annealed (i.e., diamond with a graphene layer on its surface). However, accounts exist only for acid-cleaned starting materials so far. Another reactant for the hydroxylation of the diamond surface should be water. A short report on the high temperature treatment with water vapor at 800°C with no spectroscopic details of the resulting diamond surface was reported back in 2007.[28]

A very detailed study on nanodiamond carrying hydroxyl and hydroxymethyl functions after reductive treatment was published in 2011. Combined analysis using FTIR, NMR, XPS, and zeta potential analysis showed that a large amount of OH groups is generated upon reduction using lithium aluminum hydride. Adsorbed water, however, needs to be removed by vacuum treatment in order to assess the actual surface termination properly (Fig. 8.8).[29] Depending on the initial carbonyl functions, different types and amounts of OH groups are generated on the diamond surface.

### 8.2.4 NITROGEN TERMINATION

Another important surface termination for further grafting of functional moieties is the one with nitrogen and related groups such as NH and $NH_2$. For many nanoparticles, nitrogen termination can be achieved using small organic molecules such as cysteamine, etc., which bind to the surface and exhibit $NH_2$ groups in close proximity.

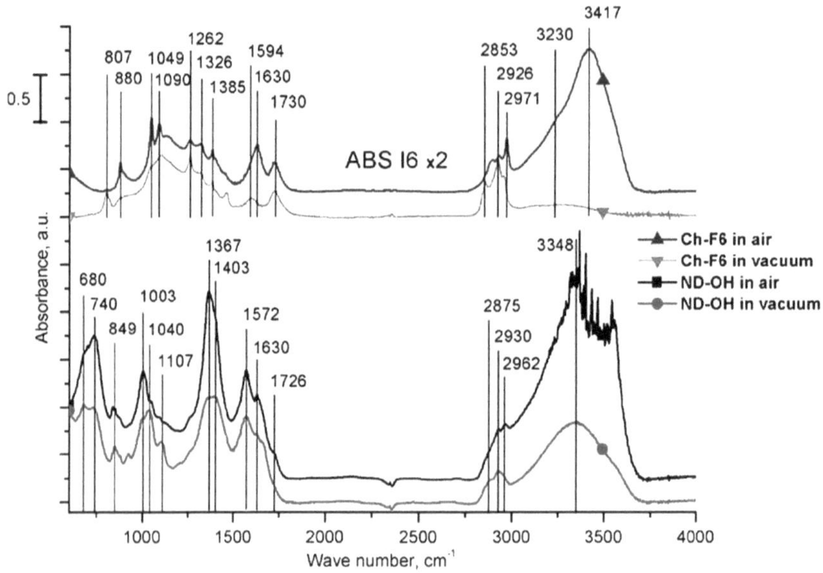


**FIGURE 8.8**

FTIR spectra of pristine (sample name Ch-F6) and OH-terminated nanodiamond before and after the treatment in vacuum. As can be seen from the IR spectra, a strong adsorption of water occurs on nanodiamond terminated with hydrophilic groups such as carboxyl, carbonyl, and hydroxyl groups. Only for the nanodiamond after reduction, the vacuum drying does not remove the OH signals completely, indicating the presence of covalently bound OH groups (with permission from ACS; Shenderova, O., et al. J. Phys. Chem. C 2011, 115 (39), 19005–19011)[29]

For diamond, however, it should be possible to directly bind the nitrogen to the particle surface as the carbon–nitrogen bond is typically rather stable, and many functional groups containing nitrogen are known in the field of organic chemistry.

So far, the ultimate goal – a homogeneously nitrogen-terminated nanodiamond surface – has not been achieved. In the following, the development of methods and workaround options will be discussed.

From a theoretical point of view, direct nitrogen termination of the diamond surface could be done either with pure nitrogen, NH, or $NH_2$ moieties. Nitrile groups bound to the diamond lattice could be considered as well. However, these represent a functional group themselves and the nitrile carbon is not part of the diamond lattice. It would be more appropriate to call this a CN termination.

Fig. 8.9 shows conceivable structures for nitrogen termination. However, the bonding situation depends strongly on the orientation of the lattice facet that is terminated. Barnard and coworkers reported on clear differences regarding the stability of different nitrogen terminations on the [111], [100], and [110] facets of diamond.[30] According to the calculations it should be possible though to achieve selective func-

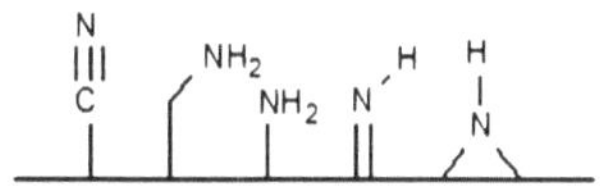


**FIGURE 8.9**

Possible bonding schemes for nitrogen on the diamond surface

tionalization with $NH_2$ groups. Stability of the resulting aminated diamond surface in oxidizing and humid environments is however an unsolved issue. As Popov and colleagues pointed out in 2011 already, ammonia plasma treatment of UNCD leads to the formation of $NH_2$ groups on the diamond surface as indicated by an XPS peak for the N1s at 399.1 eV.[31] However, the aminated UNCD kept under ambient conditions at least partially reacts with water from the surrounding air replacing $NH_2$ against OH (as shown by XPS investigation and contact angle measurements). This is most likely due to a nucleophilic substitution of protonated $NH_2$ (a rather good leaving group) by water with subsequent deprotonation. This reaction should occur obviously also on nanodiamond particles with comparable surface structure and curvature (crystallite size of UNCD is comparable to particle size in typical ND samples). The authors also mentioned a detrimental effect of ammonia plasma treatment on the integrity of the $sp^3$ lattice as indicated by XPS data that suggest an increase of $sp^2$ carbon after the plasma reaction.

An early attempt for the grafting of $NH_2$ groups was reported for microcrystalline diamond that was submitted to photochemical chlorination first, followed by the reaction of ammonia under photochemical conditions (Fig. 8.10).[32] This approach led not only to a surface exhibiting amino groups but also OH functions, most likely introduced by post-treatment substitution. Other attempts to directly terminate nanodiamond with nitrogen were carried out using nitrogen plasma and binary mixtures of $N_2$ and $H_2$ in the plasma.[33] In general, it was found that pure nitrogen had a stronger effect on the surface structure than plasma treatment in the presence of hydrogen. The latter most likely executes a healing effect by saturating surface dangling bonds generated in the plasma bombardment. With substrate temperature below 800°C, the surface was covered with a layer of $C–N_xH_y$, but no clear-cut evidence for $NH_2$ groups could be found. At higher substrate temperatures, the surface carried mainly nitrile groups. These could be formed by annealing of the abovementioned $C–N_xH_y$ groups. True $NH_2$ termination of the diamond surface has not been achieved so far, though.

However, a multitude of alternative methods for the establishment of amino groups on the diamond surface have been developed. Early reports on aminosilanization (see Section 8.3.2.1) were followed by more sophisticated approaches. An obvious approach is the amidation of surface carboxyl groups using diamines. As the carbonyl reactivity of carboxylic acids is rather low, either acid chlorides or active esters need to be formed intermediately. In order to obtain high surface loadings with distant $NH_2$ groups, a prior oxidation step is recommended. However, the direct use

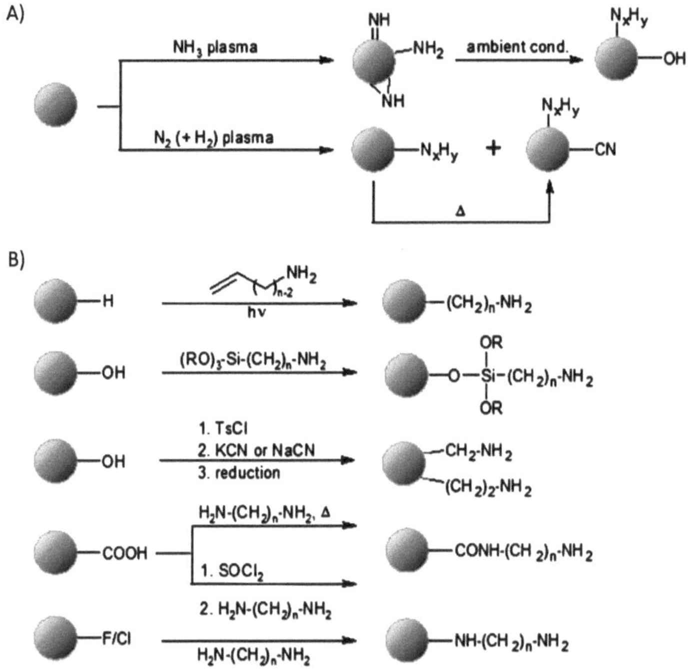


**FIGURE 8.10**

Methods for the attachment of amino groups onto nanodiamond: (A) direct methods; (B) indirect methods for the attachment of amino groups

of the already existing COOH groups can be sufficient, too, as shown in a paper by Basiuk et al.[34] In a solvent-free reaction they baked pristine nanodiamond in the presence of aliphatic and aromatic diamines as well as amine-containing polymers such as $H_2N$–$PEG_{2000}$–$NH_2$ and PEI. Mochalin et al. used thionyl chloride to form the required acid chloride on nanodiamond before reacting it with ethylene diamine.[35] The same scheme was followed by other groups using either ethylene diamine as well or diamines with longer alkyl chains.[36] Another option is the nucleophilic replacement of halogen atoms by amino derivatives. This approach had been reported by Khabashesku in the past[37] and was further used by Chinese researchers who functionalized nanodiamond first with fluorine in a rather mild reaction scheme and subsequently replaced the fluorine atoms by ethylene diamine moieties.[38] Chlorine can be equally replaced by short and long diamines.[39] The introduction of terminal nitrile groups followed by their reduction leads to $CH_2NH_2$ moieties that can be either attached directly to the diamond surface (see below) or via a linker.[40]

A very stable grafting of aminated alkyl chains is possible by using hydrogenated starting material for the photochemical attachment of terminal aminoalkenes. This method has been extensively used for the functionalization of diamond films, but can be used for nanodiamond as well.[41] One example is the modification of hydrogenated detonation diamond with 10-amino dec-1-ene via the grafting of the trifluoroacetic acid protected amino alkene and subsequent deprotection.[42]

The amino groups so far closest to the surface have been established by a classic organochemical approach. Shenderova and coworkers first reduced carbonyl groups on the diamond surface to alcohol functions that were then transformed into tosyl groups – a much better leaving group. In a nucleophilic substitution these tosyl groups were replaced by nitrile groups. Reduction of the latter leads to $CH_2NH_2$ groups that are placed either directly on the diamond lattice (if the alcohol was formed from a carbonyl moiety) or are separated by one more $CH_2$ groups from the actual diamond surface (if the alcohol group was obtained from a COOH group). These primary amino groups can be used for all types of coupling reactions such as amidations as they are not hindered by the diamond and are sterically accessible.[26b]

In summary, the direct amination of nanodiamond has not been achieved so far in a completely homogeneous manner, but a broad choice of indirect methods for the establishment of amino groups near the diamond surface is available.

### 8.2.5 SULFUR TERMINATION

Reports on the functionalization of nanodiamond with thiol groups or any other sulfur containing functional groups are relatively scarce (Fig. 8.11). The direct functionalization of the diamond surface using a photochemical reaction of elemental sulfur was reported for diamond films as well as for particles by Nakamura et al. The resulting surface carries not only single sulfur atoms, but also chains of several sulfur atoms due to the opening of the $S_8$ ring of sulfur.[43]

Oxygenated sulfur functional groups have been introduced by thermolysis of dibenzyldisulfide in the presence of oxidized nanodiamond.[44] The exact structure of the surface termination remains unclear, though.

Sulfur functional groups such as thiols, thioethers, sulfones, etc., can also be immobilized at the end of more complex linkers such as aromatic moieties bound by C–C coupling reactions to the diamond surface (see Section 8.3.3.1 for arylation) or as terminal groups of silanes (see Section 8.3.2.1 for silanization). We have recently succeeded in reacting a benzannulated cyclobutene carrying thioether groups onto the surface of nanodiamond. The thioethers can be further transformed into sulfones by the oxidation with, e.g., *meta*-chloroperbenzoic acid (MCPBA).[40]

An approach using thiourea was published by Hsu et al.[45] They used hydroxylated nanodiamond and reacted it with thiourea under acidic conditions. It is proposed that cationic sites at the diamond surface induced by the protonation of the OH groups and their subsequent cleavage react with the sulfur of the urea derivative and finally yield thiolated nanodiamond surfaces. The resulting product, without doubt, contains sulfur; however, the actual bonding situation is not fully clear. From a mechanistic point of view, the formation of a non-planar tertiary bridgehead cation is not favored. However, the certainly present hydroxymethyl groups could be easily converted into the respective thiols as the method works best for primary and secondary alcohols.

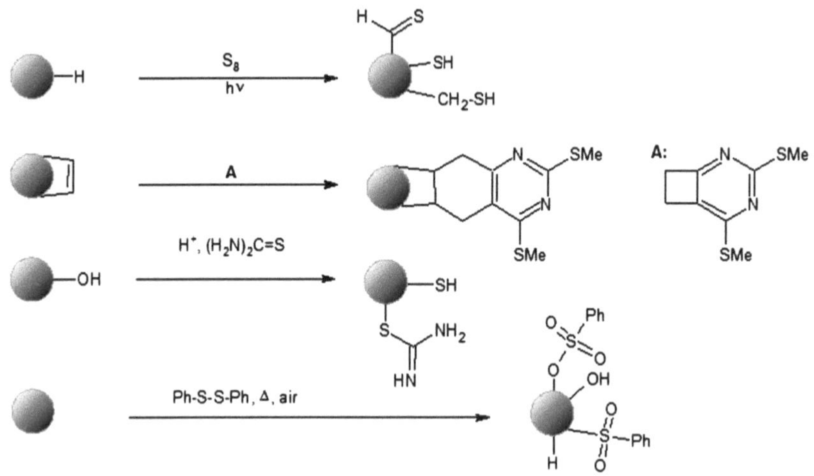


**FIGURE 8.11**

Sulfur functionalization of nanodiamond

Further research will be needed to develop this interesting method into a fully controllable technique.

## 8.2.6 HALOGENATION

Due to the versatile reactivity of halogenated carbon atoms, the halogenation of the diamond surface is an attractive goal. Many reports on different techniques have been published on the termination of diamond surfaces with different halogens. The halogen atom can terminate apical positions or positions in the sidewall of the adamantane units (a $CH_2$ moiety when H-terminated). The latter would, however, require a second halogen or another binding partner (e.g., hydrogen). This is not impossible, but the sensitivity against nucleophilic attack under ambient conditions, i.e., in the presence of water molecules, is further enhanced for $CX_2$ units. Halogenated nanodiamond should be therefore kept under inert conditions, namely when higher halogens are bound to the surface.

Fluorination is the most covered technique by now (Fig. 8.12). This has to do with the improved stability of the resulting fluorinated diamond surface, e.g., against nucleophilic attack by water molecules (F is a rather bad leaving group). The classical approach using elemental fluorine under high temperature was reported by Khabashesku already in 2004.[46] To improve the fluorination efficiency, the process was carried out at 150–500°C in the presence of hydrogen. Most likely, HF is formed in situ and represents the actual fluorination agent. A similar treatment using a fluorine stream was reported by Ivanov et al. for the fabrication of fluorinated nanodiamond for tribological applications.[47] They also used $SF_4$ to obtain surface fluorination under milder conditions. In this case, however, only functional groups will be replaced by fluorine

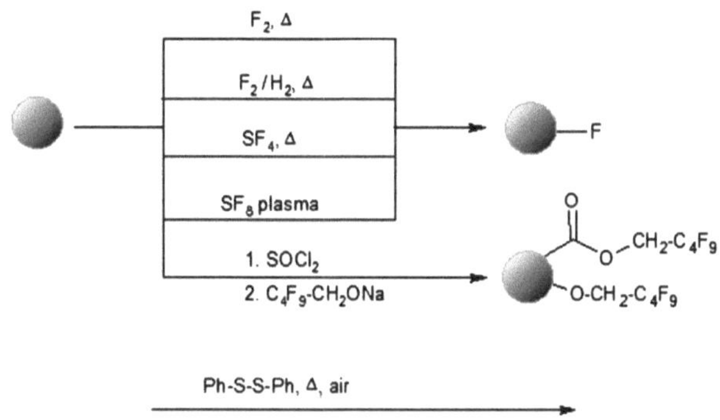


**FIGURE 8.12**

Fluorination methods suitable for nanodiamond

atoms, the substitution of surface bound hydrogen is not possible. Wang et al. used fluorine gas at 350°C and 3.4 kPa for the fluorination of nanodiamond.[38] A multi-step fluorination method resulting in significant reduction of agglomerate size was reported by Huang et al. They used a closed reactor kept at 80 kPa and 200°C for the individual fluorination steps.[48] A detailed study on the nature of C–F bonds after the fluorination of different types of nanodiamond preparation revealed that mainly OH groups are replaced by fluorine and the C–H bonds are much less attacked. Furthermore, $sp^2$ carbon in the sample is etched due to the formation of volatile $C_xF_y$ species.[49] Results were confirmed using NEXAFS, NMR, and FTIR data. $SF_6$ plasma was also used for the fluorination of the diamond surface resulting in 14.5 at.% surface loading.[50] The fluorinated surface showed a rather high stability in air (more than 2 months), but was easily hydrolyzed when treated with 0.05 M NaOH.

Indirect fluorination was reported using perfluoroalkyl radicals. The attachment of perfluorobutyl groups increased the hydrophobicity of the resulting materials.[51] The attachment of these groups is, however, not homogeneous; oxygen bridges are formed with the perfluoroalkyl moieties, which is due to oxygenated initial surface groups such as COOH and C=O.

Chlorination of nanodiamond is a much more challenging task than fluorination. This is due to reduced reactivity of chlorine compared to fluorine, the larger diameter of the chlorine atoms and the increased sensitivity against nucleophilic attack (chlorine is a much better leaving group than fluorine). Only a few reports are available on chlorinated nanodiamond. Submicron diamond was already chlorinated in 2004 by a photochemical process starting with hydrogenated diamond powder.[32] It was placed in a quartz tube filled with $Cl_2$ gas and irradiated at 254 nm. The authors subsequently aminated the diamond surface without exposing the chlorinated diamond powder to air in order to avoid any aging by hydrolysis. Thermal chlorination using $Cl_2$ gas at 500°C was reported by Spitsyn et al.[11] The same authors also mentioned the use of $CCl_4$ at 350°C for the surface chlorination of detonation nanodiamond. However, detailed information on the resulting structures is unfortunately not available. Panich et al. investigated the chlorinated detonation diamond obtained by chlorine treatment in a flow reactor at 800°C using NMR and XPS. They found a coverage of 4.18 at.% with chlorine, indicating only a partial removal of the already existing surface groups. Additionally, they observed the formation of graphitic carbon on the surface of the particles which is most likely due to the high temperature during chlorine treatment.[52] As was shown, detonation diamonds can be graphitized already at temperatures around 750°C.[53,54] The use of $CCl_4$ plasma or the photochemical grafting of Cl to nanodiamond under visible light irradiation was reported by Lisichkin et al.[50] They found that the wet-chemical chlorination resulted in slightly higher loadings (12 vs. 8 at.%). Neither method yielded a fully chlorinated surface, though. Experiments on the hydrolytic stability revealed a higher sensitivity for hydrolysis under basic pH with increased sensitivity of chlorinated ND compared to fluorinated ND in ambient air.

Accounts for bromination of nanodiamond are scarce. The attempts typically lead to a mixed Br and OH termination when starting with oxygen terminated nanodiamond. Surface annealed and H-terminated nanodiamond will also not lead to a fully brominated surface due to sterical hindrance (bromine is significantly larger than chlorine).[55] Compared to chlorinated diamond surfaces, bromine termination is even more sensitive to hydrolysis. Garcia and coworkers reported on the partial bromination of hydroxylated nanodiamond obtained by reaction with Fenton reagent.[56] They used NBS (*N*-bromosuccinimide) in $CCl_4$ as a mild bromination agent that can form $Br_2$ as well as HBr in case proton donors are available.

In general, the availability of well-defined halogenated surfaces decreases from fluorine over chlorine to bromine, while sterical hindrance and sensitivity against hydrolysis increase. Iodination of the nanodiamond surface has not been reported so far.

### 8.2.7 ANNEALING

As discussed in the introduction, the surface of diamond is typically covered with an atomic layer of surface atoms or groups. This is due to the fact that whatever the exposed crystal facet the surface carbon atoms possess one or two bonding sites that would remain unsaturated when no coverage existed. What happens when this surface termination is removed, e.g., by harsh thermal treatment?

On diamond film, the phenomenon called surface reconstruction has been widely studied. Depending on the crystal orientation, different structural features are obtained. However, a certain $\pi$-bond character of the bonding situation on the diamond surface is observed in all these cases. Hence, one can call this process surface graphitization, keeping in mind that the amount of $sp^2$ may vary strongly from a few small patches of non-diamond carbon on the particle surface to extended and even multilayer graphene-like and graphitic structures. An excellent review on all the processes involved in phase transitions of nanodiamond has been given by Kuznetsov et al.[57] In this chapter the focus will be on recent developments in the practical execution of thermal surface annealing as it is necessary for further surface transformations (see below).

Diamond in bulk is already a metastable modification with 0.02 eV more energy per atom than the thermodynamically stable graphite.[57] Only a high activation barrier prevents it from transforming spontaneously into graphite under ambient conditions or even elevated temperatures. However, the picture changes when nanoscale particles are regarded. The smaller the size of the particle, the more the surface energy is increasing, and thus the possibility of phase transformations is enhanced. Nanodiamond can be transformed partially or completely using thermal treatment, high energy irradiation (electrons or other particles), or catalytic processes.[57]

For practical application, the thermal treatment in vacuum plays a major role as it is easily carried out in a tubular vacuum furnace and gives reproducible results for practical amounts of material. The required temperature for the onset of the $sp^2$ carbon formation in the case of detonation diamond was determined by different methods. Petit et al. observed initial formation of small amounts of $sp^2$ carbon on detonation diamond at 750°C by XPS.[53] Above 900°C significant graphitization also concerning the diamond core started, which proceeded from the surface towards the inside of the particle. Similar observations were made probing the existence of $sp^2$ carbon by the occurrence of cycloaddition reactions, i.e., Diels–Alder reactions, on the annealed surface. No reactivity towards dienes was observed for detonation diamond annealed below 750°C.[54] Above this temperature initially a dependence of surface loading on annealing temperature was reported with saturation of the loading around 900°C – indicating that no further accessible $sp^2$ carbon is formed at the surface. Detailed NMR studies revealed that the initial $sp^2$ carbon content of carefully purified detonation diamond is around 6% representing less than 30% of the particle surface. Annealing at 800°C generates an $sp^2$ content of ~23%, already more than an atomic layer around each particle.[58] For very high vacuum, Yeganeh et al. reported the onset of graphitization at temperatures of 1000 K for previously hydrogenated detonation nanodiamond.[14]

Annealing does not necessarily need to be carried out in vacuum. Already in 1999 it was reported that annealing nanodiamond in argon leads to surface graphitization when the treatment temperature rises above 800°C.[59] Later studies corroborated these findings and mentioned that annealing in argon typically leads to the formation of more extended graphitic structures.[60] One of the reasons for this could be an etching–redeposition process that is induced by the presence of reactive gases such as $CO_2$, CO, and $H_2$ (see below).

However, these observations are strongly dependent on the particle size and the initial surface termination of the annealed nanodiamond sample. Larger and more facetted particles show a comparable behavior only at significantly higher annealing temperatures, and hydrogenated samples are more stable than oxygenated nanodiamond materials.[57] The latter react initially with desorption of different gaseous compounds such as $CO_2$, CO, $H_2$, and water vapor. For these samples, defunctionalization onset was reported at temperatures as low as 500 K.[61] Typically, significant defunctionalization is observed above 700 K without the formation of large amounts of graphitic material. The amount of ordered graphitic carbon is higher when the annealing is done in an inert gas instead of vacuum. This has to do with the fact that reactive compounds desorbed from the diamond surface upon defunctionalization are still present. They can initiate an etching–redeposition process.[57] The following reactions may play a role in this phenomenon: (a) $CO_2+C \rightarrow 2CO$; (b) $C+H_2O \rightarrow CO+H_2$; (c) $C + H_2 \rightarrow CH_4$. All these reactions are reversible. The etching occurs preferentially on high-energy structures such as nanodiamond particles and carbon onions, whereas redeposition will occur in the form of thermodynamically favored graphitic carbon. Such behavior has been observed also for the formation of carbon onions at much higher temperatures in argon.[60]

Cebik et al. suggested a graphitization scheme with four stages for detonation nanodiamond in argon atmosphere comprising the following scenarios (Fig. 8.13)[62]: Beginning at 600–700°C, the surface of the smallest particles of the sample starts to transform into $sp^2$ carbon without the formation of any ordered graphitic structures. When the temperature reaches 900°C, all particles are covered with $sp^2$ carbon, but no onion formation is yet observed. Almost complete transformation into non-diamond carbon is observed above 1000°C; however, this carbon is still disordered and not onion-like. The actual formation of carbon onions sets in only at temperatures above 1100°C. Similar results were obtained by Zang et al.[63] These authors investigated the influence of thermal annealing on the conductivity and electrochemical activity of the resulting nanodiamond samples. An earlier study had found that below and above the Debye temperature of diamond at 1910 K the graphitization mechanism differs and that the onset of removal of all functional groups is at 1370 K.[64]

Mona et al. studied the effect of thermal annealing and gas treatment of larger nanodiamond on the photoluminescence of larger nanodiamond (100 and 300 nm).[65] Although only small amounts of $sp^2$ carbon are formed, even after annealing at 900°C the photoluminescence is considerably enhanced.

For annealing in a nitrogen atmosphere, no significant graphitization of the nanodiamond sample was observed until 900°C with only a small increase in $sp^2$ content compared to the starting material, i.e., acid-cleaned detonation nanodiamond. For annealing at 1500°C in $N_2$, thorough graphitization was observed. XPS and Raman spectroscopy were used as complementary methods for this investigation.[66]

In conclusion, the behavior of each individual nanodiamond sample has to be investigated carefully before a procedure is applied to obtain a certain annealing state. Size, morphology, and surface termination of the crystallites are equally important as the chosen annealing atmosphere.

## 8.3 GRAFTING OF MORE COMPLEX MOIETIES

### 8.3.1 GENERAL CONSIDERATIONS

So far, methods for the direct modification of the immediate nanodiamond surface have been reported. However, this limits the surface diversity to those atoms that

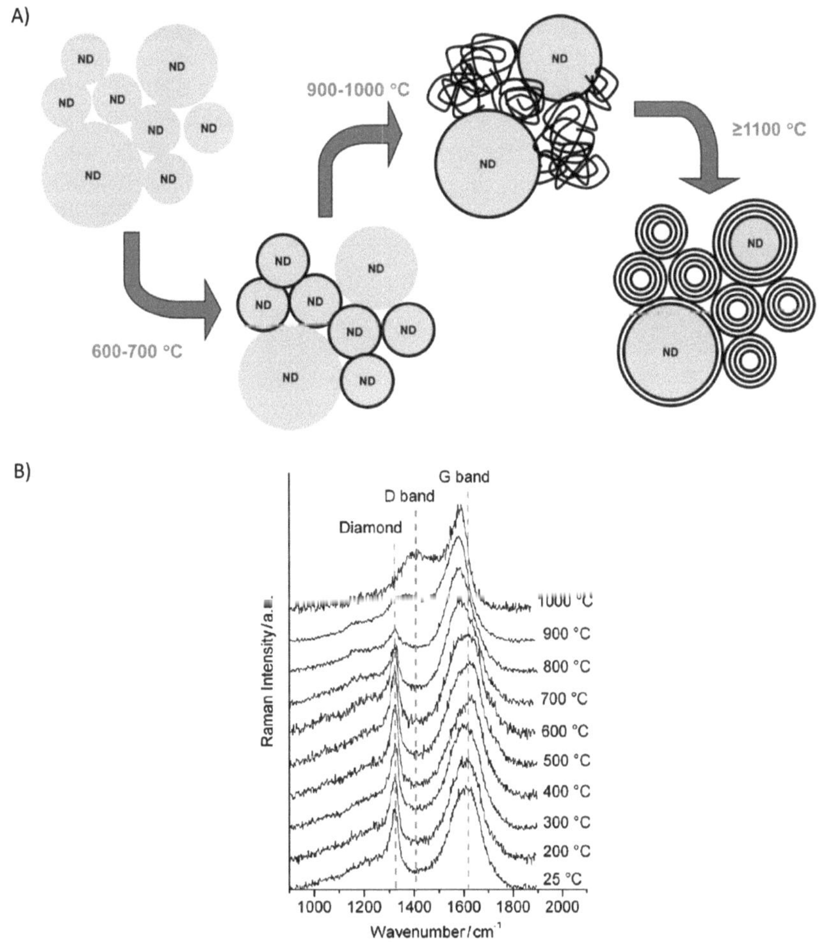


**FIGURE 8.13**

(A) Proposed four stage annealing process for nanodiamond in argon; (B) Raman spectra ($l_{exc.} = 325$ nm) of nanodiamond annealed at different temperatures in argon (© 2013 Tayor&Francis, Fig. 8, Fig. 1a from: Cebik, J., et al. Nanotechnology 2013, 24 (20), 205703[62])

can be directly attached to the surface by forming a covalent bond with the nanodiamond's carbon atoms. It was already discussed that some of the terminating groups such as halogens or amines possess limited stability, namely against hydrolysis, thus limiting their use under ambient conditions or in biological systems. Nevertheless, these moieties, and many more not accessible by direct termination of the nanodiamond surface, are very useful for the grafting, e.g., biomolecules or other organic molecules by amide formation, click chemistry, or substitution reactions.

In order to avoid the problematic properties of materials carrying these groups directly at their surface, or to establish otherwise unavailable functional groups at all, it is recommended to use inert linker structures to extend the distance between the reactive moiety and the diamond surface. Such linkers should be grafted with high stability to the nanoparticle surface and should not be reactive on their own.

Linkers can be used to modify the properties of the original nanoparticle by using either hydrophilic or hydrophobic moieties between the grafting site and the functional group at the terminal position. The linker can be also used to improve, e.g., the biocompatibility of a nanoparticle by masking its surface, an effect that also can help to prevent, e.g., non-specific interactions of this surface with functional molecules that are intended to bind to the reactive group.

When linkers with an elongated (and possibly rigid) structure are used, this can help to enable a full coverage with molecules that otherwise would impede the attachment of additional moieties due to their own size. The leftover of unreacted surface groups can be also minimized this way.

Another important advantage of the use of linkers in between the nanoparticle and the functional moiety is the prevention of agglomeration upon functionalization. Linkers can be chosen for their anti-agglomeration properties, e.g., by using oligoethylene glycol derivatives for the colloidal dispersion in aqueous media or aromatic or long-chain aliphatic hydrocarbons for a hydrophobization of the nanoparticle surface.

In recent years these strategies have been employed on nanodiamond to a large extent to produce a big variety of functional nanodiamond materials. In the following chapters the chemistry necessary for the design of complex nanodiamond conjugates will be discussed. This includes the immobilization of linker systems as well as the actual immobilization of functional molecules on the surface of nanodiamond itself or the pre-functionalized nanodiamond-linker conjugates.

## 8.3.2 LIGATION VIA HETEROATOMS

### 8.3.2.1 *Silanization*

One of the first reports on a more complex functionalization of nanodiamond used the formation of a siloxane shell on the particle surface.[26a] Trialkoxy silanes with different organic residues can be used to prepare the particle for the grafting of biomolecules, dyes, monomers, etc. (see below). There are, however, a few issues with silanization: Firstly, a multilayer shell is formed due to secondary condensation reactions. Secondly, the functional shell is not stable against hydrolysis even under slightly acidic pH or in short alcohols such as methanol and ethanol.[67] Thirdly, the condensation reactions between siloxane groups on different particles result in the formation of covalently linked aggregates that are not easily broken up (Fig. 8.14). The latter can be circumvented by using a mechanochemical approach. The beads assisted disintegration using strong ultrasound (BASD) helps to gradually functionalize the surface of the individual particles and prevent the formation of large, functionalized agglomerates.[68] Rojas et al. used silanized nanodiamond to bind an $^{18}F$ label

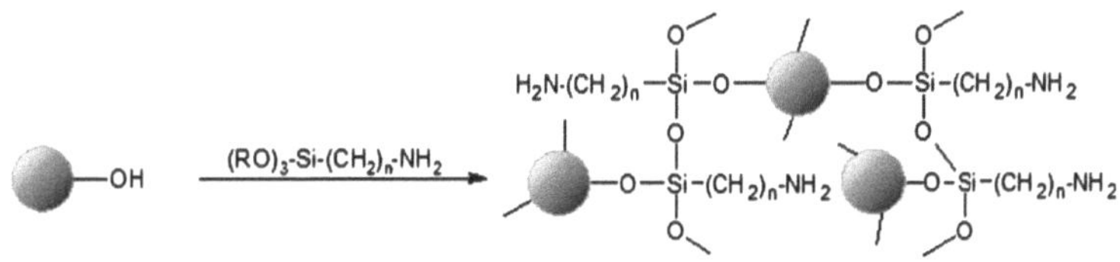


**FIGURE 8.14**

Silanization of nanodiamond can lead to a substantial increase of agglomerate size due to inter-particle condensation reactions which crosslink individual particles

that can be used to study biodistribution.[69] However, the low long-term stability of silanized nanodiamond makes it difficult to guarantee the stable attachment of the label on the nanoparticle. Silanized nanodiamond has been used for a variety of biomedical applications.[70–72] Nesterenko and coworkers have studied the effect of different silylation procedures using ICP-MS.[73] They found a highly enhanced silicon content in samples produced with efficient silylation techniques; however, even the starting materials already contain a certain amount of Si and a lot of other elements, too.

The formation of a dense silica shell starting typically with tetraethoxysilane (TEOS) can be used to stabilize nanodiamond particles and make their outer shape more regular. Either stable colloids of nanodiamond in, e.g., aqueous solutions of polyvinylpyrrolidone (PVP)[74] or the silica shell formed inside a liposome[75] have been used to ensure the formation of core–shell objects with nanodiamond inside. Although long-term stability of these conjugates is not yet known, this silica shell coating is a promising direction for the future.

#### 8.3.2.2 Esterification and etherifications

The formation of ester bonds is a very useful method for the grafting of larger moieties as it is quite flexible. The surface could either be terminated with OH or COOH groups. In the case of OH-terminated surfaces, a carboxylic acid, acid halogenide, or activated ester (or other activated acid derivative) is required. The free acid is typically not used due to reactivity issues (under the acidic conditions used for the esterification, the products of the esterification are not stable in the case of diamond, see below). In the case of COOH as the terminating group on the diamond surface, the reaction with alcohols can be improved by transforming the surface COOH first into an acid chloride or an activated ester. Also coupling agents that form the reactive species can be used. Examples include the esterification using DCC and EDC and EEDQ.[76]

Early attempts for the esterification of hydroxylated nanodiamond used acid chlorides (Fig. 8.15). Although the reaction is in general successful and does not require harsh conditions, the products exhibit limited stability, especially against hydrolysis. This is obviously due to the nature of the resulting esters being often *tert*-butyl or

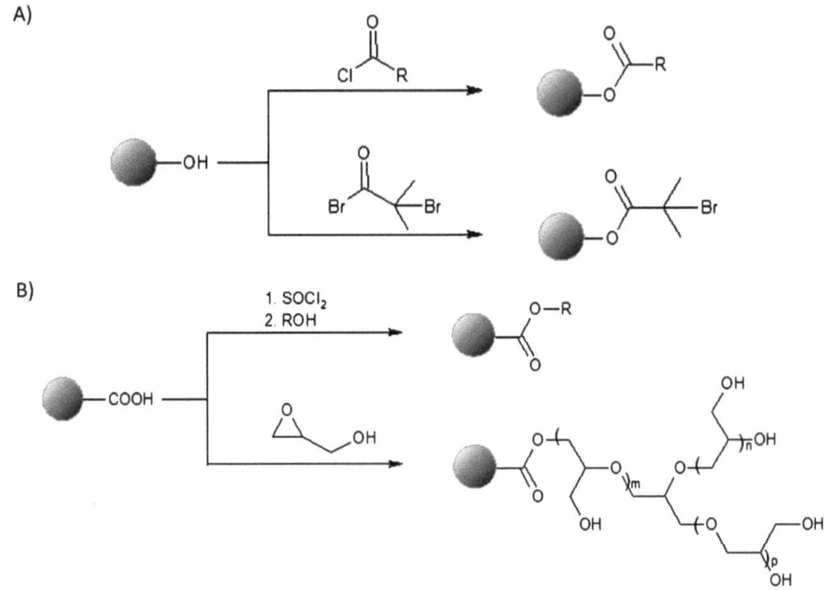


**FIGURE 8.15**

Esterification of (A) hydroxylated and (B) carboxylated nanodiamond

neopentyl esters (stemming from the reaction of tertiary OH or $CH_2OH$ bound to tertiary positions).[77]

The same reaction was used to graft 2-bromo-2-methylpropanoyl bromide onto hydroxylated nanodiamond and initiate the formation poly(ethylene glycol)methyl ether methacrylate (PEGMA) in a surface initiated atom transfer radical polymerization (ATRP).[78] This method allows high loadings as the diffusion towards the reactive sites is not hampered by already grafted polymer chains. Additionally, the monomer units could be further functionalized allowing for an extremely high functionalization degree of the resulting particles.

The reaction of OH groups on nanodiamond with 1-adamantanecarbonyl chloride led to the functionalization of nanodiamond with adamantyl moieties.[79] The interaction of the adamantyl residues with $\beta$-cyclodextrine led to highly water-soluble supramolecular conjugates.

COOH groups have been used for a long time to graft various moieties. Mostly, the formation of amides (see below) was preferred because of their higher stability against cleavage. However, a number of examples for esterification have been reported, too. Zhang et al. reacted oxidized nanodiamond first with thionyl chloride in order to form the reactive acid chloride and in a second step immobilized polyethylene glycol chains in the presence of trimethylamine (to capture the released HCl) by ester formation with the surface.[78] Although the PEGylation of the carboxylated diamond surface led to more hydrophilic and water dispersible materials, the authors

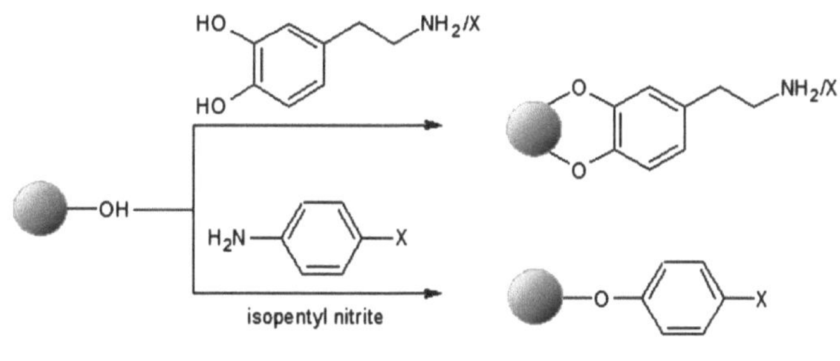


**FIGURE 8.16**

Ether formation on hydroxylated nanodiamond

mentioned that the "grafting from" approach using ATRP led to much higher degrees of functionalization (see above).

Another possibility for improved conjugate solubility was presented by Zhao et al. They used nanodiamond carrying both OH and COOH groups and reacted those with glycidol in a "grafting from" ring-opening polymerization, forming branched polyglycerol on the nanodiamond surface. The branching of the polymer increased the hydrophilicity due to the higher number of terminal OH groups and led to fully water-dispersible nanodiamond conjugates.[80] The high solubility enabled not only chromatographic purification but also the characterization with solution phase NMR methods. A similar approach was reported by Boudou et al. for the solubilization of fluorescent nanodiamond. Such conjugates may be used for bioimaging applications in the future.[81]

Ether formation can be equally achieved using hydroxylated nanodiamond. A recent, very versatile example is the use of dopamine derivatives (Fig. 8.16). The attachment of the catechol moiety proceeds smoothly without the use of additional reagents. Functionalization of the dopamine enables the subsequent grafting of other functional units.[82] When the catechol is functionalized with a triethylene glycol moiety between the aromatic ring and the functional group (here, azides), the colloidal stability of the conjugates is significantly increased without altering the overall conjugate architecture.[83] Ether formation is also observed when hydroxylated nanodiamond is reacted with aryl diazonium salts, typically generated in situ from the respective anilines.[40]

#### *8.3.2.3* ***Amidation***

Carboxylic groups on the diamond surface can be easily reacted with amines to form amides. The versatility of this approach led to numerous reports for the immobilization of functional molecules such as proteins, dyes, polymer chains, etc., on the surface of nanoscale diamond.[3] Here, some recent representative examples will be discussed to show the broad applicability of this approach.

The simplest amide of carboxylated nanodiamond can be formed when ND–COOH is first transformed into the acid chloride, e.g., by reacting it with oxalyl or thionyl chloride. In a second step this water sensitive derivative is reacted with gaseous ammonia that is bubbled into a suspension of ND–C(O)Cl in tetrahydrofurane (THF).[84]

The same strategy can be followed when grafting larger molecules with reactive amino groups or using COOH groups that are bound to the diamond surface via a spacer. However, for the coupling of stereochemically defined compounds, this approach bears the risk of side reactions or altering the stereochemistry and hence hampering the function of, e.g., amino acids and peptides. Therefore, coupling and activating agents are used to avoid the intermediate formation of the highly reactive acid chloride. The direct amidation of ND–COOH using free amines is not possible, as the concurrent acid–base reaction yields the unreactive carboxylate and ammonium salt. Badea and coworkers reported on the covalent attachment of lysine via a diaminopropane linker using a strategy with protective groups for the selective immobilization via the linker. Subsequently, the protective groups were removed and the nanodiamond obtained a polycationic nature.[85]

Recently it was reported that *O*-phosphorylethanolamine (O-PEA) was coupled to ND–COOH using 1-ethyl-3-(3-dimethylaminopropyl)carbodiimide (EDC).[71] The resulting conjugate can be used for improving the adhesion of nanodiamond on titanium alloys for implant materials due to the presence of the terminal phosphate groups.

In a similar manner Sreenivasan et al. coupled barstar (inhibitor of barnase, a bacterial ribonuclease) to nanodiamond. The barstar–barnase couple can be applied to generate extremely stable, non-covalent bonds between the object carrying the barstar moiety (here the nanodiamond) and a barnase functionalized counterpart.[86] Polylysine was grafted in a comparable manner using EDC and NHS (*N*-hydroxysuccinimide), leading to a highly functionalized nanodiamond–biopolymer conjugate. The amino groups of the protein layer were used to immobilize further functional moieties, in this case aryl boronic acids for subsequent Suzuki coupling.[87]

Another way for the specific binding of functional moieties using amidation was shown by Knapinska et al. A collagen-derived peptide was bound to a resin with its *N*-terminus being free. ND–COOH was then coupled to it using 2-(6-chloro-1H-benzotriazole-1-yl)-1,1,3,3-tetramethylaminium hexafluorophosphate (HCTU) for activation.[88] Subsequently, the peptide carrying the nanodiamond at its *N*-terminus can be released from the resin. This method avoids the unspecific grafting of the carboxylated nanodiamond to functional groups in the side chains of the peptide (Fig. 8.17).

A variation that does not use coupling agents was reported by Basiuk et al. They heated O-terminated detonation nanodiamond with different amines in a sealed tube at temperatures between 170 and 220°C. The resulting conjugates had strongly bound amines on their surface. The technique also works for polymeric amines such as poly(ethylene glycol) diamine (PEGDA), and polyethylenimine (PEI), however, the actual bonding situation is not fully known.[34]

As for the esterification or etherification, it should be possible to carry out amidations and amine formations using aminated nanodiamond and carboxylic acids or

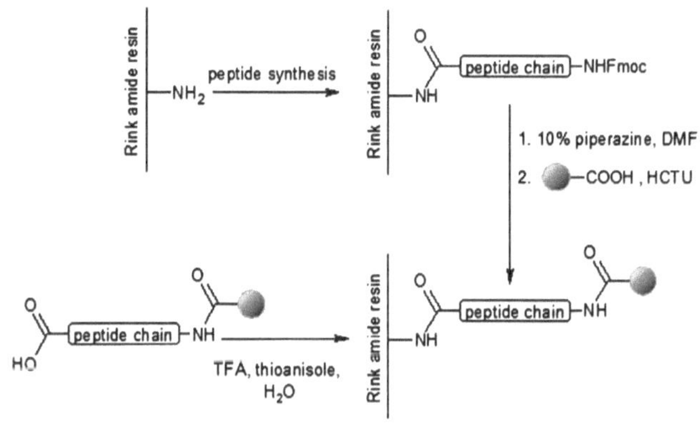


**FIGURE 8.17**

Resin-based strategy for the site-selective attachment of carboxylated nanodiamond onto the N-terminus of a peptide chain using a reaction scheme with protecting groups in the side chains of the growing peptide

alcohols, respectively. However, the availability of aminated nanodiamond is rather limited, and only a few reports have been given on these types of reactions. One option to make use of ND–$NH_2$ was discussed by Mochalin et al. Using ND–$NH_2$ (actually ND–C(O)NH–$(CH_2)_2$–$NH_2$ introduced by reacting ND–COOH with thionyl chloride and then ethylenediamine) an epoxy resin was cured resulting in significantly increased hardness, Young's modulus, and lower creep.[35]

#### *8.3.2.4* ***Phosphorylation***

The grafting of groups containing phosphorous on nanodiamond is relatively new. Only a few reports exist. All of them use oxygen-containing surface groups as anchor points for the phosphate moieties.

Presti et al. reported the reaction of phosphoryl chloride with detonation nanodiamond (Fig. 8.18) and obtained very high surface loadings of up to 1.66 mmol $g^{-1}$.[89] This loading is most likely due to the formation of a multilayer of condensed phosphate moieties (see 8.3.2.1 for similar effects on silanized samples). The phosphate shell enhances the thermal and oxidation stability of the encapsulated nanodiamond significantly. The authors also reported on the $^{31}$P NMR results for the functionalized samples – an analytical tool that is very useful for the characterization of the oxidation state of the phosphorous. Changing the phosphorylation agent to alkyl or aryl phosphonic dichloride gives rise to alkyl/aryl moieties on the surface of the phosphonated nanodiamond.[90] The use of alkyl groups with terminal functionalities allows for further grafting, e.g., of imidazolium groups without important losses in phosphorous content.[91] Phenylphosphonic acid was used to functionalize nanodiamond in a similar manner.[92] The resulting material equally possesses an enhanced thermal stability. Additionally, the authors found the grafting on nanodiamond with a mixed oxidized

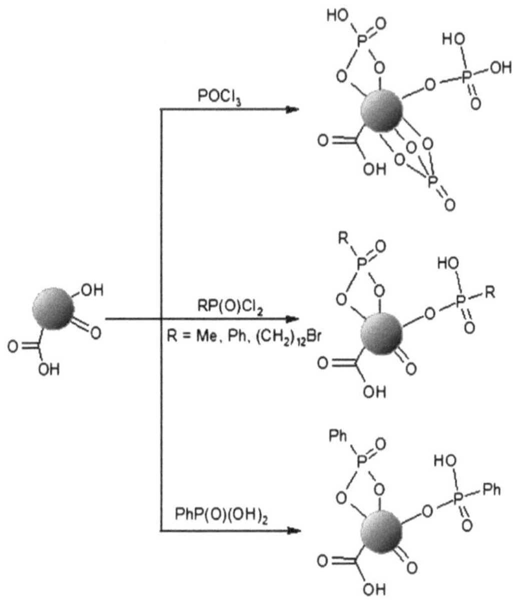


**FIGURE 8.18**

Reaction of nanodiamond surfaces with phosphorylating agents

surface proceeds more efficiently than on carboxylated nanodiamond. Most likely the presence of free OH groups and labile ether moieties enhances the reactivity of the former compared to ND–COOH. Recently, the use of nanodiamond functionalized with methylphosphonic dichloride as a reinforcing additive for polyimide films has been reported.[93] The thermal stability was significantly enhanced in composite films with only 0.5 wt.% of phosphonated nanodiamond; however, the optical transparency was dramatically reduced.

## 8.3.3 LIGATION VIA CARBON–CARBON BONDS

### *8.3.3.1 Functionalization using aryl diazonium salts*

The versatility of aryl diazonium salts has been known for a long time in organic synthesis. They can be transformed into a variety of derivatives such as aryl halogenides (including fluorides), phenols, aryl nitriles, etc. It is also possible to remove the diazonium moiety reductively to yield the native C–H bond instead, a strategy that is used for the site selective introduction of other functional groups (and then remove the directing group in a subsequent step). The reactivity of the aryl diazonium cation is due to the easy removal of molecular nitrogen, which leaves a highly reactive phenyl cation or radical behind (depending on the actual reaction conditions, see below). The cation (or radical) cannot be stabilized by conjugation with the aromatic $\pi$-system as

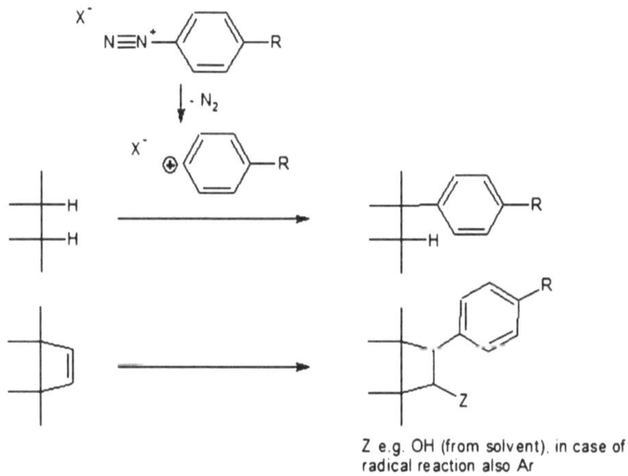


**FIGURE 8.19**

Arylation of nanodiamond with different surface terminations using diazonium salts

the empty (half filled) $sp^2$ orbital stands perpendicular to the aromatic system. The charge (the unpaired electron) is therefore highly localized at the respective carbon atom and hence the phenyl cation (radical) is highly reactive.

In the past it was reported that diazonium salts can be used to functionalize carbon nanotubes. In most cases, these reactions were carried out electrochemically. Using this approach, the reactive species is not a phenyl cation but the respective radical as one electron per unit is transferred to the aromatic species.[94] In a similar way boron-doped or ultrananocrystalline, and therefore conductive, diamond films can be functionalized.[95] However, this is difficult when undoped material and nanoparticles are to be functionalized. Here, the wet-chemical method without an electric bias has to be applied (Fig. 8.19). The actual mechanism is not completely understood to date, both cationic and radical species might be involved.[3,96] One issue with this type of reaction is the occurrence of azo coupling side reactions when high concentrations and acidic conditions are used. Typically, this can be monitored using absorption spectroscopy as the resulting azo compounds are usually strongly colored. Under suitable reaction conditions this concurrent side reaction can be largely avoided.

One of the first reports on arylation of hydrogenated nanodiamond was by Yeap et al., where 4-bromo and 4-nitro benzenediazonium tetrafluoroborates were used to modify the surface of diamond nanoparticles previously hydrogenated in a thermal process.[15] Girard et al. reported on the same procedure for hydrogenated detonation diamond.[16]

Surface annealed nanodiamond exhibits an atomic structure that is similar to the situation on nanotubes and fullerenes. It should be hence possible to not only modify hydrogenated nanodiamond but also graphitized and surface annealed nanodiamond using diazonium salts. It was reported that diazonium salts generated in situ from the respective aniline derivatives readily undergo reactions with annealed nanodiamond, leading to nanodiamond derivatives that can be dispersed in different media depending on the functional groups bound to the aryl moiety.[97,98] It is also possible to use isolated diazonium salts (typically, the tetrafluoroborates are used) to bind aryl moieties onto nanodiamond.

When the nanodiamond is neither hydrogenated nor surface annealed, the bonding between the aryl moiety and the diamond surface typically occurs via heteroatom bridges. Mangeney et al. reported on the ultrasonic grafting of isolated 4-nitrobenzenediazonium tetrafluoroborate onto O-terminated HTHP nanodiamond.[99] Hydroxylated nanodiamond could be functionalized with 4-aminobenzyl(*N*-Boc)amine using isopentyl nitrite for the in situ generation of the diazonium salt.[40] Loh and coworkers found that carboxylated nanodiamond reacts with aryl diazonium salts when Mohr's salt (ammonium iron(II) sulfate hexahydrate) is added. It acts as an electron donor and helps to generate the phenyl radicals.[15] The reactivity can be also enhanced by a mechanochemical process. Simultaneous ultrasound wet-milling and reaction with a suitable diazonium salt results in an efficient surface functionalization and, at the same time, very good colloidal dispersion of the resulting particles.[68] Besides the generation of phenyl species due to the strong sonication, the surface of the diamond particles is altered as well and either freshly generated surface dangling bonds or $\pi$-bonds formed due to the milling play a role in this process.

#### 8.3.3.2 Formation of cyclic structures on the diamond surface

The concept of carbon–carbon bond formation can be extended to systems where more than one bond is formed upon grafting of an organic moiety. Such reactions are typically not only entropically favored, but also enhance the stability of the resulting conjugate – provided the cyclic structure is not too strained. In the past several ring-forming reactions have been carried out on different carbon nanomaterials such as the Diels–Alder reaction, the Prato reaction, or the grafting of bromomalonates in the Bingel–Hirsch reaction. However, there are only a few reports on these versatile reactions for the functionalization of nanodiamond. One of the reasons is the fact that $sp^2$ carbon is required for these reactions to work properly. As described in Sect. 8.2.7, the controlled formation of a defined amount of graphene-like carbon on the diamond surface is a challenging task. Several layers of carbon are in most cases detrimental for the envisaged applications of the nanodiamond conjugate; hence a single layer or less of $sp^2$ carbon is required.

Already in 2007 Qi et al. reported on the reaction of butadiene with reconstructed diamond film surfaces and the influence of this surface modification on the electron affinity.[100] In 2011 the Diels–Alder reaction on annealed nanodiamond was reported.[54] The surface could be functionalized using *ortho*-quinodimethanes as the diene, as soon as the nanodiamond had been annealed at least at 750°C (Fig. 8.20). This temperature corresponds well to the conditions reported for initial $sp^2$ carbon formation from XPS analysis. Conventional dienes were, however, not suitable for the grafting. Dienes such as cyclopentadiene and anthracene gave poor results most likely due to strain-induced retro-Diels–Alder reaction. The use of *ortho*-quinodimethanes

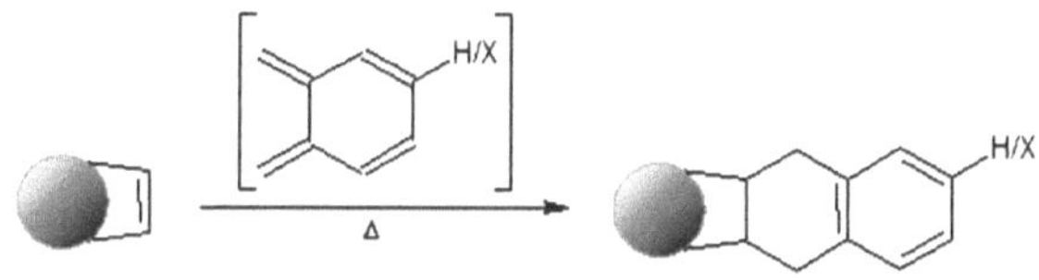


**FIGURE 8.20**

The Diels–Alder reaction on nanodiamond

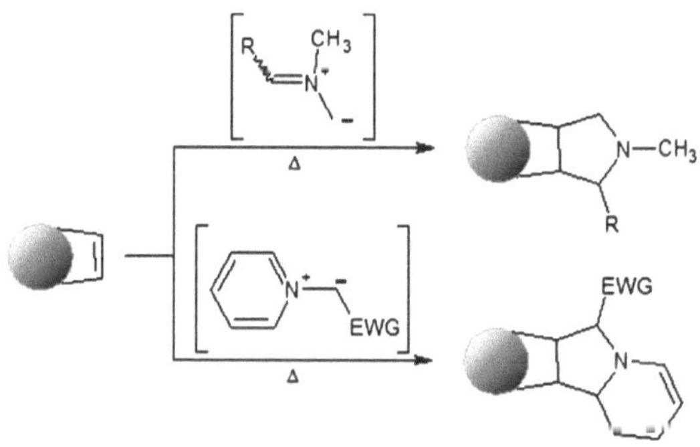


**FIGURE 8.21**

The Prato reaction on nanodiamond using conventional azomethine ylides (top) and pyridinium ylides (bottom); EWG, electron withdrawing group, e.g., CN, COOH

was introduced by Müllen and coworkers for the functionalization of the even more curved surface of $C_{60}$.[101] The formation of the aromatic ring upon grafting prevents the backward reaction. With an analogous approach, heterocycles can be grafted onto the nanodiamond surface. 2,3-bis(bromomethyl)-5,6-dicyanopyrazine was used to form the required quinodimethane in situ and to establish pyrazine molecules with two nitrile groups on the nanodiamond surface.[40] These nitrile groups can be used as versatile moieties for further functionalization, e.g., by reducing them to highly reactive primary amino groups. Another approach to *ortho*-quinodimethanes is the ring-opening of benzocyclobutenes.[40] As it does not require further reagents for the formation of the diene, this method is free of side products and hence very promising in those cases where the benzocyclobutene is easily available.

Besides the formation of six-membered rings, the formation of pyrrolidine rings can be achieved on annealed nanodiamond using the Prato reaction.[102] The required azomethine ylides can be easily obtained from the reaction of sarcosine with different aldehydes or from pyridine derivatives (Fig. 8.21). It was, however, shown that residual carbonyl species on the surface of partially annealed nanodiamond can react with the azomethine ylides as well. Such side reaction forms different heterocycles on the diamond surface. Depending on the envisaged application, the removal of these groups by hydrogenation or reduction prior to annealing should be considered.

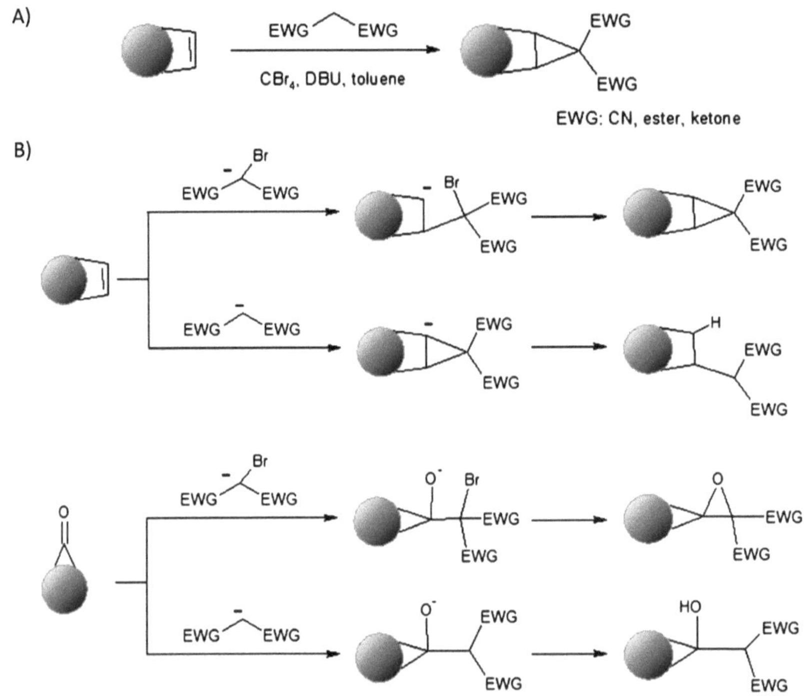


**FIGURE 8.22**

Bingel–Hirsch reaction on nanodiamond (A), comparison of reactions of bromomalonates with $\pi$-bonds and carbonyl groups on the nanodiamond surface (B). The epoxide resulting from the reaction of a bromomalonate with a surface carbonyl group will be opened upon aqueous workup to the corresponding diol

For fullerenes the cyclopropanation using bromomalonates as reagents under basic conditions is a well-known procedure.[103] As the $\pi$-bonds on annealed nanodiamond possess a bent structure and hence an enhanced reactivity, too, it should be possible to graft bromomalonates in a similar way as on fullerenes. And indeed, the Bingel–Hirsch reaction is a versatile tool for the functionalization of nanodiamond with a large variety of groups using the respective malonates or bromomalonates (Fig. 8.22A).[104] The reaction has been studied in detail, and it turned out that the reactive carbanionic species can also react with carbonyl functions on the diamond surface. Their nucleophilicity allows even the functionalization of surface oxidized nanodiamond without significant amounts of graphitic carbon. As shown in Fig. 8.22B, the resulting structures contain additional hydroxy functions stemming from the carbonyl oxygen.

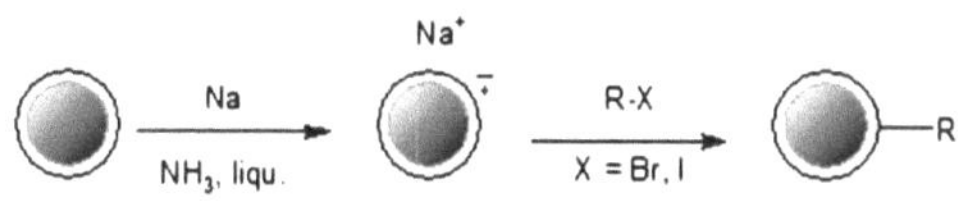


FIGURE 8.23

Reaction of annealed nanodiamond with solvated electrons and subsequent alkylation

#### 8.3.3.3 Miscellaneous C–C bond formations

Besides the already discussed reactions for C–C bond formation, several other approaches have been reported.

Billups and coworkers treated annealed detonation nanodiamond with sodium in liquid ammonia to form a kind of "nanodiamond salt" which then readily reacts with alkyl halides to form alkylated nanodiamond. When these alkyl chains carry a terminal carboxyl group the nanodiamond becomes water soluble in the form of small aggregates (Fig. 8.23).[105] Similar treatment was already reported in 2007 by Korolkov et al., when H-terminated nanodiamond was first treated with butyl lithium before being alkylated with alkyl halides.[106] Another very versatile method for the grafting of more complex moieties onto the surface of annealed nanodiamond is the use of maleimide.[107] Maleimides can be easily grafted even onto biomolecules, and the double bond of the maleimide enables addition reactions with the diamond surface. It remains, however, unclear what actual bonding situation results from this approach. A [2 + 2] thermal cycoaddition seems rather unlikely – a stepwise mechanism needs to be considered. Also the involvement of heteroatoms cannot be excluded.

The addition of carbenes also leads to the direct functionalization of diamond surfaces. For films this has been reported in the past.[108] Nanodiamond particles have, however, not been functionalized using this approach.

Radicals generated from polyoxyethylene macro azo-initiator (Azo-POE) were used to form bonds directly with the nanodiamond surface.[109] The radical is trapped by the diamond surface. However, the exact nature of the treated nanodiamond was not given. Therefore the grafting could occur either by C–C bond formation (if C–H or $sp^2$ carbon is involved) or grafting onto heteroatoms such as oxygen when other surface groups are present. Besides the thermal generation of radicals that react with the nanodiamond surface the photochemical production of such radicals has been reported. Ahmad et al. used UV light to generate propan-2-ol-2-yl radicals from 2-hydroxy-2-methyl-1-phenyl-1-propanone which were then grafted onto the surface of dispersed submicron diamond (d = 500 nm) in acetone.[110] The dispersibility of such hydroxyalkylated submicron diamond particles in organic solvents improved. Again, the nature of the surface of the initial diamond material is unknown and so is the actual bonding situation after the grafting step.

### 8.3.4 SECONDARY GRAFTING ON PRE-FUNCTIONALIZED NANODIAMOND

#### 8.3.4.1 General considerations

Typically, the immobilization of larger functional moieties is not directly done on a specifically terminated nanodiamond surface (Fig. 8.24). This has several reasons: first, it is often rather difficult to selectively functionalize a certain position in large (bio)molecules with a suitable anchor group that can directly undergo reactions with surface atoms on the diamond surface. Second, already immobilized moieties may cover further anchor points on the surface by undergoing non-covalent interactions with the diamond (e.g., hydrogen bond formation with OH or COOH groups). Third, highly efficient coupling reactions like the class of "click reactions" require groups such as azides, alkynes, or thiols that cannot be easily introduced directly onto the diamond surface.

Hence, the application of pre-functionalized nanodiamond has become the standard for the immobilization of larger functional moieties. In doing so, the linker unit can additionally provide some beneficial functions, such as modulating hydrophilicity/hydrophobicity and dispersibility, acting as a stiff spacer to prevent unspecific non-covalent interactions, or presenting the anchor groups in larger distances from each other and hence enabling higher loadings. With different terminal groups on the linkers available after the prefunctionalization step an orthogonal grafting is possible, too. The latter had been demonstrated using nanodiamond functionalized by arylation using diazonium salt coupling. The prefunctionalized nanodiamond carried not only benzoeic acid moieties that were available for amide formation but also alkyne groups that could be used for triazole formation with an azide.[98] Azide–alkyne coupling was also used for the immobilization of highly reactive TEMPO (2,2,6,6-tetramethylpiperidinyl-*N*-oxyl) onto nanodiamond by clicking an alkyne derivative of the radical onto nanodiamond after silanization of ND-OH with 3-bromopropyltrichlorosilane and substitution of Br against azide.[111]

Biomolecules are often available with azide tags, too (see below). Other easily accessible derivatives include activated esters and thiols. Such moieties enable selective grafting onto pre-functionalized nanodiamond and do not pose problems with residual side products or catalysts in the product. Such contaminations represent a major concern when investigating, e.g., the biological activity of nanodiamond derivatives. Observations may originate from contaminants and not the diamond itself.

The following chapters will show a selection of representative modifications of nanodiamond. Many more examples are available in the literature but would go beyond the limit of this account. They use very similar grafting techniques to those shown here and have expanded the world of nanodiamond derivatives in their own right.

#### 8.3.4.2 Grafting of peptides and proteins

Peptides are one of the most relevant classes of biomolecules and have been immobilized on various nanostructures and surfaces in the past. Such conjugates can be used for sensing and therapeutic applications both in vitro and in vivo. It is therefore very

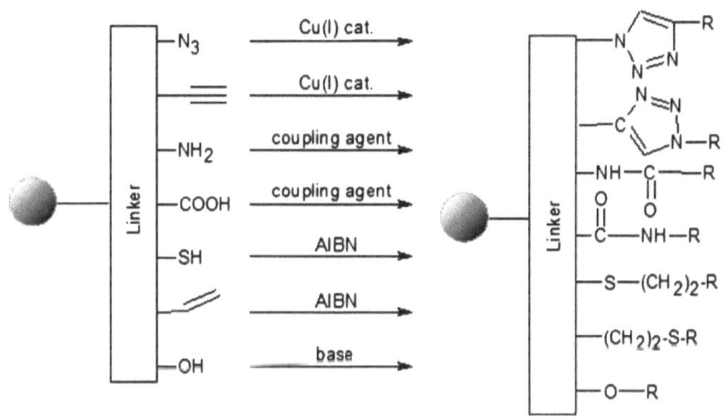


**FIGURE 8.24**

Methods for subsequent grafting of more complex moieties onto prefunctionalized nanodiamond

attractive to combine the unique properties of nanodiamond (such as biocompatibility, small size, inherent fluorescence, etc., see Chapters 2, 7, 12 and 16 for details), with proteins and peptides.

Although it is in principle possible to build a peptide starting with a covalently linked amino acid on the nanodiamond surface,[26a] this approach is not viable for larger peptide sequences. Especially the native folding of the peptide chain will most likely not be achieved using such a "grafting from" approach, not to speak of the tedious synthetic procedures when the peptide sequence exceeds the length of a few tens of amino acids. Therefore, entire peptide structures need to be immobilized in one step. Obviously, this raises the question of site selective immobilization.

In noncovalent immobilization strategies this is one of the major drawbacks. On the other hand, with the individual interactions being not too strong, an equilibrium between connected and unconnected bonding sites may occur. Nonetheless, the noncovalent immobilization of peptides and proteins onto nanodiamond is very popular due to easy synthetic procedures and the still high functionality of the resulting conjugates in many cases.

For fluorescent nanodiamond, the adsorption of bovine serum albumin and $\alpha$-lactalbumin has been achieved. These diamond nanoparticles possess a carboxylated surface with negative zetapotential and can easily undergo mainly electrostatic interactions and hydrogen bonding with proteins.[112] The resulting conjugates were stable in colloidal solution and small enough to use them in superresolution imaging in cells by STED microscopy.

Cheng and coworkers reported on the adsorption of lysozyme on nanodiamond of different type and origin.[113] The function of the protein was best preserved when 100 nm HTHP diamond with a carboxylated surface was used. Other nanodiamonds (5 nm detonation type and 50 nm HTHP) showed larger detrimental effects on the lysozyme activity. The authors attributed this to a possible modification of the 3D structure of the protein due to the more pronounced non-covalent interactions on the smaller nanodiamonds that in addition contain larger fractions of non-diamond carbon in their surface. Lysozyme was also shown to be stably adsorbed on "modified" detonation nanodiamond (no details on modification given, most likely the detonation diamond has been thoroughly acid cleaned and hence exhibits a high number of oxygen containing surface groups such as COOH and OH).[114] In aqueous suspension no desorption was detected. However, removal was possible using BTP buffer solution (containing 1,3-bis(tris(hydroxymethyl)methylamino)propane). The subsequent assessment of its activity revealed no irreversible changes to the protein structure. In the same study the authors compared covalently linked lysozyme that was attached via a benzoquinone linker. Although the activity of the ND–lysozyme conjugate was reduced to about one half, it was more resistant against proteolytic attack by pepsin.

A similar reduction in protein efficacy was reported in another study where bone morphogenetic growth factor 2 (BMP2) was either physisorbed or covalently linked onto nanodiamond using the benzoquinone approach.[115] Here, the 3D structure of the covalently bound protein was apparently too strongly modified to still enable the proper function of the factor.

Carboxylated nanodiamond can also be used to form highly biocompatible multilayer non-covalent coatings with proteins.[116] The authors found that by simply adsorbing different polycationic peptides a multilayer structure of a strongly adsorbed inner shell and a reversibly adsorbed out layer was formed. With amphiphilic peptides stable colloidal dispersions could be obtained that are very promising for biomedical applications as they exhibit a positive zetapotential.

As mentioned before, covalent grafting of proteins and peptides is possible as well. The most obvious ligation method seems to be the grafting of proteins via amide formation. Many reports have been given in the past on this technique. One recent example used the formation of a sulfo-NHS ester on fluorescent nanodiamond and coupled this pre-functionalized fND with a monoclonal mouse antibody to target actin filaments.[117] Apparently, in this case the unselective grafting onto one of the $NH_2$ groups of the protein (either the *N*-terminus or, most likely, one of the lysine side chains) does not induce problems for the antibody function. Another example was reported by Weng et al.[118] They functionalized fluorescent ND–COOH with transferrin by stirring the components in phosphate saline buffer in the presence of *N*-(3-dimethylaminopropyl)-*N*′-ethyl-carbodiimide hydrochloride.

More specific grafting is possible when functional groups not occurring in peptides are used for the ligation. The most prominent is probably the alkyne–azide click ligation as both azides and terminal alkynes are not found naturally in peptide sequences (Fig. 8.25). They can, however, be rather easily introduced in a site-specific manner. Recent reports on "click-immobilization" of peptides have been given by Cigler and coworkers and Komatsu and coworkers. While the latter group used the formation of azides on polyglycerol coated nanodiamond by a substitution strategy for the ligation of a peptide terminated with propargyl glycine,[119] the group of P. Cigler grafted a cyclic RGD peptide (with RGD referring to the amino acid se-

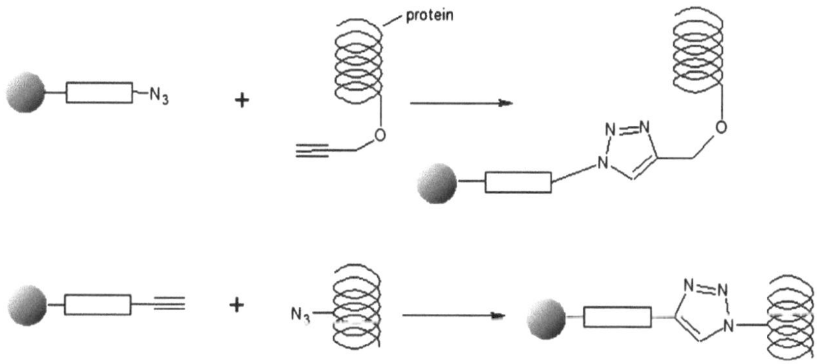


**FIGURE 8.25**

Covalent grafting of peptides onto nanodiamond using artificial anchor groups and thereby increasing the site selectivity of the grafting. Clickable groups can be either attached to one of the termini of the peptide chain (top) or introduced in the side chain (bottom), e.g., by integrating non-natural amino acids during the synthesis of the peptide. Substitution in already existing peptide side chains (e.g., the amino groups in lysines) can also be used, leads however to less specific results

quence) functionalized with a substitutional azide group in the lysine side chain onto fluorescent nanodiamond carrying a methacrylated silica layer that was subsequently coated with a copolymer produced from *N*-(2-hydroxypropyl)methacrylamide and *N*-propargylacrylamide.[120] Such conjugates are candidates for the specific targeting of tumor cells.

An interesting method for the selective grafting of peptide chains onto nanodiamond was reported by Knapinska et al.[88] They first synthesized the respective peptide on a solid phase and reacted the *N*-terminus with carboxylated nanodiamond before releasing the peptide from the resin. Using this approach non-specific interactions can be reduced to a minimum and highly selective grafting can be achieved as only the terminal $NH_2$ group can be addressed easily.

#### 8.3.4.3 Grafting of saccharides

Saccharides are another important class of biomolecules that have been grafted onto nanodiamond. This class of compounds is relevant for a variety of processes such as bacterial infections, biofilm formation, cell-cell recognition, and many more. Therefore, a variety of ligation techniques have been reported to stably immobilize saccharides onto the surface of diamond. The resulting conjugates can be used either for detection and sensing or for therapeutic approaches.

One of the first reports on covalent saccharide immobilization was given by Kuznetsov et al. They used fluorinated detonation nanodiamond and submitted it to substitution using lithium monosucrate.[121] However, the rate of fluorine replacement remained rather low as the C–F bonds on diamond do not show a reactivity comparable to single walled carbon nanotubes (where the replacement was highly efficient as shown in the same report).

One of the most versatile ligation methods is the so-called Click chemistry, where the reaction partners undergo bond formation in a highly efficient manner under mild conditions (moderate temperature, aqueous conditions) without the formation of side products and the need of complex catalysts, etc.[122] The archetypical click reaction unarguably is the formation of triazoles from alkynes with azides using a copper catalyzed reaction. This grafting approach has also been used to immobilize saccharides on pre-functionalized nanodiamond (Fig. 8.26).[123] The authors immobilized azide terminated trimeric thiosugar clusters on alkyne functionalized nanodiamond. It is equally possible to use the "clickable" moieties in the opposite direction. As had been shown for fluorescent dyes and model systems,[98] the diamond prefunctionalization with an azide-terminated linker enables the grafting of saccharides carrying alkyne groups.[124] The linker molecules carrying the azide or alkyne groups can be immobilized via ester or amide formation, Diels–Alder reactions, diazonium salt grafting, or the formation of ethers between surface OH groups and dopamine.[83]

Before the grafting of saccharides using triazole formation was reported, another click-type reaction had been used for the immobilization of mannose and glucose as well as a trimeric mannose cluster.[125] Thiol groups on nanodiamond generated by reduction of sulfonic acids were used as binding sites. These react readily with alkenes in a so-called thiol-ene reaction. The resulting nanodiamond-sugar conjugates were highly water soluble, and selective recognition of mannose-specific bacteria was achieved. Very similar results were also achieved using the formation of thiourea moieties at the grafting site.[126] This approach further facilitates the pre-functionalization of nanodiamond as only $NH_2$ groups are required (which can be introduced directly with the linker or can be easily generated, e.g., by reducing nitrile functions). On the saccharide side the easily available thiocyanate groups are required. The latter can be installed by a simple nucleophilic substitution under mild conditions or by, e.g., the reaction of a peracetylated saccharide with trimethylsilyl isothiocyanate in the presence of a Lewis acid such as $SnCl_4$.[127]

Using the saccharide functionalized nanodiamonds, a variety of assays have been developed that are described in the reports mentioned above.

#### 8.3.4.4 Grafting of other biomolecules

Besides peptides and saccharides the grafting of other classes of biomolecules is of importance. Some of these compounds can be used as drugs or for the targeting of specific cell types. Also in sensing the grafting of DNA sequences or affinity tags can be useful. Some recent examples will be shown here that present a cross-section of the available methods and compound classes.

DNA can be either grafted directly onto nanodiamond by using non-covalent binding (e.g., by electrostatic interaction with positively charged nanodiamond particles) or interacting with prefunctionalized nanodiamond via a linker unit. The latter strategy has been used by Zhang et al.[128] They coated fluorescent nanodiamond with a non-covalently linked protein layer that carried biotin moieties at its outer surface.

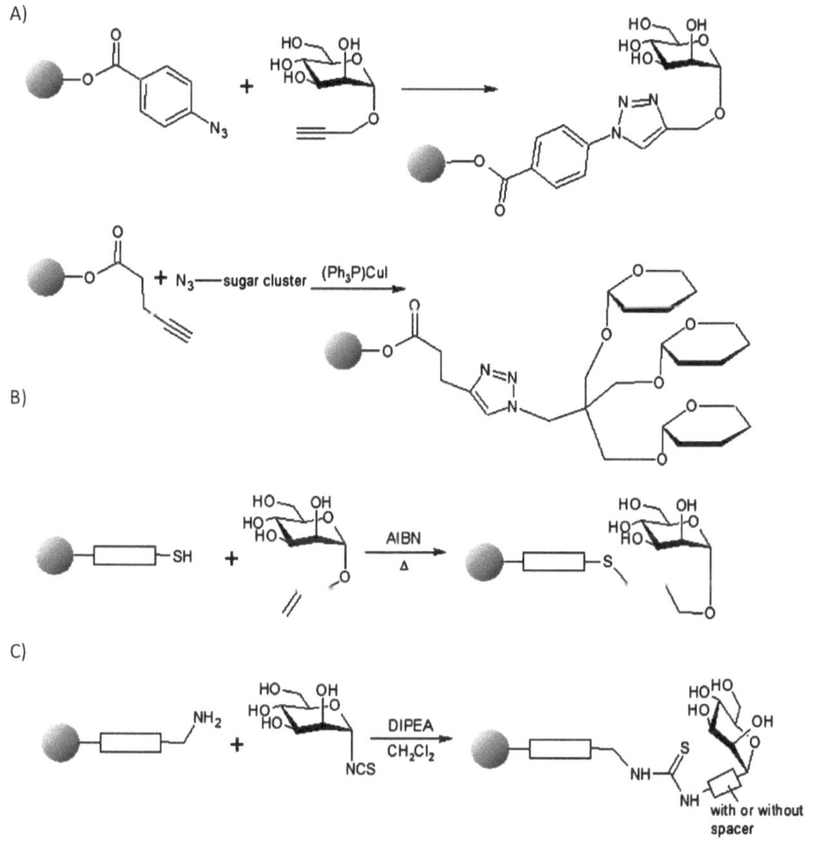


**FIGURE 8.26**

Immobilization of saccharides onto nanodiamond using different linker structures such as (A) azide–akyne click chemistry, (B) the thiol-ene reaction, and (C) thiourea formation

By using DNA origami structures that had avidin units as the arranging structural elements, the remaining binding pockets of these avidins were used to attach nanodiamond in a selective manner onto the DNA nanostructures.

Peptide nucleic acids (PNA) were also immobilized on nanodiamond. They can be used to reliably recognize DNA motifs and for targeting of these structures.[36a] By binding the PNA to carboxylated diamond via a tailored amidation reaction, the covalent PNA conjugate could be produced efficiently. As PNA binds more strongly to DNA and RNA than DNA itself, highly selective starting applications can be envisaged.

Targeting can be also achieved using smaller molecules such as folic acid.[36b] First, oxidized nanodiamond was reacted with PEG-diamine and then terminated with folate. The resulting nanodiamond conjugate was loaded with doxorubicin. The au-

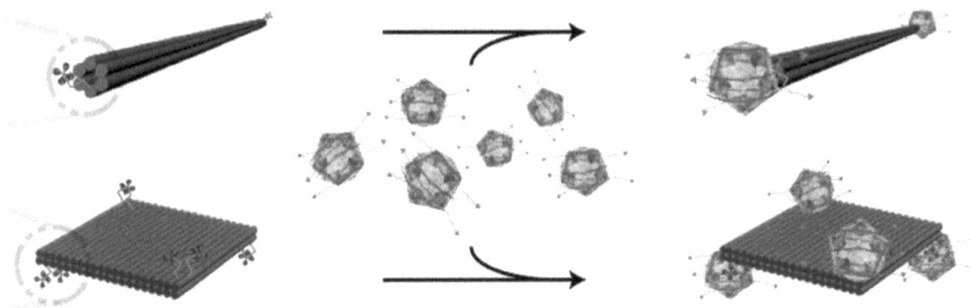

**FIGURE 8.27**

DNA origami with nanodiamonds (reprinted with permission from: Zhang, T., et al. J. Am. Chem. Soc. 2015, 137, 9776–9779,[128] © 2015 American Chemical Society)

thors observed then a more specific release of the drug when folate was present that addresses the folate receptors which are more abundant on tumor cells.

In a similar approach transferrin–doxorubicin adducts were covalently linked onto carboxylated nanodiamond using the classic EDC/NHS strategy.[129] The novelty in this approach was to load the drug onto the protein prior to its immobilization on the diamond surface. The authors observed a specific interaction with the transferrin receptors and hope to use this for targeted drug delivery in the future. So far, the efficacy of doxorubicin was not significantly enhanced by the formation of ND–Tf–Dox conjugates. One of the issues the authors discussed was the non-specific interaction of proteins with the functionalized diamond surface. Such non-covalent interaction could alter the surface properties and the ability for drug release tremendously. In order to overcome these difficulties, further research on protein interactions with diamond of different surface structure is absolutely necessary.

Proteins can also be used to facilitate the conjugation of other biomolecules to nanodiamond. Cytochrome P450 was used to enhance the amount of oxidized groups on nanodiamond that then reacted with the anti-IgG.[130]

#### 8.3.4.5 *Grafting of dyes, labels, and drugs*

Besides the immobilization of biomolecules on nanodiamond, the grafting of other functional moieties has been investigated in great detail. One of the reasons for this strong interest is the possibility to combine the unique properties of the nanodiamond particles with the functionality of the grafted unit. These can be either dyes and other labels or drugs and targeting moieties.

Fluorescent dyes can, e.g., be used for the study of aggregation phenomena as well as energy transfer between dye molecules. Maitra et al. reported on the covalent grafting of an oligo(phenylenevinylene) (OPV) using the formation of acid chlorides on the diamond surface and subsequent amide formation with an aromatic amino group of the chromophore.[131] The fluorescence was strongly concentration-dependent in an unexpected manner. With increasing concentration, the green fluorescence of the chromophore became strongly enhanced in the agglomerates that are formed at concentrations above $2 \times 10^{-2}$ mg mL$^{-1}$. An interparticle energy transfer from the diamond cores to the chromophores of neighboring particles was identified as the origin of this behavior. Other reasons such as enhanced emission by aggregation of OPV or quenching of the ND fluorescence by aggregation have been ruled out by control experiments.

Another report using two independently attached chromophores has been reported by Meinhardt et al.[98] TAMRA and Oregon Green, two commonly used fluorescent dyes for bioimaging applications have been immobilized on nanodiamond via an aromatic linker system with either an alkyne (for click chemistry) or a carboxylic group (for amide formation) in the *para* position to the grafting site. Such systems can be used as a model for orthogonal immobilization, e.g., of drugs (attached on the cleavable amide site) and targeting moieties (attached to the stable triazole ring formed in the click reaction). In this case, the emission of the two dyes could be excited separately or jointly. Although in principle possible from a spectroscopic point of view, an efficient excitation energy transfer from Oregon Green to TAMRA was not observed. Only a small shoulder of TAMRA emission can be detected when Oregon Green is excited.

Förster resonance energy transfer (FRET) between a luminescent lattice defect and a dye immobilized on the surface of silanized nanodiamond has been achieved by Wrachtrup and coworkers.[72] Negatively charged NV centers act as the donor and a fluorescent dye as the acceptor. Transfer efficiencies of up to 86% have been observed. Such systems could be used for FRET-based scanning probes in the future. However, the absorption of the NV center is not very strong, and the resulting luminescence is typically not very intense.

An approach for the enhancement of the NV emission using organic fluorophores as "harvesting antennas" has been presented.[132] In this case the chromophores act as the donor and the NV center as the acceptor in the FRET based system. Bovine serum albumin labeled with fluorescein isothiocyanate (FITC) was attached to the surface of fluorescent nanodiamond carrying NV centers. The authors showed an efficient enhancement in brightness of the NV(-) emission due to the energy transfer. In the future, a controlled attachment using linkers with defined geometry will allow for the tunable interaction between the chromophore and the lattice defect.

Chromophores attached to the surface of different types of nanodiamond have been used as well for the imaging of these nanodiamonds in cells or small organisms. However, such imaging is always limited by the depth of penetration of the excitation light (and the transparence of the tissue for the emitted light). Therefore, non-luminescent labels are highly attractive for the study of interactions with organs deep in the body or for the study of biodistribution of nanodiamond conjugates.

Besides the termination of nanodiamond with tritium[133,134] the grafting of $^{18}$F and $^{125}$I labeled structures has been reported for nanodiamond. While the latter was incorporated in a protein that was adsorbed on pre-functionalized nanodiamond,[135] $^{18}$F was introduced in a covalent manner: hydroxylated nanodiamond was first silanized with 3-aminopropylsilane groups and the radiotracer was grafted as *N*-succinimidyl 4-[$^{18}$F]fluorobenzoate. The biodistribution of the conjugate was studied using positron emission tomography (PET).[69] However, it is known from previous experiments that nanodiamond conjugates with a shell generated by silanization is not stable against hydrolysis for extended periods of time under physiological conditions (the more acidic the conditions, the less stable). It remains therefore unclear whether the recorded $^{18}F$ signal is only due to unaltered conjugates or detached $^{18}F$-containing moieties, too.

For many applications the use of a radiolabel is at least inconvenient and other methods such as MRI are favored. In medical applications the so-called contrast agents help to enhance resolution and to decrease required data collection times. Nanoparticles decorated with such contrast agents can be used not only to study the biodistribution of the particle itself but also improve the properties of the spin label.

One of the first reports was on the immobilization of a gadolinium(III)chelate complex using amide formation as the grafting step. The resulting conjugate not only worked as a spin label but significantly reduced the longitudinal relaxation time $T1$ of water protons and gave a ten-fold increase of relaxivity per gadolinium ion compared to the free Gd(III) complex.[136] More recent examples include the grafting of the Gd(III)-chelate onto a polyglycerol shell coating of the nanodiamond.[137] In this approach a large number of Gd ions can be immobilized on one diamond particle. Furthermore, the polyglycerol shell improves the dispersibility of the aggregates in physiological solutions such as PBS (phosphate buffered saline). The gadolinium can, however, also be immobilized directly on the diamond surface. In this case, carboxylic acids on the nanodiamond surface act as ligands.[138] It remains to be investigated whether these complexes that are formed directly at the diamond surface are sufficiently stable under physiological conditions which is needed for any type of bioapplications.

One of the major bioapplications for any kind of nanoparticle is the delivery of drugs. Several examples of therapeutic agents immobilized on nanodiamond have been already described in the previous chapters (such as saccharides with antibacterial activity). Here only a few more examples for the covalent grafting of drugs or drug precursors are discussed to show the broad applicability of surface functionalized nanodiamond. More examples can be found in Chapter 16 on applications of nanodiamond.

An antiviral agent was immobilized using azide-terminated nanodiamond and alkyne-functionalized phenylboronic acids such as 4-[(1-oxo-4-pentyn-1-yl)amino]-phenylboronic acid. This led to the formation of conjugates that act as "borono lectins" and hinder the entry of hepatite C viruses into liver cells.[139]

Azide-alkyne click coupling was also used to graft nitroxyl radicals (i.e., TEMPO) onto prefunctionalized nanodiamond. In this case the diamond was first silanized, and the azide was introduced in a subsequent step.[111] Nitroxyl radicals play a role as scavengers of reactive oxygen species and hence could be envisaged as useful in the protection against tumor development and neurodegenerative diseases.[140]

Besides such functional anticancer drugs, other small molecules can be used to treat tumors. Carbon monoxide is such a molecule that can be used to affect tumor cells. It is toxic but if released in a site-selective manner, it could be of high interest as

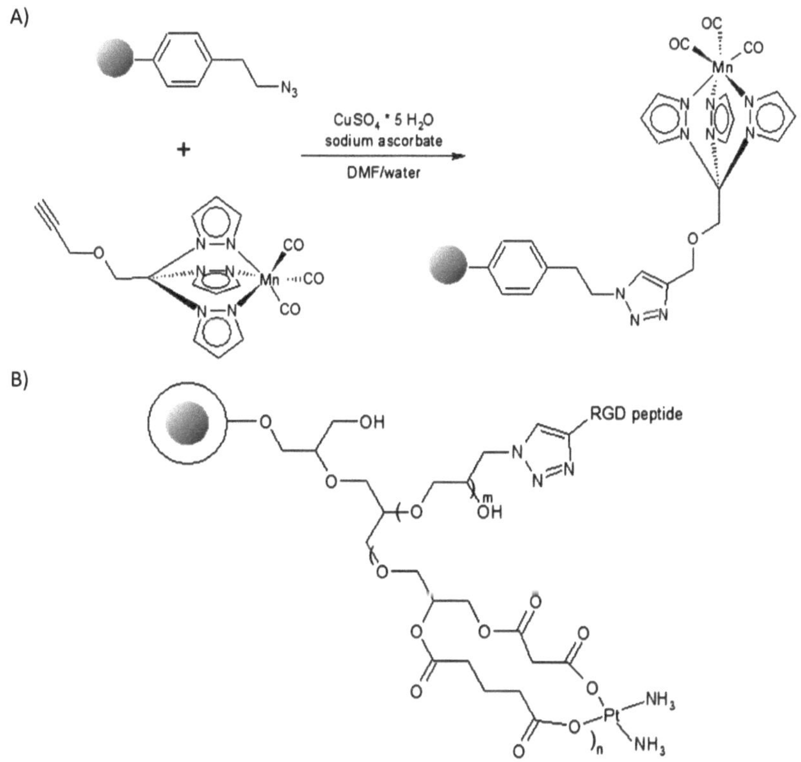


**FIGURE 8.28**

Immobilization of antitumor agents: (A) a Photo-CORM that releases cytotoxic CO upon irradiation with visible light and (B) a platinum complex derived from cisplatin, a well-known antitumor agent (the RGD peptide is used for targeting)

no adverse effects due to resulting side products are to be expected. CO is normally presented in the form of transition metal complexes that can release CO when triggered, e.g., with visible light or other stimuli. A manganese based photochemically CO releasing complex (Photo-CORM) has been immobilized onto arylated nanodiamond using click chemistry (Fig. 8.28A).[141] CORM does not lose its activity, and loading of the nanodiamond particles with the photocleavable moiety is done via click chemistry.

Other antitumor agents that have been immobilized onto nanodiamond in recent years include cisplatin, among others. When grafted onto nanodiamond coated with a hydrophilic polymer, the toxicity and hence the antitumor activity was enhanced by an order of magnitude.[142] Antitumor activity induced by platinum complexes was also observed for nanodiamond prefunctionalized with polyglycerol and grafted with the Pt compound in the periphery (Fig. 8.28B). By adding a targeting peptide, i.e., the RDG peptide, specific targeting of certain cell types even in the presence of other cells is possible.[143]

#### 8.3.4.6 Grafting of polymer components

Nanoparticles have been used as fillers of polymers for long time. They can modify the mechanical, electrical, thermal, and chemical properties of the resulting composite to a large extent. The earliest example for a carbon-based filler is rubber, where basically carbon black (which, in fact, consists of carbon nanoparticles) is used to mechanically reinforce the rubber of car tires. Additionally, these nanoparticles also improve the heat conductivity and act as a pigment.

Nanodiamond is also an attractive material for the synthesis of composite materials.[144] Due to its unique properties such as hardness, heat conductivity, high electrical resistivity, low friction coefficient, and chemical inertness, it can be used as a filler for many polymers. Especially the rich surface chemistry (see above) makes it possible to tailor the interaction of the diamond nanoparticles and the polymer matrix, both for non-covalent interactions (i.e., wetting properties) and covalent bonding. The latter ensures a strong interaction between the matrix and the filler and helps to evenly disperse the nanoparticles in the polymer.

Good dispersion is a requirement for the efficient use of nanoparticles in polymer composites. Only well-dispersed nanoparticles ensure the formation of a large volume fraction of an interphase between the nanoparticles and the matrix. This interphase is the place, where the actual modification of the polymer's properties takes place. The smaller the particles, the higher the fraction of interphase in the composite, and hence the more significant the change of properties.

Besides the shape of the particles (spherical particles generate more interphase than similarly sized nanoobjects of other shapes, the actual size of the filler objects is crucial for the observable effects. Therefore, nanodiamond with particle sizes of less than 5 nm are extremely promising for the synthesis of high performance composite materials. However, the dispersion of these particles needs to be perfect. If agglomerates of much larger size are present in the polymer matrix, they will act as defects in the polymer structure (Fig. 8.29). Instead of improving the polymer's properties, they would then actually deteriorate them. Therefore, the dispersion and efficient prevention of reagglomeration, e.g., by covalent crosslinking with the polymer matrix, are of utmost importance for the desired performance of the composites.

Various polymer composites are known. Starting with simple mixing of the components, without any further treatment via dispersion of tailor-made nanoparticles in a suitable polymer matrix (leading to strong noncovalent interaction and hence good wetting properties) to covalent composites, many different types of nanodiamond polymer composites for various purposes have been investigated. For covalent composites, there are in principle several modes of bonding, such as the star-shaped, pearl necklace, and pendant chain composites or the highly crosslinked composites with nanoparticles acting as the nodes in the polymer network (Fig. 8.30).

In the following section representative examples for composite materials containing nanodiamond will be given. More information can be found in the review articles

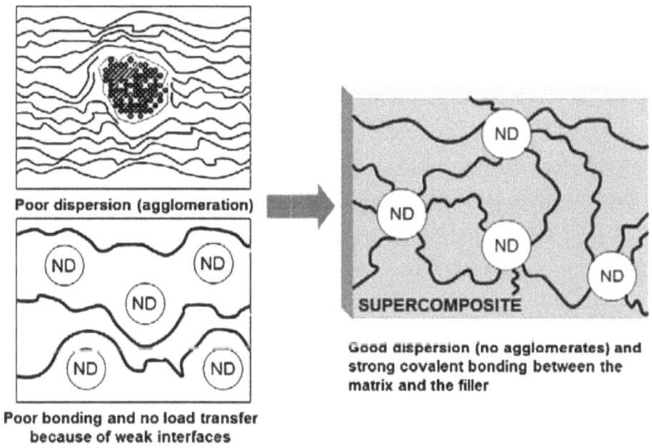


**FIGURE 8.29**

Polymer composites with agglomerates incorporated in the matrix (top left) and well-dispersed primary particles (bottom left). The agglomerates disturb the interaction between the polymer chains and introduce defects into the bulk material. They are not able to induce the desired "nanofiller effect". On the other hand, well dispersed nanoparticles form a large fraction of interphase with the matrix and can induce the change of properties to the full extent. However, the full effect of well-dispersed nanodiamond particles can be only achieved when strong interactions such as covalent bonds exist between the matrix and the nanofiller (from: Mochalin, V. N., Gogotsi, Y. Diam. Relat. Mater. 2015, 58, 161–171[144])

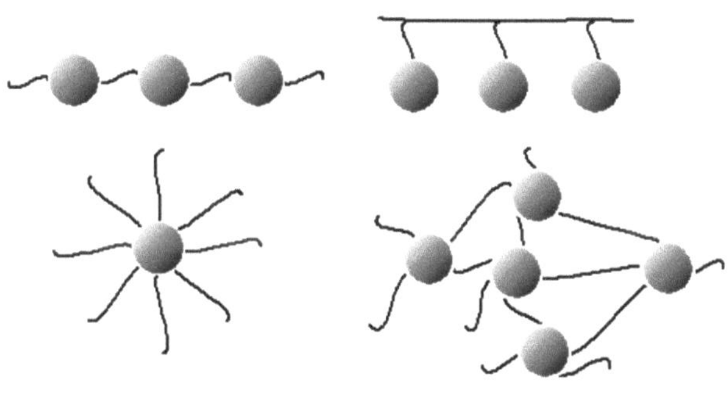

**FIGURE 8.30**

Various covalently bound composites of polymers with spherical nanoparticles such as nanodiamond

by Mochalin and Gogotsi,[144] Ullah et al.,[145] Jabeen et al.[146], and Badamshina et al.[147] A book chapter dealing with the surface chemistry for the synthesis of nanodiamond polymer composites summarizes also work done before 2011.[148]

As mentioned above, the use of nanoparticles as nodes in the polymer network (i.e., as crosslinkers) promises strong improvement especially of mechanical properties of the resulting composites. Neitzel et al. showed already in 2012 that epoxy resins can be cured using amine-functionalized nanodiamond.[149] The latter provides the amino groups for the curing of the polymer. By improving the dispersion of the nanoparticles in the matrix significant increases in mechanical strength and hardness can be achieved. All polymer chains that are linked to nanodiamond particles are influenced by this bonding – the volume fraction of the interphase increases, leading to very good mechanical properties already at rather low nanodiamond loading. Such materials are attractive for bioapplications (e.g., screws for implants, implant materials, etc.) due to their strength and biocompatibility.

The curing agent does not need to be replaced completely by aminated nanodiamond. Guo et al. reported on the addition of triethylenetetramine functionalized nanodiamond to a typical epoxy resin mixture. Curing was completed used triethylenetetramine.[150] The obtained composites showed improved fracture toughness and storage modulus. It was also reported that nanodiamond treated with ammonium bicarbonate at 100°C (claiming that $NH_2$ groups are formed at the diamond surface) can be used to form epoxy composites. In this work stirred media milling helped to disperse the nanodiamond in the respective solvent and hence improved the dispersion of the nanoparticles in the polymer matrix. As shown by the authors, improved dispersion leads to improved mechanical properties.[151]

Besides the immobilization of curing agent on the diamond surface, one can also use typical diamond surface groups such as hydroxyl functions or carboxyl groups for covalent interactions with side chain groups of the polymer or other residual functional groups of the polymer chain. Pauzauskie and coworkers have incorporated oxygen terminated nanodiamond into resorcin–formaldehyde gels and eventually formed a three-dimensional monolithic aerogel with a surface of almost 600 $m^2\ g^{-1}$.[152] Such highly porous composites could be useful for the storage of small molecules, in catalysis or energy storage.

Star-shaped conjugates of nanodiamond with polymeric chains are a very common way of producing composites. In this case crosslinking is not possible (limited to the ends of very long polymer chains). Several approaches have been reported. One can start with an initiator molecule on the surface of the diamond nanoparticle and use methods such as atom transfer radical polymerization (ATRP) or reversible addition–fragmentation chain transfer (RAFT) to build the polymer chains. Similarly monomers or suitable groups for the attachment of monomers can be immobilized on the diamond surface. And thirdly, already existing polymer chains with reactive endgroups can be bound to the nanoparticle. In the latter approach steric hindrance of subsequent grafting by chains already attached to the nanoparticle, and possible blocking of active surface area by covering it in a noncovalent manner, cannot be excluded. That is why most of the reports so far use the continuous growth of the polymer on the diamond surface.

When grafting of monomer units onto the diamond surface is chosen, these can be directly incorporated into the growing polymer chains or network. Depending on the concentration of the monomer-functionalized nanodiamond in the polymer solution, either heavily crosslinked networks or chains with ND incorporated in the polymer

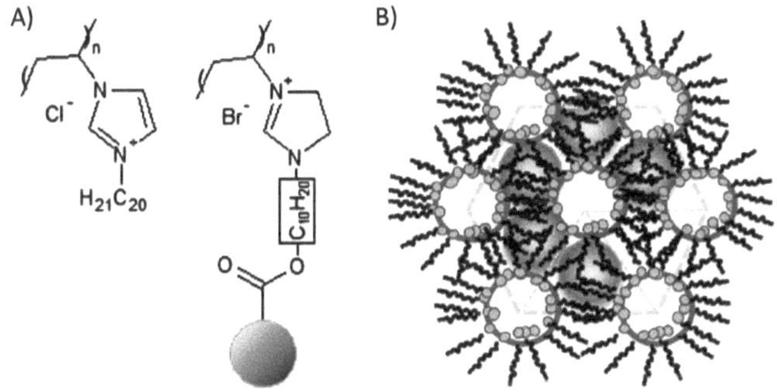


**FIGURE 8.31**

Nanodiamond composite functionalized with an ionic liquid polymer: (A) structure, (B) columnar structure incorporating the nanodiamond particles (reprinted with permission from: Ringstrand, B. S., et al. Macromol. Rapid Commun. 2016, 37, 1155–1167[155])

sequence will be formed (some crosslinking will obviously occur even at very low ND concentrations). For the perfect dispersion of the monomer–ND building blocks, perfect colloidal stability of the functionalized nanoparticles is very important. For methacrylate moieties it was shown that simultaneous beads assisted sonication and surface silanization with the acrylate-terminated silane derivative leads to fully soluble, monomer-functionalized nanodiamond particles that can be then incorporated in the PMMA without any agglomeration.[68] Another acrylate-based incorporation of nanodiamond was reported recently by Alishiri et al.[153] Nanodiamond functionalized with hydroxyethyl methacrylate was covalently incorporated into an acrylate terminated polyurethane, a promising material for hard tissue engineering applications. The functionalized ND agglomerates (no primary particles in this case) were located in the mechanically less stable regions of the mother polymer, thereby enhancing the mechanical properties to a great extent.

The milling assisted deagglomeration was later used by Hsin et al. to directly polymerize monomers with the help of a radical initiator on the surface of the appearing diamond nanoparticles.[154] The resulting composites showed interesting tribological properties.

In a similar approach 3-decyl-1-vinylimidazolium bromide was covalently attached onto carboxylated nanodiamond via ester formation and the resulting conjugate added to 3-decyl-1-vinylimidazolium chloride which is then polymerized using UV light as radical initiator. The resulting lyotropic mesophase was characterized and highly ordered structures with hexagonal columns have been described after the polymerization (Fig. 8.31).[155]

Besides the already mentioned crosslinking of epoxy resins, the immobilization of different epoxy monomers has been shown to yield composites with unique properties. Bisphenol A epoxy monomers[156] or glycidol[157] has been used for the growth of polymer brushes from the nanodiamond surface. The resulting conjugates can be used, e.g., for thermal applications due to their impact on thermal conductivity or for mechanically more resistant epoxy resins.

Polymer-functionalized nanodiamond is also a really attractive material for bioapplications. It was reported that polymer coated nanodiamond produced by RAFT synthesis and subsequently functionalized with cisplatin (a well-known antitumor drug) showed significantly higher toxicity levels against tumor cells than cisplatin on "naked" nanodiamond.[142] Other bioapplications include the production of implant materials or synthetic replacements for tissues. Further examples will be given below.

The grafting-from approach can be also used to start the polymerization from the diamond surface without immobilizing special moieties for these techniques. As long as suitable reactive surface groups are present these can be used to attach the first monomeric unit and to continue the polymerization from that one. Sun et al. reported on the activation of hydroxylated nanodiamond by deprotonation using sodium hydride, thus generating an in situ initiator for an anionic polymerization to polylactide (PLA).[158] Also ring-opening polymerization can be started from suitable functional groups on the nanodiamond surface. In this way, ND modified with polylactic acid (PLLA) or with ε-caprolactone has been produced.[159]

Electrically conducting polymers have been immobilized, too. Polyaniline has been grafted electrochemically onto nanodiamond, and the resulting conjugates were tested for their application as electrodes and capacitors.[160] PANI derivatives of nanodiamond have also been useful for sensing applications, e.g., the detection of nitrite in water.[161]

In contrast to the composites discussed above, non-covalent composites are not held together by actual chemical bonding. The interaction of the polymer matrix and the filler nanoparticles, in our case nanodiamond, is merely a non-covalent interaction. These interactions include electrostatic and hydrophobic interactions as well as further van der Waals interactions and hydrogen bonding. This diversity of possible attractive (and repulsive) forces makes the surface termination of the nanofiller extremely important. Phase segregation, dewetting, and agglomeration of the nanoparticles are detrimental to the properties of the composite material (e.g., by reducing the actual interphase and creating voids or cluster regions in the matrix structure). Therefore, for each polymer matrix the optimal diamond surface needs to be found in order to obtain a very homogeneous distribution of fully separated nanodiamonds in the polymer. In the following paragraphs a selection of recent examples for non-covalent nanodiamond–polymer composites for various applications will be given.

As with covalent composites non-covalent composites containing epoxy resins containing nanodiamond are abundant in the literature. Neitzel et al. compared the performance of "as received" detonation nanodiamond (i.e., exhibiting a mixed oxidized surface termination) and aminated detonation nanodiamond as nanofiller in epoxy composites containing Epon828 (diglycidyl ether of bisphenol A).[162] They found that both nanofillers improved the mechanical properties of the polymer matrix, however, the aminated particles showed significantly better values due to the stronger, most likely partially covalent interaction with the matrix. Besides the Young's modulus increase, the wear resistance of the polymer was improved, too.

A series of different concentrations of acid-treated detonation nanodiamond were investigated in a composite with an epoxy matrix.[163] The authors found that beyond a certain concentration (here >0.7%) the influence on the mechanical properties such as storage modulus, hardness, and loss modulus was inverted, and lower values were observed again. Further addition then led to an increase in mechanical resistance again. It was discussed that agglomeration might be the reason for the decrease with higher concentration. However, the fact that further increase in the ND amount leads to stronger effects again is not fully explainable by this. Agglomeration should be even more prevalent in these cases, and most likely small aggregates or nanodiamonds are incorporated into the polymer matrix. The authors argue though that an increase in matrix–filler interface leads to the improvement and is outplaying agglomeration. However, in another study it was also reported that even rather large agglomerates of ~100 nm show up to 2.5 times higher Young's modulus compared to neat epoxy.[164] If the interaction of the aggregate surface – which is, in fact, the surface of the outer individual diamond nanoparticles – with the matrix is generating a large interphase these interphases overlap in the composite and form an interconnected network and thereby lead to strong changes in the mechanical performance.

Besides epoxy resins as the matrix, also polyesters are very popular components of nanodiamond composites. Several of these have been identified as not only biocompatible but also biodegradable. This makes them attractive for medical applications such as surgery (e.g., suture material), bone and tissue engineering (as scaffolds could be bioresorbed after, e.g., growth of new bone or cartilage).

The most prominent example is polylactide, a biodegradable thermoplastic.[165] For many of the intended applications, the mechanical properties of PLA are, however, too weak, and the polymer requires reinforcement. This can be efficiently done using nanodiamond as the nanofiller. It can be either used to start the polymerization of the PLA from the surface of the nanodiamond (producing core–shell objects or star-shaped composites) as mentioned above or simply interact with the polymer matrix in a non-covalent manner with its surface groups. Gogotsi and coworkers reported on the production of PLA composites with detonation nanodiamond for bone engineering. They used ND functionalized with octadecylamine and poly(L-lactic acid, PLLA).[165a] This termination of the nanofiller particles resulted in an even distribution in the matrix and hence to a significant improvement of the mechanical properties while maintaining the biocompatibility of the PLA. The same group also investigated the mineralization capacity of the PLA–ND composite (here also PLLA) and found an increase to the polymer alone.[165b] This opens perspectives for these materials in regenerative medicine.

PLA modified with other polymers has been reported to be beneficial for the performance of implant materials. Poly(L-lactide)-co-($\varepsilon$-caprolactone), in short poly(LLA-co-CL), is one of the most promising candidates. When modifying it with fully dispersed detonation nanodiamond particles by simply adsorbing them on the surface of a porous scaffold structure produced from the copolymer, they attach via non-covalent interaction and improve the cell proliferation on the scaffold material.[166] Furthermore, cell differentiation, an important process in the formation of new bone, was enhanced.

It was later shown, that the surface termination of the nanodiamond particles has a tremendous effect on the mechanical properties of the poly(LLA-co-CL)-ND composite when the particles are not simply adsorbed on the scaffold surface but are already present during polymerization.[167] Although some phase segregation was observed for pristine detonation nanodiamond (due to colloidal instability in the reaction medium used for the polymerization), most likely by unfavorable interactions with CL parts of the copolymer, benzoquinone and especially PLA-prefunctionalized DND particles could be homogeneously incorporated. The latter showed also significantly improved mechanical properties such as a six times higher *E* modulus. The PLA chains bound to the nanoparticle surface increase the interphase between the particles and the matrix to a much bigger extent than a small functional moiety such as benzoquinone.

Another copolymer, poly(lactide-co-glycolide), in short PLGA, modified with nanodiamond showed beneficial properties when seeded with mesenchymal stem cells such as very good biocompatibility and increased hardness.[168] These composites are equally promising for bone tissue engineering.

Obviously, the mechanical improvement of polyesters is not limited to biomedical polymers – any kind of polyester can be modified with suitably terminated nanodiamond. Poly(ethylene terephthalate) has been modified with different types of carbonaceous nanofillers, among them acid terminated nanodiamond.[169] Nanodiamond was incorporated using melt mixing with a commercial PET sample. Both the addition of ND and graphene nanoplatelets increased the crystallinity of the polymer and the thermal stability of the composite. However, tensile properties differed when ND was applied alone or in combination with the nanographene platelets.

Other composites with hydrophilic groups either in the polymer chain, such as for polycarbonates, or as pendants on the chain like in poly(methylmethacrylate) (PMMA) can be easily modified non-covalently with suitably terminated nanodiamond. Jee et al. reported on investigations of these interactions using ND terminated with COOH, OH, $NH_2$, or amide groups and the resulting mechanical properties of the polycarbonate and PMMA composites of these functionalized nanodiamond materials.[170] Especially amine and amide groups were observed to impart beneficial properties to the composite. A shape stabilized phase change material (SSPCM) consisting of PMMA and polyethyleneglycol as the phase change material was modified with nanodiamond, leading to enhanced thermal conductivity and volume resistivity.[171] Such composites are attractive for energy storage applications.

Polyimide has been also modified with different amounts of ND ranging up to 50 wt.% of functionalized nanodiamond.[172] In this case, ND carrying a derivative of benzoeic acid showed an increase in thermostability of dielectric properties of composites for high energy density capacitors. Phosphorylated nanodiamond was also shown to improve the thermal properties of polyimide composites.[93] The combustion in air occurs 100 K above the combustion for the neat polymer. However, the originally transparent polymer became increasingly opaque with higher ND contents.

A recent example for polyurethane composites with nanodiamond was reported by Xia et al.[173] A polyurethane/epoxy interpenetrating network was modified using nanodiamond. In this case covalent and non-covalent interactions play a role for the effect of the nanofiller in the matrix. A review on polyurethane (PU) composites with different carbon nanoobjects summarizes further work on using nanofillers to improve the properties of PU.[146]

Other studies reported on the incorporation of nanodiamond in different conjugated polymers such as polythiophene or polypyrrole.[174] Functionalization led to improved interactions between the matrix and the nanodiamond particles and helped to prevent agglomeration.

Polar groups in the side chains of the polymer can also help to improve the interactions between the matrix and the diamond nanoparticles. Sordi et al. reported on the production of polyester composites using sulfonated liquid crystalline polyesters with different monomers.[175] Glass transition temperatures were increased when the interaction between the filler and the polymer was strengthened by the sulfonate groups.

Polyvinyl alcohol (PVA) is another polymer with polar groups attached to the chain. They can be modified with hydrophilic nanodiamond (e.g., pristine ND with mixed oxygenated surface or ND-COOH). The incorporation of nanodiamond into PVA leads to very good reinforcement of the nanocomposite compared to the empty polymer.[176] Other authors reported on the increase of the refractive index of PVA upon incorporation of nanodiamond.[177]

Composites produced from poly(phenylene isophthalamide) from the aramide class of polar polymers and nanodiamond can be used for separation membranes, e.g., for methanol and methyl acetate mixtures.[178]

A rather unique non-covalent incorporation of nanodiamond into a polymer was proposed by Cha et al.[179] Ligand exchange reactions of ferrocenes attached to the polymer backbone with aromatic rings on the diamond surface were suggested as the mode of attachment. However, the polar interaction of oxygenated surface groups either with the iron centers or with the polymer itself seem more likely as no special treatment for the establishment of extended $\pi$-systems has been carried out with the nanodiamond applied in this study. From XPS and Raman measurements, it is known that $sp^2$ carbon on nanodiamond occurs typically in a non-ordered fashion and not as arrays of aromatic rings.

There are many more examples in the literature for nanodiamond–polymer composites including those with biopolymers based on non-covalent interactions between the nanofiller and the matrix that would deserve reference in this chapter. Space, however, is limited. They are all based on the same set of non-covalent interactions and requirements, such as homogeneous dispersion, prevention of agglomeration, suitable wetting properties, etc. In general, it can be concluded that the covalent or non-covalent incorporation of suitable nanodiamond particles into polymer composites induces a multitude of beneficial effects. Mechanical properties such as tensile strength, hardness wear resistance, and storage modulus are improved, thermal conductivity and stability are enhanced, and the biocompatibility is conserved or even improved. Tailor-made nanodiamond composites are promising candidates for future, large-scale applications if the interactions between the composites' components are fully understood and controllable, and if agglomeration and phase segregation can be completely prevented.

#### 8.3.4.7 Miscellaneous

In the preceding chapters a broad variety of surface modifications for nanodiamond has been presented, ordered by the initial surface termination or the class of immobilized moieties. There is, however, much more to find in the literature that cannot be easily categorized into one of the preceding sections. Several of these examples will be discussed in this chapter.

Quite often these examples include the use of hybrid structures where nanodiamond is only one of the functional components. For instance, functionalized nanodiamond has been used as a component in organic solar cells that are based on the classic poly(3-hexylthiophene) (P3HT)-fullerene bulk heterojunction model.[180] Carboxylated nanodiamond forms a composite with the polymer matrix and acts as an efficient charge transporter. It is expected that the application of $C_{60}$-decorated nanodiamond will lead to enhanced performance of polymer-based photovoltaic cells.

Another hybrid system consisting of nanodiamond and gold/silver particles has been presented by Cheng et al.[181] The functionalized PEG-coated nanodiamond with linkers containing thiol groups that enable interaction with gold/silver nanoparticles. Subsequent addition of thiolated transferrin then modified the surface of these hybrid covalent aggregates with a functional moiety that allows improved uptake of the nanoobjects into living cells. It turned out that the cytotoxic effect of these functional nanoparticles was higher than for the control systems. Therefore, further research into such nanodiamond hybrid systems seems to be a promising approach.

Colloidosomes as another class of hybrid systems of nanodiamond with additional components were reported by Maas and coworkers.[182] The termination of detonation nanodiamond either with a high number of carboxylic groups leading to a strongly negative zetapotential or the formation of a strongly positive surface enabled the interaction of the respective nanodiamond materials with differently charged emulsifiers such as stearic acid in the case of carboxylated ND and stearyl amine in the case of hydrogen-treated ND. The resulting colloidosomes showed a high shell stability.

## 8.4 FUTURE DIRECTIONS AND CHALLENGES

As can be seen in the previous sections the surface chemistry of nanodiamond has reached a certain level of maturity. A broad variety of transformations is available to synthetize tailor-made nanodiamond conjugates for any desired application.

However, several challenges remain:

- It has not been possible so far to synthetize 1:1 conjugates of nanodiamond with functional moieties in a reliable, non-statistical way. This would be very important for various imaging and sending applications.
- Particles with an asymmetric surface termination, such as Janus particles, have also not been produced in a controlled way. Such particles could be beneficial in polymer composites, as seeds in CVD growth and many other applications.
- A rather trivial, but nonetheless not easy to meet requirement, is the 100% loading of all reactive sites on the nanodiamond surface with functional units. In most cases, reactive sites are blocked by already bound moieties, especially in the case of larger biomolecules.
- Equally, the site specific immobilization of functional molecules (e.g., only at apices of facetted particles or only on 111 facets, etc.) would be of high interest.
- Regarding the potential biomedical applications of nanodiamond, there are still certain obstacles that need to be overcome before an actual application can be envisaged. Most of these concern the stability against agglomeration in physiological media and serum, however, the prevention of non-specific interactions of proteins, e.g., from the serum need to be controlled as well.

Nanodiamond surface chemistry has reached a state where establishing a general protocol for the functionalization of different types of diamond would be highly desirable. An openly available toolbox would help to investigate materials with reproducible properties. Such standardization is one of the next steps on the way to industrial and clinical application of functionalized nanodiamond.

## REFERENCES


1. (a) Gunawan, M. A.; Hierso, J.-C.; Poinsot, D.; Fokin, A. A.; Fokina, N. A.; Tkachenko, B. A.; Schreiner, P. R. Diamondoids: Functionalization and Subsequent Applications of Perfectly Defined Molecular Cage Hydrocarbons. *New J. Chem.* **2014**, *38* (1), 28–41;
(b) Clay, W. A.; Dahl, J. E. P.; Carlson, R. M. K.; Melosh, N. A.; Shen, Z.-X. Physical Properties of Materials Derived From Diamondoid Molecules. *Rep. Prog. Phys.* **2015**, *78* (1), 016501;
(c) Bostedt, C.; Landt, L.; Moller, T.; Dahl, J. E.; Carlson, R. M. K. Diamondoids. In Barnard, A. S., Guo, H., Eds., *Nature's Nanostructures*; Pan Stanford Publishing Pte. Ltd.: Singapore, 2012, pp. 169–194;
(d) Schwertfeger, H.; Fokin, A. A.; Schreiner, P. R. Diamonds Are a Chemist's Best Friend. Diamondoid Chemistry Beyond Adamantane. *Angew. Chem. Int. Ed.* **2008**, *47* (6), 1022–1036.
2. (a) Lai, L.; Barnard, A. S. Tuning the Electron Transfer Properties of Entire Nanodiamond Ensembles. *J. Phys. Chem. C* **2014**, *118* (51), 30209–30215;
(b) Barnard, A. S.; Per, M. C. Size and Shape Dependent Deprotonation Potential and Proton Affinity of Nanodiamond. *Nanotechnology* **2014**, *25* (44), 445702;

(c) Yuan, T.; Larsson, K. Theoretical Study of Size Effects on Surface Chemical Properties for Nanoscale Diamond Particles. *J. Phys. Chem. C* **2014**, *118* (45), 26061–26069;
(d) Larsson, K. Surface Chemistry of Diamond. In *Novel Aspects of Diamond;* Top. Appl. Phys., Vol. 121, 2015, pp. 53–83;
(e) Lai, L.; Barnard, A. S. Anisotropic Adsorption and Distribution of Immobilized Carboxyl on Nanodiamond. *Nanoscale* **2014**, *6* (23), 14185–14189;
(f) Paci, J. T.; Man, H. B.; Saha, B.; Ho, D.; Schatz, G. C. Understanding the Surfaces of Nanodiamonds. *J. Phys. Chem. C* **2013**, *117* (33), 17256–17267.

3. Krueger, A.; Lang, D. Functionality Is Key: Recent Progress in the Surface Modification of Nanodiamond. *Adv. Funct. Mater.* **2012**, *22* (5), 890–906.
4. (a) Yang, J. H. C.; Teii, K. Mechanism of Enhanced Wettability of Nanocrystalline Diamond Films by Plasma Treatment. *Thin Solid Films* **2012**, *520* (21), 6566–6570;
(b) Pinzari, F.; Ascarelli, P.; Cappelli, E.; Mattei, G.; Giorgi, R. Wettability of HF-CVD Diamond Films. *Diam. Relat. Mater.* **2001**, *10* (3–7), 781–785;
(c) Park, Y.-S.; Son, H.-G.; Kim, D.-H.; Oh, H.-G.; Lee, D.-S.; Kim, M.-H.; Lim, K.-M.; Song, K.-S. Microarray of Neuroblastoma Cells on the Selectively Functionalized Nanocrystalline Diamond Thin Film Surface. *Appl. Surf. Sci.* **2016**, *361*, 269–276.
5. (a) Liu, F. B.; Wang, J. D.; Liu, B.; Li, X. M.; Chen, D. R. Effect of Electronic Structures on Electrochemical Behaviors of Surface-Terminated Boron-Doped Diamond Film Electrodes. *Diam. Relat. Mater.* **2007**, *16* (3), 454–460;
(b) Brown, N.; Hod, O. Controlling the Electronic Properties of Nanodiamonds via Surface Chemical Functionalization: A DFT Study. *J. Phys. Chem. C* **2014**, *118* (10), 5530–5537;
(c) Ristein, J.; Riedel, M.; Ley, L.; Takeuchi, D.; Okushi, H. Band Diagrams of Intrinsic and p-Type Diamond With Hydrogenated Surfaces. *Phys. Status Solidi A: Appl. Res.* **2003**, *199* (1), 64–70;
(d) Speranza, G.; Torrengo, S.; Filippi, M.; Minati, L.; Vittone, E.; Pasquarelli, A.; Dipalo, M.; Kohn, E. In Situ Thermal Treatment of UV-Oxidized Diamond Hydrogenated Surface. *Surf. Sci.* **2010**, *604* (9–10), 753–761;
(e) Ristein, J. Surface Science of Diamond: Familiar and Amazing. *Surf. Sci.* **2006**, *600* (18), 3677–3689;
(f) Lin, I.-N.; Koizumi, S.; Yater, J.; Koeck, F. Diamond Electron Emission. *MRS Bull.* **2014**, *39* (6), 533–541.
6. Schmidlin, L.; Pichot, V.; Comet, M.; Josset, S.; Rabu, P.; Spitzer, D. Identification, Quantification and Modification of Detonation Nanodiamond Functional Groups. *Diam. Relat. Mater.* **2012**, *22*, 113–117.
7. (a) Angus, J. C.; Argoitia, A.; Gat, R.; Li, Z.; Sunkara, M.; Wang, L.; Wang, Y. Chemical Vapor Deposition of Diamond. *Phil. Trans. R. Soc. A* **1993**, *342*, 195–208;
(b) Butler, J. E.; Mankelevich, Y. A.; Cheesman, A.; Ma, J.; Ashfold, M. N. R. Understanding the Chemical Vapor Deposition of Diamond: Recent Progress. *J. Phys.: Condens. Matter* **2009**, *21* (36), 364201;
(c) Williams, O. A. Nanocrystalline Diamond. *Diam. Relat. Mater.* **2011**, *20* (5–6), 621–640.
8. (a) Szunerits, S.; Nebel, C. E.; Hamers, R. J. Surface Functionalization and Biological Applications of CVD Diamond. *MRS Bull.* **2014**, *39*, 517–524;
(b) Krueger, A. Diamond Nanoparticles: Surface Modifications and Applications. In Sarin, V., Ed., Comprehensive Hard Materials, Vol. 3; Elsevier: Oxford, UK, 2014, pp. 379–406.

9. Burton, J. W. Reduction of C=X to CH2. In Knochel, P., Molander, G. A., Eds., Comprehensive Organic Synthesis, Vol. 8, 2nd edition; 2014, pp. 446–478.
10. Chenneberg, L.; Goddard, J.-P.; Fensterbank, L. Reduction of Saturated Alcohols and Amines to Alkanes. In Knochel, P., Molander, G. A., Eds., Comprehensive Organic Synthesis, Vol. 8, 2nd edition; Elsevier: Oxford, UK, 2014, pp. 1011–1030.
11. Spitsyn, B. V.; Denisov, S. A.; Skorik, N. A.; Chopurova, A. G.; Parkaeva, S. A.; Belyakova, L. D.; Larionov, O. G. *Diam. Relat. Mater.* **2010**, *19*, 123–127.
12. Williams, O. A.; Hees, J.; Dieker, C.; Jager, W.; Kirste, L.; Nebel, C. E. *ACS Nano* **2010**, *4*, 4824–4830.
13. Mitev, D. P.; Townsend, A. T.; Paull, B.; Nesterenko, P. N. Screening of Elemental Impurities in Commercial Detonation Nanodiamond Using Sector Field Inductively Coupled Plasma-Mass Spectrometry. *J. Mater. Sci.* **2014**, *49* (10), 3573–3591.
14. Yeganeh, M.; Coxon, P. R.; Brieva, A. C.; Dhanak, V. R.; Šiller, L.; Butenko, Yu. V. *Phys. Rev. B* **2007**, *75*, 155404.
15. Weng, S. Y.; Chen, S.; Loh, K. P. *Langmuir* **2009**, *25*, 185–191.
16. Girard, H. A.; Arnault, J.-C.; Perruchas, S.; Saada, S.; Gacoin, T.; Boilot, J.-P.; Bergonzo, P. *Diam. Relat. Mater.* **2010**, *19*, 1117–1123.
17. Girard, H. A.; Petit, T.; Perruchas, S.; Gacoin, T.; Gesset, C.; Arnault, J. C.; Bergonzo, P. *PCCP* **2011**, *13*, 11481.
18. Lai, L.; Barnard, A. S. *Nanoscale* **2011**, *3*, 2566–2575.
19. Nguyen, T.-T.-B.; Chang, H. C.; Wu, V. W.-K. *Diam. Relat. Mater.* **2007**, *16*, 872–876.
20. Bogatyreva, G. P.; Marinich, M. A.; Oleinik, G. S.; Bazalii, G. A.; Bilochenko, V. A.; Panova, A. N. *J. Superhard Mater.* **2011**, *33*, 208–216.
21. Shenderova, O.; Koscheev, A.; Zaripov, N.; Petrov, I.; Skryabin, Y.; Detkov, P.; Turner, S.; Van Tendeloo, G. Surface Chemistry and Properties of Ozone-Purified Detonation Nanodiamonds. *J. Phys. Chem. C* **2011**, *115*, 9827–9837.
22. Rakha, S. A.; Ali, N.; Haleem, Y. A.; Alam, F.; Khurram, A. A.; Munir, A. *J. Mater. Sci. Technol. (Shenyang, China)* **2014**, *30* (8), 753–758.
23. Osswald, S.; Yushin, G.; Mochalin, V.; Kucheyev, S. O.; Gogotsi, Y. *J. Am. Chem. Soc.* **2006**, *128*, 11635–11642.
24. Gaebel, T.; Bradac, C.; Chen, J.; Say, J. M.; Brown, L.; Hemmer, P.; Rabeau, J. R. *Diam. Relat. Mater.* **2012**, *21*, 28–32.
25. (a) Pichot, V.; Comet, M.; Fousson, E.; Baras, C.; Senger, A.; Le Normand, F.; Spitzer, D. *Diam. Relat. Mater.* **2008**, *17*, 13–22;
(b) Osswald, S.; Havel, M.; Mochalin, V.; Yushin, G.; Gogotsi, Y. *Diam. Relat. Mater.* **2008**, *17*, 1122–1126.
26. (a) Krüger, A.; Liang, Y.; Jarre, G.; Stegk, J. *J. Mater. Chem.* **2006**, *16*, 2322–2328;
(b) Ciftan Hens, S.; Cunningham, G.; Tyler, T.; Moseenkov, S.; Kuznetsov, V.; Shenderova, O. *Diam. Relat. Mater.* **2008**, *17*, 1858–1866.
27. (a) Martín, R.; Alvaro, M.; Herance, J. R.; García, H. *ACS Nano* **2010**, *4*, 65–74;
(b) Jirásek, V.; Lukeš, P.; Kozak, H.; Artemenko, A.; Člupek, M.; Čermák, J.; Rezek, B.; Kromka, A. *RSC Adv.* **2016**, *6*, 2352–2360.
28. Gordeev, S. K.; Korchagina, S. B. *J. Superhard Mater.* **2007**, *29*, 124–125.
29. Shenderova, O.; Panich, A. M.; Moseenkov, S.; Hens, S. C.; Kuznetsov, V.; Vieth, H.-M. *J. Phys. Chem. C* **2011**, *115* (39), 19005–19011.
30. Lai, L.; Barnard, A. S. *J. Phys. Chem. C* **2011**, *115*, 6218–6228.
31. Koch, H.; Kulisch, W.; Popov, C.; Merz, R.; Merz, B.; Reithmaier, J. P. *Diam. Relat. Mater.* **2011**, *20*, 254–258.

32. Sotowa, K.-I.; Amamoto, T.; Sobana, A.; Kusakabe, K.; Imato, T. *Diam. Relat. Mater.* **2004**, *13*, 145–150.
33. Stanishevsky, A. V.; Walock, M. J.; Catledge, S. A. *Appl. Surf. Sci. B* **2015**, *357*, 1403–1409.
34. Basiuk, E. V.; Santamaria-Bonfil, A.; Meza-Laguna, V.; Gromovoy, T. Yu.; Alvares-Zauco, E.; Contreras-Torres, F. F.; Rizo, J.; Zavala, G.; Basiuk, V. A. *Appl. Surf. Sci.* **2013**, 324–334.
35. Mochalin, V. N.; Neitzel, I.; Etzold, B. J. M.; Peterson, A.; Palmese, G.; Gogotsi, Y. *ACS Nano* **2011**, *5*, 7494–7502.
36. (a) Gaillard, C.; Girard, H. A.; Falck, C.; Paget, V.; Simic, V.; Ugolin, N.; Bergonzo, P.; Chevillard, S.; Arnault, J. C. *RSC Adv.* **2014**, *4*, 3566–3572;
(b) Dong, Y.; Cao, R.; Li, Y.; Wang, Z.; Li, L.; Tian, L. *RSC Adv.* **2015**, *5*, 82711–82716.
37. Liu, Y.; Gu, Z.; Margrave, J. L.; Khabashesku, V. N. *Chem. Mater.* **2004**, *16*, 3924–3930.
38. Wang, Y.; Huang, H.; Zang, J.; Meng, F.; Dong, L.; Su, J. *Int. J. Electrochem. Sci.* **2012**, *7*, 6807–6815.
39. Yakovlev, R. Yu.; Kulakova, I. I.; Leonidov, N. B.; Lisichkin, G. V. *Mendeleev Commun.* **2012**, *22*, 213–214.
40. Jarre, G.; Heyer, S.; Memmel, E.; Meinhardt, T.; Krueger, A. *Beilstein J. Org. Chem.* **2014**, *10*, 2729–2737.
41. Nichols, B. M.; Butler, J. E.; Russell, J. N. Jr.; Hamers, R. J. *J. Phys. Chem. B* **2005**, *109*, 20938–20947.
42. Hopper, A. P.; Dugan, J. M.; Gill, A. A.; Fox, O. J. L.; May, P. W.; Haycock, J. W.; Claeyssens, F. *Biomed. Mater.* **2014**, *9*, 045009.
43. (a) Nakamura, T.; Ohana, T.; Hagiwara, Y.; Tsubota, T. *Phys. Chem. Chem. Phys.* **2009**, *11* (4), 730–734;
(b) Nakamura, T.; Ohana, T. *Jpn. J. Appl. Phys.* **2012**, *51* (8, Pt. 1), 085201;
(c) Nakamura, T.; Ohana, T. *Trans. Mater. Res. Soc. Jpn.* **2013**, *38* (3), 415–418.
44. Su, S.; Wei, J.; Zhang, K.; Qiu, J.; Wang, S. *Colloid Polym. Sci.* **2015**, *293* (4), 1299–1305.
45. Hsu, M.-H.; Chuang, H.; Cheng, F.-Y.; Huang, Y.-P.; Han, C.-C.; Chen, J.-Y.; Huang, S.-C.; Chen, J.-K.; Wu, D.-S.; Chu, H.-L.; Chang, C.-C. *ACS Appl. Mater. Interfaces* **2014**, *6*, 7198–7203.
46. Liu, Y.; Gu, Z.; Margrave, J. L.; Khabashesku, V. N. *Chem. Mater.* **2004**, *16*, 3924.
47. Ivanov, M. G.; Ivanov, D. M.; Pavlyshko, S. V.; Petrov, I.; Vargas, A.; McGuire, G.; Shenderova, O. *Fuller. Nanotub. Carbon Nanostructures* **2012**, *20*, 606–610.
48. Huang, H.; Wang, Y. H.; Zang, J. B.; Bian, L. Y. *Appl. Surf. Sci.* **2012**, *258*, 4079–4084.
49. Zagrebina, E. M.; Generalov, A. V.; Klyushin, A. Yu.; Simonov, K. A.; Vinogradov, N. A.; Dubois, M.; Frezet, L.; Martensson, N.; Preobrajenski, A. B.; Vinogradov, A. S. *J. Phys. Chem. C* **2015**, *119*, 835–844.
50. Lisichkin, G. V.; Kulakova, I. I.; Gerasimov, Y. A.; Karpukhin, A. V.; Yakovlev, R. Yu. *Mendeleev Commun.* **2009**, *19*, 309–310.
51. Voznyakovskii, A. P.; Kalinin, A. V.; Mokeev, M. V.; Agibalova, L. V.; Vlasova, E. N. *Russ. J. Appl. Chem.* **2012**, *85* (7), 1090–1094.
52. Panich, A. M.; Sergeev, N. A.; Olszewski, M.; Froumin, N.; Dideykin, A. T.; Sokolov, V. V.; Vul, A. Ya. *J. Nanosci. Nanotechnol.* **2015**, *15*, 1030–1036.
53. Petit, T.; Arnault, J.-C.; Girard, H. A.; Sennour, M.; Bergonzo, P. *Phys. Rev. B* **2011**, *84*, 233407.
54. Jarre, G.; Liang, Y.; Betz, P.; Lang, D.; Krueger, A. *Chem. Commun.* **2011**, *47*, 544–546.

55. Tiwari, A. K.; Goss, J. P.; Briddon, P. R.; Wright, N. G.; Horsfall, A. B.; Rayson, M. J. *Phys. Status Solidi A* **2012**, *209*, 1703–1708.
56. Martín, R.; Concepcion Heydorn, P.; Alvaro, M.; Garcia, H. *Chem. Mater.* **2009**, *21*, 4505–4514.
57. Kuznetsov, V. L.; Butenko, Y. V. Diamond Phase Transitions at Nanoscale. In Shenderova, O. A., Gruen, D. M., Eds., *Ultrananocrystalline Diamond;* 2nd edition; Elsevier Ltd.: Oxford, UK, 2012, pp. 181–244.
58. Panich, A. M.; Shames, A. I.; Sergeev, N. A.; Olszewski, M.; McDonough, J. K.; Mochalin, V. N.; Gogotsi, Y. *J. Phys.: Condens. Matter* **2013**, *25*, 245303.
59. Chen, J.; Deng, S. Z.; Chen, J.; Yu, Z. X.; Xu, N. S. *Appl. Phys. Lett.* **1999**, *74*, 3651–3653.
60. Zeiger, M.; Jaeckel, N.; Weingarth, D.; Presser, V. *Carbon* **2015**, *94*, 507–517.
61. Butenko, Yu. V.; Kuznetsov, V. L.; Paukshtis, E. A.; Stadnichenko, A. I.; Mazov, I. N.; Moseenkov, S. I.; Boronin, A. I.; Kosheev, S. V. *Fuller. Nanotub. Carbon Nanostructures* **2006**, *14*, 557–564.
62. Cebik, J.; McDonough, J. K.; Peerally, F.; Medrano, R.; Neitzel, I.; Gogotsi, Y.; Osswald, S. *Nanotechnology* **2013**, *24* (20), 205703.
63. Zang, J.; Wang, Y.; Bian, L.; Zhang, J.; Meng, F.; Zhao, Y.; Ren, S.; Qu, X. *Electrochim. Acta* **2012**, *72*, 68–73.
64. Butenko, Y. V.; Kuznetsov, V. L.; Chuvilin, A. L.; Kolomiichuk, V. N.; Stankus, S. V.; Khairulin, R. A.; Segall, B. *J. Appl. Phys.* **2000**, *88*, 4380–4388.
65. Mona, J.; Tu, J.-S.; Kang, T.-Y.; Tsai, C.-Y.; Perevedentseva, E.; Cheng, C.-L. *Diam. Relat. Mater.* **2012**, *24*, 134–138.
66. Xie, F. Y.; Xie, W. G.; Gong, L.; Zhang, W. H.; Chen, S. H.; Zhang, Q. Z.; Chen, J. *Surf. Interface Anal.* **2010**, *42*, 1514–1518.
67. Krueger, A.; Ozawa, M.; Jarre, G.; Liang, Y.; Stegk, J.; Lu, L. *Phys. Status Solidi A* **2007**, *204*, 2881–2887.
68. Liang, Y.; Ozawa, M.; Krueger, A. *ACS Nano* **2009**, *3*, 2288–2296.
69. Rojas, S.; Gispert, J. D.; Martin, R.; Abad, S.; Menchon, C.; Pareto, D.; Victor, V. M.; Alvaro, M.; Garcia, H.; Herance, J. R. *ACS Nano* **2011**, *5*, 5552–5559.
70. (a) Prabhakar, N.; Näreoja, T.; von Haartman, E.; Sxen Karaman, D.; Jiang, H.; Koho, S.; Dolenko, T. A.; Hänninen, P. E.; Vlasov, D. I.; Ralchenko, V. G.; Hosomi, S.; Vlasov, I. I.; Sahlgren, C.; Rosenholm, J. M. *Nanoscale* **2013**, *5*, 3713–3722;
(b) von Haartman, E.; Jiang, H.; Khomich, A. A.; Zhang, J.; Burikov, S. A.; Dolenko, T. A.; Ruokolainen, J.; Gu, H.; Shenderova, O. A.; Vlasov, I. I.; Rosenholm, J. M. *J. Mater. Chem. B* **2013**, *1*, 2358–2366.
71. Goncalves, J. P. L.; Shaikh, A. Q.; Reitzig, M.; Kovalenko, D. A.; Michael, J.; Beutner, R.; Cuniberti, G.; Scharnweber, D.; Opitz, J. *Beilstein J. Org. Chem.* **2014**, *10*, 2765–2773.
72. Tisler, J.; Reuter, R.; Laemmle, A.; Jelezko, F.; Balasubramanian, G.; Hemmer, P. R.; Reinhard, F.; Wrachtrup, J. *ACS Nano* **2011**, *5*, 7893–7898.
73. Duffy, E.; Mitev, D. P.; Thickett, S. C.; Townsend, A. T.; Paull, B.; Nesterenko, P. N. *Appl. Surf. Sci. A* **2015**, *357*, 397–406.
74. Rehor, I.; Slegerova, J.; Kucka, J.; Proks, V.; Petrakova, V.; Adam, M.-P.; Treussart, F.; Turner, S.; Bals, S.; Sacha, P.; Ledvina, M.; Wen, A. M.; Steinmetz, N. F.; Cigler, P. *Small* **2014**, *10*, 1106–1115.
75. Bumb, A.; Sarkar, S. K.; Billington, N.; Brechbiel, M. W.; Neuman, K. C. *J. Am. Chem. Soc.* **2013**, *135*, 7815–7818.

76. Siengalewicz, P.; Mulzer, J.; Rinner, U. Synthesis of Esters and Lactones. In Knochel, P., Molander, G. A., Eds., Comprehensive Organic Synthesis, Vol. 6, 2nd edition; Elsevier: Oxford, UK, 2014, pp. 355–410.
77. Krueger, A.; Boedeker, T. *Diam. Relat. Mater.* **2008**, *17*, 1367–1370.
78. Zhang, X.; Fu, C.; Feng, L.; Ji, Y.; Tao, L.; Huang, Q.; Li, S.; Wei, Y. *Polymer* **2012**, *53*, 3178–3184.
79. Liu, M.; Xu, D.; Wang, K.; Deng, F.; Wan, Q.; Zeng, G.; Huang, Q.; Zhang, X.; Wei, Y. *RSC Adv.* **2015**, *5*, 96983–96989.
80. Zhao, L.; Takimoto, T.; Ito, M.; Kitagawa, N.; Kimura, T.; Komatsu, N. *Angew. Chem. Int. Ed.* **2011**, *50*, 1388–1392.
81. Boudou, J.-P.; David, M.-O.; Joshi, V.; Eidi, H.; Curmi, P. A. *Diam. Relat. Mater.* **2013**, *38*, 131–138.
82. Barras, A.; Lyskawa, J.; Szunerits, S.; Woisel, P.; Boukherroub, R. *Langmuir* **2011**, *27*, 12451–12457.
83. Khanal, M.; Turcheniuk, V.; Barras, A.; Rosay, E.; Bande, O.; Siriwardena, A.; Zaitsev, V.; Pan, G.-H.; Boukherroub, R.; Szunerits, S. *Langmuir* **2015**, *31*, 3926–3933.
84. Dhanak, V. R.; Butenko, Yu. V.; Brieva, A. C.; Coxon, P. R.; Alves, L.; Siller, L. *J. Nanosci. Nanotechnol.* **2012**, *12*, 3084–3090.
85. Kaur, R.; Chitanda, J. M.; Michel, D.; Maley, J.; Borondics, F.; Yang, P.; Verrall, R. E.; Badea, I. *Int. J. Nanomed.* **2012**, *7*, 3851–3866.
86. Sreenivasan, V. K. A.; Ivukina, E. A.; Deng, W.; Kelf, T. A.; Zdobnova, T. A.; Lukash, S. V.; Veryugin, B. V.; Stremovskiy, O. A.; Zvyagin, A. V.; Deyev, S. M. *J. Mater. Chem.* **2011**, *21*, 65–68.
87. Xu, G.; Zhang, W.; Wei, L.; Lu, H.; Yang, P. *Analyst* **2013**, *138*, 1876–1885.
88. Knapinska, A. M.; Tokmina-Roszyk, D.; Amar, S.; Tokmina-Roszyk, M.; Mochalin, V. N.; Gogotsi, Y.; Cosme, P.; Terentis, A. C.; Fields, G. B. *Biopolymers* **2015**, *104*, 186–195.
89. Presti, C.; Alauzun, J. G.; Laurencin, D.; Mutin, P. H. *Chem. Mater.* **2013**, *25*, 2051–2055.
90. Presti, C.; Thankamony, A. S. L.; Alauzun, J. G.; Mutin, P. H.; Carnevale, D.; Lion, C.; Vezin, H.; Laurencin, D.; Lafon, O. *J. Phys. Chem. C* **2015**, *119*, 12408–12422.
91. Presti, C.; Alauzun, J. G.; Laurencin, D.; Mutin, P. H. *Langmuir* **2014**, *30*, 9239–9245.
92. Wahab, Z.; Foley, E. A.; Pellechia, P. J.; Anneaux, B. L.; Ploehn, H. J. *J. Colloid Interface Sci.* **2015**, *450*, 301–309.
93. Beyler-Cigil, A.; Cakmakci, E.; Kahraman, M. V. *Polym. Compos.* **2015**, *37*, 2285–2292.
94. Bahr, J. L.; Yang, J.; Kosynkin, D. V.; Bronikowski, M. J.; Smalley, R. E.; Tour, J. M. *J. Am. Chem. Soc.* **2001**, *123*, 6536–6542.
95. Wang, J.; Firestone, M. A.; Auciello, O.; Carlisle, J. A. *Langmuir* **2004**, *20*, 11450–11456.
96. Mohamed, A. A.; Salmi, Z.; Dahoumane, S. A.; Mekki, A.; Carbonnier, B.; Chehimi, M. M. *Adv. Colloid Interface Sci.* **2015**, *225*, 16–36.
97. Liang, Y.; Meinhardt, T.; Jarre, G.; Ozawa, M.; Vrdoljak, P.; Schoell, A.; Reinert, F.; Krueger, A. *J. Colloid Interface Sci.* **2011**, *354*, 23–30.
98. Meinhardt, T.; Lang, D.; Dill, H.; Krueger, A. *Adv. Funct. Mater.* **2011**, *21*, 494–500.
99. Mangeney, C.; Qin, Z.; Dahoumane, S. A.; Adenier, A.; Herbst, F.; Boudou, J.-P.; Pinson, J.; Chehimi, M. M. *Diam. Relat. Mater.* **2008**, *17*, 1881–1887.
100. Qi, D.; Liu, L.; Gao, X.; Ouyang, T.; Chen, S.; Loh, K. P.; Wee, A. T. S. *Langmuir* **2007**, *19*, 9722–9727.
101. Belik, P.; Guegel, A.; Kraus, A.; Walter, M.; Muellen, K. *J. Org. Chem.* **1995**, *60*, 3307–3310.

102. Lang, D.; Krueger, A. *Diam. Relat. Mater.* **2011**, *20*, 101–104.
103. (a) Bingel, C. *Chem. Ber.* **1993**, *126*, 1957–1959;
(b) Bingel, C.; Schiffer, H. *Liebigs Ann.* **1995**, 1551–1553;
(c) Camps, X.; Hirsch, A. *J. Chem. Soc. Perkin Trans.* **1997**, *1*, 1595–1596.
104. Betz, P.; Krueger, A. *Chem. Phys. Chem.* **2012**, *13*, 2578–2584.
105. Kuznetsov, O.; Sun, Y.; Thaner, R.; Bratt, A.; Shenoy, V.; Wong, M. S.; Jones, J.; Billups, W. E. *Langmuir* **2012**, *28*, 5243–5248.
106. Korolkov, V. V.; Kulakova, I. I.; Tarasevich, B. N.; Lisichkin, G. V. *Diam. Relat. Mater.* **2007**, *16*, 2129–2132.
107. El Tall, O.; Hou, Y.; Abou-Hamad, E.; Raja, I. U.; Hedhili, M. N.; Peng, W.; Mahfouz, R.; Bakr, O. M.; Beaujuge, P. M. *Chem. Mater.* **2014**, *26*, 2766–2769.
108. Wang, H.; Griffiths, J.-P.; Egdell, R. G.; Moloney, M. G.; Foord, J. S. *Langmuir* **2008**, *24*, 862–868.
109. Cha, I.; Hashimoto, K.; Fujiki, K.; Yamauchi, T.; Tsubokawa, N. *Mater. Chem. Phys.* **2014**, *143*, 1131–1138.
110. Ahmad, M. N.; Zia, K. M.; Nadeem, S.; Farooq, T.; Anium, M. N.; Shah, A. T.; Khurshid, S.; Shehzad, K.; Zuber, M.; Yang, W. *Asian J. Chem.* **2013**, *25*, 9840–9844.
111. Romanova, E. E.; Akiel, R.; Cho, F. H.; Takahashi, S. *J. Phys. Chem. A* **2013**, *117*, 11933–11939.
112. Tzeng, Y.-K.; Faklaris, O.; Chang, B.-M.; Kuo, Y.; Hsu, J.-H.; Chang, H.-C. *Angew. Chem. Int. Ed.* **2011**, *50*, 2262–2265.
113. Perevedentseva, E.; Cai, P.-J.; Chiu, Y.-C.; Cheng, C.-L. *Langmuir* **2011**, *27*, 1085–1091.
114. Mogilnaya, O.; Bondar, V. *Micro Nanosyst.* **2012**, *4*, 41–47.
115. Suliman, S.; Xing, Z.; Wu, X.; Xue, Y.; Pedersen, T. O.; Sun, Y.; Døskeland, A. P.; Nickel, J.; Waag, T.; Lygre, H.; Finne-Wistrand, A.; Steinmüller-Nethl, D.; Krueger, A.; Mustafa, K. *J. Control. Release* **2015**, *197*, 148–157.
116. Swiecicki, J.-M.; Tailhades, J.; Lepeltier, E.; Chassaing, G.; Lavielle, S.; Mansuy, C. *Colloids Surf. A* **2013**, *431*, 73–79.
117. Bradac, C.; Volz, T.; Say, J. M.; Rastogi, I. D.; Cordina, N. M.; Brown, L. *J. Biophotonics* **2016**, *9*, 296–304.
118. Weng, M.-F.; Chang, B.-J.; Chiang, S.-Y.; Wang, N.-S.; Niu, H. *Diam. Relat. Mater.* **2012**, *22*, 96–104.
119. Zhao, L.; Nakae, Y.; Qin, H.; Ito, T.; Kimura, T.; Kojima, H.; Chan, L.; Komatsu, N. *J. Org. Chem.* **2014**, *10*, 707–713.
120. Slegerova, J.; Hajek, M.; Rehor, I.; Sedlak, F.; Stursa, J.; Hruby, M.; Cigler, P. *Nanoscale* **2015**, *7*, 415–420.
121. Kuznetsov, O. V.; Pulikkathara, M. X.; Lobo, R. F. M.; Khabasheskua, V. N. *Russ. Chem. Bull.* **2010**, *59*, 1495–1505.
122. Kolb, H. C.; Finn, M. G.; Sharpless, K. B. *Angew. Chem. Int. Ed.* **2001**, *40*, 2004–2021.
123. (a) Khanal, M.; Larsonneur, F.; Raks, V.; Barras, A.; Baumann, J.-S.; Martin, F. A.; Boukherroub, R.; Ghigo, J.-M.; Ortiz Mellet, C.; Zaitsev, V.; Garcia Fernandez, J. M.; Beloin, C.; Siriwardena, A.; Szunerits, S. *Nanoscale* **2015**, *7*, 2325–2335;
(b) Siriwardena, A.; Khanal, M.; Barras, A.; Bande, O.; Mena-Barragan, T.; Mellet, C. O.; Garcia Fernandez, J. M.; Boukherroub, R.; Szunerits, S. *RSC Adv.* **2015**, *5*, 100568–100578.
124. Barras, A.; Martin, F. A.; Bande, O.; Baumann, J.-S.; Ghigo, J.-M.; Boukherroub, R.; Beloin, C.; Siriwardena, A.; Szunerits, S. *Nanoscale* **2013**, *5*, 2307–2316.
125. Hartmann, M.; Betz, P.; Sun, Y.; Gorb, S. N.; Lindhorst, T. K.; Krueger, A. *Chem. Eur. J.* **2012**, *18*, 6485–6492.

126. Fessele, C.; Wachtler, S.; Chandrasekaran, V.; Stiller, C.; Lindhorst, T. K.; Krueger, A. *Eur. J. Org. Chem.* **2015**, 5519–5525.
127. Kühne, M.; Györgydeák, Z.; Lindhorst, T. K. *Synthesis* **2006**, *6*, 949–951.
128. Zhang, T.; Neumann, A.; Lindlau, J.; Wu, Y.; Pramanik, G.; Naydenov, B.; Jelezko, F.; Schüder, F.; Huber, S.; Huber, M.; Stehr, F.; Högele, A.; Weil, T.; Liedl, T. *J. Am. Chem. Soc.* **2015**, *137*, 9776–9779.
129. Wang, Z.; Tian, Z.; Dong, Y.; Li, L.; Tian, L.; Li, Y.; Yang, B. *Diam. Relat. Mater.* **2015**, *58*, 84–93.
130. Su, S.; Wang, S.; Qiu, J. *Sci. Adv. Mater.* **2014**, *6*, 202–207.
131. Maitra, U.; Jain, A.; George, S. J.; Rao, C. N. R. *Nanoscale* **2011**, *3*, 3192–3197.
132. Fudala, R.; Raut, S.; Maliwal, B. P.; Zerda, T. W.; Gryczynski, I.; Simanek, E.; Borejdo, J.; Rich, R.; Akopova, I.; Gryczynski, Z. *Curr. Pharm. Biotechnol.* **2014**, *14*, 1127–1133.
133. Girard, H. A.; El-Kharbachi, A.; Garcia-Argote, S.; Petit, T.; Bergonzo, P.; Rousseau, B.; Arnault, J.-C. *Chem. Commun.* **2014**, *50*, 2916–2918.
134. Badun, G. A.; Chernysheva, M. G.; Yakovlev, R. Yu.; Leonidov, N. B.; Semenenko, M. N.; Lisichkin, G. V. *Radiochim. Acta* **2014**, *102*, 941–946.
135. Purtov, K. V.; Petunin, A. I.; Burov, A. E.; Puzyr, A. P.; Bondar, V. S. *Nanoscale Res. Lett.* **2010**, *5*, 631–636.
136. Manus, L. M.; Mastarone, D. J.; Waters, E. A.; Zhang, X.-Q.; Schultz-Sikma, E. A.; MacRenaris, K. W.; Ho, D.; Meade, T. J. *Nano Lett.* **2010**, *10*, 484–489.
137. Zhao, L.; Shiino, A.; Qin, H.; Kimura, T.; Komatsu, N. *J. Nanosci. Nanotechnol.* **2015**, *15*, 1076–1082.
138. Osipov, V. Yu.; Aleksenskiy, A. E.; Takai, K.; Vul, A. Ya. *Phys. Solid State* **2015**, *57*, 2314–2319.
139. Khanal, M.; Vausselin, T.; Barras, A.; Bande, O.; Turcheniuk, K.; Benazza, M.; Zaitsev, V.; Teodorescu, C. M.; Boukherroub, R.; Siriwardena, A.; Dubuisson, J.; Szunerits, S. *ACS Appl. Mater. Interfaces* **2013**, *23*, 12488–12498.
140. Chonpathompikunlert, P.; Yoshitomi, T.; Han, J.; Toh, K.; Isoda, H.; Nagasaki, Y. *Ther. Deliv.* **2011**, *2*, 585–597.
141. Doerdelmann, G.; Meinhardt, T.; Sowik, T.; Krueger, A.; Schatzschneider, U. *Chem. Commun.* **2012**, *48*, 11528–11530.
142. Huynh, V. T.; Pearson, S.; Noy, J.-M.; Abboud, A.; Utama, R. H.; Lu, H.; Stenzel, M. H. *ACS Macro Lett.* **2013**, *2*, 246–250.
143. Zhao, L.; Xu, Y.-H.; Qin, H.; Abe, S.; Akasaka, T.; Chano, T.; Watari, F.; Kimura, T.; Komatsu, N.; Chen, X. *Adv. Funct. Mater.* **2014**, *24*, 5348–5357.
144. Mochalin, V. N.; Gogotsi, Y. *Diam. Relat. Mater.* **2015**, *58*, 161–171.
145. Ullah, M.; Kausar, A.; Siddiq, M.; Subhan, M.; Abid Zia, M. *Polym.-Plast. Technol. Eng.* **2015**, *54*, 861–879.
146. Jabeen, S.; Kausar, A.; Muhammad, B.; Gul, S.; Farooq, M. *Polym.-Plast. Technol. Eng.* **2015**, *54*, 1379–1409.
147. Badamshina, E.; Estrin, Y.; Gafurova, M. *J. Mater. Chem. A* **2013**, *1*, 6509–6529.
148. Neitzel, I.; Mochalin, V.; Gogotsi, Y. In Shenderova, O. A., Gruen, D. M., Eds., *Ultrananocrystalline Diamond;* 2nd ed.; 2012, pp. 421–456.
149. Neitzel, I.; Mochalin, V. N.; Niu, J.; Cuadra, J.; Kontsos, A.; Palmese, G. R.; Gogotsi, Y. *Polymer* **2012**, *53*, 5965–5971.
150. Guo, H.; Sheng, H.; Peng, X.; Yu, X.; Naito, K.; Qu, X.; Zhang, Q. *Polym. Compos.* **2014**, *35*, 2144–2149.
151. Khan, M.; Shahzad, N.; Xiong, C.; Zhao, T. K.; Li, T.; Siddique, F.; Ali, N.; Shahzad, M.; Ullah, H.; Rakha, S. A. *Diam. Relat. Mater.* **2016**, *61*, 32–40.

152. Manandhar, S.; Roder, P. B.; Hanson, J. L.; Lim, M.; Smith, B. E.; Mann, A.; Pauzauskie, P. J. *J. Mater. Res.* **2014**, *29*, 2905–2911.
153. Alishiri, M.; Shojaei, A.; Abdekhodaie, M. J. *RSC Adv.* **2016**, *6*, 8743–8755.
154. Hsin, Y. L.; Chu, H.-Y.; Jeng, Y.-R.; Huang, Y.-H.; Wang, M. H.; Chang, C. K. *J. Mater. Chem.* **2011**, *21*, 13213–13222.
155. Ringstrand, B. S.; Seifert, S.; Podlesak, D. W.; Firestone, M. A. *Macromol. Rapid Commun.* **2016**, *37*, 1155–1167.
156. Zhang, H. H.; Liu, Y. T.; Wang, R.; Yu, X. Y.; Qu, X. W.; Zhang, Q. X. *Chin. Chem. Lett.* **2011**, 485–488.
157. Branson, B. T.; Beauchamp, P. S.; Beam, J. C.; Lukehart, C. M.; Davidson, J. L. *ACS Nano* **2013**, *7*, 3183–3189.
158. Sun, Y.; Olsén, P.; Waag, T.; Krueger, A.; Steinmüller-Nethl, D.; Albertsson, A.-C.; Finne-Wistrand, A. *Part. Part. Syst. Charact.* **2015**, *32*, 35–42.
159. Liu, R.; Zhao, F.; Yu, X.; Naito, K.; Ding, H.; Qu, X.; Zhang, Q. *Diam. Relat. Mater.* **2014**, *50*, 26–32.
160. Zang, J.; Wang, Y.; Zhao, X.; Xin, G.; Sun, S.; Qu, X.; Ren, S. *Int. J. Electrochem. Sci.* **2012**, *7*, 1677–1687.
161. Gopalan, A. I.; Lee, K.-P.; Komathi, S. *Biosens. Bioelectron.* **2011**, *26*, 1638–1643.
162. Neitzel, I.; Mochalin, V.; Bares, J. A.; Carpick, R. W.; Erdemir, A.; Gogotsi, Y. *Tribol. Lett.* **2012**, *47*, 195–200.
163. Rakha, S. A.; Raza, R.; Munir, A. *Polym. Compos.* **2013**, *34*, 811–818.
164. Haleem, Y. A.; Liu, D.; Chen, W.; Wang, C.; Hong, C.; He, Z.; Liu, J.; Song, P.; Yu, S.; Song, L. *Comp. Part B* **2015**, *78*, 480–487.
165. (a) Zhang, Q.; Mochalin, V. N.; Neitzel, I.; Knoke, I. Y.; Han, J.; Klug, C. A.; Zhou, J. G.; Lelkes, P. I.; Gogotsi, Y. *Biomaterials* **2011**, *32*, 87–94;
(b) Zhang, Q.; Mochalin, V. N.; Neitzel, I.; Hazeli, K.; Niu, J.; Kontsos, A.; Zhou, J. G.; Lelkes, P. I.; Gogotsi, Y. *Biomaterials* **2012**, *33*, 5067–5075.
166. Xing, Z.; Pedersen, T. O.; Wu, X.; Xue, Y.; Sun, Y.; Finne-Wistrand, A.; Hellem, S.; Kloss, F. R.; Waag, T.; Krueger, A.; Steinmüller-Nethl, D.; Mustafa, K. *Tissue Eng.* **2013**, *19*, 1783–1791.
167. Sun, Y.; Finne-Wistrand, A.; Waag, T.; Xing, Z.; Yassin, M.; Yamamoto, A.; Mustafa, K.; Steinmüller-Nethl, D.; Krueger, A.; Albertsson, A.-C. *Macromol. Mater. Eng.* **2015**, *300*, 436–447.
168. Brady, M. A.; Renzing, A.; Douglas, T. E. L.; Liu, Q.; Wille, S.; Parizek, M.; Bacakova, L.; Kromka, A.; Jarosova, M.; Godier, G.; Warnke, P. H. *J. Nanosci. Nanotechnol.* **2015**, *15*, 1060–1069.
169. Bistričić, L.; Borjanović, V.; Leskovac, M.; Mikac, L.; McGuire, G. E.; Shenderova, O.; Nunn, N. *J. Polym. Res.* **2015**, *22*, 39.
170. Jee, A.-Y.; Lee, M. *J. Nanosci. Nanotechnol.* **2011**, *11*, 533–536.
171. Zhang, L.; Zhu, J.; Zhou, W. *Adv. Mater. Res.* **2012**, *347–353* (Pt. 5), 2764–2767.
172. Wang, D. H.; Fillery, S. P.; Durstock, M. F.; Dai, L.; Vaia, R. A.; Tan, L.-S. *Adv. Mater. Res.* **2013**, *785–786*, 410–416.
173. Xia, S.; Liu, Y.; Zhang, L.; Wang, D.; Zou, W.; Peng, J.; Cao, S. *J. Appl. Polym. Sci.* **2014**, *131*, 40244.
174. Ashraf, R.; Kausar, A.; Siddiq, M. *Iran. Polym. J.* **2014**, *23*, 531–545.
175. Sordi, D.; De Ruijter, C.; Orlanducci, S.; Picken, S. J.; Sudhoelter, E. J. R.; Terranova, M. L.; de Smet, L. C. P. M.; Dingemans, T. J. *J. Polym. Sci. A* **2011**, *49*, 1079–1087.
176. Morimune, S.; Kotera, M.; Nishino, T.; Goto, K.; Hata, K. *Macromolecules* **2011**, *44*, 4415–4421.

177. Ogata, T.; Yagi, R.; Nakamura, N.; Kuwahara, Y.; Kurihara, S. *ACS Appl. Mater. Interfaces* **2012**, *4*, 3769–3772.
178. Avagimova, N. V.; Toikka, A. M.; Polotskaya, G. A. *Pet. Chem.* **2015**, *55*, 276–282.
179. Cha, I.; Shirai, K.; Fujiki, K.; Yamauchi, T.; Tsubokawa, N. *Diam. Relat. Mater.* **2011**, *20*, 439–444.
180. Lau, X. C.; Desai, C.; Mitra, S. *Sol. Energy* **2013**, *91*, 204–211.
181. Cheng, L.-C.; Chen, H. M.; Lai, T.-C.; Chan, Y.-C.; Liu, R.-S.; Sung, J. C.; Hsiao, M.; Chen, C.-H.; Her, L.-J.; Tsai, D. P. *Nanoscale* **2013**, *5*, 3931–3940.
182. Maas, M.; Bollhorst, T.; Zare, R. N.; Rezwan, K. *Part. Part. Syst. Charact.* **2014**, *31*, 1067–1071.

CHAPTER

# HR-EELS investigations of hydrogenated nanodiamond films

# 9

**Shaul Michaelson, Alon Hoffman**
*Schulich Faculty of Chemistry, Technion, Haifa, Israel*

**Contents**



## 9.1 INTRODUCTION

High resolution electron energy loss spectroscopy (HR-EELS) detects meV losses of a highly monochromatic low energy electron beam scattered from a solid surface and thus probes vibrational modes localized within a few Å of the surface region.[1] It is highly sensitive to the 1–2 upper atomic monolayers and therefore provides unique information about physicochemical processes that occur on the very surface area, such as physical and chemisorption,[2] dissociative adsorption,[3] homo- and hetero-epitaxial deposition,[4] etc. The properties and associated practical applications of diamond films are significantly modified by their surface structure and composition.[5–9] The chemical characterization of the very surface region of diamond films is thus most important. For example, a monolayer of adsorbed hydrogen atoms induces negative electron affinity (NEA)[10] and a p-type surface conductivity,[6] while the last phenomenon is strongly affected by the presence of water adsorbents or $sp^2$ carbon.[11–13] The structure of the upper monolayer is critically important in homo-epitaxial diamond deposition, where crystal imperfections give rise to formation of unwanted non-epitaxial defects, hillocks[14–16] and twin boundaries.[17] Also, diamond surface transfer doping is a matter of the upper monolayer of the adsorbed species.

On the other hand, the chemical and physical characterization of the uppermost surface atomic layers of diamond films presents a great challenge. Common surface sensitive techniques (X-ray photoelectron spectroscopy (XPS), Auger electron spectroscopy (AES), electron energy loss spectroscopy (EELS), etc.) usually probe a few nanometers rather than the upper atomic surface layer.[18] The study of diamond films consisting of grain size in nanometric scale requires the use of quite complicated transmission electron microscopy usually combined with X-ray diffraction analysis. The quest of additional complementary techniques providing information about the grain size, grain boundaries, and film surface is of high importance.

HR-EELS was extensively applied in the past to study different aspects of adsorbed molecules and atoms on different carbon film surfaces.[3,19–23] This technique offers a major advantage over optical vibrational techniques such as Raman or IR-spectroscopy since it is intrinsically surface-sensitive. In addition to the vibrational properties of the substrate, surface low-energy electron scattering provides insights into the sample conduction-band density-of-states (DOS). In this review we present HR-EELS results of well-defined polycrystalline diamond films from micrometer grain size (few hundreds of nm) down to ultrananocrystalline diamond films with grain size of 2–5 nm which were deposited by different CVD methods.

## 9.2 ASSIGNMENT OF DIAMOND HR-EELS PEAKS

In this section we summarize the basic HR-EELS peak assignment based on the reduced mass shift related vibrations induced by isotopic exchange, by means of ion induced modifications of diamond surface and high grain boundary-to-bulk atomic ratio associated to nanodiamond films. Once the peak origin, its spectral position, and sensitivity are established, basic research on physicochemical properties can be conducted.

### 9.2.1 MODIFICATIONS INDUCED BY ISOTOPIC EXCHANGE

In this section the nature of the different peaks (C–C related or C–H related) appearing in the HR-EEL spectra of hydrogenated polycrystalline diamond films has been established.[19,21] It was achieved by analysis of the modifications of the HR-EEL spectra induced by the isotopic exchange of H by D and $^{12}C$ by $^{13}C$. This exchange was performed by using an appropriate gas mixture for the CVD growth of the films: (i) complete exchange of D by H (using $CD_4 + D_2$) or partial exchange (using $CH_4 + D_2$ or $CH_4 + D_2$) instead of ($CH_4 + H_2$) and (ii) replacement of $^{12}C$ atoms by $^{13}C$ (using $^{13}CH_4$+$H_2$ instead of $^{12}CH_4$+$H_2$). The completely exchanged samples were exposed to ex-situ MW-H (MW-D) plasma to ensure their H (D) termination. All samples were characterized by HR-EEL, XPS, and electronic EELS (plasmon scattering).

Hot filament (HF) chemical vapor deposited polycrystalline diamond films were grown using methane/hydrogen ratio 1/99, pressure 50 Torr, filament (Re) temperature 2000 °C, and silicon substrate temperature 800 °C.[24] The films were grown from gas mixtures of the following composition: $^{12}CH_4 + H_2$, $^{12}CD_4 + H_2$, $^{12}CH_4 + D_2$, $^{12}CD_4 + D_2$, $^{13}CH_4 + H_2$.[25] Hydrogenation was done ex situ in a commercial Astex CVD microwave reactor (Astex model AX5010) at a sample temperature of 800 °C and hydrogen (deuterium) pressure of 50 Torr.

The HR-EELS chamber is equipped with a Delta 0.5 spectrometer (VSI-SPECS) consisting of a double monochromator and a single analyzer housed in an ultra-high vacuum (UHV) system with base pressure of $\sim 8 \times 10^{-10}$ Torr. All spectra were recorded at room temperature in the specular geometry with an incident angle of 55° from the surface normal, incident energy of 5 eV, and full width half maximum (FWHM) of 5–8 meV and up to loss energies of 700 meV. The HR-EELS chamber is attached to a second UHV system equipped with various surface analytical techniques allowing in situ surface modification and characterization. The surface phase composition (plasmon losses) was analyzed in the second analytical chamber by electronic EELS with a primary electron energy of $E_p = 500$ eV and FWHM of 0.7 eV. The hydrogenated diamond surface was ion beam damaged by irradiation with 500 eV Ar ions. The chemical composition of the surface was monitored by X-ray photoelectron spectroscopy (XPS, not shown).

The HR-EEL spectra of the films deposited from the different gas mixtures are given in Fig. 9.1. The $^{12}CH_4 + H_2$ ($^{12}CD_4 + D_2$) film was exposed ex situ to MW-H (MW-D) plasma for 20 min. The two other (partially exchanged samples) were measured as-grown. Prior to the HR-EELS measurements, the samples were annealed at 450 °C under UHV conditions. The different vibrational modes are enumerated on top of Fig. 9.1 for clarity. Modes #2, #4, and #6 are present in all spectra, indicating that they are associated with C–C vibrations. Peaks #1 and #3 appear only in spectra 1C and 1D of the $CH_4 + D_2$ and $CD_4 + D_2$ films, while peaks #5 and #7 appear only in spectra 1A and 1B of the $^{12}CH_4 + H_2$ and $^{12}CD_4 + H_2$ films.[19]

Fig. 9.2 gives the HR-EEL spectra of the films deposited from the $^{12}CH_4 + H_2$, $^{12}CD_4 + D_2$, and $^{13}CH_4 + H_2$ gas mixtures. The main vibrational modes are enumerated on the plot in the same way as in Fig. 9.1. Note that the exchange of $^{12}C$ by $^{13}C$ shifts peaks #2, #4, #6, and #8 in Fig. 9.2A ($^{12}CH_4 + H_2$) to lower loss energies in Fig. 9.2B ($^{13}CH_4 + H_2$) by $\Delta$, $2\Delta$, $3\Delta$, and $4\Delta$, respectively, where $\Delta = 5$ meV. Mode #7 is nearly unaffected by the exchange of $^{12}C$ by $^{13}C$ (Figs. 9.2A and 9.2B) but it disappears when hydrogen is replaced by deuterium (Fig. 9.2C ($^{12}CD_4 + D_2$) compared to 9.2A ($^{12}CH_4 + H_2$)).

A comparison of the spectra of Fig. 9.1 and Fig. 9.2 indicates that peaks #2, #4, #6, and #8 remained intact under substitution of hydrogen by deuterium. These peaks

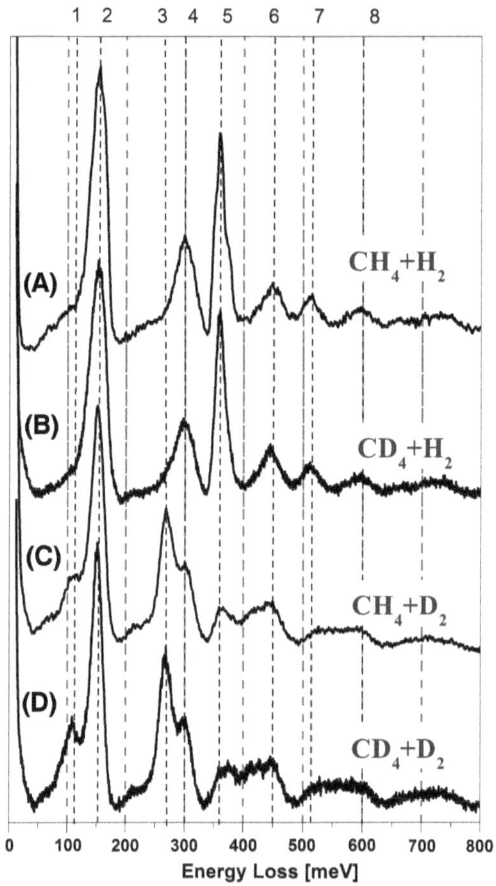


**FIGURE 9.1**

HR-EEL spectra of HF-CVD diamond films deposited from different isotopic gas mixtures. The different loss features are enumerated on the plot. The sample deposited from $CH_4 + H_2$ was treated in MW-H plasma for 20 min whereas sample deposited from $CD_4 + D_2$ gas mixture was treated in MW-D plasma for the same time

were, however, slightly shifted by replacing $^{12}C$ by $^{13}C$ (Fig. 9.2B), indicating that they are associated with C–C modes only and not with C–H modes. In contradiction, peaks #5 and #7 disappear, and new peaks (#1 and #3) appear upon exchanging hydrogen with deuterium (Figs. 9.1 and 9.2). Peaks #5 and #7 remain intact by replacing $^{12}C$ with $^{13}C$ (Fig. 9.2B). This indicates that peaks #5 and #7 are associated with C–H modes. We can thus discuss three families of peaks: (i) pure C–C peaks, (ii) pure C–H peaks, and (iii) coupling of C–C and C–H peaks.

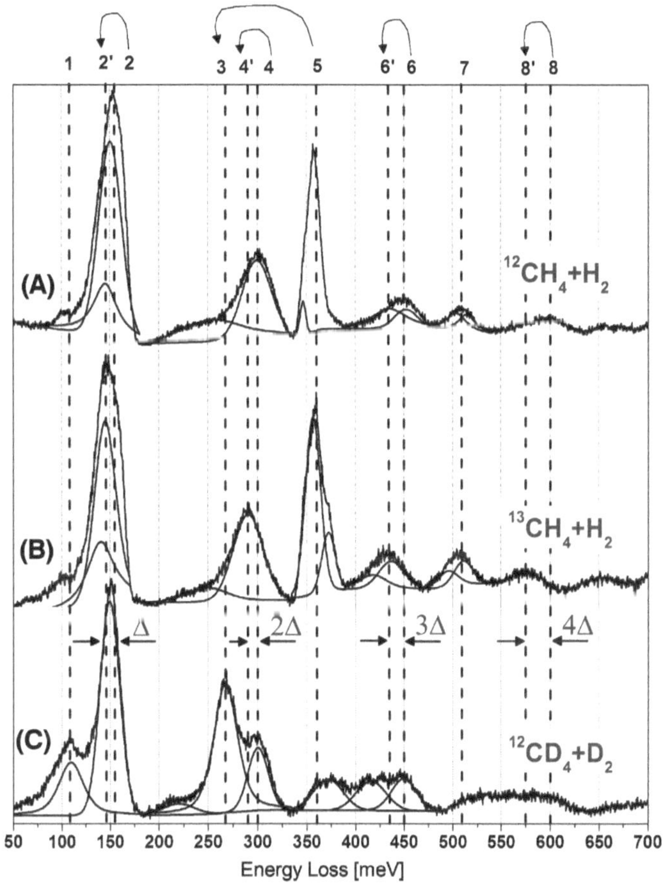


**FIGURE 9.2**

Three HR-EEL spectra of diamond films deposited from isotopic gas mixtures with peak fitting. The peak enumeration is similar as in Fig. 9.1. Each peak was fitted to two possible contributions of C–C and C–H modes

## 9.2.2 FINGERPRINTS OF WELL-DEFINED DIAMOND SURFACE CRYSTALLINITY

In order to better understand the origin of the pure C–C peaks (150, 300, 450, and 600 meV) Ar irradiation was performed on a well-defined polycrystalline diamond surface. Fig. 9.3 shows the HR-EEL spectra of a diamond film deposited from $^{12}CH_4 + H_2$ gas mixture following ex situ MW hydrogenation for ~20 min as is (Fig. 9.3A), and after bombardment by 500 eV Ar ~$1 \times 10^{15}$ ions/cm$^2$ (Fig. 9.3B). Note the drastic decrease in intensity of the 300, 450, and 600 meV peaks. At this point it is worth noting the absence of the overtones #4, #6, and #8 on the HR-EEL

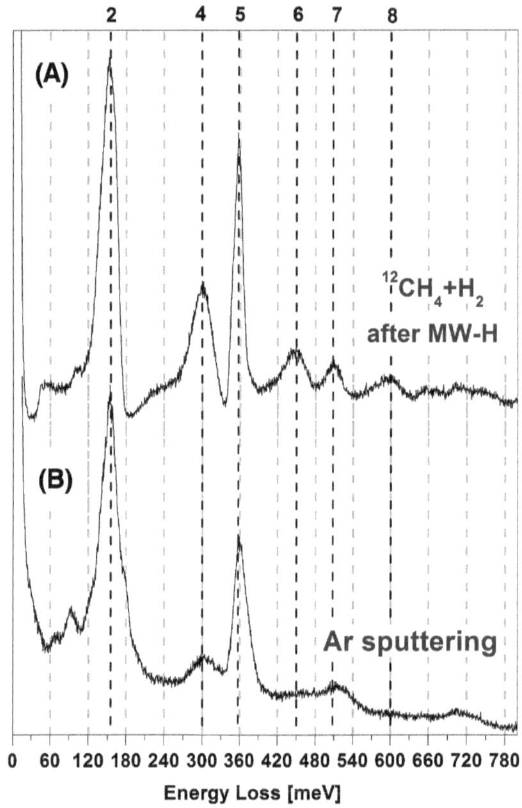


**FIGURE 9.3**

HR-EEL spectra of HF-CVD diamond films deposited from $^{12}CH_4 + H_2$ gas mixture following ex situ MW-H treatment for 30 min (A) and Ar bombardment for 15 min at ion energy of 500 eV (B). Note the strong decrease in relative intensity of the 300, 450, and 600 meV positioned features following the ion bombardment

spectrum of amorphous carbon films reported in the literature and in our previous works. Mode #2 is nevertheless present and prominent. On the other hand, these overtones are observed on the well-defined (100), (111), and polycrystalline diamond surfaces. The electronic loss spectra of these films (using a primary electron energy of 500 eV) are shown in Fig. 9.4. Prominent diamond bulk and surface plasmons can be identified in spectrum 9.4A of the MW-H terminated surface, while the Ar-irradiated film (9.4B) displays a broad feature centered near 30 eV loss energy with no surface plasmon, which is associated with the ion beam amorphized diamond surface.

These results show that the intensity of these three overtone HR-EEL peaks (300, 450, and 600 meV) decreases most significantly after the diamond surfaces have been irradiated with low energy Ar ions (Fig. 9.3). In this case the characteristic diamond

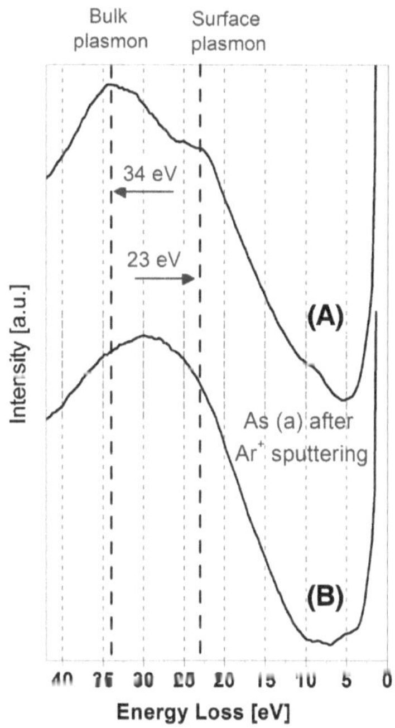


**FIGURE 9.4**

Electronic EEL spectra at primary electron energy $E_p = 500$ eV of diamond samples deposited from $^{12}CH_4 + H_2$ gas mixture (A) and irradiated by $Ar^+$ ions with the dose of $1 \times 10^{15}$ ions/cm$^2$ and primary energy of 500 eV (B). Note the disappearance of diamond surface (23 eV) and bulk (34 eV) plasmons onto Ar-irradiated surface

loss features (the surface and bulk diamond plasmon, spectrum 9.4A) are replaced by amorphous carbon losses (Fig. 9.4C). These modes should be therefore associated with well-defined diamond surfaces rather than with amorphous carbon surfaces. We thus conclude that appearance of the three modes, #4, #6, and #8, at 300, 450, and 600 meV in the HR-EEL spectrum of diamond should be attributed to higher losses of the slow electron onto diamond optical phonon and may serve as a fingerprint of well-defined diamond surfaces. Independent observation of these features on MW hydrogen treated single crystal (100) diamond surface[26,27] confirms this claim.

### 9.2.3 GRAIN BOUNDARIES CONTRIBUTIONS TO HR-EEL SPECTRUM

In this section we apply HR-EEL spectroscopy to study the hydrogen bonding configuration on nanodiamond film surfaces.[28] We compare spectra of HF CVD polycrystalline diamond films (discussed in previous sections) with the spectra of micro-wave deposited nanodiamond films. The nanocrystalline diamond film was deposited by microwave (MW) CVD at a frequency of 2.45 GHz using a microwave power of

1200 W and total gas pressure of 200 Torr onto an Si substrate at 800 °C.[29] Prior to deposition, the silicon substrate was seeded mechanically in solution with diamond powder. The gas mixture composition was 1.4% $CH_4$, 1% $H_2$, and 97.6% Ar (growth rate 2.7 μmℏ), resulting in a grain size in the 10–30 nm range.

Fig. 9.5 shows the HR-EEL spectra of the sub-microcrystalline diamond film as a function of annealing temperature. The shape of the spectrum (Fig. 9.5(A)) is dominated by five main peaks positioned at 155, 300, ~360, 450, and ~510 meV, discussed in previous sections. As may be seen in Fig. 9.5(A), the general shape of the spectra is nearly unaffected up to annealing to 800 °C, while annealing to 1000 °C results in elimination of hydrogen associated features (360 and 510 meV peaks), and a new peak appears near 90 meV. This mode was associated with C=C dimer localized on bare hydrogen surface.

The detailed peak fitting procedure of the C–H stretching HR-EELS mode (Fig. 9.5(B)) reveals at least three different contributions at ~350, ~360, and ~373 meV. The first two are most likely associated with hydrogen adsorption onto diamond (111) and (100) surfaces,[30] while the last one may be attributed to $sp^2$ hybridized carbon (symmetric olefinic or aromatic $sp^2$ CH stretch).[31,32] Upon annealing to 800 °C, the maximum peak position is slightly shifted to smaller loss energies (from 359 to 357 meV), while the $sp^2$ associated mode at ~373 meV is still prominent. Based on the peak fitting procedure, the ratio of 373 meV mode area to the total C–H stretching area is 0.255 (in the case of 800 °C annealing). According to thermal stability studies of different kinds of hydrogenated carbon matrixes, annealing to a temperature of ~600 °C results in desorption of hydrogen atoms alongside with hydrocarbon species independently of the method used for amorphous carbon film preparation. This allows us to attribute the 373 meV HR-EELS mode to hydrogen bonded to $sp^2$ hybridized carbon positioned at diamond grain surface and boundary, which may be stable up to these high temperatures. This suggestion may be confirmed by HR-EELS studies of the nanodiamond film discussed below.

Figs. 9.6(A)–(C) show the HR-EEL spectra of the nanodiamond film (grain size of 10–30 nm). Detailed analysis of C–H stretching mode is shown in Figs. 9.6(D)–(F). Fig. 9.6(A) corresponds to the as-deposited sample followed by in situ annealing to 520 °C. It is dominated by mixed C–C stretching and C–H bending mode at 155 meV, the overtones of diamond optical phonon at 300 and 450 meV, C–H stretching vibration at 360–375 meV, and mixed C–H associated mode at ~515 meV. Note the strong splitting of the C–H stretching mode in this spectrum: the two maxima are positioned at ~360 and 373 meV, reflecting hydrogen bonding to $sp^3$ and $sp^2$ carbon correspondingly (Fig. 9.6(D)). Subsequent annealing to 800 °C does not change the general shape of the spectrum (Fig. 9.6(B)), while the relative intensity of two different C–H stretching modes is somewhat changed (Fig. 9.6(E)). This change may be associated with thermal induced desorption of hydrocarbon species. Based on the peak fitting procedure, the ratio of 373 meV mode area to the total C–H stretching area is 0.431 (Fig. 9.6(E)).

The as-deposited nanodiamond film was exposed ex situ to MW activated hydrogen plasma (MW-H) for 10 min. This treatment is known to increase the surface

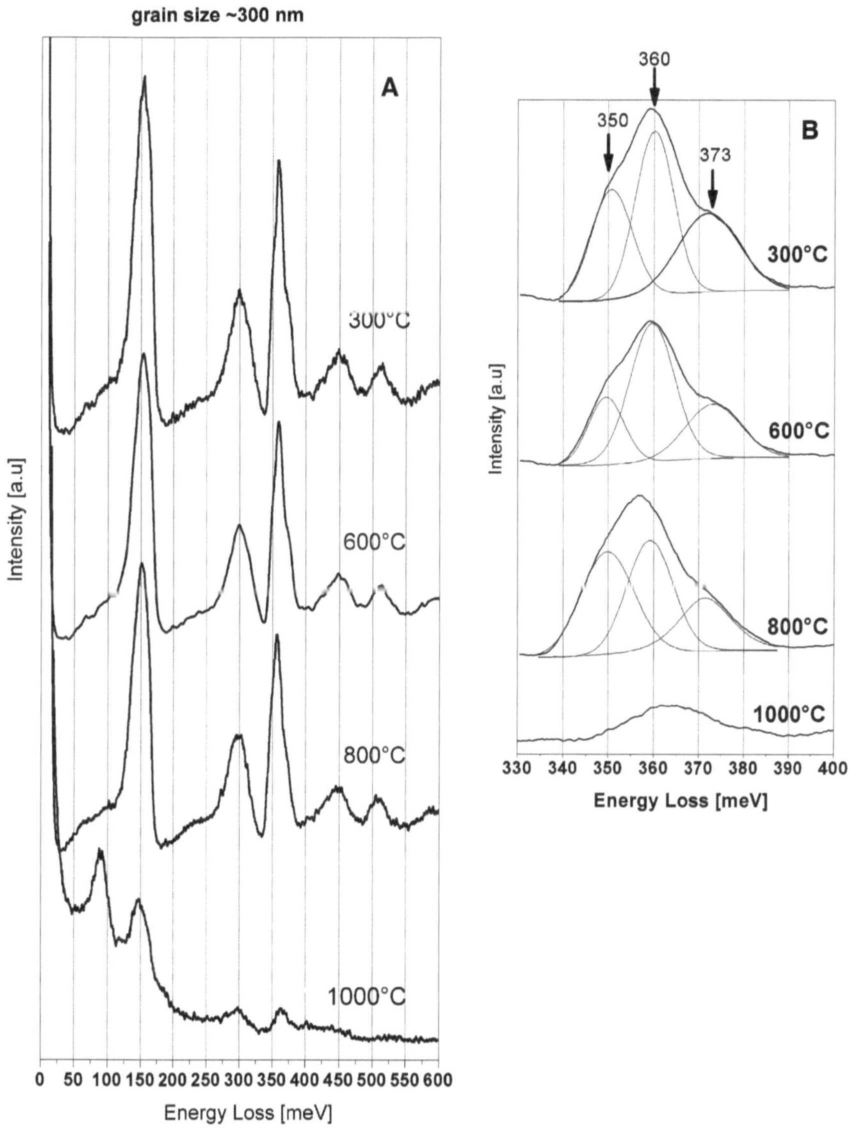


**FIGURE 9.5**

(A) HR-EEL spectra of as-deposited HF CVD films as a function of annealing temperature. The primary electron energy was 5 eV and FWHM of 7–8 meV. The shape of the spectrum is nearly unchanged up to $T_A = 800\,°C$, while $T_A = 1000\,°C$ results in nearly complete hydrogen desorption (the absence of strong 360 meV peak) alongside with appearance of C=C dimer mode at 92 meV. (B) Detailed analysis of HR-EELS C–H stretching mode of as-deposited HF CVD diamond film. Peak fitting procedure reveals three different contributions at ~350, ~360, and 373 meV energy loss

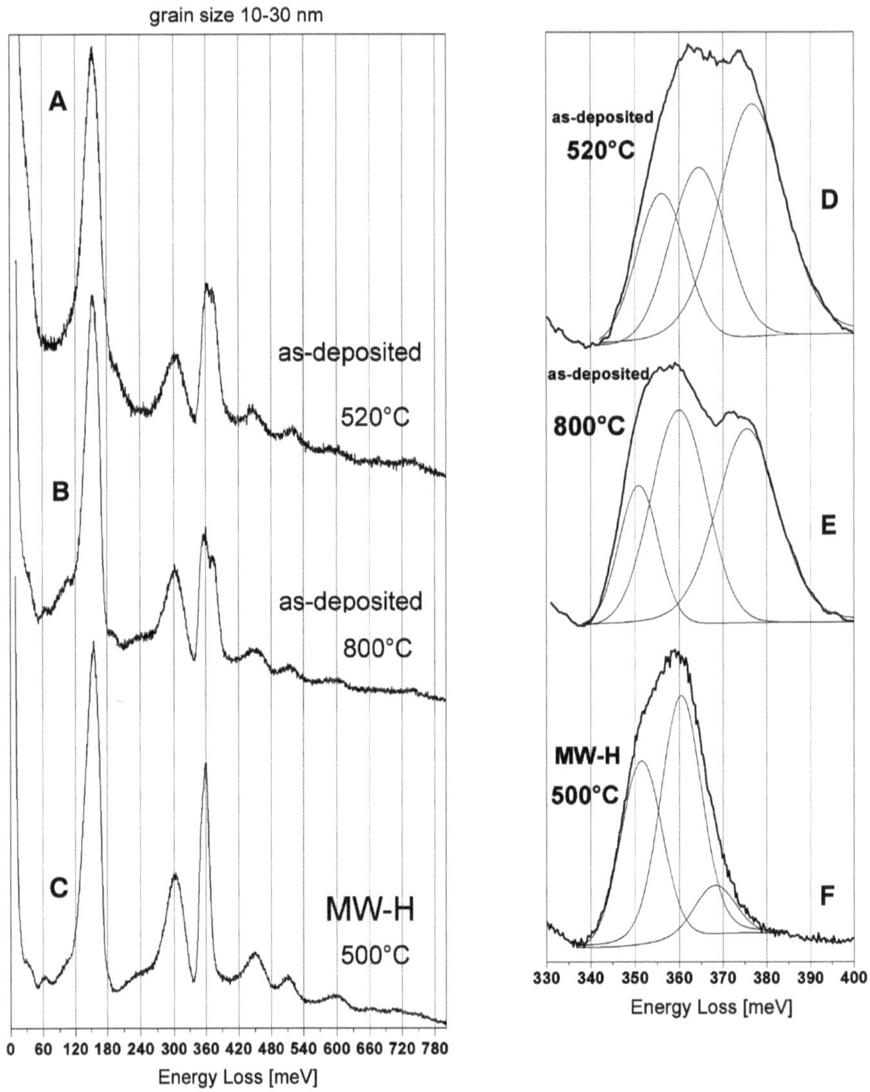


**FIGURE 9.6**

HR-EEL spectra of MW CVD deposited nanodiamond films with 10–30 nm grain size. (A) As-deposited sample in situ annealed to 520 °C. (B) As-deposited sample followed subsequent in situ annealing to 800 °C. (C) As-deposited sample followed ex situ MW hydrogenation for 10 min and consequent in situ annealing to 500 °C. (D)–(F) Detailed analysis of HR-EELS C–H stretching mode (data originated from spectra (A)–(C), respectively. Note the splitting of C–H stretching mode of as-deposited nanodiamond sample due to strong contribution of 373 meV mode. The last mode disappears following ex situ MW-H treatment (spectra (C) and (F)) smoothing alongside with preferential etching of $sp^2$ carbon. The HR-EEL spectrum of ex situ MW-H treated and in situ annealed to 500 °C nanodiamond sample is shown in Fig. 9.6(C). The splitting of C–H stretching vibrations disappears and now, this mode is centered at 360 meV (Fig. 9.6(F)) and the 373 meV mode's intensity is strongly reduced. Exposure to MW activated hydrogen results in preferential etching of $sp^2$ hybridized carbon resulting in a relative increase of the $sp^3$ hybridized carbon on the film's surface.

Now, we compare the HR-EELS analysis of as-deposited MW CVD nanodiamond (Fig. 9.6) with previously discussed as-deposited hot filament (HF) CVD submicron crystalline samples (Fig. 9.5). As can be seen in Fig. 9.5(B), the HR-EEL spectrum of as-deposited HF CVD films with ~300 nm grains also displays contributions at ~350, ~360, and 373 meV within the C–H stretching modes which are stable up to 800 °C. Obviously, the ratio of grain boundaries in the case of nanodiamond films (10–30 nm grains) exceeds that of submicron crystallites (~300 nm grains). Since the contribution of the 373 meV mode is much more prominent in the case of nanometric grains, this mode may be associated with hydrogen bonded to $sp^2$ hybridized carbon positioned on the surface and grain boundary of the nanodiamond film. This is a very exciting feature. In our studies of hydrogen adsorption on the ion irradiated diamond surfaces, we concluded that nearly all kinds of hydrogenated carbon decompose near ~600 °C. This is in complete agreement with other studies of thermal stability of diamondlike carbon which undergoes erosion/destabilization at temperature about 600 °C independently of deposition process.[33,34] The present results show that $sp^2$ hybridized carbon is stable up to temperatures above 800 °C. This finding allows one to distinguish the hydrogenated $sp^2$ hybridized carbon bonded to diamond grains from chemisorbed (or physisorbed) molecules on polycrystalline diamond surface by simple annealing procedure.

It is very interesting to note at this stage the possible correlation between the 373 meV HR-EELS mode and the well-known Raman 1140 $cm^{-1}$ peak of nanodiamond. It was shown in the works of many authors,[35–37] including us,[38] that the 1140 $cm^{-1}$ Raman peak is attributed to C–H bending mode of $sp^2$ carbon and positioned most likely in the diamond grain boundary. Both peaks (Raman 1140 $cm^{-1}$ and HR-EELS ~373 meV mode) are attributed to ($sp^2$)-carbon–hydrogen vibration. Both peaks are stable up to elevated temperatures where hydrogen bonded to amorphous carbon undergoes desorption. Both peaks are strongly enhanced in the case of nanodiamond grains. Therefore, our results strongly suggest that the ~373 meV mode can be attributed to the stretching of the same C–H vibration, which contributes to the 1140 $cm^{-1}$ Raman peak (currently attributed to $\nu_1$ mode of trans-polyacetylene (t-PA); the bending of $sp^2$–C–H). Both peaks may serve as indication of nanocrystalline character of diamond films. However, due to different probing depths of Raman and HR-EEL spectroscopies (tens or hundreds of nanometers in the case of Raman and 1–2 nm in the case of HR-EELS) quantitative comparison of these two peaks is impossible.

We conclude, therefore, that hydrogen bonding to surfaces and grain boundaries of nanodiamond films display a clear contribution to the HR-EELS at 373 meV which

**Table 9.1** Assignment of basic HR-EELS peaks based on isotopic exchange studies

| Diamond surface | HR-EEL position | Present assignment |
|---|---|---|
| Hydrogenated diamond (C-12) surface | ~155 meV | Mixed $^{12}C$–H bend and $^{12}C$–$^{12}C$ stretch modes |
| | 300 meV | First overtone of $^{12}C$ diamond optical phonon |
| | 351–375 meV | ($sp^3/sp^2$) $^{12}C$–H stretching mode |
| | 450 meV | Second overtone of $^{12}C$ diamond optical phonon |
| | 510 meV | Overlapping of both $^{12}C$–H bend and $^{12}C$–H stretching modes and $^{12}C$–$^{12}C$ stretch and $^{12}C$–H bend modes |
| | 600 meV | Third overtone of C(12) diamond optical phonon |
| Deuterated diamond (C-12) surface | ~110 meV | $^{12}C$–D bending modes |
| | 150 meV | $^{12}C$–$^{12}C$ stretching modes |
| | 260–280 meV | $^{12}C$–D stretching modes |
| | 300 meV | First overtone of $^{12}C$ diamond optical phonon |
| | 420 meV | Coupling of $^{12}C$–$^{12}C$ stretch and $^{12}C$–D stretch modes |
| | 450 meV | Second overtone of $^{12}C$ diamond optical phonon |
| Hydrogenated diamond (C-13) surface | 148 meV | Mixed $^{13}C$–H bend and $^{13}C$–$^{13}C$ stretch modes |
| | 290 meV | First overtone of $^{13}C$ diamond optical phonon |
| | 350–375 meV | ($sp^3/sp^2$) $^{13}C$–H stretching mode |
| | 435 meV | Second overtone of $^{13}C$ diamond optical phonon |
| | 508 meV | Overlapping of both $^{13}C$–H bend and $^{13}C$–H stretching modes and $^{13}C$–$^{13}C$ stretch and $^{13}C$–H bend modes |
| | 580 meV | Third overtone of C(13) diamond optical phonon |

differs from bonding to the microcrystalline diamond surfaces whose main contribution is centered around 360 meV. The ~373 meV mode is stable up to elevated temperatures of ~800 °C where all kinds of hydrogenated amorphous carbon are decomposed. The intensity of this mode is drastically decreased following ex situ treatment in MW activated hydrogen plasma, most likely due to preferential etching of $sp^2$ carbon bonded to diamond grain surfaces and boundaries. Our assignments of diamond-related HR-EELS peaks are summarized in Table 9.1.

## 9.3 NANOSIZE EFFECTS IN THE HR-EEL SPECTRA OF NANODIAMOND

In this section we summarize our HR-EELS studies of hydrogenated diamond films composed of diamond grains of varying size of ~300, ~20, and ~5 nm.[39] We found that the relative intensities of the HR-EELS modes associated with C–C lattice phonons are comparably larger than modes associated with the C–H surface vibrational modes in the case of nano-crystalline compared to micro-crystalline diamond films. In order to explain these effects, the fundamental excitation mechanisms leading to these differences were assessed by measuring the excitations functions of the different films and their electron, reflectivity. It is suggested that these effects are associated with the larger probability for inelastic scattering of electrons by lattice phonons mediated by scattering with internal grain boundaries and defects in the case of nano-crystalline diamond films.

The investigated films consist of diamond films with ~5, ~20, and ~300 nm grain size grown by three different kinds of chemical vapor deposition (CVD) methods onto silicon substrates:

1. Hot filament (HF) CVD diamond films were grown from a hydrogen-rich gas mixture ($CH_1/H_2$ ratio was 1/99) at a substrate temperature of 800 °C for 1 h, as described in previous sections.[24]

2. dc glow discharge (GD) CVD films were grown from hydrogen-rich plasma ($CH_4/H_2$ ratio was 9/91) at substrate temperature of 880 °C and total gas pressure of 20 Torr.[40–42] The mechanism underlying the formation of nanodiamond films deposited by dc GD CVD is completely different from HF CVD described above.[40] No pretreatment procedure is needed to induce diamond nucleation on silicon substrate. The nanodiamond nucleation and growth occur from energetic particle bombardment in the sub-surface region of the film resulting in the upper surface region consisting of hydrogenated amorphous carbon. The film evolves on top of a hydrogenated carbonaceous precursor containing a mixture of an amorphous and graphitic carbon formed during the first ~20 min of deposition (~300 nm thickness). When the precursor density reaches ~3 g/cm$^3$, nanodiamond particles precipitate and grow to a final size of ~5 nm, as was estimated in transmission electron microscopy (TEM) and X-ray diffraction experiments. The total sp$^3$ content of the film grown for 1 h reaches ~80% of total carbon atoms, while hydrogen concentration are in the range of 15–20 at.%. The nanodiamond particles are embedded in hydrogenated amorphous carbon matrix.

3. Microwave (MW) CVD nanodiamond films were grown from a hydrogen-poor gas mixture consisting of 1.4% CH4 and 98.6% Ar.[29] In this case, the growth rate is 0.96 μm/h and diamond grain size ~20 nm. The films were grown on Si substrates maintained at 800 °C using a microwave power of 1200 W and total gas pressure of 200 Torr. Prior to deposition, the silicon substrates were seeded mechanically in a solution with diamond powder.

To ensure that the three diamond surfaces are hydrogen terminated, the as-deposited films were exposed to hydrogen microwave plasma for ~10 min at a substrate temperature of ~800 °C. It was determined[41] that this processes results in fully hydrogenated surfaces and etching of most non-diamond carbon from the surface of the films. A schematic cartoon of the surface conditions of the two main kinds of films that are investigated in present work are shown in Fig. 9.7 for convenience.

The HR-EEL spectra of nanodiamond films are different from those of micron grain size films in the relative intensities of C–C related modes relative to C–H vibrational modes. Fig. 9.8 shows a comparison between the HR-EEL spectra measured for the three kinds of diamond films grown for 1 h by different methods: HF CVD (grains ~300 nm), dc GD CVD (grains ~5 nm), and MW CVD (grains 3–20 nm). Since the first strong peak at ~155 meV is composed of mixed C–C and C–H modes, we should compare only pure C–C vibration modes with pure C–H modes, namely, overtones of diamond optical phonon at 300 meV with C–H stretching mode at 362 meV. From the spectra shown in Fig. 9.8, for the HF CVD (grains ~300 nm) diamond film the relative ratio of intensity of 362 meV mode is about two times more intense than the 300 meV mode. For the nanodiamond films deposited by MW CVD (grains ~20 nm) and dc GD CVD (grains ~5 nm) similar intensities were measured for these two modes. These results clearly show that the HR-EELS modes associated with C–C lattice phonons are relatively more intense than modes associated with the C–H surface vibrational modes in the case of nanocrystalline compared to microcrystalline diamond films.

Additional investigations of this effect was performed by means of low energy electron excitation function, reported therein.[39] In addition to the vibrational properties of the substrate, low-energy electron scattering provides insights into the sample conduction-band density-of-states (DOS).[43,44] When the intensity of the elastic peak is followed as a function of the incident electron energy, the global electron elastic reflectivity of the sample is probed. When inelastic excitation functions of selected vibrational modes and associated multiple losses are considered, this is the DOS characterizing the environment of the probed species and the surface versus lattice character of the excited vibrational modes which are probed. To further assess the reported effect, HR-EEL spectra were measured as a function of incident electron energy alongside with excitation functions and electron elastic reflectivity for the diamond films consisting of ~300 and ~5 nm grain size, and the results were compared. Detailed explanation of this technique is behind the scope of this review, and below we provide the main conclusions, derived therein.[39]

In the case of nanodiamond sample, the peaks associated to the total electron reflectivity were broader than for the microcrystalline films. This is probably associated to the shorter ranger order in the former which is expected to result in band broadening. The larger overall intensity of the elastic reflectivity in the nanocrystalline film is most likely due to the shorter scattering length of low energy electrons in this material which results in a higher yield of elastically backscattered electrons. The shorter scattering length is associated with scattering at defects and grain boundaries within the nanocrystalline diamond films whose density is much larger than for the microcrystalline films. For the microcrystalline diamond, the excitation function is

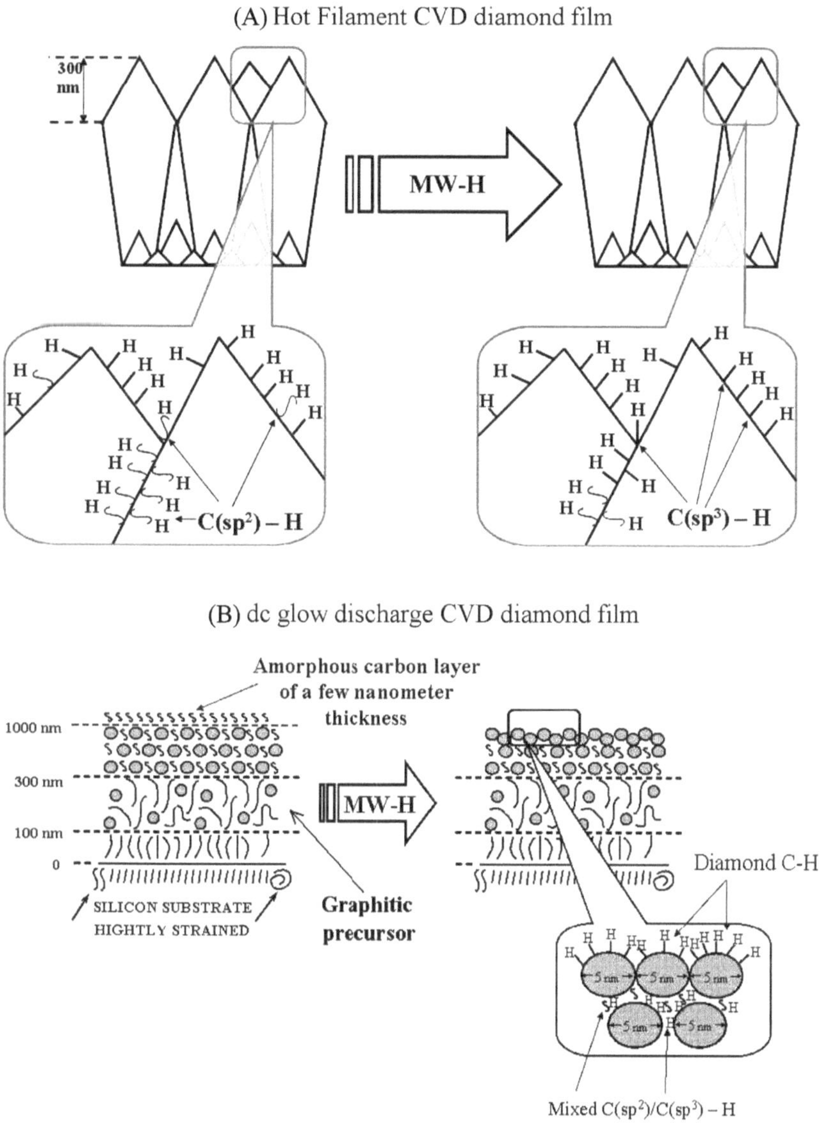


**FIGURE 9.7**

Schematic cartoon demonstrating the difference in film structure and the effect produced by exposure of as-deposited films to microwave (MW) activated hydrogen. (A) Hot filament CVD film before and after MW hydrogenation. The film has a columnar structure with surface protruded average grain size of ~300 nm. Hydrogenation results in preferential etching of $sp^2$ hybridized carbon. (B) dc glow-discharge nanodiamond films before and after MW hydrogenation. The H atoms etch the upper amorphous carbon layer and result in exposure of hydrogenated diamond crystallites with average size of ~5 nm

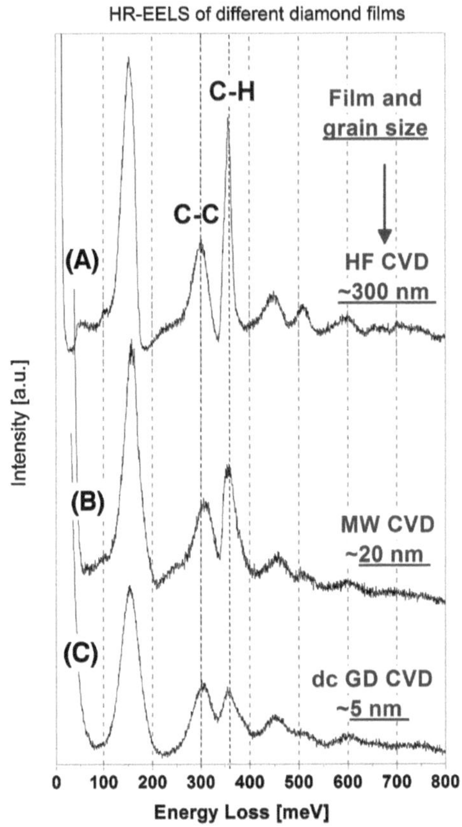


**FIGURE 9.8**

Comparison of HR-EEL spectra of three different diamond films which underwent ex situ microwave hydrogen treatment: (A) HF CVD film grown for 1 hour (grain size ~300 nm); (B) MW CVD sample (grain size ~20 nm); (C) nanodiamond film grown by dc GD method (grain size ~5 nm). Note the different ratio of C–C associated modes relative to C–H vibrations

dominated by band structure effects whereas for the nanodiamond it is dominated by a C–H local resonance.

Our results clearly show that the vibrational modes associated with surface modes are relatively more intense than bulk modes in the case of microcrystalline as compared to nanocrystalline films. This effect is believed to be due to the shorter elastic mean free path in the nanograined films. This shorter elastic mean free path is associated with an enhanced probability for inelastic scattering of low energy electrons within the crystal field. For low energy electrons, as these used in HR-EELS studies, the main channel for energy loss is by scattering with impurities, internal grain boundaries mediating the excitation of lattice phonons. The density of these is obviously larger in the case of nanodiamond films as compared to the microcrystalline one. It is, therefore, reasonable to expect that the probability of scattering by lattice phonon is larger for the nanocrystalline than for the microcrystalline films due to the larger density of internal grain boundaries in the former. It is suggested that these effects are associated with the larger probability for inelastic scattering of electrons by lattice phonons mediated by scattering with internal grain boundaries and defects in the case of nanocrystalline diamond films.[39]

## 9.4 HYDROGENATION AND THERMAL STABILITY OF NANO- AND MICROCRYSTALLINE DIAMOND FILMS STUDIED BY HR-EELS

The thermal stability of hydrogen atoms bonded to the diamond matrix is a very important issue. Nearly in all cases diamond films deposition and surface characterization are carried out in different experimental chambers, therefore the studied films are exposed to ambient conditions to a varying degree. In situ vacuum annealing procedure is a very simple way to remove ambient contamination from the diamond surface, which makes it possible to distinguish between hydrocarbon contamination and hydrogen directly bonded to the diamond matrix. Our studies reveal that thermal desorption of ambient contaminations occurs at least at 500–600 °C and results in irreversible changes in the diamond surface.[45] In order to obtain bare diamond surfaces annealing, procedure to at least 1000 °C must be performed. The questions of surface reconstruction, hydrogen desorption induced defects, and impact of grain size on the degree of diamond graphitization are still open.

### 9.4.1 THERMAL STABILITY OF HYDROGENATED NANODIAMOND SURFACE

In this section we present our study[46] of the effect of diamond grain size upon: (i) ex situ hydrogenation by exposure of the diamond surfaces to microwave (MW) hydrogen plasma, (ii) annealing the diamond surfaces at 1000 °C, and (iii) in situ hydrogenation of the as annealed and bared surfaces by atomic hydrogen. Three different kinds of carbon films were investigated in this study: (i) polycrystalline diamond films with ~300 nm grain size ("submicron" diamond films) deposited by hot filament (HF) CVD, (ii) polycrystalline diamond films with ~5 nm grain size ("nanodiamond" films) deposited by dc glow discharge (dc GD) CVD, and (iii) highly oriented pyrolitic graphite (HOPG) sample that was used as a model $sp^2$ bonded carbon material reference. The hydrogen bonding and near surface phase composition following the different processes were probed by high resolution electron energy loss spectroscopy (HR-EELS) and electronic EELS. The impact of diamond crystalline size on the shape of HR-EEL spectrum was reported.

Two kinds of polycrystalline diamond films were used in the present study. These were deposited by the conventional HF and dc GD CVD methods, discussed in Section 9.3. The evolution mechanism underlying the film formation is different in both cases and was extensively investigated by us in previous years.[40,47] In Figs. 9.9(A) and 9.9(D), the HR-EEL spectra of the dc GD CVD and HF CVD diamond films are respectively shown, followed by ex situ MW hydrogenation and in situ annealing to 600 °C. The shape of these spectra was extensively discussed in the previous sections. Comparison of HR-EEL spectra of fully hydrogenated nanodiamond (Fig. 9.9(A)) and submicron (Fig. 9.9(D)) surfaces reveals different intensity ratio of the pure C–H to C–C vibration features. Since the first strong peak at ~155 meV is composed of mixed C–C and C–H modes, we should compare only pure C–C vibration modes with pure C–H modes, namely, overtones of the diamond optical phonon at ~300 meV with the C–H stretching mode at ~360 meV. In the case of HF CVD (grain size ~300 nm) the relative ratio of intensity of the ~360 meV mode with respect to the intensity of the ~300 meV phonon overtone is ~1.9. In the case of dc GD CVD nanodiamond films (grains ~5 nm) this ratio is ~0.9. The difference between the HR-EEL spectrum of nanodiamond and microcrystalline diamond surface is speculated to be associated with the shorter elastic mean free path of the scattered electrons in the former (cf. Section 9.3). It should be emphasized at this stage that while the two kinds of diamond films investigated in this work are quite different in their microstructure and composition, their surface regions differ mainly by diamond grain size and density of grain boundaries.

Figs. 9.9(B) and 9.9(E) show HR-EEL spectra of bare nano- and submicron-diamond surface following in situ annealing to 1000 °C. Desorption of hydrogen is evidenced by the drastic decrease in the C–H stretching mode intensity at ~360 meV. The obvious difference in the shape of these two spectra can be seen. The HR-EEL spectrum of the submicron-diamond film surface (~300 nm grain size, Fig. 9.9(E)) displays features associated to the bare diamond surface, namely C–C stretching vibration at ~150 meV and C=C dimer stretching near ~90 meV. The intensity of the first overtone of the diamond optical phonon at ~300 meV is strongly reduced following 1000 °C annealing (Fig. 9.9(E)), while its second overtone completely vanished from the spectrum. In contrast to the submicron-diamond, the spectrum of bare nanodiamond surface (~5 nm grain size, Fig. 9.9(B)) displays a nearly featureless spectrum that is similar to the spectrum of the clean HOPG surface, shown in Fig. 9.9(G). The HR-EEL spectrum of graphite is known to be featureless when recorded at specular mode. The continuous background of these two spectra may be related to the excitation of graphitic $\pi \rightarrow \pi*$ interband transitions.[48] Both vertical and longitudinal graphitic phonons are not distinguished in specular reflection experiments.[49] The electronic EEL spectra (plasmon losses, not shown) upon annealing at 1000 °C of the nanodiamond and submicron-diamond surfaces display graphitic $\pi$ electron plasmon losses at ~6 eV.[50]

We conclude that thermal hydrogen desorption results in diamond surface reconstruction and phase transformation, which occurs differently for each of the diamond films investigated in the present work. While for the case of ~300 nm grains film

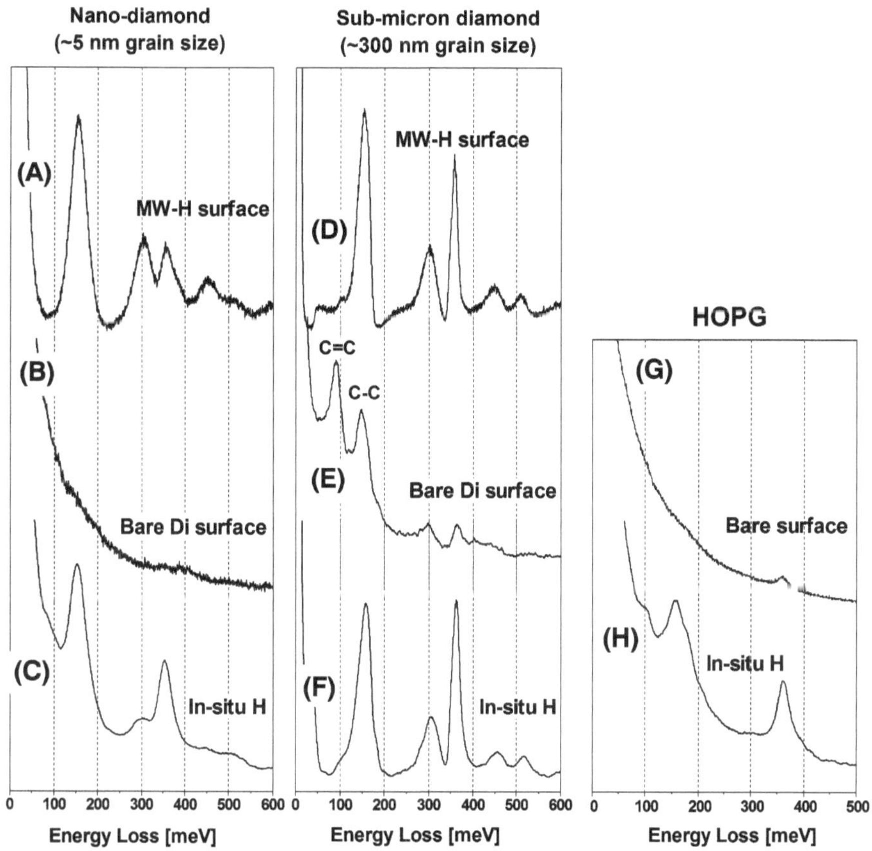


**FIGURE 9.9**

(A)–(C) HR-EEL spectra of dc GD CVD nanodiamond, (D)–(F) HF CVD submicron diamond, and (G)–(H) HOPG films: (A) dc GD CVD diamond film followed ex situ MW hydrogenation and in situ annealing to 600 °C, (B) the (A) sample followed in situ annealing to 1000 °C, (C) the (B) sample followed subsequent in situ exposure to atomic hydrogen, (D) HF CVD diamond sample followed ex situ MW hydrogenation and in situ annealing to 600 °C, (E) the (D) sample followed in situ annealing to 1000 °C, (F) sample as in (E) followed subsequent in situ exposure to atomic hydrogen, (G) HR-EEL spectrum of the clean HOPG surface annealed to 275 °C, and (H) in situ hydrogenation of the clean HOPG surface

hydrogen desorption results in one monolayer of C=C dimers on the film surface, in the case of ~5 nm grains film graphitization occurs more deeply toward the subsurface region of the diamond crystallite. These results are in agreement with other studies of thermal graphitization of diamond films composed of different grain sizes.[5,51–54] Butenko et al.[51] studied the density decrease (i.e., graphitization) in diamond films composed of 4.2, 4.7, and ~360 nm grain size as a function of thermal annealing by means of $\gamma$-ray densitometer. It was shown that in the case of the 4.2 and 4.7 nm diamond grains the film's density decreases more severely (starting at ~1100 °C) than in the case of ~360 nm diamond grains, for which graphitization occurs only above the Debay temperature of $T_D = 1910\,°C$.[51] Baidakova et al.[52] showed by X-ray diffraction analysis that nanodiamond of 4.4 nm size prepared by the detonation technique graphitizes at ~930 °C. These experimental findings are in agreement with the theoretical study of Kwon et al.[53] who elucidated the existence of an additional phase diagram boundary between the nanodiamond and graphite for grain size up to 15 nm. On the other hand, Raman spectroscopy studies provided by Herchen et al.[54] show that single crystal type IIa diamond displays a diamond-graphite phase transition only for $T_{ph} > 1900\,°C$.

We suggest that the effect observed in the present study may be correlated with the general tendency of decreasing the phase transition temperature for nanometer particles compared to the bulk material. Such a decrease in a phase transition temperature $T_{ph}$, for instance, at the melting point, was observed for metal clusters,[55–59] metal compounds,[60] as well as for the second order phase transition of ferroelectrics.[61] This decrease in $T_{ph}$ is usually attributed to increase in Gibbs free energy of nanocrystal due to the contribution of high energy surface atoms, which results in a reduced thermodynamics stability of the nanocrystals. A theory based on the Laplace and the Gibbs–Duhem equations[62] predicts a melting temperature depression ΔT of the form $\Delta T_{ph} = T_{bulk} - T(r)$, where $T(r)$ depends on the ratio of the surface/bulk atoms, surface tension, and related parameters.[56,57,59] In addition, Gamarnik proposed[63,64] that the diamond–graphite phase transition depends not only on the difference in free energies between the phases, but also on the activation energy of the phase transition, which can be less for smaller diamond particles because of larger interatomic distances.[64] Therefore, a decrease in the activation energy in smaller metastable diamond particles results in decreasing of the diamond–graphite phase transition temperature. This is expected to cause the preferable graphitization of smaller diamond particles.

Therefore the observed difference between HR-EEL spectra of submicron- and nanodiamond films following 1000 °C annealing may be associated with different degrees of surface graphitization: H desorption from submicron grains results in nearly one monolayer of C=C dimers due to rearrangement of the carbon dangling bonds, while in the case of nanometric grains surface-to-bulk graphitization may take place within the few monolayers depth.

### 9.4.2 IN SITU HYDROGEN ADSORPTION: DIAMOND PHASE RECOVERY

Figs. 9.9(C) and 9.9(F) show HR-EEL spectra of bare nano- and submicron-diamond film surfaces, which were subjected to in situ hydrogenation. H atoms terminate unsaturated carbon bonds providing $sp \rightarrow sp^2$ and $sp^2 \rightarrow sp^3$C hybridization change. In our previous work we showed[19,21] that the appearance of diamond optical phonon overtones at ~300, ~450, and 600 meV are characteristic solely to well-defined diamond surface. Here we suggest that the relative intensity of the ~360 meV peak (pure C–H mode) to the ~300 meV peak (pure C–C mode) may be qualitatively recognized as a degree of surface recovery. Comparison of the HR-EEL spectra shown in Figs. 9.9(D) and 9.9(F) of submicron-diamond surface reveals that the intensity ratio of 360/300 peaks increases from ~1.9 in the case of well-defined MW hydrogenated surface (Fig. 9.9(D)) to ~2.2 following in situ annealing and subsequent hydrogenation (Fig. 9.9(F)). A detailed HR-EELS study of submicron films showed[65] that the diamond surface underwent irreversible changes following in situ 1000 °C annealing, which cannot be recovered by subsequent in situ exposure to atomic hydrogen. Similarly, the surface of nanodiamond films underwent distinguished irreversible changes, shown in Figs. 9.9(A) and 9.9(C). While the ~360/300 HR-EELS peaks ratio is ~0.9 in the case of MW hydrogenated nanodiamond (Fig. 9.9(A)), this ratio increases to ~1.4 in Fig. 9.9(C). In such a case, the submicron diamond surface recovery (Fig. 9.9(F)) is higher than that of nanodiamond film (Fig. 9.9(C)). Most likely, this is a consequence of a higher degree of graphitization of the nanodiamond surface compared to submicron one. For comparison, in Fig. 9.9(H) we show the HR-EEL spectrum of the clean HOPG surface following the same in situ exposure to atomic H. The spectrum is dominated by a C–H stretching mode at 360 meV (predominantly, $sp^3$ carbon) and a broad band centered at 160 meV, which may be attributed to C–C stretch and C–H bending vibrations. A small shoulder may be distinguished at ~100 meV, which may be attributed to a vertical phonon mode. No peaks at ~300 or ~450 meV loss energy were observed.

Summarizing this section, the effect of annealing and hydrogenation of polycrystalline diamond film consisting of different grain sizes by HR-EEL spectroscopy was reported:

1. UHV annealing of fully hydrogenated films to 1000 °C results in thermal hydrogen desorption and diamond phase degradation. In the case of ~5 nm crystallites, thermal graphitization occurs more severely, than in the case of submicron ~300 nm grains. We associate this result with a size effect where high energy surface atoms reduce the thermodynamic stability of nanocrystals.

2. In situ adsorption of hydrogen atoms gives rise to reappearance of diamond features for both kinds of film, while the extent of the surface recovery is lower in the case of nanodiamond film. Re-emergence of bulk diamond features evidences rather partial than full nanodiamond graphitization at present conditions.

## 9.5 HR-EELS STUDIES OF NANO-DIAMOND PHASE FORMATION BY BEN ON 3C-SIC(100)

Deposition of thin and smooth nanocrystalline diamond films requires a high degree of control of the nucleation stage.[66] In this section we present the successive steps of the bias-enhanced nucleation (BEN) process as studied ex situ by HR-EELS on 3C-SiC(100).

The BEN process is an in situ pretreatment for diamond nucleation.[67,68] It consists in applying a negative bias voltage to the substrate with respect to the grounded reactor walls during the $CH_4/H_2$ plasma exposure. This pretreatment is technically easy to perform. Nevertheless, the mechanisms induced on the substrate surface (ion bombardment, etching, carbon deposition) and the plasma modifications are rather complex. Indeed, the produced effects not only depend on the applied bias voltage, but also on the plasma parameters (pressure, gas composition, microwave power, etc.), the surface substrate properties, and the geometry of the reactor. In fact, biasing the substrate exposed to a microwave plasma leads to the formation of a plasma sheath just above the substrate which is the result of ion bombardment from the plasma and electronic emission from the surface. The relevant characteristics of the plasma sheath, like ion energy distribution and electronic emission, are not measurable, but they are linked to process parameters (pressure, nature of the substrate, temperature, geometry, etc.). It is thus essential to control the whole process for reproducible results.

In the BEN studies a single-domain n-type 3C-SiC thin film epitaxially grown (3-off axis) on a Si(100) substrate was used.[66] Reconstructed surfaces were prepared using a silicon source and a $C_2H_4$ gas line, connected to the chamber via a precision leak valve. Plasma treatments were performed in a home-designed MP-CVD reactor equipped with a 2.45 GHz–2 kW SAIREM microwave generator. The sample holder was placed onto a molybdenum disk in order to minimize residual contamination from the support during the plasma discharge. Each sample underwent two successive CVD steps. In the first step surfaces were exposed to a 3% $CH_4/H_2$ gas mixture for 5 or 10 min to stabilize the plasma parameters and the substrate temperature. In the second step (nucleation), a negative bias voltage was applied to the substrates thus inducing a bombardment of the surface by $C_xH_y^+$ species. A growth step was then applied to some samples to get thin nano-crystalline diamond films.

The surface analysis was carried out by ex situ HR-EELS of the as prepared SiC surface after annealing at 573 K to desorb contaminants. The HR-EELS spectra recorded at this stage show clearly that a very thin carbonaceous layer composed of fully saturated $CH_x$ species is formed on the 3C-SiC(100) substrate (Fig. 9.10). In addition to the 117 and 234 meV bands associated with the optical phonon loss and its first overtone characteristic of SiC(100), 150–180 and 360 meV bands are observed. The broad band at 150–180 meV is associated with the mixed C–H bending C–C stretching modes of $CH_x$ surface species. The 360 meV peak is characteristic of $sp^3$–C–H stretching.

After the nucleation step, diamond growth leads to the appearance of the additional carbon C (1s) binding state located at +0.5 eV with respect to the diamond component observed by XPS.[66] According to previous XPS work performed on diamond surfaces, this contribution can be assigned to polyhydrides formation ($CH_x$, $x > 1$).[69] Ex situ HR-EELS investigations performed on 3C-SiC samples after BEN confirm this assignation. The HR-EEL spectrum, measured after annealing at 573 K,

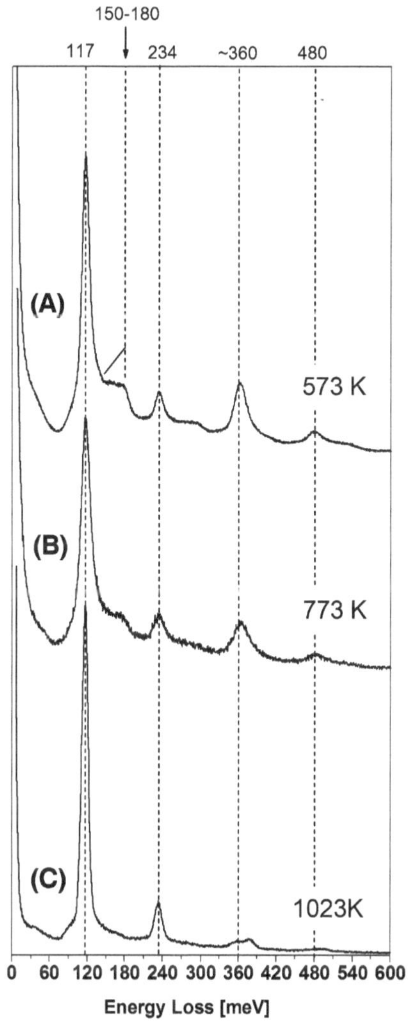


**FIGURE 9.10**

(A)–(C) HR-EELS of 3C-SiC(100) after the first CVD step as a function of annealing temperature

reveals peaks at 155, 300, 360, 450, and 510 meV, associated with the diamond structure (Fig. 9.11). The broad peak centered at 155 meV is composed of several modes, the diamond optical phonon and C–H bending. The high energy loss side of this peak is associated with $CH_x(x > 1)$ bending modes. These species are removed after a thermal annealing at 773 K. The weakly bonded $CH_x(x > 1)$ species are incorporated into the diamond crystals by a sub-plantation process. This mechanism

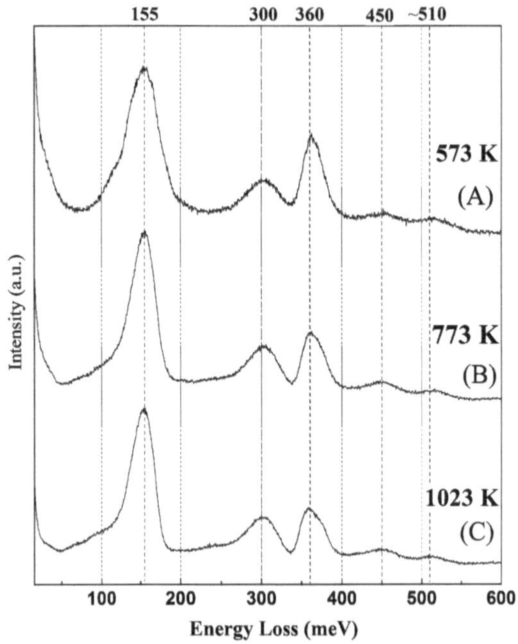


**FIGURE 9.11**

HR-EEL spectra of 3C-SiC(100) following the biasing step as a function of annealing temperature

is induced by the $C_xH_y$ bombardment taking place during BEN, resulting in high hydrocarbon species which are weakly bonded to the SiC surface.

HR-EELS investigations performed on areas that were not covered during the CVD process underline the high inertness of the 3C-SiC(100) surface under BEN conditions.[70] The shape and relative intensities of the HR-EELS peaks are evidence of the contribution of diamond-associated features and a very thin carbonaceous layer on top of the 3C-SiC substrate. After annealing at 773 K, the changes in the HR-EEL spectra reveal the desorption of the carbonaceous layer which most likely covered the 3C-SiC surface onto which diamond nucleation did not take place. After annealing at 1023 K the HR-EEL spectrum displays peaks associated with high-quality hydrogenated nanodiamond and well-defined SiC surfaces The area uncovered by the nanodiamond film exposes the SiC surface to the plasma environment, but only a very thin carbonaceous layer of relatively low thermal stability was deposited, and this desorbed upon annealing at 773 K. These results emphasize that diamond nucleation and growth is preferred under the biasing conditions. If the diamond phase is not formed, no other solid carbon films can be formed. The 3C-SiC(100) surface undergoes very little BEN damage. 3C-SiC exhibits a very different behavior compared to silicon for which a sub-plantation mechanism has been invoked to explain diamond nucleation during BEN.[71] The associated ion bombardment leads to a strong modification of the Si substrate on several tens of nanometers. Diamond nuclei are formed within an amorphous carbon matrix. Here, taking into account the weak surface modifications, a different nucleation scheme must be proposed, probably involving the mobility of carbon species at the surface. Indeed, their surface diffusion may be increased by the deposited energy under ionic bombardment.

Unlike silicon, the carbon-terminated 3C-SiC surface demonstrates a very high inertia under the BEN process. HR-EELS results clearly show that the first plane of this surface is preserved even after BEN, revealing a very low level of damage. Moreover, a significant enhancement of the diamond nucleation density starting from this surface stoichiometry has been observed. This last result opens the way to the use of 3C-SiC(100) as a substrate to get a more pronounced diamond nucleation effect by controlling its extreme surface. This is very helpful in achieving thinner nanocrystalline diamond films in the future with a controlled diamond/SiC interface.

## REFERENCES


1. Ibach, H.; Mills, D. L. Electron Energy Loss Spectroscopy and Surface Vibrations. 1982.
2. Henderson, M. A.; Mitchell, G. E.; White, J. M. The Chemisorption of Methyl Halides (Cl, Br and I) on Pt(111). *Surf. Sci.* **1987**, *184* (1–2), L325–L331.
3. Michaelson, S.; Chandran, M.; Zalkind, S.; Shamir, N.; Akhvlediani, R.; Hoffman, A. Dissociative Adsorption of Molecular Deuterium and Thermal Stability onto Hydrogenated, Bare and Ion Beam Damaged Poly- and Single Crystalline Diamond Surfaces. *Surf. Sci.* **2015**, *642*, 16–21.
4. Aizawa, T.; Ando, T.; Kamo, M.; Sato, Y. High-Resolution Electron-Energy-Loss Spectroscopic Study of Epitaxially Grown Diamond (111) and (100) Surfaces. *Phys. Rev. B* **1993**, *48* (24), 18348–18351.
5. Pate, B. The Diamond Surface: Atomic and Electronic Structure. *Surf. Sci.* **1986**.
6. Maier, F.; Riedel, M.; Mantel, B.; Ristein, J.; Ley, L. Origin of Surface Conductivity in Diamond. *Phys. Rev. Lett.* **2000**.
7. Lurie, P.; Wilson, J. The Diamond Surface: I. The Structure of the Clean Surface and the Interaction with Gases and Metals. *Surf. Sci.* **1977**.
8. van der Weide, J.; Zhang, Z.; Baumann, P. Negative-Electron-Affinity Effects on the Diamond (100) Surface. *Phys. Rev. B* **1994**.
9. Cui, J.; Ristein, J.; Ley, L. Electron Affinity of the Bare and Hydrogen Covered Single Crystal Diamond (111) Surface. *Phys. Rev. Lett.* **1998**.
10. Cui, J.; Ristein, J.; Ley, L. Electron Affinity of the Bare and Hydrogen Covered Single Crystal Diamond (111) Surface. *Phys. Rev. Lett.* **1998**, *81* (2), 429–432.
11. Szameitat, M.; Jiang, X.; Beyer, W. Influence of Adsorbates on the Surface Conductivity of Chemical Vapor Deposition Diamond. *Appl. Phys. Lett.* **2000**.
12. Riedel, M.; Ristein, J.; Ley, L. Recovery of Surface Conductivity of H-Terminated Diamond after Thermal Annealing in Vacuum. *Phys. Rev. B* **2004**.
13. Foord, J.; Lau, C.; Hiramatsu, M. Influence of the Environment on the Surface Conductivity of Chemical Vapor Deposition Diamond. *Diam. Relat.* **2002**.
14. Bauer, T.; Schreck, M. High Growth Rate Homoepitaxial Diamond Deposition on off-Axis Substrates. *Diam. Relat.* **2005**.

15. Bogdan, G.; Nesládek, M.; D'Haen, J. Growth Characterization of Near-atomically Flat, Thick Homoepitaxial CVD Diamond Films. *Phys. Status Solidi A, Appl. Res.* **2005**.
16. Achard, J.; Tallaire, A.; Sussmann, R. The Control of Growth Parameters in the Synthesis of High-Quality Single Crystalline Diamond by CVD. *J. Cryst.* **2005**.
17. Butler, J. E.; Oleynik, I. A Mechanism for Crystal Twinning in the Growth of Diamond by Chemical Vapour Deposition. *Philos. Trans. R. Soc., Math. Phys. Eng. Sci.* **2008**, *366* (1863), 295–311; discussion 311.
18. Vickerman, J. C.; Gilmore, I. S. *Surface Analysis – The Principal Techniques*. 2nd Edition; 2009.
19. Michaelson, S.; Lifshitz, Y.; Hoffman, A. High Resolution Electron Energy Loss Spectroscopy of Hydrogenated Polycrystalline Diamond: Assignment of Peaks through Modifications Induced by Isotopic Exchange. *Diam. Relat. Mater.* **2007**, *16* (4–7), 855–860.
20. Thoms, B. D.; Russell, J. N.; Pehrsson, P. E.; Butler, J. E. Adsorption and Abstraction of Hydrogen on Polycrystalline Diamond. *J. Chem. Phys.* **1994**, *100* (11), 8425–8431.
21. Michaelson, S.; Hoffman, A.; Lifshitz, Y. Determination of Vibrational Modes in Electron Energy Loss Spectroscopy of Polycrystalline Diamond Surfaces by Isotopic Exchange. *Appl. Phys. Lett.* **2006**, *89* (22), 223112–223113.
22. Bobrov, K.; Shechter, H.; Folman, M.; Hoffman, A. Deuterium Adsorption-Desorption from Diamond (100) Single Crystal Surfaces Studied by TPD. *Diam. Relat. Mater.* **1998**, *7* (2–5), 170–176.
23. Lee, S.-T.; Lin, Z.; Jiang, X. CVD Diamond Films: Nucleation and Growth. *Mater. Sci. Eng., R Rep.* **1999**, *25* (4), 123–154.
24. Akhvlediani, R.; Lior, I.; Michaelson, S.; Hoffman, A. Nanometer Rough Sub-Micrometer-Thick and Continuous Diamond Chemical Vapor Deposition Film Promoted by a Synergetic Ultrasonic Effect. *Diam. Relat. Mater.* **2002**, *11* (3–6), 545–549.
25. Ternyak, O.; Michaelson, S.; Tkach, L.; Akhvlediani, R.; Hoffman, A. The Impact of Gas Isotopic Exchange on the Growth Rate and Hydrogen (deuterium) Bonding within CVD Diamond Films. *Phys. Status Solidi* **2007**, *204* (9), 2839–2846.
26. Thoms, B. D.; Butler, J. E. HREELS Scattering Mechanism from Diamond Surfaces. *Phys. Rev. B* **1994**, *50* (23), 17450–17455.
27. Thoms, B. D.; Butler, J. E. HREELS and LEED of: The 2 × 1 Monohydride Dimer Row Reconstruction. *Surf. Sci.* **1995**, *328* (3), 291–301.
28. Michaelson, S.; Ternyak, O.; Hoffman, A.; Williams, O. A.; Gruen, D. M. Hydrogen Bonding at Grain Surfaces and Boundaries of Nanodiamond Films Detected by High Resolution Electron Energy Loss Spectroscopy. *Appl. Phys. Lett.* **2007**, *91* (10), 103104.
29. Zhou, D.; Gruen, D. M.; Qin, L. C.; McCauley, T. G.; Krauss, A. R. Control of Diamond Film Microstructure by Ar Additions to CH[sub 4]/H[sub 2] Microwave Plasmas. *J. Appl. Phys.* **1998**, *84* (4), 1981–1989.
30. Aizawa, T.; Ando, T.; Yamamoto, K.; Kamo, M.; Sato, Y. Surface Vibrational Studies of CVD Diamond. *Diam. Relat. Mater.* **1995**, *4* (5–6), 600–606.
31. Lee, S.; Apai, G. Surface Phonons and CH Vibrational Modes of Diamond (100) and (111) Surfaces. *Phys. Rev. B* **1993**.
32. Ristein, J.; Stief, R. T.; Ley, L.; Beyer, W. A Comparative Analysis of a-C:H by Infrared Spectroscopy and Mass Selected Thermal Effusion. *J. Appl. Phys.* **2008**, *84* (1998), 3836–3847.
33. Benlahsen, M.; Racine, B.; Zellama, K.; Turban, G. On the Hydrogen Incorporation, Intrinsic Stress and Thermal Stability of Hydrogenated Amorphous Carbon Films Deposited

from an Electron Cyclotron Resonance Plasma. *J. Non-Cryst. Solids* **2001**, *283* (1–3), 47–55.

34. Wild, C.; Koidl, P. Thermal Gas Effusion from Hydrogenated Amorphous Carbon Films. *Appl. Phys. Lett.* **1987**, *51* (19), 1506.
35. Ferrari, A. C.; Robertson, J. Raman Spectroscopy of Amorphous, Nanostructured, Diamond-like Carbon, and Nanodiamond. *Philos. Trans. R. Soc., Math. Phys. Eng. Sci.* **2004**, *362* (1824), 2477–2512.
36. Kuzmany, H.; Pfeiffer, R.; Salk, N.; Günther, B. The Mystery of the 1140 $cm^{-1}$ Raman Line in Nanocrystalline Diamond Films. *Carbon N.Y.* **2004**, *42* (5–6), 911–917.
37. Ferrari, A.; Robertson, J. Origin of the 1150-$cm^{-1}$ Raman Mode in Nanocrystalline Diamond. *Phys. Rev. B* **2001**.
38. Michaelson, S.; Ternyak, O.; Hoffman, A.; Lifshitz, Y. Hydrogen Incorporation Processes in Nanodiamond Films Studied by Isotopic Induced Modifications of Raman Spectra. *Appl. Phys. Lett.* **2006**, *89* (13), 131918.
39. Hoffman, A.; Lafosse, A.; Michaelson, S.; Bertin, M.; Azria, R. Nano Size Effects in the High Resolution Electron Energy Loss Spectra and Excitation Function of Hydrogenated Diamond Films. *Surf. Sci.* **2008**, *602* (18), 3026–3032.
40. Lifshitz, Y.; Meng, X. M.; Lee, S. T.; Akhveldiany, R.; Hoffman, A. Visualization of Diamond Nucleation and Growth from Energetic Species. *Phys. Rev. Lett.* **2004**, *93* (5), 056101.
41. Michaelson, S.; Hoffman, A. Hydrogen Bonding, Content and Thermal Stability in Nano-Diamond Films. *Diam. Relat. Mater.* **2006**, *15* (48), 486–497.
42. Lifshitz, Y.; Köhler, T.; Frauenheim, T.; Guzmann, I.; Hoffman, A.; Zhang, R. Q.; Zhou, X. T.; Lee, S. T. The Mechanism of Diamond Nucleation from Energetic Species. *Science* **2002**, *297* (5586), 1531–1533.
43. Lafosse, A.; Hoffman, A.; Bertin, M.; Teillet-Billy, D.; Azria, R. Density-of-States Effect on Surface and Lattice Vibrational Modes in Hydrogenated Polycrystalline Diamond. *Phys. Rev. B* **2006**.
44. Lafosse, A.; Billy, D.; Guillotin, J.; Le Coat, Y.; Azria, R.; Laikhtman, A.; Hoffman, A. Role of Electronic Band Structure and Resonances on Electron Reflectivity and Vibrational Excitation Functions: The Case of Hydrogenated Diamond. *Phys. Rev. B* **2003**.
45. Michaelson, S.; Hoffman, A. Ambient Contamination of Poly-Crystalline Diamond Surfaces Studied by High-Resolution Electron Energy Loss Spectroscopy and X-Ray Photoelectron Spectroscopy. *Diam. Relat. Mater.* **2008**, *17* (6), 920–924.
46. Michaelson, S.; Akhvlediani, R.; Hoffman, A. Hydrogenation and Thermal Stability of Nano- and Microcrystalline Diamond Films Studied by Vibrational Electron Spectroscopy. *J. Appl. Phys.* **2008**, *104* (8), 083527.
47. Michaelson, S.; Ternyak, O.; Hoffman, A.; Lifshitz, Y. Correlation between Diamond Grain Size and Hydrogen Retention in Diamond Films Studied by Scanning Electron Microscopy and Secondary Ion Mass Spectroscopy. *Appl. Phys. Lett.* **2007**, *90* (3), 031914.
48. Annett, J. F.; Palmer, R. E.; Willis, R. F. Surface Dielectric Response of a Semimetal: Electron-Energy-Loss Spectroscopy of Graphite. *Phys. Rev. B* **1988**, *37* (5), 2408–2414.
49. Zecho, T.; Güttler, A.; Sha, X.; Jackson, B.; Küppers, J. Adsorption of Hydrogen and Deuterium Atoms on the (0001) Graphite Surface. *J. Chem. Phys.* **2002**, *117* (18), 8486.
50. Diebold, U.; Preisinger, A.; Schattschneider, P.; Varga, P. Angle Resolved Electron Energy Loss Spectroscopy on Graphite. *Surf. Sci.* **1988**, *197* (3), 430–443.

51. Butenko, Y. V.; Kuznetsov, V. L.; Chuvilin, A. L.; Kolomiichuk, V. N.; Stankus, S. V.; Khairulin, R. A.; Segall, B. Kinetics of the Graphitization of Dispersed Diamonds at "low" Temperatures . *J. Appl. Phys.* **2000**, *88* (7), 4380.
52. Baidakova, M.; Siklitsky, V.; Vul, A. Y. Ultradisperse-Diamond Nanoclusters Fractal Structure and Diamond–Graphite Phase Transition. *Chaos Solitons Fractals* **1999**, *10* (12), 2153–2163.
53. Kwon, S. J.; Park, J.-G. Theoretical Analysis of the Graphitization of a Nanodiamond. *J. Phys. Condens. Matter* **2007**, *19* (38), 386215.
54. Herchen, H.; Cappelli, M. A. First-Order Raman Spectrum of Diamond at High Temperatures. *Phys. Rev. B* **1991**, *43* (14), 11740–11744.
55. Buffat, P.; Borel, J.-P. Size Effect on the Melting Temperature of Gold Particles. *Phys. Rev. A* **1976**, *13* (6), 2287–2298.
56. Zhang, M.; Efremov, M. Y.; Schiettekatte, F.; Olson, E. A.; Kwan, A. T.; Lai, S. L.; Wisleder, T.; Greene, J. E.; Allen, L. H. Size-Dependent Melting Point Depression of Nanostructures: Nanocalorimetric Measurements. *Phys. Rev. B* **2000**, *62* (15), 10548–10557.
57. Qi, W. H.; Wang, M. P. Size and Shape Dependent Melting Temperature of Metallic Nanoparticles. *Mater. Chem. Phys.* **2004**, *88* (2–3), 280–284.
58. Meng, Q.; Zhou, N.; Rong, Y.; Chen, S.; Hsu (Xu Zuyao), T. Y.; Zuyao, X. Size Effect on the Fe Nanocrystalline Phase Transformation. *Acta Mater.* **2002**, *50* (18), 4563–4570.
59. Qi, W. H. Size Effect on Melting Temperature of Nanosolids. *Physica B, Condens. Matter* **2005**, *368* (1–4), 46–50.
60. Zhu, K.-R.; Zhang, M.-S.; Hong, J.-M.; Yin, Z. Size Effect on Phase Transition Sequence of $TiO_2$ Nanocrystal. *Mater. Sci. Eng. A* **2005**, *403* (1–2), 87–93.
61. Zhong, W. L.; Wang, Y. G.; Zhang, P. L.; Qu, B. D. Phenomenological Study of the Size Effect on Phase Transitions in Ferroelectric Particles. *Phys. Rev. B* **1994**, *50* (2), 698–703.
62. Defay, R.; Prigogine, I.; Bellemans, A.; Everett, D. *Surface Tension and Adsorption*. 1966.
63. Gamarnik, M. Y. Size-Related Stabilization of Diamond Nanoparticles. *Nanostruct. Mater.* **1996**, *7* (6), 651–658.
64. Gamarnik, M. Y. Size Changes of Lattice Parameters in Ultradisperse Diamond and Silicon. *Phys. Status Solidi* **1990**, *161* (2), 457–462.
65. Lafosse, A.; Bertin, M.; Michaelson, S.; Azria, R.; Akhvlediani, R.; Hoffman, A. Surface Defects Induced by in-Situ Annealing of Hydrogenated Polycrystalline Diamond Studied by High Resolution Electron Energy Loss Spectroscopy. *Diam. Relat. Mater.* **2008**, *17* (6), 949–953.
66. Arnault, J. C.; Saada, S.; Delclos, S.; Rocha, L.; Intiso, L.; Polini, R.; Hoffman, A.; Michaelson, S.; Bergonzo, P. Surface Science Contribution to the BEN Control on Si(100) and 3C-SiC(100): Towards Ultrathin Nanocrystalline Diamond Films. *Chem. Vap. Depos.* **2008**, *14* (7–8 spec. iss.), 187–195.
67. Wolter, S.; Stoner, B.; Glass, J. Textured Growth of Diamond on Silicon via in Situ Carburization and Bias-enhanced Nucleation. *Appl. Phys.* **1993**.
68. Stoner, B.; Ma, G.; Wolter, S.; Glass, J. Characterization of Bias-Enhanced Nucleation of Diamond on Silicon by Invacuo Surface Analysis and Transmission Electron Microscopy. *Phys. Rev. B* **1992**.
69. Graupner, R.; Maier, F.; Ristein, J.; Ley, L.; Jung, C. High-Resolution Surface-Sensitive C 1 S Core-Level Spectra of Clean and Hydrogen-Terminated Diamond (100) and (111) Surfaces. *Phys. Rev. B* **1998**, *57* (19), 12397–12409.

70. Arnault, J.-C.; Saada, S.; Michaelson, S.; Hangaly, N. K.; Akhvlediani, R.; Hoffman, A. Bias Enhanced Diamond Nucleation onto 3C–SiC(100) Surfaces Studied by High Resolution X-Ray Photoelectron and High Resolution Electron Energy Loss Spectroscopies. *Diam. Relat. Mater.* **2008**, *17* (3), 377–382.
71. Lifshitz, Y.; Kasi, S.; Rabalais, J. Subplantation Model for Film Growth from Hyperthermal Species: Application to Diamond. *Phys. Rev. Lett.* **1989**, *62* (11), 1290–1293.

## ACKNOWLEDGEMENTS

This review was carried out with the financial support of the European Community FP7-Energy Project ProME$^3$ThE$^2$US$^2$ "Production Method of Electrical Energy by Enhanced Thermal Electron Emission by the Use of Superior Semiconductors", Grant Agreement No. 308975, website: www.prometheus-energy.eu. A.H. would like to acknowledge the Technion-IIT fund for the promotion of research.

CHAPTER

# Surface potential of nanodiamonds investigated by KPFM

10

**Bohuslav Rezek*, Stepan Stehlik†**

*Faculty of Electrical Engineering, Czech Technical University, Prague, Czech Republic* Institute of Physics, Czech Academy of Sciences, Prague, Czech Republic†*

## Contents



## 10.1 NANODIAMONDS AND ELECTRICAL CHARGE

Nanoparticles of diverse materials and sizes represent an emerging type of materials that find broad use in commercial products as well as attract attention in fundamental research with a perspective of novel effects and phenomena. In this respect,

**Nanodiamonds**

diamond nanoparticles (mostly denoted as nanodiamonds, NDs, in similarity to nano-iron, nanosilver, etc.) are gaining considerably increasing interest due to their specific properties and wide range of possible applications.[1,2]

Formation of NDs by detonation has been firstly reported in 1961.[3] Nowadays, NDs of various kinds such as typically detonation nanodiamonds (DNDs) or high pressure high temperature (HPHT) NDs are available in large quantities although their surface and material quality may differ.[4] Applicability of NDs ranges from biology (fluorescence imaging,[5] drug delivery[6]) and chemical sensors (Surface Acoustic Wave sensors,[7] microelectromechanical sensors[8]) to single photon sources[9,10] and application of NDs as reflectors in low energy neutron physics.[11] In all these applications, NDs offer unique properties such as biocompatibility, various possibilities of surface functionalization,[12] presence of functional impurities and defects ($NV^-$ and other color centers), radiation hardness, etc. Partially graphitized NDs with fullerene-like surface reconstruction or fully graphitized NDs known as onion-like carbon (OLC) are perspective nanomaterials for supercapacitors.[13,14] NDs are also indispensable for chemical vapor deposition of nanocrystalline diamond where they serve as seeds for growth on diverse substrate materials and morphologies.[15–17] Further details on NDs fabrication and specific applications are provided in Chapter 2 or Chapters 14 to 18, respectively.

Numerous analytical techniques are used to characterize properties of NDs and nanoparticles, in general. Atomic structure can be directly visualized by High Resolution Transmission Electron Microscopy (HRTEM, see Chapter 3)[18] or Scanning Transmission Electron Microscopy (STEM),[4] composition and surface chemistry can be characterized by X-ray Photoelectron Spectroscopy (XPS) or infrared and Raman spectroscopies[19] (for details see Chapter 4), particle size by Dynamic Light Scattering (DLS)[20] or Atomic Force Microscopy (AFM).[4] It is often advantageous to correlate methods with spatial resolution high enough to characterized individual nanoparticles (TEM, AFM), yet still able to provide some degree of statistical information,[21] with methods providing more or less averaged information (XPS, FTIR, Raman, DLS), which can be sometimes misleading due to heterogeneity of the ND samples.[4,22]

One of the key factors for many of the aforementioned properties and applications is electrical charge of the NDs. So-called zeta potential is the mostly used quantity to characterize electrostatic charge of nanoparticles. It is well established from colloidal chemistry, from which also its name stems.[23] There are various charging mechanisms giving rise to zeta potential, such as dissociation of surface chemical groups, their dipoles and electron affinity differences. Zeta potential is defined as an electrical potential at the boundary of bulk solution and a fixed water layer, which envelops the nanoparticle and which is retained during the nanoparticle movements in solution. Zeta potential is measured by applying an electric field across the colloid and monitoring nanoparticle migration velocity, typically by DLS. Measurement of zeta potential on NDs as a function of pH showed that oxidized NDs have negative zeta potential within the full range of pH while graphitized NDs or H-NDs have positive zeta potential up to $pH = 12$.[20,24] Zeta potential has high importance for mutual nanoparticle interactions in solutions. It influences, for instance, the stability of NDs' colloidal solutions and assembly of NDs on surfaces.[25] For further details see Chapter 11. However, zeta potential is not the charge or work function of the nanoparticle itself.

Yet, electrical charge, potential, and work function of the nanoparticle are equally important. They influence, for instance, the adsorption of proteins[26] or drugs,[27] penetration of cell membrane,[28] luminescence,[29,30] and electron emission.[20,31] Here we show that Kelvin probe and in particular Kelvin probe force microscopy (KPFM) can be used for detailed analysis of electrical potential of NDs on microscopic level and it can facilitate understanding of effects related with the nanodiamond electrical charge.

## 10.2 KELVIN PROBE FORCE MICROSCOPY (KPFM)

### 10.2.1 FROM MACROSCOPIC MEASUREMENTS TO MICROSCOPY

The original Kelvin probe method dates back to 1861 when Lord Kelvin performed experiments with two plates of different metals forming a parallel plate capacitor and electroscope. A scheme of this original setup is shown in Fig. 10.1A. Already by then, well prior to the discovery of the electron in 1905, he detected charging of the plates when the metals were different, provided correct interpretation (albeit in different words) and sorted metals by this effect. The fundamental principle of the charging is based on equilibration of work functions, as shown in Fig. 10.1B. Due to the difference in the work function ($\Phi$) of the probe and sample, a contact potential difference (CPD) arises as

$$V_{CPD} = \frac{\Delta\Phi}{e} = \frac{(\Phi_{sample} - \Phi_{tip})}{e}. \tag{10.1}$$

The work function is defined as

$$\Phi = E_{vac} - E_F, \tag{10.2}$$

where $E_{vac}$ is energy of vacuum level and $E_F$ represents the Fermi energy of electrons.

In principle, if the work function of one material is known (a probe) one can determine the work function of the studied sample. Nowadays, such macroscopic Kelvin probe measurement is performed by a vibrating electrode (in the form of fork or tip), and it became a useful tool for material characterization in air or ultra-high vacuum.[32,33]

Kelvin Probe Force Microscopy (KPFM) miniaturizes this kind of measurement by taking advantage of sharp conductive tips on cantilevers employed in atomic force microscopy (AFM) and by using the AFM system itself to precisely control XYZ position above the sample under study.[34,35] An example of a KPFM system in our

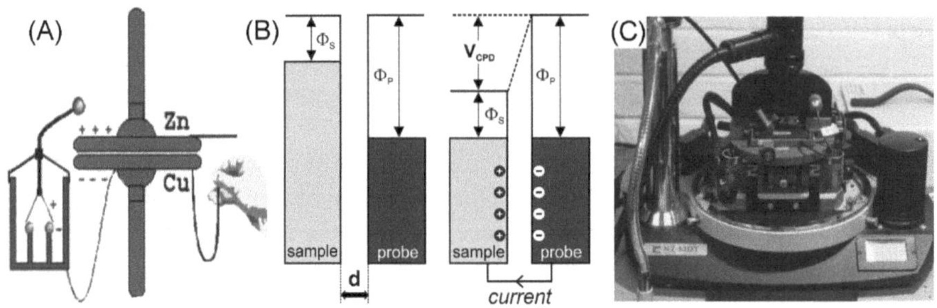


**FIGURE 10.1**

(A) Schematic drawing of original Lord Kelvin's systems (used with permission from kelvinprobe.info). (B) Principle mechanism of Kelvin probe showing two materials with different work functions prior and after electrical connection while still having a narrow gap in between them. (C) Photograph of modern microscopic Kelvin probe based on atomic force microscopy system including coaxial illumination by macro-optics as well as sideways illumination by optical fiber

laboratory is shown in Fig. 10.1C. When CPD is modulated by an external alternating voltage, the resulting alternating electrostatic force gives rise to mechanical oscillations of AFM cantilever. A feedback loop is used to compensate CPD by external dc voltage so that the oscillations cease. Thereby, microscopic mapping of the surface potential is possible and it can be directly correlated with microscopic surface morphology obtained by AFM.[36]

For quantitative determination of the work function, the work function of the KPFM tip needs to be independently calibrated. We found practical to use a layer of gold deposited on the sample directly or at least on a reference substrate. The metal work function can be calibrated by X-ray (XPS) or ultraviolet (UPS) photoelectron emission spectroscopies[37] or by total photoelectron yield spectroscopy (TPYS)[36] using Fowler's photoelectron emission theory. In addition, as TPYS measures the energy gap between occupied states and the vacuum level, the surface Fermi level is not affecting the results in a complex way as in XPS/UPS. Electron affinity of diamond with various surface terminations can be thereby deduced in a straightforward way.[38] By complementing CPD measurements in KPFM with electron affinity and work function calibration from TPYS, one can construct a surface and subsurface energetic band configuration by using relatively straightforward guideline scheme.[39] Note that the metal work function calibration by XPS or TPYS should be done in natural (air adsorbed) condition for deducing diamond surface work function from KPFM data in air.[36] On diamond and other materials such as graphene, adsorbates from surrounding environment influence electronic properties of the material. Removal of adsorbates by e.g. annealing in ultra-high vacuum (UHV) thus provides a clean surface but also a material with completely different electronic properties. Therefore, it is essential to perform KPFM and other analysis in application relevant state of the sample.

Based on the different work function of hydrogenated and oxidized diamond surface, KPFM clearly distinguished areas with different surface termination, although topography showed no difference.[36,40] Moreover, KPFM was used to recognize covalent grafting of organic molecules to the diamond surface by the change of surface work function.[41]

KPFM does not detect only work function though. In principle, the obtained signal is a superposition of contact potential difference (CPD) due to difference between KPFM tip and sample work functions with any additional electrostatic charge in the system which modulates the CPD signal. Thereby, one can characterize charge trapping in nanocrystalline diamond,[42] evaluate Schottky contact charging and photovoltages in diamond photodiodes,[39] monitor photovoltage kinetics in diamond-organic heterostructures,[41] or detect photovoltages on pristine nanodiamonds.[43]

KPFM measurements can thus provide valuable insight into material properties, surface chemistry, as well as into the various electronic and opto-electronic effects with high spatial resolution.

## 10.2.2 SPATIAL RESOLUTION LIMIT

It may seem that one needs very sharp AFM tips to accurately measure surface potential/CPD of nm-sized particles. That is quite the opposite though. Theoretical calculations[44] showed that the observed potential is a locally weighted average over all potentials present on the sample surface and thus lead to the conclusion that good spatial resolution in potential maps is obtained by long, slender, and slightly blunt tips. These calculations are in agreement with basic physics that larger capacity arises from larger electrodes at closer distance.

We evidenced this effect also experimentally, directly on NDs by using a two-pass KPFM regime. Fig. 10.2A shows potential contrast of scattered NDs as a function of KPFM measurement height during the second pass. The measurement height (or more precisely so-called lift height) is given with respect to the height during AFM topography scan in the first pass. Negative values thus correspond to closer approach towards the sample surface during KPFM. This is possible because in a typical tapping (intermittent contact) mode topography measurement the tip oscillation amplitude is on the order of tens of nanometers while in KPFM the electrostatically induced oscillations are an order of magnitude lower, typically only a few nm. If the tip in KPFM approaches the surface too close though, it may occasionally touch the surface. This leads to instabilities during KPFM measurement (due to mechanical disturbance of electrostatically induced weak oscillations of the cantilever as well as due to a shortcut between the tip and sample), which are visible as short streaks (local instabilities) or completely smeared lines (overall instability).

There are various single-pass KPFM methods,[45] which make the measurement faster and compensate tip–sample potential difference during topography scan (thereby possibly reducing topographic artifacts induced by electrostatic forces).[46,47] On the other hand, the two-pass regime of KPFM enables finding optimal condition for the best, yet stable contrast by adjusting the measurement height in the second

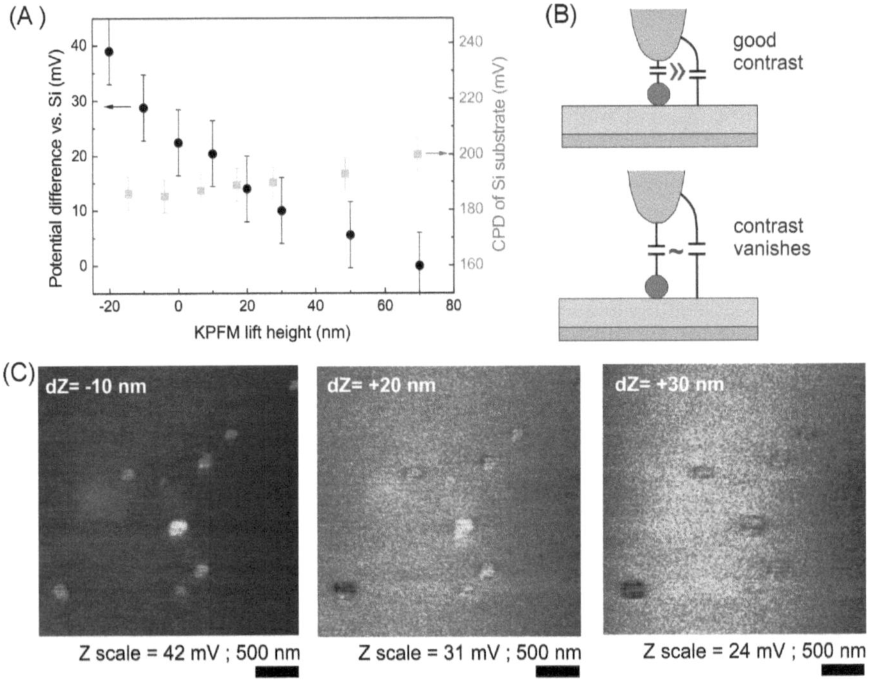


**FIGURE 10.2**

(A) Potential contrast on top of scattered nanodiamonds as a function of KPFM measurement height (lift height) during the second pass. The measurement height is given with respect to height during AFM topography scan in the first pass. Negative values thus correspond to further approach towards the sample surface during KPFM. (B) Scheme showing interplay of local and stray capacitance with impact on potential contrast of nanodiamonds. (C) KPFM images as a function of measurement height. Edge artifacts (dark circles) around the nanoparticles became noticeable at low contrast (larger tip–nanodiamond distance). Data were acquired with a PtCr-coated Si AFM tip on plasma-oxidized HPHT NDs with 50 nm nominal size deposited on Si substrate. The error bars were taken as RMS of CPD values on the substrate

pass with respect to surface topography obtained in the first pass. This is in particular useful for KPFM measurements of objects in sizes comparable or larger than the tip radius (>5 nm, i.e. the sample is not atomically flat) when measurement of topography and potential often requires different settings for optimal results.

Lowering the distance to surface in the second pass increases local capacitance between the KPFM tip and nanoparticle, thereby reducing possible effects of stray capacitance and increasing the potential contrast of NDs versus the substrate. This effect is schematically shown in Fig. 10.2B and practically in Fig. 10.2C where KPFM images for different measurement height are shown. At the smallest measurement distance (lift height of −10 nm) the bright potential of NDs is well pronounced. When the tip distance increases (lift height of +50 nm), the NDs are barely visible in the potential image. In fact, they are noticeable only due to edge artifacts[48] (dark circles) around the NDs, which arise due to finite size and geometry of the AFM tip and NDs.

Considering the effects of stray capacitance and edge artifacts, one may wonder what the ultimate spatial resolution of KPFM is. The answer is simple – down to atomic scale. Atomically resolved KPFM measurements on a Si surface with 7 × 7 reconstruction were demonstrated by frequency-modulated KPFM (FM-KPFM) in ultrahigh vacuum.[49]

## 10.3 SURFACE POTENTIAL OF NANODIAMONDS

### 10.3.1 DEPOSITION OF NANODIAMONDS ON SUBSTRATES

For characterization of surface potential, it is preferable that NDs are evenly distributed on the substrate. Deposition from colloidal solution is typically used. It is influenced by surface chemistry and zeta potential of both the NDs and the substrate. For example, deposition of hydrogenated NDs (H-NDs) on a silicon substrate is advantageous[25] due to positive zeta potential of H-NDs and negative zeta potential of Si substrate (which is covered by native $SiO_x$ layer) close to neutral $pH$. The opposite zeta potentials lead to electrostatic attraction between the H-NDs and Si substrate. On the other hand, deposition of oxidized NDs (O-NDs) on Si substrates is hindered due to electrostatic repulsion between O-NDs and Si substrate, both having negative zeta potential close to neutral $pH$. In order to overcome this issue, one can use several approaches. One is a cationic polyelectrolyte coating of a substrate.[15] Another is $pH$ adjustment of the O-ND colloid to $pH \sim 3$ at which the O-NDs still keep negative zeta potential but Si substrate turns to positive values due to protonation of surface Si–OH groups. Although this approach has been applied to citrate-stabilized Au nanoparticles,[50] we found it applicable to O-NDs, too. Ultrasonic agitation can be used to overcome the electrostatic barriers between the NDs and a substrate and promote at least low density deposition even at the same zeta potentials.[4] Also electrophoretic deposition can be used to promote deposition of NDs (both positive and negative ZP) on a substrate with good electrical conductivity such as Au or graphite.[51] Nevertheless, we observed that the resulting ND density on substrate is determined by the dilution rather than by specific deposition method. In order to coat the substrates with scattered NDs of various surface terminations (useful not only for KPFM but also as etching masks,[52] electrochemically active sites,[53] single photon sources,[54] etc.), we found most practical to use ultrasonic agitation or drop casting of highly diluted colloid. Typical ND colloids are made with concentration on the order of 1 mg/mL. To obtain reasonably scattered NDs on substrate we used 1:10 to 1:100 dilution of the initial colloidal solution. Suitable concentration thus corresponds to about 1–10 μg/mL.

## 10.3.2 INFLUENCE OF SIZE ON NANODIAMOND POTENTIAL

For characterization of size and morphology of NDs on substrates, atomic force microscopy can provide statistically relevant data down to 1 nm particles[4] and the data can be directly correlated with surface potential measurements by KPFM or with other regimes such as phase images,[55,56] electrical current measurements[22] or even with intentional in situ such as local electrostatic charging.[57,58]

Fig. 10.3A shows schematic drawing of the AFM/KPFM measurement setup. The substrate has an electrode deposited on top. We mostly used Au thin film (50 nm) sputtered on n-doped Si wafer, yet other metals can be applied as well. Compared to thermally evaporated Au, sputtered Au film provides smoother and featureless surface which is more suitable for characterization of nanoparticles. The NDs are depicted as individual nanoparticles or nanoparticle clusters of various sizes. This corresponds to typical situation for detonation NDs. For HPHT nanodiamonds the clustering is less pronounced and size distribution is wider[59] due to various sizes of NDs after crushing and milling of the initial HPHT microcrystals (see Chapter 2).

### *10.3.2.1 Surface potential on scattered nanodiamonds*

Fig. 10.3B shows a typical AFM topography of NDs deposited on the substrate; in this case oxidized (air-annealed) DNDs on Au substrate. Single NDs of 5 nm height as well as clusters with height up to 50 nm are observed. Note that lateral dimensions of NDs are overestimated by 20–40 nm as the AFM image is influenced by convolution of the nanodiamond and tip shape as they are of comparable dimensions. The size of NDs or ND clusters can be characterized more precisely from the AFM height. Of course, this method may not provide accurate values for highly asymmetric nanoparticles or clusters, and neither does DLS. Nevertheless, we will see further below that the KPFM potential is determined mostly by the nanodiamond/cluster thickness and not by its lateral size.

Corresponding KPFM map of the same NDs is shown in Fig. 10.3C. NDs are well resolvable in the KPFM image as dark spots. One thing is immediately noticeable, the potential is not constant and it seems to vary with the size. Characteristic value of surface potential of a particular type of NDs is thus difficult to define. We have proposed that one can still use a methodology[20] which is schematically shown in Fig. 10.3D. The height of ND or ND cluster is obtained as the difference between the utmost height of the ND (identified from the AFM scan) and the background height in close ND vicinity (yet outside the range of possible edge artifacts). KPFM potential values are acquired at the same positions. Potential at the ND top can be taken as absolute value for direct comparison or it can be subtracted from the background potential, whereby potential difference vs. substrate is obtained.

Using such data obtained on Si and Au substrates with scattered oxidized (air annealed) detonation NDs, Fig. 10.3E plots surface potential of NDs as a function of their height. We can see that the height dependence of the potential shows a clear trend that can be fitted well by a linear regression in the range of recorded heights. It is also noticeable that the linear trend is similar on both Si and Au substrate and that

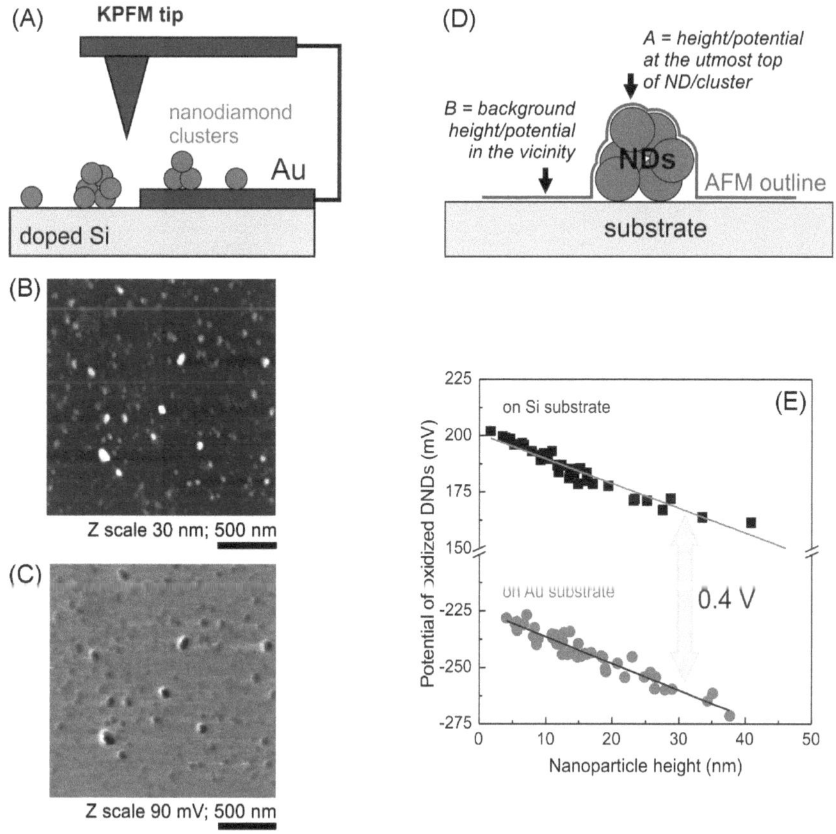


**FIGURE 10.3**

(A) Scheme of KPFM on individual nanodiamonds and nanodiamond clusters.
(B) Topography of oxidized (air-annealed) DNDs on Si substrate. (C) Corresponding map of surface potential by KPFM under optimized measurement parameters in the second pass.
(D) Schematic depiction of methodology for evaluation of surface potential of NDs.
(E) Nanodiamond surface potential plotted as function of nanoparticle height by using the methodology. Data were acquired on Si and Au substrates with scattered DNDs annealed (oxidized) in air. The graph was adapted with permission from: Stehlik, S., et al. Langmuir 2013, 29, 1634–1641[20]

the same NDs exhibit up to 0.4 V potential difference depending on the substrate on which they are deposited.

Albeit it was reported only recently for NDs, a height dependence of surface potential for nanoscale objects is generally not a new finding. Seemingly similar linear size-dependence of potential against substrate was observed for Pt clusters deposited on $TiO_2$ surfaces.[60] However, in our case the trend is not linear as a function of cluster area but as a function of particle height. If we assume that the height of nanoparticles

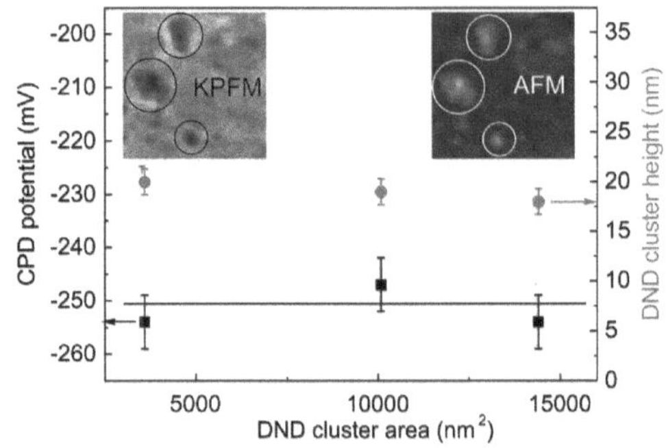


**FIGURE 10.4**

KPFM potential contrast on detonation nanodiamond clusters of similar height yet of significantly different lateral size. Lateral size is characterized as an area obtained from AFM topography image. The inset shows potential map (left) and topography image (right) with circles denoting the analyzed clusters. The data were acquired on oxidized (air-annealed) DNDs. The error bars were estimated from RMS values on the substrate

(nanoparticle clusters more precisely) can be used as an estimate of interface area diameter, the trend changes to nonlinear as a function of the effective area. Nevertheless, we tried to use the dipole-per-area model of Sasahara et al.,[60] but we were not able to obtain a good fit. In fact, the dipole-per-area model is similar to averaging of potential in KPFM measurement: the larger the Pt cluster area, the more the potential diverges from the background. Thus, the unsuccessful fit can be considered as another indication that the observed KPFM contrast is not dominated by the spatial averaging effects in the case of nanodiamonds.

We have also noticed that the KPFM potential is determined mostly by the nanodiamond particle/cluster thickness on the substrate and not by its lateral size. This is shown in Fig. 10.4. It shows several NDs clusters which have different area yet almost the same height. The graph clearly shows that the potential of the clusters is independent of their area. This is another indication that spatial averaging (or in other words stray capacitance) is not determining the observed trends.

Dependence of KPFM potential on the height was reported also on silicon nanocrystals embedded in $Si/SiO_2$ and measured in air[58] as well as on NaCl layers on Cu(111) substrates[61] and InAs quantum dots on GaAs(001) substrates[62] measured in ultra-high vacuum (UHV). For InAs quantum dots the height dependence was attributed to quantum size effect where the amount of charges accumulated in the QD is determined by the confinement energy levels in the QD of various size.[62] This effect was limited to sizes below 7 nm on InAs and saturated above this size. However, for nanodiamonds the quantum confinement effects are expected below 2 nm, which is not the case here. Even if the potential was due to a property of individual NDs inside the clusters, it should not change with the cluster size.

The work function can change as a function of thickness as well. For NaCl films, the work function changed by as much as 0.6 eV before saturation at 4 monolayer thickness.[61] For metals, the work function changed by up to 0.1 eV until 10 nm thickness.[63] However, for nanodiamonds the effect seems not to saturate at least up to 50 nm size. Thus it is most likely not related to the thickness-dependent work function.

Moreover, the height dependence was observed also for HPHT nanodiamond ensembles in the size range of 5 to 30 nm.[19] The potential differences and slope of the potential-height dependence were comparable for detonation and HPHT nanodiamonds. Thus the observed effect is not due to some specific features of detonation NDs and it does not matter whether the nanoparticle is made of a single crystal diamond or represents a cluster of smaller nanodiamonds.

A critical question may still arise whether the observed size dependence is possibly due to some effect of stray capacitance from the background. The stray capacitance may lead to potential averaging that is known to influence KPFM measurements across material boundaries or nanoscale features.[44,48] The potential contrast may be reduced by as much as 50% due to the stray capacitance of the AFM tip and cantilever.[64]

As shown in Fig. 10.2A, an increase in the tip–ND separation up to 70 nm during KPFM measurement indeed reduced the potential difference vs. substrate until it almost disappeared due to diminishing local tip–ND capacitance. However, this effect leads only to overall decrease in the potential difference of ND vs. substrate and it did not influence (qualitatively) the observed trend of potential as a function of size. This is illustrated by the KPFM image series as a function of measurement height in Fig. 10.2C. Control experiments with KPFM in the frequency-modulated regime, which is sensitive to force gradients and thereby enables in principle much higher spatial resolution,[49] provided similar (even lower) potential differences than those in Fig. 10.2.

Nevertheless, whatever the selected KPFM lift height, with increasing height of NDs the distance of the tip to background electrode increases and the stray capacitance should decrease, thereby possibly increasing contribution of local tip-NDs capacitance. Fig. 10.2A shows the results of the experiment where the background potential as a function of KPFM measurement height was measured. Unlike the potential of NDs, the change of background potential with height is negligible (<0.2 mV/nm). This indicates that the decrease in stray capacitance as a function of tip height cannot explain the linear trend observed for nanodiamonds.

#### *10.3.2.2* ***Surface potential on nanodiamond layers***

Another way to avoid contribution of substrate material to stray capacitance is to measure continuous nanodiamond layers. Relatively homogeneous, about 40 nm thick ND layers can be formed by drop-casting of several tens of μl of an ND colloid with concentration of about 1 mg/ml. We took advantage of the fact that the deposited ND layers do not have uniform thickness. The KPFM potential on local maxima in topography is correlated with the layer height. The full height was determined by making a scratch in the layer to expose underlying Au or Si substrate and subtracting the mean height of the substrate from the relative height profile of the ND layer. As shown in

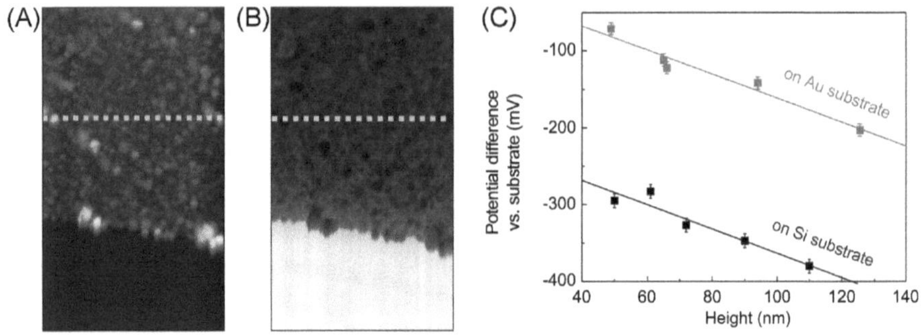


**FIGURE 10.5**

KPFM of continuous nanodiamond layers deposited by drop-casting from a colloidal solution on substrates. (A) Topography of ND layer on Si with the substrate exposed in the bottom part. (B) Map of surface potential by KPFM in the second pass under optimized measurement parameters. The line indicates where the height–potential values were taken. (C) Nanodiamond surface potential plotted as a function of local layer height. Lines correspond to linear regression fit of the data. The data were acquired on Si and Au substrates (grounded for reference) covered with hydrogenated DNDs

Fig. 10.5, the linear trend of ND potential vs. height is confirmed also in the case of ND layers. It remains linear up to about 120 nm thickness. Further study is needed to elucidate why the trends on Au and Si substrate do not overlap in spite of subtraction of the substrate potential.

All the data on potential–height dependence thus show that the stray capacitance has negligible influence in KPFM on NDs. Further analysis of subtle stray capacitance effects on NDs may be done, for instance, by deconvolution procedures that are able to account for the stray capacitance effect when the tip shape is known or estimated.[65]

### 10.3.3 INFLUENCE OF SUBSTRATE ON NANODIAMOND POTENTIAL

By analyzing surface potential of nanodiamonds as a function of size, one can notice another feature: while the trend of potential is similar for different substrate materials, the potential itself is not. Fig. 10.3 shows that the very same nanodiamonds exhibit very different potential values when deposited on Si and Au substrates. The difference is about 0.4 V, which corresponds to the difference of the work function between n-Si and Au. As we have already excluded that stray capacitance would dominate the KPFM potential, only one explanation remains: the nanodiamonds must be significantly electrostatically charged due to interaction with the substrate.

The gained amount of charge for the nanodiamond can be approximated by a simple capacitance equation, $Q = C*U$. Assuming typical nanoscale capacitance on the order of $10^{-18}$ F (attofarads)[66] and using the same capacitance on both surfaces due to same nanodiamonds, $dQ = 0.4*10^{-18}/e = 2.5$ electrons (where unit electron charge is $e = 1.602 * 10^{-19}$ C). Thus few electrons per nanodiamond particle can be gained by interaction with a substrate.

The data in Fig. 10.3 were obtained on oxidized NDs. Oxidized diamond is highly resistive and exhibits surface states able to retain electrical charge.[67] The charging of oxidized NDs is thus possible, too. Similar effect was observed also for conductive nanodiamonds with hydrogenated or graphitized surfaces and even for gold nanoparticles, though covered by surface stabilizing groups such as citrates.[20,57]

The effect of different charging on different substrates has been also corroborated by differences in electron emission[20] and photoluminescence decay.[29] More details about the impact of substrate-induced charging on other properties of nanodiamonds are described later below.

#### 10.3.3.1 Intentional electrostatic charging

Further analysis and experiments with intentional charging by AFM tip confirmed that oxidized and also hydrogenated and graphitized nanodiamonds as well as gold nanoparticles are able to retain electrostatic charge.[57] The charge retention time was estimated to be at least 60 s based on time-resolved KPFM experiments.[57] Fig. 10.6 illustrates how the potential contrast of oxidized NDs changes after intentional charging by the given voltage on the AFM tip during the topography acquisition in the first pass along each line scan. Data were obtained in air on grounded Au substrate to avoid the possibility of substrate charging which is well known on Si wafers. The images show clear inversion of NDs' potential contrast on the substrates as a function of charging polarity. The plot of potential vs. nanodiamond height (based on the methodology described above) confirms consistent effect irrespective of the ND height. The induced potential difference is as much as 160 mV by using charging voltage ±4 V on 20 nm high ND clusters.

It is noticeable that the slopes turn to more negative for applied positive voltage, and vice versa. This means that the charging proceeds actually via the substrate. In another words, if a positive bias voltage is applied on the Si tip, the nanoparticle gains a negative charge through the negatively charged substrate. This occurs during the first pass without any compensation of potential difference between the tip and sample. If the charge is effectively trapped at a nanoparticle it is then accordingly resolved with negative or positive contrast against the substrate. The effect differs from conventional contact charging methods where the polarity of retained charge is the same as the charging polarity such as on nanocrystalline diamond films,[42] fluorocarbon films[68] or silicon quantum dots.[47] The above observation also means that a possible charging effect via tip–ND contact during tapping/semicontact AFM regime is negligible.

The effect is observed, albeit not so pronounced, on hydrogenated or graphitized NDs as well. Lower charging efficiency is most likely due to higher electrical conductivity of hydrogenated or graphitized NDs.[22] Transfer doping of H-NDs has been confirmed in agreement with the same effect on bulk diamond.[69,70] Also graphite is a relatively good electrical conductor, and it was shown that even a few or single graphene layer on DNDs may provide noticeable electrical conductivity.[71]

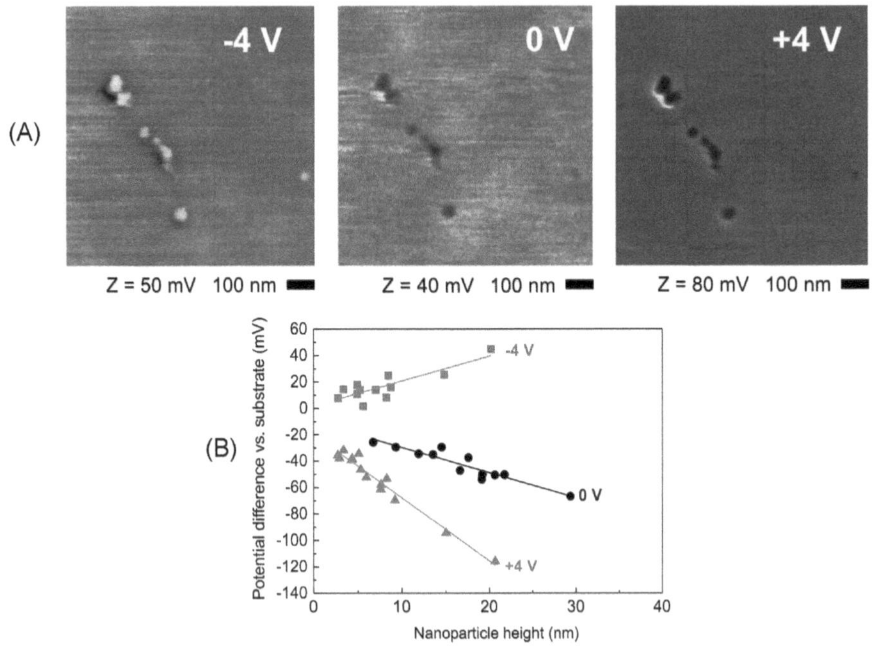


**FIGURE 10.6**

KPFM on nanodiamonds after intentional charging during the first pass, i.e. topography acquisition across each line scan. Data were obtained in air on nanodiamonds oxidized by air annealing (O-DND) on grounded Au substrate. The substrate potential was subtracted by the described methodology, thus the displayed data correspond to potential difference vs. the given substrate

## 10.3.4 INFLUENCE OF NANODIAMOND SURFACE CHEMISTRY ON SURFACE POTENTIAL

Surface condition (including various surface treatments, chemical reactions, adsorbates, contaminations) affects also the work function and thereby the potential at a surface, often quite considerably. These changes arise from the formation of electric dipoles at the surface, which change the energy for electrons to reach the vacuum level. The control and characterization of the state of the surface (chemical, structural, electronic) is thus fundamental for understanding and applying properties of nanoparticles and other nanostructures.

To assess the effect of surface condition of NDs on their potential, one can use well-established methods to modify the surface of the same NDs by various chemical moieties. The most common are hydrogenation (by microwave plasma, high-temperature annealing in hydrogen, or even r.f. plasma in air or solution),[19,72,73] oxidation (by ozone treatment, oxygen plasma, or annealing in ambient air),[4,74]

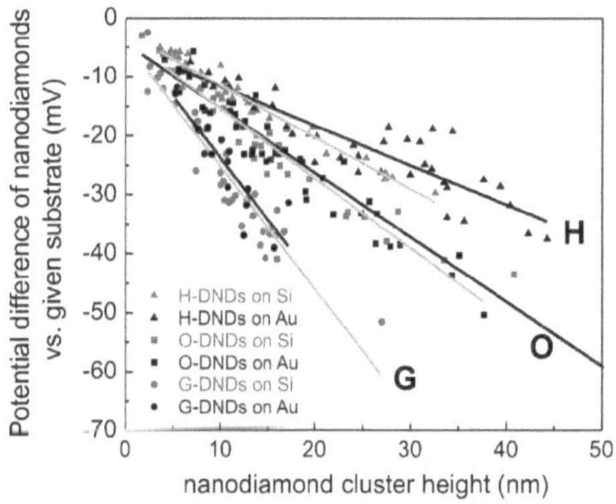


**FIGURE 10.7**

KPFM on nanodiamonds after various treatments: hydrogenation in plasma ($H$), oxidation by annealing in air ($O$), graphitization by annealing in vacuum ($G$). Data were obtained on Si and Au substrates by Si tip. The substrate potential was subtracted by the above described methodology, thus the displayed data correspond to potential difference vs. the given substrate. The different surface chemistry of NDs is resolved in potential irrespective of the employed substrate material. Adapted with permission from: Stehlik, S., et al. Langmuir 2013, 29, 1634–1641[20]

graphitization (by annealing in vacuum, see Chapter 8).[75] Other modifications are, of course, also possible, like fluorination[76] or amination.[77,78]

Fig. 10.7 shows KPFM data on nanodiamonds after various treatments: hydrogenation in plasma (H), oxidation by annealing in air (O), graphitization by annealing in vacuum (G). Data were obtained on Si and Au substrates by Si tip. The substrate potential was subtracted by the methodology described above, thus the displayed data correspond to potential difference vs. the given substrate. One can see that the surface chemistry indeed modulates the surface potential of the same nanodiamonds.

The effect is consistent across various heights of NDs. However, it is also evident that without collecting the potential data as a function of height, the effect of the surface chemistry might not be distinguished as for a given potential value one can find NDs with each modification. The effect of surface chemistry becomes clear only when the data is compared for NDs of the same height.

For H and O termination, the effect of surface chemistry is in agreement with bulk diamond; for oxygen-terminated surfaces the work function (and thus potential) is lower than for hydrogen-terminated surfaces by as much as 300 mV.[36,79] The only difference is an order of magnitude which is lower (within 50 mV) in the case of NDs. The reason is most likely the additional charging of NDs on substrates, as discussed above.

More negative potential for graphitized NDs may be due to a larger work function of graphitic patches on diamond[80] or due to higher charge storage capacity of graphitic materials (as shown e.g. on grain boundaries in nanocrystalline diamond films[42] or onion-like carbons in supercapacitors[13]).

Note that the different surface chemistry of NDs is resolved in potential irrespective of the employed substrate material. Thus the charging of NDs due to substrate does not obscure the effects of surface chemistry. We assume that the potential differences between the NDs with various surface terminations are determined mostly by their work functions.

Further KPFM experiments performed on variously modified NDs and correlated with FTIR spectroscopy confirmed the chemistry driven differences. In fact, based on the above data, KPFM was able to disclose that the C–H vibrations observed by FTIR do not necessarily correspond to H-terminated diamond but are rather related to hydrogenated amorphous carbon.[19,73] Thus without correlation with KPFM, the FTIR could lead to false conclusions.

### 10.3.5 INFLUENCE OF PROBING TIP ON NANODIAMOND POTENTIAL

In KPFM the contact potential difference (CPD) between the tip and studied material primarily arises from the difference between their work functions (modulated by any other electrostatic charge and photovoltage in the sample) and it is then compensated by application of an external voltage to the tip or to the sample during KPFM measurement. Thus the material of the tip should not change the relative differences between objects or areas having different work function in the KPFM image. However, this is not valid for nanoparticles.

Experiments with KPFM tips made of highly n-doped silicon with and without PtCr coating showed that both Si and PtCr coated tips can resolve similar height dependence of ND potential, its accommodation to various substrates, as well as its modulation due to different surface chemistry (H, O, G).[57] One could only notice that the Pt-coated tip is more prone to contamination and instability. The main difference was that the potentials were overall negative for Si tips and overall positive for Pt-coated tips on both Si and Au substrates. A similar effect was observed also for Au nanoparticles on Si and Au substrates.[57] Since this phenomenon was observed also on other substrates such as Pt or Ni, we assume that the presence of a native silicon oxide layer on the Si substrate does not play any significant role here.

On the other hand, both types of KPFM tip correctly recorded the potential change across the Au–Si interface or potential of other larger objects such as organic patches.[57] Thus the ND contrast inversion is not due to any artifact of the KPFM system, and the Au–Si interface can be used as a reasonable standard to verify correct KPFM operation. Of course, other reference samples, such as Pt/Au patterns, can be used depending on availability. In addition, as mentioned above, Au work function can be independently calibrated by photoelectron spectroscopy and enables us to deduce the absolute value of the surface work function for the sample.

The effect of KPFM tip material on ND potential can be explained as follows. When the sample and the tip are connected via an external circuit in KPFM, the contact potential difference arises. Due to a difference in the work function, the Si tip (lower work function than that of Au) gains a positive charge while the gold substrate and NDs gain a negative charge. For Pt-coated tip (higher work function than that

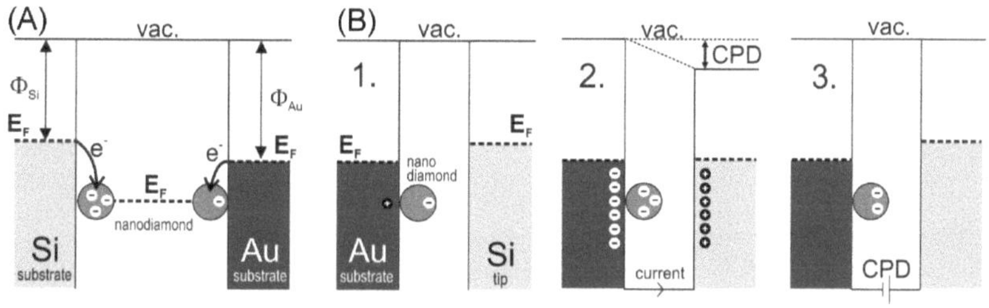


**FIGURE 10.8**

(A) Schematic model of inherent nanodiamond charging on different substrates (Si and Au in this case). (B) Schematic model of nanodiamond charging due to interaction with KPFM tip (Si in this case). For the Pt tip the charge polarity is the opposite due to a higher work function. The scheme is simplified by showing only the work function, thus for Si it does not show valence and conduction bands with energetic gap

of Au) the polarity is the opposite. This happens as soon as the tip and sample are connected in the AFM system via the external circuit, i.e. already during the first pass when the topography is acquired and CPD is not compensated in any way. In that context it is similar to the intentional charging described above. If some of the charge is retained by NDs, KPFM detects dark, negative potential in the case of Si tip and bright, positive potential for Pt-coated tip. This is a specific property of these nanoscale objects as for larger objects (e.g. organic patches or metal thin films) the potential contrast inversion does not occur. Fig. 10.8 schematically summarizes the effects of substrate and KPFM tip materials on inherent charging of nanodiamonds.

It should be emphasized that the observed effect is related not only to the KPFM setup. The KPFM tip actually represents a general interface material to study properties of nanoparticles in a complete electronic circuit (substrate, nanoparticle, tip, and external connections). The results show that such a circuit represents a complex system where the nanoparticle potential is significantly influenced by electronic communication among all involved components.

### 10.3.6 INFLUENCE OF ADSORBED WATER ON NANODIAMOND POTENTIAL

Another general factor is water. Water interaction with nanodiamonds is fundamental for both the technology and applications. However, it is still not fully understood, and explanations of various phenomena are debated. For instance, formation of nanophase water on DNDs was reported, but without studying the role of surface chemistry.[81] A recent study showed that the orientation of water molecules in the solvation shell depends on the NDs' zeta potential which reflects surface chemistry to a certain degree.[82] Giant dielectric permitivity of DNDs was also attributed to the interaction of adsorbed water with functional groups on nanodiamond, yet further investigation is needed to elucidate its actual mechanism.[83]

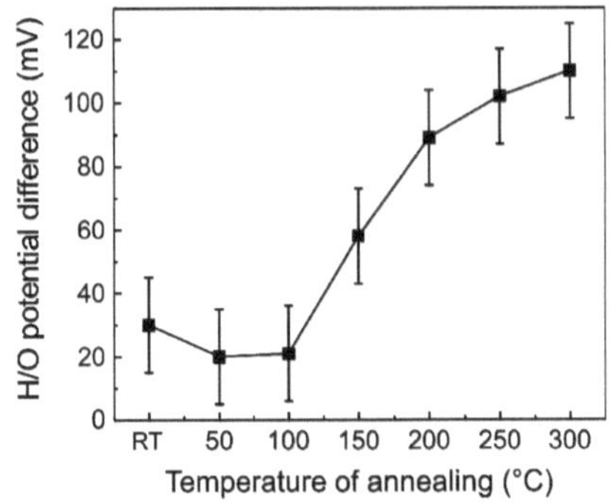


**FIGURE 10.9**

Difference of KPFM potential between H-DNDs and O-DNDs on HOPG substrate as a function of the annealing temperature in an oxygen and water free glovebox

Water molecules and ions that are inherently adsorbed on surfaces in air may decrease the overall CPD contrast in KPFM under ambient conditions, i.e. screen the sample potential.[84] This is due to a dipole of water molecules and charges of ions. The screening was, for instance, pronounced on doped patterns on Si wafers in an environment with high relative humidity (RH > 50%). At relatively low humidity (RH ~ 30%), diamond and metal surfaces exhibited reasonable potential contrast, in agreement with expected work function differences.[36] Yet, geometric constraints as well as specific functional groups can lead to a different structure of surface-bound water on nanoparticles compared to bulk water properties.

In order to study the effect of adsorbed water on the CPD of H-DNDs and O-DNDs, an annealing sequence in the glovebox was performed from room temperature to 300°C with a 50°C step.[85] At each temperature, the samples were annealed for 30 min and immediately measured by KPFM in the glovebox. The KPFM data showed a large shift of the CPD by as much as 0.5 V during the annealing sequence for both H-DNDs and O-DNDs. This indicates a strong screening effect of the adsorbed water on NDs at room temperature. Fig. 10.9 illustrates the resulting potential difference between H-DNDs and O-DNDs on HOPG as a function of the annealing temperature. It shows that H/O difference remains more or less stable up to 100°C and then increases rapidly up to 250°C where it starts to saturate. This may be explained by a thicker water layer as well as a different water structure and binding on H-DNDs compared to O-DNDS, as suggested by FTIR and thermal analysis.[85] Thus KPFM can reveal different water desorption efficiency related to different water–nanodiamond interactions on NDs with different surface chemistry.

However, a change of the electronic properties of diamond itself due to the removal of surface water should be considered as well. As the Fermi level is pinned at the surface states of oxidized diamond,[86] the shift of the CPD during the annealing sequence can be attributed only to gradual desorption of water. In the case of H-DNDs, water desorption can change the adsorbate-induced transfer doping equilibrium and thereby influence the surface electronic properties including the work function.[87] Thermal analysis indicated that the annealing up to 300°C leads to the complete desorption of water from both H-DNDs and O-DNDs.[85] It remains unclear though whether the strongly bound counter anions that are related with the transfer doping of H-diamond also desorb in this temperature range. KPFM measurement showed a gradual change of potential with increasing temperature on both H-DNDs and O-DNDs rather than abrupt decrease in the work function expected after counter ion desorption.[87] The work function of H-DNDs also remained close to that of O-DNDs (within 150 mV) all the time. It is thus plausible to assume that a change of KPFM potential contrast corresponds mainly to a reduced potential screening. Nevertheless, measurements of photoelectron spectroscopy[70] or surface conductivity during the water desorption from NDs would be useful to ascertain this issue.

## 10.4 IMPACT ON OTHER NANODIAMOND PROPERTIES

The observed effect of nanodiamond charging due to substrate may have an impact also on other properties such as photoluminescence, chemical reactions, electronic charge transport, etc. So far the effect was noticed in the case of electron emission, photoluminescence, and photovoltage. These effects are illustrated in Fig. 10.10.

### 10.4.1 IMPACT ON ELECTRON EMISSION FROM NANODIAMONDS

Nanodiamonds possess interesting features for electron emission.[31] Different potentials of NDs on different substrates should give rise to different electron emission properties. This effect was indirectly probed by scanning electron microscopy with a secondary electron (SE) detector.[20] Fig. 10.10A shows SE on a layer of oxidized DNDs for two different accelerating voltages 2 kV and 0.5 kV. The images as well as intensity histogram show that the very same NDs exhibit systematically different SE intensity on different substrates. This is in agreement with previous SEM measurements at 10 kV.[20] Penetration depth of electron beam is about 80 nm for 2 kV and 10 nm for 0.5 kV in our case. The SE signal thus arises mainly from the DNDs layer, and not from the underlying substrate. The darker contrast on Si vs. Pt may be explained by a higher negative charge of NDs on Si (in agreement with the scheme in Fig. 10.8) which blocks emission of secondary electrons from the NDs. Although the SEM results can be considered only in a qualitative manner, they directly evidence a charge exchange between the NDs and the substrate, and independently support the results obtained by KFM.

### 10.4.2 IMPACT ON LIGHT EMISSION FROM NANODIAMONDS

The effect of substrate on electronic and photoluminescence (PL) properties of nanostructures such as graphene oxide[88] or $MoS_2$[89] has been recently demonstrated. Similarly, an inherent charge transfer between NDs and the substrate was found to shift

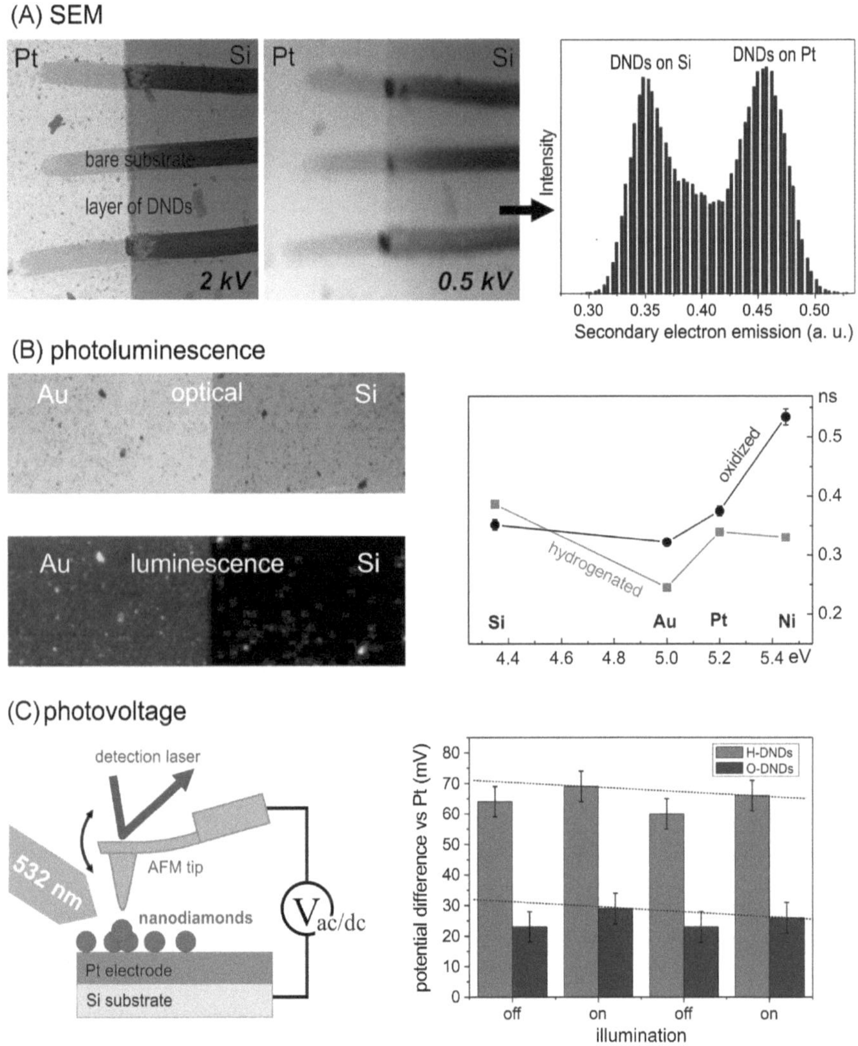


**FIGURE 10.10**

(A) Secondary electron emission images and intensity histogram on a layer of oxidized DNDs for different accelerating voltages, 2 kV and 0.5 kV. (B) Difference in photoluminescence intensity of hydrogenated DNDs on Au and Si surface (due to different reflectivity) and non-uniform change of photoluminescence decay kinetics as a function of substrate work function. (C) Photovoltage measurements performed by KPFM on oxidized and hydrogenated DNDs on the Pt substrate according to the depicted experimental configuration. The dashed line indicates a shift of photovoltage during illumination cycling the PL spectra and change PL lifetimes of nanodiamonds.[29] The PL changes depend on whether NDs are hydrogen or oxygen terminated, which is another indication of the charging effects. Fig. 10.10B illustrates a difference in PL intensity of NDs on Au and Si surface (due to different reflectivity actually) and non-uniform change of PL decay kinetics as a function of the substrate work function (adjusted by use of different metals).

There are two possible mechanisms of how the PL can be affected by the substrate: (i) charge transfer from photo-generated excitons from DNDs to substrate and (ii) trapped charge on DNDs due to the work function difference. The charge transfer from excitons is unlikely as it is typically associated with a power law PL decay.[90] The influence of charge trapping is more likely. The surface plays a much larger role due to considerable surface-to-volume ratio (~15% on 5 nm DNDs). The large dipole moment of PL centers is extremely sensitive to strain and mobile charges.[91] The involvement of charge trapping at the surface or defect states is corroborated also by shortened PL decays. The charge is mostly retained in diamond surface states (including states arising due to surface oxidation, imperfect hydrogen termination, $sp^2$ phase, and dangling bonds) as indicated by microscopic and electronic transport studies.[42,67] The persistent electrostatic charge due to substrate interaction can thus explain different PL decay times as well as PL spectral shifts for the same NDs on different substrates.

### 10.4.3 PHOTOVOLTAGE ON NANODIAMONDS

The energy of absorbed light can be transformed not only back to photoluminescence but also to photovoltage. Fig. 10.10C shows photovoltage measurements performed by KPFM on DNDs on the Pt substrate according to the depicted experimental configuration. The graph shows the resulting surface potentials of H-DNDs and O-DNDs in the dark and under illumination (green 532 nm laser) vs. bottom Pt electrode, i.e. the potential of Pt substrate is subtracted from the potential of DNDs. The potential changes as the illumination is repeatedly switched on and off. For both types of DND the resulting photovoltage is reproducibly around 5 mV. One can also notice a slight shift in the potential as the illumination cycling proceeds (indicated by the dashed line). This could be related to a persistent trapping of some of the photo-generated charge on the nanodiamonds. Similar long-term effect has been observed on organic blends for photovoltaic applications.[92]

Detection of photovoltage (albeit small) by KPFM on individual DNDs is surprising from several aspects. First of all, diamond is a wide bandgap semiconductor, transparent to visible light in the ideal case. However, surface states and non-diamond carbon on the surface of NDs provide visible light absorption, well know from optical experiments.[90] KPFM thus provides a direct proof that this absorption can be transformed to photovoltage on nanosized diamonds. In addition, KPFM setup is symmetric in terms of contacts: Pt substrate, spherical particle, and Pt-coated tip. Thus photovoltage should totally cancel out. The nonzero photovoltage can thus arise due to irregularities in the DNDs shape, surface termination, and work function of Pt on the tip and substrate. Partial compensation of photovoltage by AFM laser diode (650 nm) is also possible, as observed, for instance, in the case of photocurrents on silicon.[93] Nevertheless, the observed photovoltage indicated that nanodiamonds may become miniature energy sources. Further local (photo)current measurement on NDs may shed more light on the photovoltage effects.[22]

## 10.5 CONCLUSIONS

We have seen that the surface potential of nanodiamonds reflects many features and phenomena on this exciting nanoscale material. Kelvin probe force microscopy was introduced as a tool for measurements of surface potential even on individual or a few nanodiamonds. Specific challenges such as spatial resolution, stray capacitance, and dependence of surface potential on nanodiamond size were analyzed and discussed. Based on this analysis, measurement of linear size dependence of potential can be taken as a universal characterization methodology for NDs (and also for other nanoparticles). It was similar for detonation ND clusters (composed of multiple nanodiamonds) and HPHT single crystal nanodiamonds. In addition, the slope of the linear trend can be used to resolve various surface chemical moieties on diamond. This has been found particularly useful for accurate interpretation of FTIR spectra after nanodiamond chemical modifications.

KPFM further revealed that electrically resistive as well as conductive nanodiamonds are prone to charge trapping due to the interaction with the substrate material they are deposited on. This leads to pronounced change of their potential (up to 0.4 V difference observed) which they assume according to the substrate. The effect was observed also on gold nanoparticles. The charge trapping was further studied and confirmed by intentional voltage application via KPFM tip. The significant charging of nanodiamonds and other nanoparticles in dependence on a substrate thus seems to be a real and general effect. It fits well to a broader picture of nanoscale materials where doping and bandgap of two-dimensional materials (such as graphene or $MoS_2$) is to a large extent determined by a substrate.[94,95] The different charge of nanodiamonds on different substrates has been also manifested by impact on other properties such as electron emission, photoluminescence, and photovoltage.

## REFERENCES

1. Schrand, A. M.; Hens, S. A. C.; Shenderova, O. A. Nanodiamond Particles: Properties and Perspectives for Bioapplications. *Crit. Rev. Solid State Mater. Sci.* **2009**, *34*, 18–74.
2. Mochalin, V. N.; Shenderova, O.; Ho, D.; Gogotsi, Y. The Properties and Applications of Nanodiamonds. *Nat. Nanotechnol.* **2012**, *7*, 11–23.
3. DeCarli, P. S.; Jamieson, J. C. Formation of Diamond by Explosive Shock. *Science* **1961**, *133*, 1821–1822.

4. Stehlik, S.; Varga, M.; Ledinsky, M.; Jirasek, V.; Artemenko, A.; Kozak, H.; Ondic, L.; Skakalova, V.; Argentero, G.; Pennycook, T.; et al. Size and Purity Control of HPHT Nanodiamonds Down to 1 nm. *J. Phys. Chem. C* **2015**.
5. Takimoto, T.; Chano, T.; Shimizu, S.; Okabe, H.; Ito, M.; Morita, M.; Kimura, T.; Inubushi, T.; Komatsu, N. Preparation of Fluorescent Diamond Nanoparticles Stably Dispersed Under a Physiological Environment Through Multistep Organic Transformations. *Chem. Mater.* **2010**, *22*, 3462–3471.
6. Ho, D., Ed. *Nanodiamonds*; Springer US: Boston, MA, 2010.
7. Chevallier, E.; Scorsone, E.; Girard, H. A.; Pichot, V.; Spitzer, D.; Bergonzo, P. Metalloporphyrin-Functionalised Diamond Nano-Particles as Sensitive Layer for Nitroaromatic Vapours Detection at Room-Temperature. *Sens. Actuators B, Chem.* **2010**, *151*, 191–197.
8. Ahmad, R. K.; Parada, A. C.; Hudziak, S.; Chaudhary, A.; Jackman, R. B. Nanodiamond-Coated Silicon Cantilever Array for Chemical Sensing. *Appl. Phys. Lett.* **2010**, *97*, 093103.
9. Gruber, A.; Dräbenstedt, A.; Tietz, C.; Fleury, L.; Wrachtrup, J.; von Borczyskowski, C. Scanning Confocal Optical Microscopy and Magnetic Resonance on Single Defect Centers. *Science* **1997**, *276*, 2012–2014.
10. Treussart, F.; Jacques, V.; Wu, E.; Gacoin, T.; Grangier, P.; Roch, J.-F. Photoluminescence of Single Colour Defects in 50 nm Diamond Nanocrystals. *Physica B, Condens. Matter* **2006**, *376–377*, 926–929.
11. Nesvizhevsky, V.; Cubitt, R.; Lychagin, E.; Muzychka, A.; Nekhaev, G.; Pignol, G.; Protasov, K.; Strelkov, A. Application of Diamond Nanoparticles in Low-Energy Neutron Physics. *Materials* **2010**, *3*, 1768–1781.
12. Krueger, A.; Lang, D. Functionality Is Key: Recent Progress in the Surface Modification of Nanodiamond. *Adv. Funct. Mater.* **2012**, *22*, 890–906.
13. Pech, D.; Brunet, M.; Durou, H.; Huang, P.; Mochalin, V.; Gogotsi, Y.; Taberna, P.-L.; Simon, P. Ultrahigh-Power Micrometre-Sized Supercapacitors Based on Onion-like Carbon. *Nat. Nanotechnol.* **2010**, *5*, 651–654.
14. Portet, C.; Yushin, G.; Gogotsi, Y. Electrochemical Performance of Carbon Onions, Nanodiamonds, Carbon Black and Multiwalled Nanotubes in Electrical Double Layer Capacitors. *Carbon* **2007**, *45*, 2511–2518.
15. Girard, H. A.; Perruchas, S.; Gesset, C.; Chaigneau, M.; Vieille, L.; Arnault, J.-C.; Bergonzo, P.; Boilot, J.-P.; Gacoin, T. Electrostatic Grafting of Diamond Nanoparticles: A Versatile Route to Nanocrystalline Diamond Thin Films. *ACS Appl. Mater. Interfaces* **2009**, *1*, 2738–2746.
16. Kozak, H.; Kromka, A.; Babchenko, O.; Rezek, B. Directly Grown Nanocrystalline Diamond Field-Effect Transistor Microstructures. *Sens. Lett.* **2010**, *8*, 482–487.
17. Kromka, A.; Rezek, B.; Remes, Z.; Michalka, M.; Ledinsky, M.; Zemek, J.; Potmesil, J.; Vanecek, M. Formation of Continuous Nanocrystalline Diamond Layers on Glass and Silicon at Low Temperatures. *Chem. Vap. Depos.* **2008**, *14*, 181–186.
18. Turner, S.; Shenderova, O.; Da Pieve, F.; Lu, Y.; Yücelen, E.; Verbeeck, J.; Lamoen, D.; Van Tendeloo, G. Aberration-Corrected Microscopy and Spectroscopy Analysis of Pristine, Nitrogen Containing Detonation Nanodiamond. *Phys. Status Solidi A* **2013**, *210*, 1976–1984.
19. Kozak, H.; Artemenko, A.; Čermák, J.; Švrček, V.; Kromka, A.; Rezek, B. Oxidation and Reduction of Nanodiamond Particles in Colloidal Solutions by Laser Irradiation or Radio-Frequency Plasma Treatment. *Vib. Spectrosc.* **2016**, *83*, 108–114.

20. Stehlik, S.; Petit, T.; Girard, H. A.; Arnault, J.-C.; Kromka, A.; Rezek, B. Nanoparticles Assume Electrical Potential According to Substrate, Size, and Surface Termination. *Langmuir* **2013**, *29*, 1634–1641.
21. Stehlik, S.; Miliaieva, D.; Varga, M.; Kromka, A.; Rezek, B. Size Decrease of Detonation Nanodiamonds by Air Annealing Investigated by AFM. *MRS Adv.* **2016**, 1–6, FirstView.
22. Čermák, J.; Kozak, H.; Stehlík, Š.; Švrček, V.; Pichot, V.; Spitzer, D.; Kromka, A.; Rezek, B. Microscopic Electrical Conductivity of Nanodiamonds after Thermal and Plasma Treatments. *MRS Adv.* **2016**, 1–7, FirstView.
23. Delgado, A. V.; González-Caballero, F.; Hunter, R. J.; Koopal, L. K.; Lyklema, J. Measurement and Interpretation of Electrokinetic Phenomena. *J. Colloid Interface Sci.* **2007**, *309*, 194–224.
24. Williams, O. A.; Hees, J.; Dieker, C.; Jäger, W.; Kirste, L.; Nebel, C. E. Size-Dependent Reactivity of Diamond Nanoparticles. *ACS Nano* **2010**, *4*, 4824–4830.
25. Hees, J.; Kriele, A.; Williams, O. A. Electrostatic Self-Assembly of Diamond Nanoparticles. *Chem. Phys. Lett.* **2011**, *509*, 12–15.
26. Aramesh, M.; Shimoni, O.; Ostrikov, K.; Prawer, S.; Cervenka, J. Surface Charge Effects in Protein Adsorption on Nanodiamonds. *Nanoscale* **2015**, *7*, 5726–5736.
27. Chow, E. K.; Zhang, X.-Q.; Chen, M.; Lam, R.; Robinson, E.; Huang, H.; Schaffer, D.; Osawa, E.; Goga, A.; Ho, D. Nanodiamond Therapeutic Delivery Agents Mediate Enhanced Chemoresistant Tumor Treatment. *Sci. Transl. Med.* **2011**, *3*, 73ra21.
28. Nangia, S.; Sureshkumar, R. Effects of Nanoparticle Charge and Shape Anisotropy on Translocation Through Cell Membranes. *Langmuir* **2012**, *28*, 17666–17671.
29. Stehlik, S.; Ondic, L.; Berhane, A. M.; Aharonovich, I.; Girard, H. A.; Arnault, J.-C.; Rezek, B. Photoluminescence of Nanodiamonds Influenced by Charge Transfer from Silicon and Metal Substrates. *Diam. Relat. Mater.* **2016**, *63*, 91–96.
30. Petráková, V.; Taylor, A.; Kratochvílová, I.; Fendrych, F.; Vacík, J.; Kučka, J.; Štursa, J.; Cígler, P.; Ledvina, M.; Fišerová, A.; et al. Luminescence of Nanodiamond Driven by Atomic Functionalization: Towards Novel Detection Principles. *Adv. Funct. Mater.* **2012**, *22*, 812–819.
31. Zhirnov, V. V.; Shenderova, O. A.; Jaeger, D. L.; Tyler, T.; Areshkin, D. A.; Brenner, D. W.; Hren, J. J. Electron Emission Properties of Detonation Nanodiamonds. *Phys. Solid State* **2004**, *46*, 657–661.
32. Van Aken, B. B.; Bakker, K. J.; Heijna, M. C. R.; Reid, D.; Baikie, I.; Soppe, W. J. In Situ Measurements of Surface (photo)voltage of Roll-to-Roll Deposited Thin Film Silicon Solar Cells. *Phys. Status Solidi A* **2010**, *207*, 682–685.
33. Baikie, I. D. Kelvin Probe information site. http://www.kelvinprobe.info/index.htm.
34. Nonnenmacher, M.; O'Boyle, M. P.; Wickramasinghe, H. K. Kelvin Probe Force Microscopy. *Appl. Phys. Lett.* **1991**, *58*, 2921–2923.
35. Sadewasser, S., Glatzel, T., Eds. *Kelvin Probe Force Microscopy;* Springer Series in Surface Sciences, Vol. 48; Springer Berlin Heidelberg: Berlin, Heidelberg, 2012.
36. Rezek, B.; Sauerer, C.; Nebel, C. E.; Stutzmann, M.; Ristein, J.; Ley, L.; Snidero, E.; Bergonzo, P. Fermi Level on Hydrogen Terminated Diamond Surfaces. *Appl. Phys. Lett.* **2003**, *82*, 2266–2268.
37. Baer, Y. The Natural Energy Scale for XPS Spectra of Metals. *Solid State Commun.* **1976**, *19*, 669–671.
38. Takeuchi, D.; Riedel, M.; Ristein, J.; Ley, L. Surface Band Bending and Surface Conductivity of Hydrogenated Diamond. *Phys. Rev. B* **2003**, *68*, 041304.

39. Čermák, J.; Koide, Y.; Takeuchi, D.; Rezek, B. Spectrally Dependent Photovoltages in Schottky Photodiode Based on (100) B-Doped Diamond. *J. Appl. Phys.* **2014**, *115*, 053105.
40. Rezek, B.; Nebel, C. E. Kelvin Force Microscopy on Diamond Surfaces and Devices. *Diam. Relat. Mater.* **2005**, *14*, 466–469.
41. Rezek, B.; Čermák, J.; Kromka, A.; Ledinský, M.; Hubík, P.; Mareš, J. J.; Purkrt, A.; Cimrová, V.; Fejfar, A.; Kočka, J. Synthesis, Structure, and Opto-Electronic Properties of Organic-Based Nanoscale Heterojunctions. *Nanoscale Res. Lett.* **2011**, *6*, 238.
42. Verveniotis, E.; Kromka, A.; Rezek, B. Controlling Electrostatic Charging of Nanocrystalline Diamond at Nanoscale. *Langmuir* **2013**, *29*, 7111–7117.
43. Rezek, B.; Stehlik, S.; Kromka, A.; Arnault, J.-C.; Weis, M.; Jakabovic, J. Visible Light Photodiodes and Photovoltages from Detonation Nanodiamonds. *MRS Adv.* **2016**, 1–5, FirstView.
44. Jacobs, H. O.; Leuchtmann, P.; Homan, O. J.; Stemmer, A. Resolution and Contrast in Kelvin Probe Force Microscopy. *J. Appl. Phys.* **1998**, *84*, 1168–1173.
45. Li, G.; Mao, B.; Lan, F.; Liu, L. Practical Aspects of Single-Pass Scan Kelvin Probe Force Microscopy. *Rev. Sci. Instrum.* **2012**, *83*, 113701.
46. Sadewasser, S.; Carl, P.; Glatzel, T.; Lux-Steiner, M. C. Influence of Uncompensated Electrostatic Force on Height Measurements in Non-Contact Atomic Force Microscopy. *Nanotechnology* **2004**, *15*, S14.
47. Shimizu, N.; Ikeda, M.; Yoshida, E.; Murakami, H.; Miyazaki, S.; Hirose, M. Charging States of Si Quantum Dots as Detected by AFM/Kelvin Probe Technique. *Jpn. J. Appl. Phys.* **2000**, *39*, 2318–2320.
48. Rezek, B.; Garrido, J. A.; Stutzmann, M.; Nebel, C. E.; Snidero, E.; Bergonzo, P. Local Oxidation of Hydrogenated Diamond Surfaces for Device Fabrication. *Phys. Status Solidi A* **2002**, *193*, 523–528.
49. Sadewasser, S.; Jelinek, P.; Fang, C.-K.; Custance, O.; Yamada, Y.; Sugimoto, Y.; Abe, M.; Morita, S. New Insights on Atomic-Resolution Frequency-Modulation Kelvin-Probe Force-Microscopy Imaging of Semiconductors. *Phys. Rev. Lett.* **2009**, *103*, 266103.
50. Kolíbal, M.; Konečný, M.; Ligmajer, F.; Škoda, D.; Vystavěl, T.; Zlámal, J.; Varga, P.; Šikola, T. Guided Assembly of Gold Colloidal Nanoparticles on Silicon Substrates Prepatterned by Charged Particle Beams. *ACS Nano* **2012**, *6*, 10098–10106.
51. Schmidlin, L.; Pichot, V.; Josset, S.; Pawlak, R.; Glatzel, T.; Kawai, S.; Meyer, E.; Spitzer, D. Two-Dimensional Nanodiamond Monolayers Deposited by Combined Ultracentrifugation and Electrophoresis Techniques. *Appl. Phys. Lett.* **2012**, *101*, 253111.
52. Gao, F.; Nebel, C. E. Diamond-Based Supercapacitors: Realization and Properties. *ACS Appl. Mater. Interfaces* **2015**.
53. Holt, K. B. Electrochemistry of Nanodiamond Particles. In *Nanodiamond;* Williams, O. A., Ed., RSC Nanoscience and Nanotechnology; 2014, pp. 128–150, Chapter 6.
54. Aharonovich, I.; Castelletto, S.; Simpson, D. A.; Stacey, A.; McCallum, J.; Greentree, A. D.; Prawer, S. Two-Level Ultrabright Single Photon Emission From Diamond Nanocrystals. *Nano Lett.* **2009**, *9*, 3191–3195.
55. Williams, O. A.; Douhéret, O.; Daenen, M.; Haenen, K.; Ōsawa, E.; Takahashi, M. Enhanced Diamond Nucleation on Monodispersed Nanocrystalline Diamond. *Chem. Phys. Lett.* **2007**, *445*, 255–258.
56. Kozak, H.; Kromka, A.; Ukraintsev, E.; Houdkova, J.; Ledinsky, M.; Vaněček, M.; Rezek, B. Detecting sp2 Phase on Diamond Surfaces by Atomic Force Microscopy Phase Imaging and Its Effects on Surface Conductivity. *Diam. Relat. Mater.* **2009**, *18*, 722–725.

57. Stehlik, S.; Petit, T.; Girard, H. A.; Kromka, A.; Arnault, J.-C.; Rezek, B. Surface Potential of Diamond and Gold Nanoparticles Can Be Locally Switched by Surrounding Materials or Applied Voltage. *J. Nanopart. Res.* **2014**, *16*, 1–11.
58. Salem, M. A.; Mizuta, H.; Oda, S. Probing Electron Charging in Nanocrystalline Si Dots Using Kelvin Probe Force Microscopy. *Appl. Phys. Lett.* **2004**, *85*, 3262–3264.
59. Rehor, I.; Cigler, P. Precise Estimation of HPHT Nanodiamond Size Distribution Based on Transmission Electron Microscopy Image Analysis. *Diam. Relat. Mater.* **2014**, *46*, 21–24.
60. Sasahara, A.; Pang, C. L.; Onishi, H. Local Work Function of Pt Clusters Vacuum-Deposited on a $TiO_2$ Surface. *J. Phys. Chem. B* **2006**, *110*, 17584–17588.
61. Glatzel, T.; Zimmerli, L.; Koch, S.; Such, B.; Kawai, S.; Meyer, E. Determination of Effective Tip Geometries in Kelvin Probe Force Microscopy on Thin Insulating Films on Metals. *Nanotechnology* **2009**, *20*, 264016.
62. Yamauchi, T.; Tabuchi, M.; Nakamura, A. Size Dependence of the Work Function in InAs Quantum Dots on GaAs(001) as Studied by Kelvin Force Probe Microscopy. *Appl. Phys. Lett.* **2004**, *84*, 3834–3836.
63. Hornauer, H.; Vancea, J.; Reiss, G.; Hoffmann, H. Thickness Dependence of the Work Function in Double-Layer Metallic Films. *Z. Phys. B, Condens. Matter* **1989**, *77*, 399–407.
64. Elias, G.; Glatzel, T.; Meyer, E.; Schwarzman, A.; Boag, A.; Rosenwaks, Y. The Role of the Cantilever in Kelvin Probe Force Microscopy Measurements. *Beilstein J. Nanotechnol.* **2011**, 2, 252–260.
65. Liscio, A.; Palermo, V.; Samorì, P. Nanoscale Quantitative Measurement of the Potential of Charged Nanostructures by Electrostatic and Kelvin Probe Force Microscopy: Unraveling Electronic Processes in Complex Materials. *Acc. Chem. Res.* **2010**, *43*, 541–550.
66. Fumagalli, L.; Ferrari, G.; Sampietro, M.; Casuso, I.; Martínez, E.; Samitier, J.; Gomila, G. Nanoscale Capacitance Imaging With Attofarad Resolution Using Ac Current Sensing Atomic Force Microscopy. *Nanotechnology* **2006**, *17*, 4581.
67. Itoh, Y.; Sumikawa, Y.; Umezawa, H.; Kawarada, H. Trapping Mechanism on Oxygen-Terminated Diamond Surfaces. *Appl. Phys. Lett.* **2006**, *89*, 203503.
68. Mesquida, P.; Stemmer, A. Attaching Silica Nanoparticles From Suspension Onto Surface Charge Patterns Generated by a Conductive Atomic Force Microscope Tip. *Adv. Mater.* **2001**, *13*, 1395–1398.
69. Kondo, T.; Neitzel, I.; Mochalin, V. N.; Urai, J.; Yuasa, M.; Gogotsi, Y. Electrical Conductivity of Thermally Hydrogenated Nanodiamond Powders. *J. Appl. Phys.* **2013**, *113*, 214307.
70. Petit, T.; Girard, H. A.; Trouvé, A.; Batonneau-Gener, I.; Bergonzo, P.; Arnault, J.-C. Surface Transfer Doping Can Mediate Both Colloidal Stability and Self-Assembly of Nanodiamonds. *Nanoscale* **2013**, *5*, 8958–8962.
71. Kuznetsov, V. L.; Butenko, Y. V.; Chuvilin, A. L.; Romanenko, A. I.; Okotrub, A. V. Electrical Resistivity of Graphitized Ultra-Disperse Diamond and Onion-Like Carbon. *Chem. Phys. Lett.* **2001**, *336*, 397–404.
72. Girard, H. A.; Arnault, J. C.; Perruchas, S.; Saada, S.; Gacoin, T.; Boilot, J.-P.; Bergonzo, P. Hydrogenation of Nanodiamonds Using MPCVD: A New Route Toward Organic Functionalization. *Diam. Relat. Mater.* **2010**, *19*, 1117–1123.
73. Kromka, A.; Čech, J.; Kozak, H.; Artemenko, A.; Ižák, T.; Čermák, J.; Rezek, B.; Černák, M. Low-Temperature Hydrogenation of Diamond Nanoparticles Using Diffuse Coplanar Surface Barrier Discharge at Atmospheric Pressure. *Phys. Status Solidi B* **2015**, *252*, 2602–2607.

74. Kozak, H.; Remes, Z.; Houdkova, J.; Stehlik, S.; Kromka, A.; Rezek, B. Chemical Modifications and Stability of Diamond Nanoparticles Resolved by Infrared Spectroscopy and Kelvin Force Microscopy. *J. Nanopart. Res.* **2013**, *15*, 1–9.
75. Petit, T.; Arnault, J.-C.; Girard, H. A.; Sennour, M.; Bergonzo, P. Early Stages of Surface Graphitization on Nanodiamond Probed by X-Ray Photoelectron Spectroscopy. *Phys. Rev. B* **2011**, *84*, 233407.
76. Liu, Y.; Khabashesku, V. N.; Halas, N. J. Fluorinated Nanodiamond as a Wet Chemistry Precursor for Diamond Coatings Covalently Bonded to Glass Surface. *J. Am. Chem. Soc.* **2005**, *127*, 3712–3713.
77. Mochalin, V. N.; Neitzel, I.; Etzold, B. J. M.; Peterson, A.; Palmese, G.; Gogotsi, Y. Covalent Incorporation of Aminated Nanodiamond Into an Epoxy Polymer Network. *ACS Nano* **2011**, *5*, 7494–7502.
78. Jarre, G.; Heyer, S.; Memmel, E.; Meinhardt, T.; Krueger, A. Synthesis of Nanodiamond Derivatives Carrying Amino Functions and Quantification by a Modified Kaiser Test. *Beilstein J. Org. Chem.* **2014**, *10*, 2729–2737.
79. Tachiki, M.; Kaibara, Y.; Sumikawa, Y.; Shigeno, M.; Kanazawa, H.; Banno, T.; Soup Song, K.; Umezawa, H.; Kawarada, H. Characterization of Locally Modified Diamond Surface Using Kelvin Probe Force Microscope. *Surf. Sci.* **2005**, *581*, 207–212.
80. Cui, J. B.; Ristein, J.; Ley, L. Low-Threshold Electron Emission from Diamond. *Phys. Rev. B* **1999**, *60*, 16135–16142.
81. Korobov, M. V.; Avramenko, N. V.; Bogachev, A. G.; Rozhkova, N. N.; Ōsawa, E. Nanophase of Water in Nano-Diamond Gel. *J. Phys. Chem. C* **2007**, *111*, 7330–7334.
82. Petit, T.; Yuzawa, H.; Nagasaka, M.; Yamanoi, R.; Osawa, E.; Kosugi, N.; Aziz, E. F. Probing Interfacial Water on Nanodiamonds in Colloidal Dispersion. *J. Phys. Chem. Lett.* **2015**, *6*, 2909–2912.
83. Batsanov, S. S.; Gavrilkin, S. M.; Batsanov, A. S.; Poyarkov, K. B.; Kulakova, I. I.; Johnson, D. W.; Mendis, B. G. Giant Dielectric Permittivity of Detonation-Produced Nanodiamond Is Caused by Water. *J. Mater. Chem.* **2012**, *22*, 11166–11172.
84. Sugimura, H.; Ishida, Y.; Hayashi, K.; Takai, O.; Nakagiri, N. Potential Shielding by the Surface Water Layer in Kelvin Probe Force Microscopy. *Appl. Phys. Lett.* **2002**, *80*, 1459–1461.
85. Stehlik, S.; Glatzel, T.; Pichot, V.; Pawlak, R.; Meyer, E.; Spitzer, D.; Rezek, B. Water Interaction with Hydrogenated and Oxidized Detonation Nanodiamonds — Microscopic and Spectroscopic Analyses. *Diam. Relat. Mater.* **2016**, *63*, 97–102.
86. Maier, F.; Ristein, J.; Ley, L. Electron Affinity of Plasma-Hydrogenated and Chemically Oxidized Diamond (100) Surfaces. *Phys. Rev. B* **2001**, *64*, 165411.
87. Maier, F.; Riedel, M.; Mantel, B.; Ristein, J.; Ley, L. Origin of Surface Conductivity in Diamond. *Phys. Rev. Lett.* **2000**, *85*, 3472–3475.
88. Rani, J. R.; Lim, J.; Oh, J.; Kim, D.; Lee, D.; Kim, J.-W.; Shin, H. S.; Kim, J. H.; Jun, S. C. Substrate and Buffer Layer Effect on the Structural and Optical Properties of Graphene Oxide Thin Films. *RSC Adv.* **2013**, *3*, 5926–5936.
89. Buscema, M.; Steele, G. A.; Zant, H. S.; van der Castellanos-Gomez, J. The Effect of the Substrate on the Raman and Photoluminescence Emission of Single-Layer MoS2. *Nano Res.* **2015**, *7*, 561–571.
90. Galář, P.; Čermák, J.; Malý, P.; Kromka, A.; Rezek, B. Electrochemically Grafted Polypyrrole Changes Photoluminescence of Electronic States Inside Nanocrystalline Diamond. *J. Appl. Phys.* **2014**, *116*, 223103.

91. Greentree, A. D.; Fairchild, B. A.; Hossain, F. M.; Prawer, S. Diamond Integrated Quantum Photonics. *Mater. Today* **2008**, *11*, 22–31.
92. Čermák, J.; Rezek, B.; Cimrová, V.; Fejfar, A.; Purkrt, A.; Vaněček, M.; Kočka, J. Time-Resolved Opto-Electronic Properties of poly(3-Hexylthiophene-2, 5-Diyl): Fullerene Heterostructures Detected by Kelvin Force Microscopy. *Thin Solid Films* **2010**, *519*, 836–840.
93. Ledinský, M.; Fejfar, A.; Vetushka, A.; Stuchlík, J.; Rezek, B.; Kočka, J. Local Photoconductivity of Microcrystalline Silicon Thin Films Measured by Conductive Atomic Force Microscopy. *Phys. Status Solidi RRL – Rapid Res. Lett.* **2011**, *5*, 373–375.
94. Giovannetti, G.; Khomyakov, P. A.; Brocks, G.; Karpan, V. M.; van den Brink, J.; Kelly, P. J.; Kelly, P. J. Doping Graphene With Metal Contacts. *Phys. Rev. Lett.* **2008**, *101*, 026803.
95. Zhou, S. Y.; Gweon, G.-H.; Fedorov, A. V.; First, P. N.; de Heer, W. A.; Lee, D.-H.; Guinea, F.; Castro Neto, A. H.; Lanzara, A. Substrate-Induced Bandgap Opening in Epitaxial Graphene. *Nat. Mater.* **2007**, *6*, 770–775.

## ACKNOWLEDGEMENTS

Work on this chapter has been partially supported by the project GACR 15-01809S. Long-term collaboration on nanodiamonds with the groups of Alexander Kromka (Institute of Physics CAS, Czech Republic), Jean-Charles Arnault (CEA, LIST, Saclay France), and Ernest Mayer (University of Basel, Switzerland) is gratefully acknowledged.

CHAPTER

# Interactions with solvent

# 11

**Tristan Petit**

*Helmholtz-Zentrum Berlin für Materialien und Energie GmbH, Berlin, Germany*

**Contents**



## 11.1 INTRODUCTION

Most of nanodiamonds (NDs) applications in nanomedicine, photocatalysis or environmental science take place in liquid media. Due to their small size, down to 3 nm for the smallest ones, NDs have a large surface area on the order of 300–450 $m^2/g$. They are hence more sensitive to interactions with surrounding molecules than macroscale objects. Nanoparticles are also known to induce solvent rearrangement which might change liquid solvent properties.[1] However, our understanding of interactions of NDs with solvent molecules, which could be from chemical, electrostatic or electronic origin, remains very limited.

Aqueous environment is by far the most studied environment because of its high relevance for many applications; therefore, results presented here will mainly focus on NDs–water interactions. The adsorption of water on other nanomaterials has already been shown to have a profound impact on their chemical and catalytic activity.[2] In addition, diamond as a bulk material is known to have peculiar interactions with aqueous environment[3]; therefore, NDs–water interface worth further investigation. For the moment, there is little information of the interactions of NDs with solvents other than water, but a few examples will be mentioned when experimental data are available.

In this chapter, particular attention will be given to the experimental conditions used to probe NDs–water interactions. One of the main questions is whether results observed on NDs exposed to a few water molecules in vacuum can be extended to NDs in water dispersion where water molecules represent generally more than 99% of the total volume. In vacuum, surface chemistry modifications induced by water can be followed by exposing a clean surface to water molecules. Exposure to atmospheric moisture may already have profound effects on diamond electronic properties, as observed on bulk diamond prepared in vacuum and exposed to air.[4] For this reason, we will first detail results obtained in vacuum. Then, we will focus on experiments performed under a gaseous atmosphere, in which NDs interact with a few water layers. Modifications induced by adsorption and desorption of thin water films will provide information on the conformation of the first water hydration shells around NDs. Finally, experiments performed directly in the liquid phase will give new perspectives on the influence of water dispersion on the NDs properties, and inversely, on the impact of NDs on the hydrogen bond network of liquid water. In aqueous dispersions, intermolecular interactions between water molecules enter into play and could potentially affect the NDs–water interface.

Another point to keep in mind is that the NDs–water interactions are strongly dependent on the NDs surface chemistry. The NDs surface chemistry is extremely versatile as discussed in Chapter 8 and will condition the adsorption of molecules in the first solvation shell. We will also see that some surface functional groups can chemically react with water molecules, and a correlation between NDs–solvent interaction and specific surface chemistries is particularly relevant. A comparison of NDs–water interaction for different well-defined surface chemistries is a necessary procedure to understand the origin of unusual NDs behavior.

## 11.2 INTERACTION WITH SOLVENT MOLECULES IN VACUUM

Vacuum measurements offer the advantage that interaction with the first solvent molecules can be observed on a clean surface. To prepare a clean diamond surface, preliminary annealing at temperature up to 450°C in vacuum ensures that solvent molecules are effectively desorbed from the surface while no surface modifications are induced. The changes observed after solvent exposure in vacuum can then reliably be associated with solvent interaction. This approach has been extensively applied for bulk diamond materials. Although the reactivity of NDs toward solvent molecules is certainly enhanced by their small size, the solvent adsorption processes should be similar for bulk and nanoscale diamond. Controlled desorption of solvent molecules adsorbed on NDs surface prior to vacuum characterization can also be applied. In this case, the NDs surface is clean from adsorbed solvent molecules at the end of the process.

Another advantage of vacuum conditions is that advanced electron and X-ray surface analysis methods as well as vibrational spectroscopy can be applied. These methods are strongly sensitive to chemical bonding and electronic structure with different probe depths. Electron and X-ray spectroscopies usually probe only a few nanometers and will be extremely sensitive to surface chemistry of NDs. On the other hand, vibrational spectroscopies probe up to several micrometers and are therefore bulk-sensitive.

## INFRARED SPECTROSCOPY

Water interaction was first investigated on single crystal diamond in vacuum by infrared spectroscopy, which probes vibrational modes of diamond surface groups and of water molecules. Struck and D'Evelyn observed the formation of hydroxyl groups after exposure of a bare diamond surface to water molecules.[5] They proposed that water molecules could spontaneously dissociate on carbon dimers present on clean surfaces. Carbon dimers are formed after surface reconstruction on (100) surfaces if no hydrogen is available to saturate surface dangling bonds.[6] They proposed the following scheme:

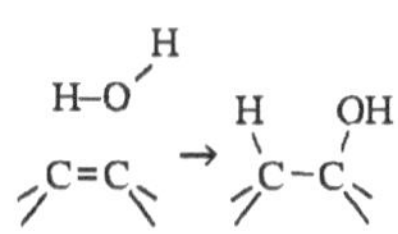


As discussed in Chapter 8, fullerene-like reconstructions (FLRs)[7] can be formed on NDs after annealing in vacuum. FLRs may have reactivity comparable to that of carbon dimers. Although such experiments have only been performed on bulk diamond yet, similar water dissociation processes could be expected on vacuum-annealed NDs.

## HIGH RESOLUTION ELECTRON ENERGY LOSS SPECTROSCOPY

Water dissociation was also reported in High Resolution Electron Energy Loss Spectroscopy (HREELS) studies.[8,9] HREELS principle was detailed in Chapter 9. The high sensitivity to different carbon bonding configuration of HREELS is valuable here. For bulk and nanocrystalline diamond, the exposure to water molecules leads to the formation of C–H and C–OH vibrational HREELS peaks, as shown in Fig. 11.1, which confirms the dissociative water adsorption on carbon dimers.

The newly formed chemical groups were stable after 300°C annealing but disappeared after 700°C annealing. Lower water dissociation was observed on deuterated surface, meaning that passivation of the diamond surface with hydrogen atoms partially prevents the dissociation of water molecules. In parallel, physical adsorption

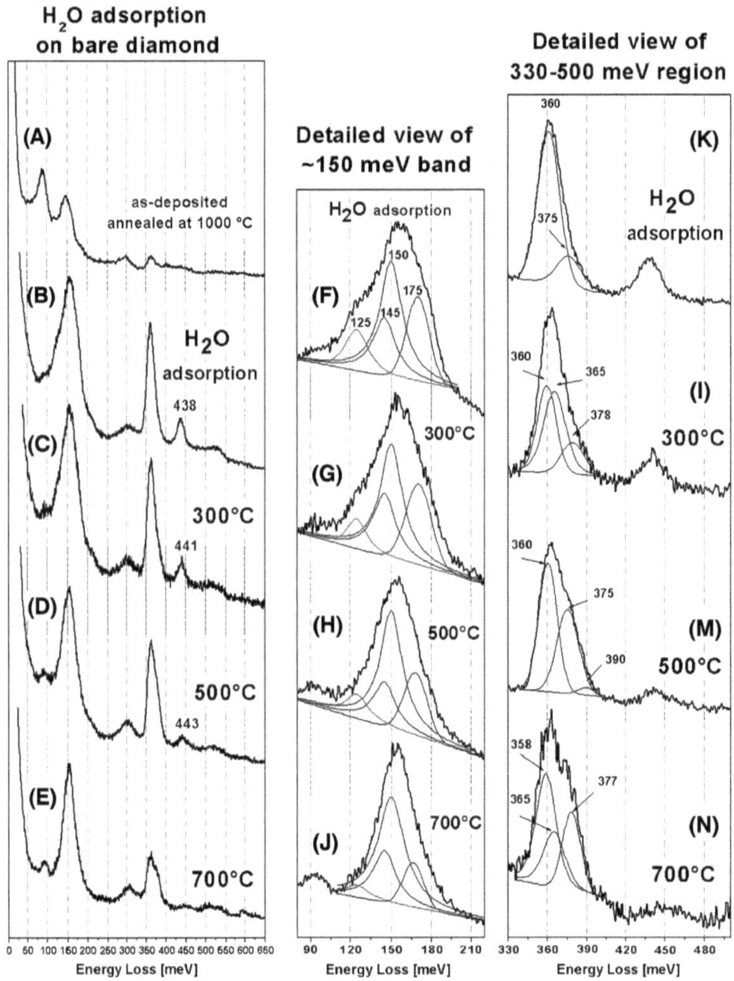


**FIGURE 11.1**

HREEL spectra of bare diamond film surface followed the exposure of the water molecules: (A) as-deposited film after 1000°C annealing; (B) in-situ exposure of the bare diamond surface to water vapor at room temperature. Note the appearance of the strong C–H stretching mode alongside with the well-defined peak at 438 meV, associated to C–OH associated mode; (C) subsequent annealing at 300°C; (D) at 500°C; and (E) at 700°C. (F)–(J) Peak fitting and detailed view of the ~150 meV centered band (data taken from (B)–(E) spectra). Fitting procedure was done for four main peaks: (1) 125 meV, associated with C–O–C vibration; (2) 145 meV, associated with C–H bending vibration; (3) 150 meV associated with diamond optical phonon, and (4) ~175 meV associated with $CH_x$ ($x = 2, 3$) bending vibration. (K)–(N) Detailed view and peak fitting of C–H stretching region. The fitting peak number and their exact position (indicated on the plot) were derived from best fitting parameters. Reprinted in adapted form from: Akhvlediani, R., et al. Surf. Sci. 2010, 604 (23–24), 2129–2138,[9] © 2010 with permission from Elsevier of water molecules was reported. Water dissociation and physisorption occur on the diamond surface, and the amount of dissociated water seems to be related to the concentration of C=C dimers. Here again, these results have been demonstrated on bulk diamond, and similar experiments remain to be performed on NDs.

### X-RAY PHOTOELECTRON SPECTROSCOPY

Water desorption from NDs, which have been first exposed to air and water, is a complementary approach which was recently applied to surface-graphitized[10] and hydrogenated NDs[11] characterized by X-ray Photoelectron Spectroscopy (XPS). In this case, NDs are dried on a substrate prior to measurements to be compatible with vacuum conditions. XPS can potentially also be applied directly in liquid, which will be discussed in Section 11.4.

For vacuum-annealed NDs, no $sp^2$ carbon signature could be detected after air exposure and water dispersion.[10] After in situ annealing at temperature up to 600°C, an $sp^2$ carbon contribution was though appearing. Surface graphitization should only appear above 700°C, and vacuum-annealed NDs are known to exhibit $sp^2$ carbons on their surface when characterized under ultra high vacuum directly after annealing.[12] It appears that exposure to air or water leads to the disappearance of the $sp^2$ signature while desorption of adsorbates in vacuum leads to the reappearance of these $sp^2$ carbons. This phenomenon was interpreted in terms of water and/or oxygen chemisorption on FLRs on NDs, following air exposure.[10]

For plasma hydrogenated NDs, a downshift of −1 eV of the carbon core-level binding energy was observed after dispersion in water.[11] This downshift vanished after in situ annealing at 300°C and was attributed to a peculiar electronic process, called surface transfer doping, first observed on hydrogenated bulk diamond surfaces.[11] During surface transfer doping, electrons from the diamond are transferred to adsorbates due to the negative electron affinity of the hydrogenated diamond surface. This electron transfer results in the formation of a hole-doped conductive layer on the diamond surface, which can be detected by XPS.[13] The adsorption of solvent molecules or other adsorbates has therefore strong impact on NDs electronic properties.

## 11.3 INTERACTION WITH SOLVENT MOLECULES UNDER GASEOUS ATMOSPHERE

In air or in a controlled gaseous atmosphere, the interaction between NDs and solvent molecules differs from vacuum conditions because a larger amount of molecules are likely to adsorb on the NDs. It remains different from the liquid environment though because intermolecular interactions between solvent molecules remain limited. Other characterization methods can be employed, which do not require vacuum conditions. Vibrational spectroscopies, such as Raman and infrared spectroscopies, are methods of choice to provide extensive information on the changes in the surface chemistry after solvent molecule exposure, but other methods will also be mentioned.

## INFRARED SPECTROSCOPY

The strong adsorption of water molecules on detonation NDs, even at low relative humidity (RH<35%), has first been evidenced by Fourier Transformed Infrared (FTIR) spectroscopy as seen in Fig. 11.2.[14] The progressive increase of OH bending and stretching modes at 1630 and 3400 $cm^{-1}$, respectively, is assigned to adsorption of water molecules.

Water molecules forming hydrogen bonds with NDs acid and basic sites can be differentiated by slight shifts in the OH stretching vibrations. Furthermore, Ji et al. observed intermolecular hydrogen bonding between neighboring water molecules, which could be explained by the formation of ordered water multilayers.[14] These intermolecular hydrogen bonds are not observed between gaseous water molecules, therefore the NDs surface behaves as a nucleation point for hydrogen-bonded water. Different amounts of adsorbed water are reported for NDs annealed in air, nitrogen, or hydrogen, suggesting surface-dependent water arrangement.

A more recent FTIR report, based on the reflection method using Attenuated Total Reflectance (ATR) crystal instead of transmission measurement with KBr pellets, has shown that water adsorption for hydrogenated NDs has a different signature than for oxidized ones.[15] The difference was attributed to hydrophobic interactions between hydrogenated NDs and water molecules. ATR-FTIR was also applied to NDs exposed to ethanol vapor up to saturation. Changes over time in the ethanol chemical vibrations in contact to the NDs were interpreted in terms of restructuration of the ethanol solvation layer.[16] FTIR therefore enables the characterization of structuration of solvent films around NDs under ambient atmosphere.

## RAMAN SPECTROSCOPY

Adsorption of water molecules on NDs affects their Raman signature. The principle of Raman spectroscopy and the characteristic features of NDs Raman spectra are detailed in Chapter 4. The feature at 1640 $cm^{-1}$ as observed in Fig. 11.3 has been attributed to O–H bending vibrations coming either from hydroxyl groups on the NDs, or from adsorbed water molecules, by Mochalin et al.[17]

This feature disappears after heating in argon atmosphere up to 300°C and does not reappear after cooling in argon (Fig. 11.3A). However, after air exposure, the Raman spectrum of a cooled sample is similar to the spectrum before heating (Fig. 11.3B). According to Mochalin et al., these observations can be interpreted as the result of strong water interaction with the ND surface after air exposure. Formation of hydroxyl groups on the surface of NDs from water dissociation or strong water physisorption is possible as discussed in Section 11.2.

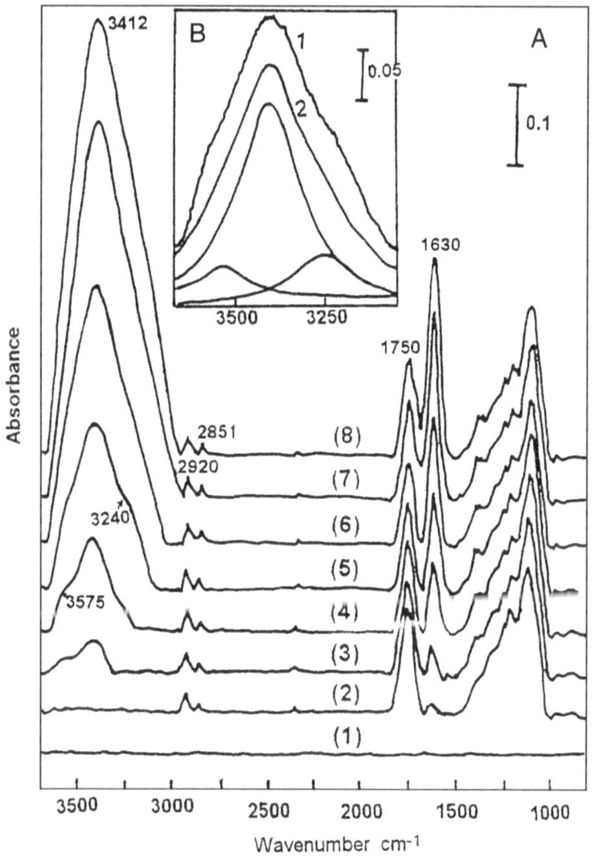


**FIGURE 11.2**

(A) FTIR spectra of KBr and NDs pellets after exposure to air over different periods of time: (1) KBr pellet heated in an in situ IR reactor cell at 140°C for 4 h and then exposed to air (<35% RH) at room temperature for 48 h; (2) NDs pellet heated as in (1) and recorded in vacuum; (3) exposed to air for 1; (4) 10; (5) 30; (6) 60; (7) 100, and (8) 140 min.
(B) Fitted spectrum of (5): 1, actual spectrum; 2, corresponding overall curve fit, remaining peaks are the calculated subbands. Reprinted in adapted form from: Ji, S., et al. Appl. Surf. Sci. 1998, 133 (4), 231–238,[14] © 1998 with permission from Elsevier

## GRAVIMETRY

Gravimetric methods, based on the measurement of mass change under different environmental conditions, can provide complementary information on the thermodynamics of solvent adsorption on NDs. Two different approaches are usually chosen. Either the sample is exposed to a progressive temperature increase under an inert atmosphere, which is called thermogravimetric analysis (TGA), or the temperature

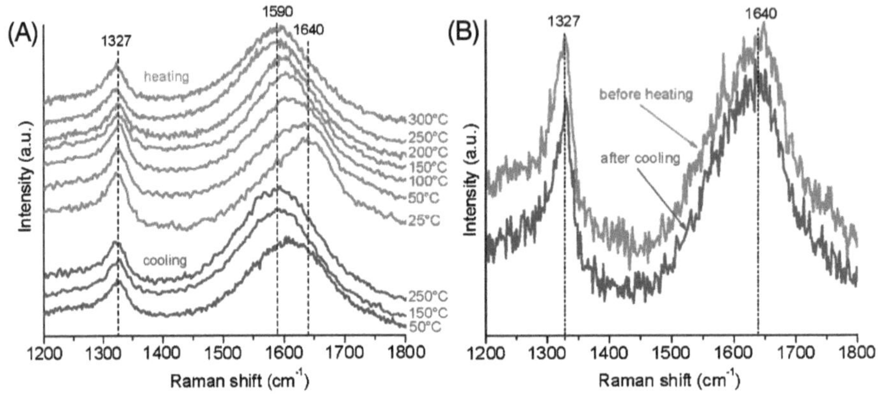


**FIGURE 11.3**

(A) UV Raman spectra of detonation NDs in Ar atmosphere at different temperatures and (B) UV Raman spectra of NDs recorded in air at room temperature before heating and after cooling. Reprinted with permission from: Mochalin, V., et al. Chem. Mater. 2009, 21 (2), 273–279,[17] © 2009 American Chemical Society

is kept constant and the sample is exposed to varying gas molecules. During these processes, the relative mass change of the sample is measured. Both approaches have been recently applied to NDs to probe their interaction with water molecules. A comparison of hydrogenated and oxidized NDs is particularly relevant to evidence a surface-dependent behavior. In Fig. 11.4A the adsorption of water molecules on NDs at room temperature, after an annealing pre-treatment at 300°C to remove adsorbents, is plotted with respect to the partial water vapor pressure.[11]

At low water vapor pressure ($P/P_0 < 0.03$), a large mass increase is observed on NDs–H while no water adsorption occurs on NDs–COOH. On the contrary, at high relative water pressure ($P/P_0 > 0.8$), a higher water adsorption rate is evidenced on oxidized NDs. The TGA measurements on oxidized and hydrogenated NDs are plotted in Fig. 11.4B.[15] In the 0–100°C region, the bulk water is desorbed while surface-bounded water is desorbed in the 100–300°C region. While most of the water molecules are desorbed at 100°C for oxidized NDs, the water desorption is strong up to 250°C for hydrogenated NDs.

From these two studies, we can conclude that water molecules adsorb at lower water vapor pressure and are desorbed at higher temperature for hydrogenated NDs than for oxidized NDs. Water molecules have therefore a stronger affinity for hydrogenated surface than oxidized ones. This is rather surprising because hydrogenated bulk diamond surfaces are known to be hydrophobic whereas oxidized ones are hydrophilic.[18] Water dissociation for hydrogenated NDs, as observed by HREELS performed in vacuum (see Section 11.2), could be a possible explanation of this strong and rapid adsorption, although a direct measurement of water dissociation in ambient conditions would be necessary to validate this hypothesis.

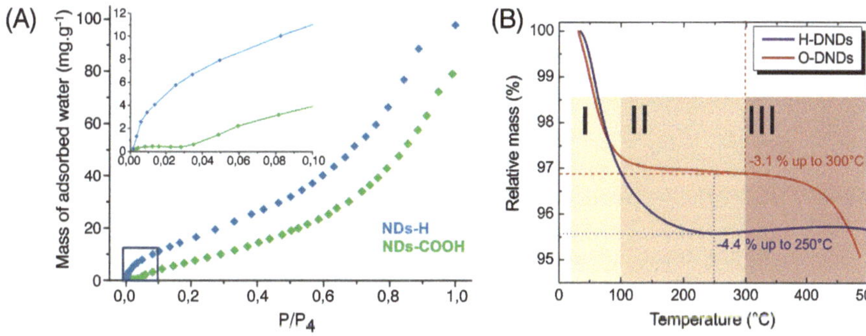


**FIGURE 11.4**

(A) Water adsorption on surface-modified nanodiamonds. Water vapor adsorption isotherms on NDs-H (blue) and NDs–COOH (green). Prior to water vapor exposure, the surface of NDs was annealed at 300°C in vacuum to remove physisorbed and weakly bounded species. Inset shows a magnified view of adsorption at low water relative pressure ($P/P_0 < 0.1$). Reproduced from: Petit, T., et al. Nanoscale 2013, 5 (19), 8958–8962,[11] by permission of The Royal Society of Chemistry. (B) TGA curves of oxidized (O-DNDs) and hydrogenated (H-DNDs) NDs. Reprinted in adapted form from: Stehlik, S., et al. Diam. Relat. Mater. 2015,[15] © 2015 with permission from Elsevier

## PERMITTIVITY

Another possibility to characterize the interaction of NDs with surrounding water molecules is the measurement of the dielectric permittivity of detonation NDs exposed to a small amount of water molecules.[19] An extremely large increase on the dielectric permittivity from $<100$ to $>10^9$ was measured on NDs exposed to water molecules when their surface was hydrogenated. The permittivity increased with higher water content, while the resistivity of the sample dropped. This phenomenon was not reproduced on oxidized NDs. The authors explained this unusual behavior by proton hopping or strong polarization of water molecules at the surface of NDs–H.[19]

# 11.4 INTERACTION WITH SOLVENT MOLECULES IN LIQUID PHASE

Probing chemical and electronic interactions between solvent molecules in vacuum and gaseous atmosphere are fundamental to better understand the impact of solvent adsorption on surface chemistry and electronic properties of NDs. However, it remains questionable if conclusions resulting from experiments on dry NDs can be directly extended to NDs in liquid dispersions. Indeed, for vacuum experiments, NDs are usually dried on a substrate and a charge transfer from the underlying substrate[20] may alter the results related to electronic states (see Chapter 10). Water molecules also have different organizations and electronic structures in the liquid and gas phases, which would potentially modify their interactions with dispersed NDs.

The investigation of NDs–solvent chemical and electronic interactions directly in the liquid phase is not an easy task because most of the analytical methods require either vacuum conditions (electron and X-ray spectroscopies) or are extremely sensitive to solvent response, screening the signature of NDs (vibrational spectroscopies).

Furthermore, measurements in the liquid phase require good colloidal stability of the NDs. The NDs stability in an aqueous dispersion is usually insured by surface charges inducing electrostatic repulsion between neighboring NDs. However, van der Waals and Coulombic forces may also enter into play,[21,22] which induces aggregation of NDs. To overcome these forces, bead-assisted milling and ultrasonication are required (see Chapter 2). We will review in this section the results currently available on the characterization of NDs in water.

## ELECTROPHORETIC MOBILITY

Before discussing advanced spectroscopy methods applied in liquids, the measurement of the electrophoretic mobility of NDs in a liquid, used to estimate the Zeta potential (ZP) of NDs, has to be mentioned. The ZP corresponds to the electrostatic potential at the slipping plane between solvent molecules bonded to the surface of the NDs and the rest of the solvent.[23] Hence it provides knowledge about the electrostatic field close to NDs and is routinely used to validate the colloidal stability of NDs.

In general, NDs dispersions with a ZP having an absolute value above 30 mV are considered stable. Note that the ZP is strongly dependent on pH and ionic strength, and should not be confused with the surface potential, defined at the nanoparticle surface. The relationship between surface chemistry and the ZP for detonation NDs at neutral pH is summarized in Table 11.1.

Depending on the surface chemistry, several hypotheses have been proposed to explain the origin of the charges (see Table 11.1). If negative ZPs are clearly induced by deprotonation of acidic groups on the NDs, the origin of the positive ZP is more controversial.[34] For hydroxylated and aminated NDs, the protonation of functional surface groups seems the most plausible interpretation, although the stability of protonated hydroxyl groups at neutral pH remains under question. For hydrogenated and graphitized NDs, no surface functional groups should be positively-charged and more complex origins have been proposed. For surface-graphitized NDs, the idea of hole formation following the adsorption of oxygen on FLRs was proposed based on XPS results.[10] On plasma-hydrogenated NDs, a different XPS signature was observed, and the surface charges may come from surface transfer doping by analogy with bulk diamond.[11] The adsorption of hydronium ions ($H_3O^+$) was also hypothesized as the origin of positive ZP on hydrogen-annealed NDs.[33] Clearly, the single measurement of ZP from electrophoretic mobility is not enough to precisely describe the surface electronic and chemical structure of dispersed NDs. It is however a method easy to

**Table 11.1** Surface-dependent Zeta potential of nanodiamonds

| Surface chemistry | Surface treatment | Zeta potential | Possible origin of Zeta potential | References |
|---|---|---|---|---|
| Carboxylated NDs | Chemical/air oxidation | − | Carboxylic groups deprotonation | 24,25 |
| Ozonated NDs | $O_3$ annealing | − | Acidic groups deprotonation | 26 |
| Hydroxylated NDs | Chemical reduction | + | Hydroxyl group protonation | 27–29 |
| Aminated NDs | Chemical amination | + | Amine protonation | 30,31 |
| Graphitized NDs | Vacuum annealing | + | Oxygen hole doping | 10,32 |
| Hydrogenated NDs | $H_2$ plasma | + | Surface transfer doping | 11 |
| | $H_2$ annealing | + | Water protonation | 33 |

implement which provides information about the electrostatic interactions between NDs and their liquid environment.

## CALORIMETRY

Differential Scanning Calorimetry (DSC) measurement is another relevant method to probe NDs–water interaction in which the amount of heat induced by temperature changes from the sample is measured. Early results on the melting/freezing of liquid water from NDs aqueous gel by Korobov et al. have revealed an unusual endothermic peak at 265 K, below the bulk water freezing peak at 273 K (Fig. 11.5).[35] This peak was attributed to strongly bonded water on NDs, a so-called nanophase of water. The different freezing temperature of this water nanophase was explained by the existence of an ice-like structure which extends over 2–3 water layers.

Korobov et al. also observed that the DSC signature of NDs strongly depend on their core and aggregate sizes.[36] A larger difference between the bulk water and nanophase melting point temperatures was observed for the smallest and well-dispersed NDs. This is a good example of the use of NDs–water interactions to learn information on NDs dispersion properties. Advanced spectroscopy methods have thought to be applied in liquid to answer the question of the impact of solvation on NDs surface chemistry and gain more knowledge about the structure of the nanophase of water.

## RAMAN SPECTROSCOPY

We have previously seen that NDs–solvent interactions can be probed by Raman spectroscopy under air through the O–H bending vibrations. In the liquid phase, the

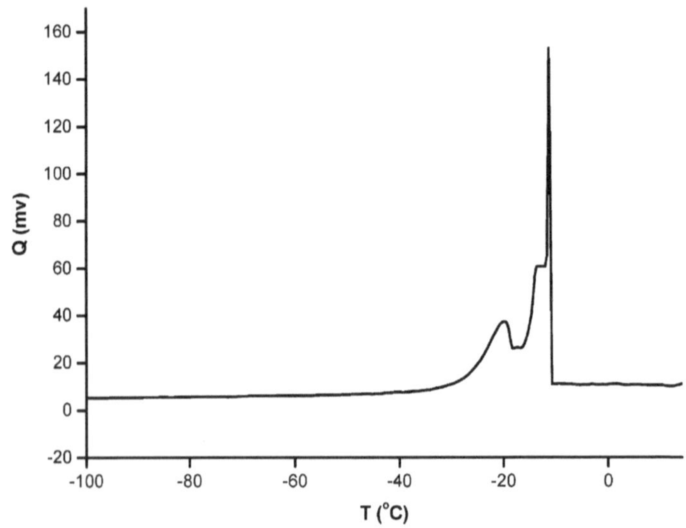


**FIGURE 11.5**

Typical DSC cooling scan of detonation nanodiamonds–water in 1:1 mixture from 293 to 173 K. The sharp peak on the right corresponds to the freezing of the bulk water. The peak on the left corresponds to the freezing of the nanophase of water. Reprinted with permission from: Korobov, M. V., et al. J. Phys. Chem. C 2007, 111 (20), 7330–7334,[35] © 2007 American Chemical Society

O–H stretching Raman mode around 3000–3500 $cm^{-1}$ appears more intense, and it was used to monitor the impact of NDs on solvent organization. Dolenko et al. have investigated modifications of fluorescence and Raman spectra for NDs with different surface terminations in water and methanol.[37,38] The authors observed that the ratio $\chi$ of the Raman bands at 3215 and 3415 $cm^{-1}$, associated to vibrations of water molecules with stronger and weaker hydrogen bonds, respectively, varies with respect to the NDs concentration (Fig. 11.6A). From this observation, the authors concluded that NDs tend to decrease the strength of the water hydrogen bonding compared to bulk water. The ratio $\chi$ was found to depend on the NDs surface chemistry, and carboxylated NDs were found to have the strongest impact on water organization (Fig. 11.6B). In addition, the overall fluorescence of the NDs dispersion seems to have an inverse correlation with the hydrogen bond weakening (Fig. 11.6B).[37]

Dolenko et al. performed similar experiments with carboxylated NDs in methanol.[38] Hydrogen bond weakening for increased NDs concentration was also observed in methanol. However, the Raman band intensity changes were found to be larger in methanol, suggesting that the hydrogen bond network is more disturbed by NDs in methanol than in aqueous dispersions.

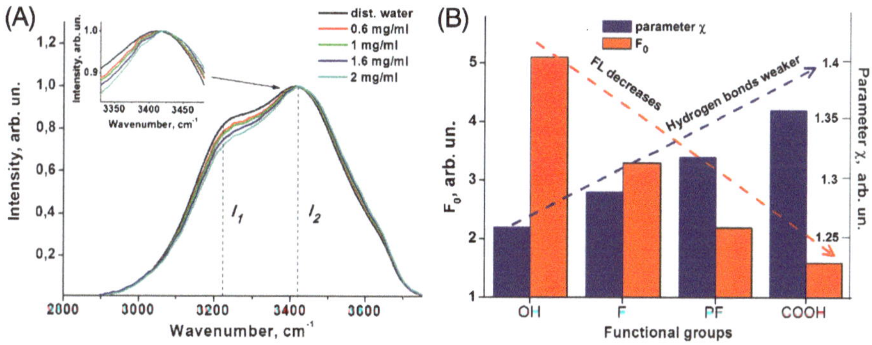


**FIGURE 11.6**

(A) Water O–H stretching bands in detonation NDs aqueous dispersions with concentrations 0.6, 1, 1.6, and 2 mg/ml. The inset shows details in the vicinity of the 3400 $cm^{-1}$ peak. (B) Dependences of the parameters $\chi$, defined as $I_2/I_1$, and $F_0$, the ratio of integral intensity of fluorescence to integral intensity of water Raman valence band, for hydroxylated (OH), fluorinated (F), polyfunctional (PF), and carboxylated (COOH) detonation NDs. Reprinted in adapted form with permission from: Dolenko, T. A., et al. J. Phys. Chem. C 2012, 116 (45), 24314–24319,[37] © 2012 American Chemical Society

## SOFT X-RAY SPECTROSCOPY

Soft X-ray spectroscopies, such as X-ray absorption (XA), X-ray emission (XE), and photoelectron spectroscopies (XPS), are advanced spectroscopy methods that probe the electronic structure of matter. With these techniques, a core electron is excited to an unoccupied electronic state by an incoming X-ray photon. Photons or electrons emitted during the filling process of the core-hole are then analyzed, which gives information on the electronic structure of the samples. XAS and XES probe the unoccupied and occupied electronic states, respectively, of the excited atom, whereas XPS measures the energy of electrons extracted from the core or valence electronic states of the atom. These methods are element-selective and can provide a high level of information about chemical environment. A detailed description of these methods is available in reference [39].

The implementation of soft X-ray spectroscopies to solutes in liquids is complex because these methods require vacuum conditions in addition to an X-ray photon source with tunable energy and high photon flux. The tunable energy is required to perform element-specific excitation. The high photon flux is necessary to have measurable signal from solutes such as nanoparticles, usually dispersed at low concentrations in a solvent. These requirements are fulfilled by third-generation synchrotron facility.[40] To probe light elements such as carbon, oxygen, or nitrogen, X-ray energies between 200 and 600 eV are necessary, which lie in the soft X-ray range. The vacuum conditions are required due to the low mean free path of X-ray photons in this energy range. Classical solvents such as water or alcohols have low vapor pressure,

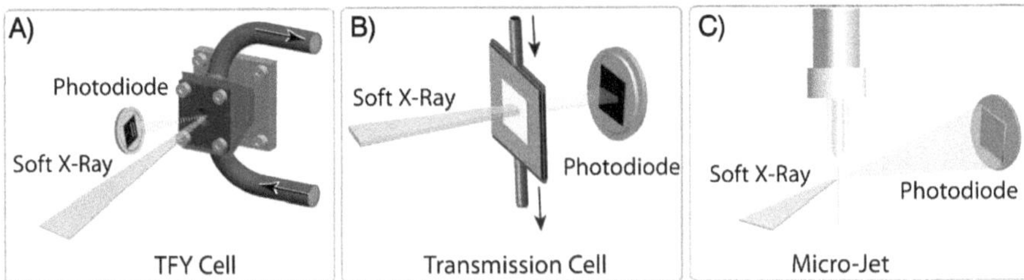


**FIGURE 11.7**

(A) Schematic of a flow-cell for detection in fluorescence yield. (B) Schematic of a flow-cell for soft X-ray spectroscopy in transmission mode. (C) Schematic of a liquid microjet for fluorescence yield. Reproduced from: Lange, K. M., Aziz, E. F. Chem. Soc. Rev. 2013, 42 (16), 6840–6859,[41] by permission of The Royal Society of Chemistry

which makes them incompatible with vacuum conditions. However, recent technical breakthroughs have offered new options to overcome this challenge.

### *X-ray absorption and emission spectroscopies*

Several approaches are possible to measure liquids with XAS and XES as illustrated in Fig. 11.7.[41] In flow cell systems which have first been introduced, the liquid sample is excited with X-rays through a thin (~100–200 nm) silicon window membrane. The X-ray total fluorescence yield (TFY) is then measured through the same membrane (Fig. 11.7A)[42,43] or the X-ray absorbance of the liquid sample is measured in transmission through a second thin membrane (Fig. 11.7B).[44,45] Another alternative is the use of high-velocity microjets, which enables the stabilization of a liquid phase in vacuum (Fig. 11.7C).[46,47] Right after measurement, the microjet is frozen in order to maintain the vacuum chamber pressure below $\sim 10^{-5}$ mbar. Microjets usually have a diameter of 10 to 50 μm, and the X-ray absorption is estimated from the TFY. These methods have been applied to pure solvent, dissolved ions, and small molecules as reviewed in reference [41]. They have also been implemented to dispersed nanoparticles[48–50] and recent results on NDs aqueous dispersions, providing new insights on the NDs–water interface, will be discussed here.[51,52]

To probe the carbon K edge of NDs dispersed in water, the use of microjet is the most adapted technique because it provides a larger photon flux and reduces sources of carbon contamination compared to membrane-based methods. Fluorescent yield measurements are also more sensitive than transmission measurements[53] to dilute samples, such as NDs in aqueous dispersions. A drawback of microjets is that a large amount of sample (hundreds of ml) of well-dispersed NDs is required for measurement, although recycling microjet systems can also be implemented.[54]

A proof of principle of carbon K edge characterization of carboxylated NDs dispersed at 1 wt.% in water has been reported in reference [51]. XAS and XES of NDs dried on a silicon substrate and flowing in a liquid microjet were compared to in-

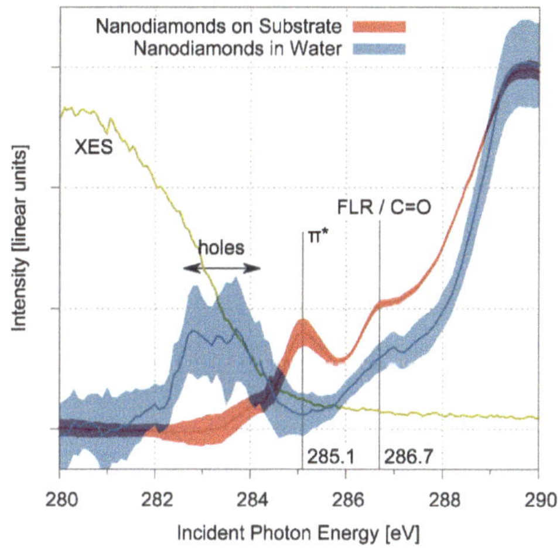


**FIGURE 11.8**

XAS pre-edge region of NDs on a substrate (red) and NDs dispersed in water at 1 wt.% (blue). XES from ND sub (green) is also plotted to indicate the occupied states in the NDs valence band. Reproduced from: Petit, T., et al. Nanoscale 2015, 7 (7), 2987–2991,[51] by permission of The Royal Society of Chemistry

vestigate the impact of solvation on NDs electronic structure. Interestingly, strong changes in the pre-edge region of the carbon K edge, where transitions related to the NDs surface chemistry enter into play, were reported. In particular, the $\pi^*$ transition at 285.1 eV, which is related to the graphitic surface reconstructions, is only observed on NDs characterized on a substrate, while a broad band feature appears at lower energy for NDs in liquid water (Fig. 11.8). This new band lies in the same energy range as the first occupied electronic states probed by XES, and was therefore attributed to holes formed in the NDs valence band.[51] These holes might originate from oxygen hole doping of graphitic reconstructions.[10] Oxygen adsorption on FLRs would explain the disappearance of the $\pi^*$ feature.

The characterization of the oxygen K edge by XAS is also of interest because it is sensitive to water molecules rather than NDs. Indeed, the small amount of oxygen on the NDs surface is negligible compared to oxygen from water molecules at NDs concentration below 10 wt.%. The XA spectrum of oxygen K edge of pure water in the gas phase is constituted of separated peaks related to the different unoccupied orbitals of oxygen atoms.[55] The spectrum of liquid water is quite different from gas phase because strong intermolecular interaction induce the formation of a band structure similar to solid semiconductors, depending on hydrogen bond network.[55,56] Hence, modification of the hydrogen bond network of water induced by NDs can be probed by looking at the oxygen K edge.

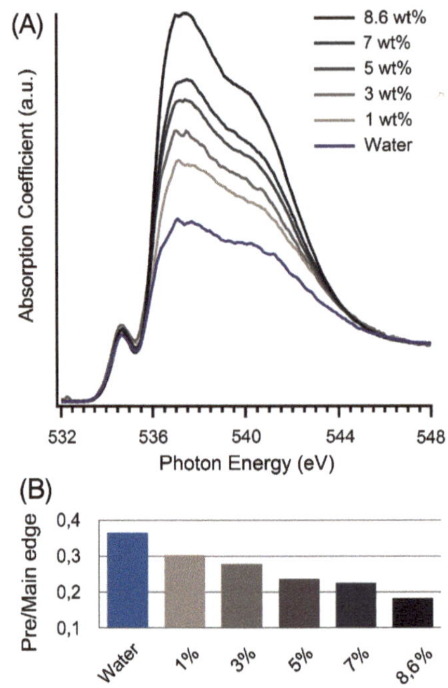


**FIGURE 11.9**

(A) XAS of oxygen K-edge from water (blue) and aqueous dispersion of NDs at different concentrations ranging from 1 to 8.6 wt.% (grey and black). (B) Pre/Main-edge ratios for the different NDs concentrations. Reprinted with permission from: Petit, T., et al. J. Phys. Chem. Lett. 2015, 6 (15), 2909–2912,[52] © 2015 American Chemical Society

To probe oxygen K edge of water molecules in an NDs dispersion, the microjet approach is not optimal anymore since fluorescence yield measurements are subject to saturation for highly concentrated samples, which is the case for oxygen atoms from water molecules in aqueous solutions. On the other hand, transmission XAS measurements performed with a transmission flow cell do not suffer from saturation effects and can be reliably used for the oxygen K edge, constituted of a pre- (535 eV), main- (538 eV) and post-edge (540 eV). This method has been previously applied to aqueous ionic solutions and water–methanol mixtures, and could evidence small changes in the water structure.[45,57,58]

Aqueous dispersions of NDs with different surface chemistries and sizes have been characterized using a transmission cell.[52] The XA spectra obtained with surface-graphitized detonation NDs are shown in Fig. 11.9. The main- and post-edges at 538 and 540 eV, respectively, increased compared to the water spectrum. These changes appear extremely large compared to ionic solutions.[58] This result was interpreted in terms of a different water organization around NDs due to their larger size compared to ions. The electronic structure of the interfacial water in the NDs solvation shell seems close to high density amorphous ice[52] which would be in agreement with calorimetry results discussed previously.

By performing XAS directly in water using a liquid microjet or a transmission flow cell, it was possible to demonstrate that aqueous solvation indeed has an impact on the NDs and solvent electronic structure. The surface chemistry of NDs was found to be altered by solvation, while the overall solvent organization was modified by the presence of NDs. Further research efforts are required to understand to which extent these observations can be reproduced with NDs having different surface chemistry.

### *X-ray photoelectron spectroscopy*

The use of liquid microjet to probe nanoparticles dispersed in water can also be applied to XPS. For example, the de/protonation of silanol groups on silica nanoparticles in a liquid microjet was used to estimate the surface potential of the nanoparticles at different pHs.[59] Since XPS is extremely surface sensitive, the probed nanoparticles are only a few nanometers from the liquid–vacuum interface and might behave differently than nanoparticles in the bulk liquid. Nevertheless, this method could provide new insights on surface potential of nanoparticles in liquid environment. XPS in a liquid has not been applied to NDs dispersions yet, but it might help understanding the origin of the Zeta potential on NDs.

## 11.5 CONCLUSION AND OUTLOOK

Interactions between NDs and solvent molecules have consequences on both the NDs surface properties and solvent organization. In water, these interactions were found to depend on the NDs surface chemistry. In particular, graphitized and hydrogenated surfaces may chemically react with the solvent, while hydroxylated and carboxylated surfaces seem less reactive, although they affect the water hydrogen bond network. Up to now, most of the published results were focused on water mainly due to its relevance for nanomedicine (see Chapter 15). However, other solvents, such as organic solvents, mineral oils, and ionic liquids, deserve more attention in the coming years since these solvents will be relevant for specific applications in lubrication (see Chapter 15), thermal fluids, and catalysis (see Chapter 18).

A broad range of advanced analysis and spectroscopy methods are available to probe NDs–solvent interactions. Close-range interactions between NDs surface and a few solvent molecules can be probed in vacuum. Measurements performed at ambient pressure in air or inert atmosphere are able to probe NDs with a few solvation shells. In the liquid phase, long-range modifications of the solvent structure induced by the NDs can also be investigated. Experiments performed in all these environments are complementary when drawing a full picture of NDs–solvent interactions.

The investigation of nanoparticle–solvent interactions in the liquid phase leads to the development of new tools to probe NDs directly in a liquid. In the future, one can envision that chemical reactions or electronic processes at the surface of NDs could be followed in operando directly in liquid media using these new spectroscopy methods. Understanding the role of a solvent on the NDs chemical reactivity and catalytic activity would help the optimization of NDs properties toward specific applications.

## REFERENCES


1. Zobel, M.; Neder, R. B.; Kimber, S. A. J. Universal Solvent Restructuring Induced by Colloidal Nanoparticles. *Science* **2015**, *347* (6219), 292–294.
2. Rubasinghege, G.; Grassian, V. H. Role(s) of Adsorbed Water in the Surface Chemistry of Environmental Interfaces. *Chem. Commun. (Camb.)* **2013**, *49* (30), 3071–3094.
3. Chakrapani, V.; Angus, J. C.; Anderson, A. B.; Wolter, S. D.; Stoner, B. R.; Sumanasekera, G. U. Charge Transfer Equilibria Between Diamond and an Aqueous Oxygen Electrochemical Redox Couple. *Science* **2007**, *318* (5855), 1424–1430.
4. Maier, F.; Riedel, M.; Mantel, B.; Ristein, J.; Ley, L. Origin of Surface Conductivity in Diamond. *Phys. Rev. Lett.* **2000**, *85* (16), 3472–3475.
5. Struck, L. M. Interaction of Hydrogen and Water With Diamond (100): Infrared Spectroscopy. *J. Vac. Sci. Technol. A Vac. Surf. Films* **1993**, *11* (4), 1992.
6. Furthmüller, J.; Hafner, J.; Kresse, G. Dimer Reconstruction and Electronic Surface States on Clean and Hydrogenated Diamond (100) Surfaces. *Phys. Rev. B* **1996**, *53* (11), 7334–7351.
7. Raty, J.-Y.; Galli, G.; Bostedt, C.; van Buuren, T. W.; Terminello, L. J. Quantum Confinement and Fullerenelike Surface Reconstructions in Nanodiamonds. *Phys. Rev. Lett.* **2003**, *90* (3), 37401.
8. Laikhtman, A.; Lafosse, A.; Le Coat, Y.; Azria, R.; Hoffman, A. Interaction of Water Vapor with Bare and Hydrogenated Diamond Film Surfaces. *Surf. Sci.* **2004**, *551* (1–2), 99–105.
9. Akhvlediani, R.; Michaelson, S.; Hoffman, A. Interaction of Water Molecules With Bare and Deuterated Polycrystalline Diamond Surface Studied by High Resolution Electron Energy Loss and X-Ray Photoelectron Spectroscopies. *Surf. Sci.* **2010**, *604* (23–24), 2129–2138.
10. Petit, T.; Arnault, J.-C. C.; Girard, H. A.; Sennour, M.; Kang, T.-Y.; Cheng, C.-L.; Bergonzo, P. Oxygen Hole Doping of Nanodiamond. *Nanoscale* **2012**, *4* (21), 6792–6799.
11. Petit, T.; Girard, H. A.; Trouvé, A.; Batonneau-Gener, I.; Bergonzo, P.; Arnault, J.-C. Surface Transfer Doping Can Mediate Both Colloidal Stability and Self-Assembly of Nanodiamonds. *Nanoscale* **2013**, *5* (19), 8958–8962.
12. Petit, T.; Arnault, J.-C.; Girard, H. A.; Sennour, M.; Bergonzo, P. Early Stages of Surface Graphitization on Nanodiamond Probed by X-Ray Photoelectron Spectroscopy. *Phys. Rev. B* **2011**, *84* (23), 233407.
13. Chen, W.; Qi, D.; Gao, X.; Wee, A. T. S. Surface Transfer Doping of Semiconductors. *Prog. Surf. Sci.* **2009**, *84* (9–10), 279–321.
14. Ji, S.; Jiang, T.; Xu, K.; Li, S. FTIR Study of the Adsorption of Water on Ultradispersed Diamond Powder Surface. *Appl. Surf. Sci.* **1998**, *133* (4), 231–238.
15. Stehlik, S.; Glatzel, T.; Pichot, V.; Pawlak, R.; Meyer, E.; Spitzer, D.; Rezek, B. Water Interaction With Hydrogenated and Oxidized Detonation Nanodiamonds – Microscopic and Spectroscopic Analyses. *Diam. Relat. Mater.* **2015**.
16. Inel, G. A.; Ungureau, E.-M.; Varley, T. S.; Hirani, M.; Holt, K. B. Solvent–Surface Interactions Between Nanodiamond and Ethanol Studied With In Situ Infrared Spectroscopy. *Diam. Relat. Mater.* **2016**, *61*, 7–13.

17. Mochalin, V.; Osswald, S.; Gogotsi, Y. Contribution of Functional Groups to the Raman Spectrum of Nanodiamond Powders. *Chem. Mater.* **2009**, *21* (2), 273–279.
18. Ostrovskaya, L.; Perevertailo, V.; Ralchenko, V.; Dementjev, A.; Loginova, O. Wettability and Surface Energy of Oxidized and Hydrogen Plasma-Treated Diamond Films. *Diam. Relat. Mater.* **2002**, *11* (3–6), 845–850.
19. Batsanov, S. S.; Gavrilkin, S. M.; Batsanov, A. S.; Poyarkov, K. B.; Kulakova, I. I.; Johnson, D. W.; Mendis, B. G. Giant Dielectric Permittivity of Detonation-Produced Nanodiamond Is Caused by Water. *J. Mater. Chem.* **2012**, *22* (22), 11166.
20. Stehlik, S.; Petit, T.; Girard, H. A.; Arnault, J.-C.; Kromka, A.; Rezek, B. Nanoparticles Assume Electrical Potential According to Substrate, Size, and Surface Termination. *Langmuir* **2013**, *29* (5), 1634–1641.
21. Xu, Q.; Zhao, X. Electrostatic Interactions Versus van Der Waals Interactions in the Self-Assembly of Dispersed Nanodiamonds. *J. Mater. Chem.* **2012**, *22* (32), 16416.
22. Lai, L.; Barnard, A. S. Interparticle Interactions and Self-Assembly of Functionalized Nanodiamonds. *J. Phys. Chem. Lett.* **2012**, *3* (7), 896–901.
23. Delgado, a. V.; González-Caballero, F.; Hunter, R. J.; Koopal, L. K.; Lyklema, J. Measurement and Interpretation of Electrokinetic Phenomena (IUPAC Technical Report). *Pure Appl. Chem.* **2005**, *77* (10), 1753–1805.
24. Gibson, N.; Shenderova, O.; Luo, T. J. M.; Moseenkov, S.; Bondar, V.; Puzyr, A.; Purtov, K.; Fitzgerald, Z.; Brenner, D. W. W. Colloidal Stability of Modified Nanodiamond Particles. *Diam. Relat. Mater.* **2009**, *18* (4), 620–626.
25. Krueger, A.; Lang, D. Functionality Is Key: Recent Progress in the Surface Modification of Nanodiamond. *Adv. Funct. Mater.* **2012**, *22* (5), 890–906.
26. Shenderova, O.; Koscheev, A.; Zaripov, N.; Petrov, I.; Skryabin, Y.; Detkov, P.; Turner, S.; Van Tendeloo, G. Surface Chemistry and Properties of Ozone-Purified Detonation Nanodiamonds. *J. Phys. Chem. C* **2011**, *115* (20), 9827–9837.
27. Krueger, A.; Liang, Y.; Jarre, G.; Stegk, J. Surface Functionalisation of Detonation Diamond Suitable for Biological Applications. *J. Mater. Chem.* **2006**, *16* (24), 2322.
28. Girard, H. A.; Petit, T.; Perruchas, S.; Gacoin, T.; Gesset, C.; Arnault, J.-C. C.; Bergonzo, P. Surface Properties of Hydrogenated Nanodiamonds: A Chemical Investigation. *Phys. Chem. Chem. Phys.* **2011**, *13* (24), 11517–11523.
29. Shenderova, O.; Panich, A. M.; Moseenkov, S.; Hens, S. C.; Kuznetsov, V.; Vieth, H. M. Hydroxylated Detonation Nanodiamond: FTIR, XPS, and NMR Studies. *J. Phys. Chem. C* **2011**, *115* (39), 19005–19011.
30. Mochalin, V. N.; Neitzel, I.; Etzold, B.; Peterson, A. M.; Palmese, G.; Gogotsi, Y. Covalent Incorporation of Aminated Nanodiamond Into an Epoxy Polymer Network. *ACS Nano* **2011**.
31. Hens, S. C.; Cunningham, G.; Tyler, T.; Moseenkov, S.; Kuznetsov, V.; Shenderova, O. Nanodiamond Bioconjugate Probes and Their Collection by Electrophoresis. *Diam. Relat. Mater.* **2008**, *17* (11), 1858–1866.
32. Liang, Y.; Meinhardt, T.; Jarre, G.; Ozawa, M.; Vrdoljak, P.; Schöll, A.; Reinert, F.; Krueger, A. Deagglomeration and Surface Modification of Thermally Annealed Nanoscale Diamond. *J. Colloid Interface Sci.* **2011**, *354* (1), 23–30.
33. Williams, O. A.; Hees, J.; Dieker, C.; Jäger, W.; Kirste, L.; Nebel, C. E. Size-Dependent Reactivity of Diamond Nanoparticles. *ACS Nano* **2010**, *4* (8), 4824–4830.
34. Mochalin, V. N.; Shenderova, O.; Ho, D.; Gogotsi, Y. The Properties and Applications of Nanodiamonds. *Nat. Nano* **2012**, *7* (1), 11–23.

35. Korobov, M. V.; Avramenko, N. V.; Bogachev, A. G.; Rozhkova, N. N.; Osawa, E. Nanophase of Water in Nano-Diamond Gel. *J. Phys. Chem. C* **2007**, *111* (20), 7330–7334.
36. Korobov, M. V.; Volkov, D. S.; Avramenko, N. V.; Belyaeva, L. A.; Semenyuk, P. I.; Proskurnin, M. A. Improving the Dispersity of Detonation Nanodiamond: Differential Scanning Calorimetry as a New Method of Controlling the Aggregation State of Nanodiamond Powders. *Nanoscale* **2013**, *5* (4), 1529–1536.
37. Dolenko, T. A.; Burikov, S. A.; Rosenholm, J. M.; Shenderova, O. A.; Vlasov, I. I. Diamond–Water Coupling Effects in Raman and Photoluminescence Spectra of Nanodiamond Colloidal Suspensions. *J. Phys. Chem. C* **2012**, *116* (45), 24314–24319.
38. Dolenko, T.; Burikov, S.; Laptinskiy, K.; Rosenholm, J. M.; Shenderova, O.; Vlasov, I. Evidence of Carbon Nanoparticle-Solvent Molecule Interactions in Raman and Fluorescence Spectra. *Phys. Status Solidi* **2015**, *212* (11), 2512–2518.
39. de Groot, F.; Kotani, A. *Core Level Spectroscopy of Solids;* Taylor & Francis: New York, 2008.
40. Attwood, D. *Soft X-Rays and Extreme Ultraviolet Radiation: Principles and Applications;* Cambridge University Press, 2007.
41. Lange, K. M.; Aziz, E. F. Electronic Structure of Ions and Molecules in Solution: A View from Modern Soft X-Ray Spectroscopies. *Chem. Soc. Rev.* **2013**, *42* (16), 6840–6859.
42. Freiwald, M.; Cramm, S.; Eberhardt, W.; Eisebitt, S. Soft X-Ray Absorption Spectroscopy in Liquid Environments. *J. Electron Spectrosc. Relat. Phenom.* **2004**, *137–140*, 413–416.
43. Guo, J.; Tong, T.; Svec, L.; Go, J.; Dong, C.; Chiou, J.-W. Soft-X-Ray Spectroscopy Experiment of Liquids. *J. Vac. Sci. Technol. A Vac. Surf. Films* **2007**, *25* (4), 1231.
44. Yang, B. X.; Kirz, J. Extended X-Ray-Absorption Fine Structure of Liquid Water. *Phys. Rev. B* **1987**, *36* (2), 1361–1364.
45. Nagasaka, M.; Hatsui, T.; Horigome, T.; Hamamura, Y.; Kosugi, N. Development of a Liquid Flow Cell to Measure Soft X-Ray Absorption in Transmission Mode: A Test for Liquid Water. *J. Electron Spectrosc. Relat. Phenom.* **2010**, *177* (2–3), 130–134.
46. Siegbahn, H.; Siegbahn, K. ESCA Applied to Liquids. *J. Electron Spectrosc. Relat. Phenom.* **1973**, *2* (3), 319–325.
47. Faubel, M.; Steiner, B.; Toennies, J. P. Photoelectron Spectroscopy of Liquid Water, Some Alcohols, and Pure Nonane in Free Micro Jets. *J. Chem. Phys.* **1997**, *106* (22), 9013.
48. Liu, H.; Yin, Y.; Augustsson, A.; Dong, C.; Nordgren, J.; Chang, C.; Alivisatos, P.; Thornton, G.; Ogletree, D. F.; Requejo, F. G.; et al. Electronic Structure of Cobalt Nanocrystals Suspended in Liquid. *Nano Lett.* **2007**, *7* (7), 1919–1922.
49. Brown, M. A.; Seidel, R.; Thürmer, S.; Faubel, M.; Hemminger, J. C.; van Bokhoven, J. A.; Winter, B.; Sterrer, M. Electronic Structure of Sub-10 nm Colloidal Silica Nanoparticles Measured by In Situ Photoelectron Spectroscopy at the Aqueous–Solid Interface. *Phys. Chem. Chem. Phys.* **2011**, *13* (28), 12720–12723.
50. Cafun, J.-D.; Kvashnina, K. O.; Casals, E.; Puntes, V. F.; Glatzel, P. Absence of $Ce^{3+}$ Sites in Chemically Active Colloidal Ceria Nanoparticles. *ACS Nano* **2013**, *7* (12), 10726–10732.
51. Petit, T.; Pflüger, M.; Tolksdorf, D.; Xiao, J.; Aziz, E. F. Valence Holes Observed in Nanodiamonds Dispersed in Water. *Nanoscale* **2015**, *7* (7), 2987–2991.
52. Petit, T.; Yuzawa, H.; Nagasaka, M.; Yamanoi, R.; Osawa, E.; Kosugi, N.; Aziz, E. F. Probing Interfacial Water on Nanodiamonds in Colloidal Dispersion. *J. Phys. Chem. Lett.* **2015**, *6* (15), 2909–2912.

53. Jaklevic, J.; Kirby, J. A.; Klein, M. P.; Robertson, A. S.; Brown, G. S.; Eisenberger, P. Fluorescence Detection of Exafs: Sensitivity Enhancement for Dilute Species and Thin Films. *Solid State Commun.* **1977**, *23* (9), 679–682.
54. Lange, K. M.; Kothe, A.; Aziz, E. F. Chemistry in Solution: Recent Techniques and Applications Using Soft X-Ray Spectroscopy. *Phys. Chem. Chem. Phys.* **2012**, *14* (16), 5331–5338.
55. Myneni, S.; Luo, Y.; Näslund, L. Å.; Cavalleri, M.; Ojamäe, L.; Ogasawara, H.; Pelmenschikov, A.; Wernet, P.; Väterlein, P.; Heske, C.; et al. Spectroscopic Probing of Local Hydrogen-Bonding Structures in Liquid Water. *J. Phys. Condens. Matter* **2002**, *14* (8), L213–L219.
56. Lange, K. M.; Aziz, E. F. The Hydrogen Bond of Water From the Perspective of Soft X-Ray Spectroscopy. *Chem. Asian J.* **2012**.
57. Nagasaka, M.; Mochizuki, K.; Leloup, V.; Kosugi, N. Local Structures of Methanol–Water Binary Solutions Studied by Soft X-Ray Absorption Spectroscopy. *J. Phys. Chem. B* **2014**, *118* (16), 4388–4396.
58. Waluyo, I.; Nordlund, D.; Bergmann, U.; Schlesinger, D.; Pettersson, L. G. M.; Nilsson, A. A Different View of Structure-Making and Structure-Breaking in Alkali Halide Aqueous Solutions Through X-Ray Absorption Spectroscopy. *J. Chem. Phys.* **2014**, *140* (24), 244506.
59. Brown, M. A.; Redondo, A. B.; Sterrer, M.; Winter, B.; Pacchioni, G.; Abbas, Z.; van Bokhoven, J. A. Measure of Surface Potential at the Aqueous-Oxide Nanoparticle Interface by XPS From a Liquid Microjet. *Nano Lett.* **2013**, *13* (11), 5403–5407.

CHAPTER

# Toxicity issues: interactions with biological species

12

**Sylvie Chevillard, Romain Grall, Jozo Delic, Vincent Paget**
*CEA DRF iRCM, Laboratory of Experimental Cancerology, Fontenay-aux-Roses, France*

## Contents



## 12.1 INTRODUCTION

Biological and medical applications of carbon nanoparticles such as fullerenes and carbon nanotubes have been extensively studied while other carbon nanoparticles, specifically nanodiamonds (NDs), are less experimented. However, recent research on the biomedical applications of NDs has increased especially in the fields of drug carrier and delivery systems (i.e. cancer therapeutics) (see Chapter 16),[1–4] potentiation of radiotherapy,[5] photoluminescent/Raman, biomarker and bioprobes,[6,7] nanosurgery,[1] orthopedy, dental field, prosthetic devices for retinal implants and implant coating,[1] analytical diagnostics,[1] and bio-chips,[8] among others.

NDs benefit from various promising properties, such as super hardness, chemical inertness, special structure/surface,[1] optical transparency, optical and luminescence spectroscopic properties (see Chapter 7).[9–11] Of particular interest is their strong near-infrared photoluminescence,[12] which makes NDs a promising alternative to the current bioimaging platforms for both in vitro and in vivo applications.[13–15] Moreover, NDs surface presents proportionally large number of surface atoms that makes them highly adapted for functionalization with numerous biomedical molecules.[16,17]

Modification of functional groups on the ND surface, for conferring specific surface charges, allows controlling colloidal stability, adsorption and release of molecules, and conjugation with biomolecules, drugs, and polymers. The polyvalence of surface treatment or ND encapsulation is very promising for improving biological tolerance and cellular uptake.

Indeed, considering the huge potential of NDs for medical applications, it is of major importance to understand hazards and risks related to NDs before practical use in clinical setting by establishing their biological behavior and effects at the levels of organisms, organs, tissues and cells. Short-term effect but also the possible long-term toxic effect of bio-persistent internalized ND should be further explored after acute, chronic, or repeated exposure according to the ways of entrance or target tissues.

## 12.2 IN VITRO BIOLOGICAL EFFECTS OF ND

In the carbon nanomaterial safety assessment, in vitro cytotoxicity test is an important research subject because it is simple, reproducible, and many human cell types could be tested for defining possible specific response according to cell function and genetic background. A large number of in vitro tests are used for screening hazards of nanoparticles for human health, most of them focused on cytotoxicity and genotoxicity by analyzing cell survival, proliferation and cell cycle, cell death, oxidative stress, DNA alterations, and repair and metabolic activity. Specific mechanisms of action can be also tested by analyzing gene or protein expression. The choice of cell types and cell lines should be adapted to the biological question (hazards after inhalation, ingestion, etc.) since previous studies suggested that the cytotoxic response to nanomaterials is quite dependent on the cell type, probably underlying the differences in uptake, retention, and intrinsic sensitivity. Most extensively, human cancer cell lines are used instead of normal cells because they are widely used in the field of toxicology, available for virtually every organ, their responses to many drugs and pollutants are mainly studied, and it is easy to get them in large quantities in reproducible conditions. For the screening of potentially toxic effects of a nanoparticle, it is recommended to test the nanoparticles on a wide range of doses on many cell lines.[18] All these tests should reduce the number of animal tests that will be further required.

### 12.2.1 IN VITRO CYTOTOXICITY

In vitro, the absence of cytotoxicity of carboxylated high pressure high temperature (HPHT) ND of different size (20, 50, and 100 nm) has been demonstrated on liver, kidney, intestine, and lung, cervical human cancer cell lines, and human fibroblasts,[6,19–22] as well as the absence of DNA alteration and Reactive Oxygen Species (ROS) generation[19] even at very high concentrations (500 μg/ml) (an example is given in Fig. 12.1).

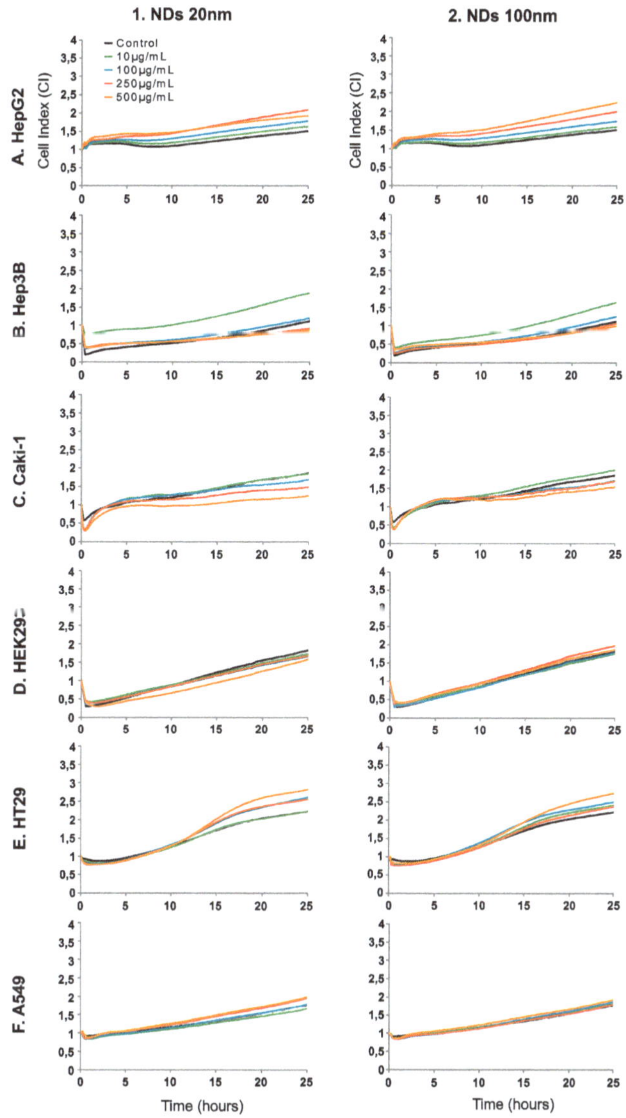


**FIGURE 12.1**

Cytotoxicity was estimated by real-time monitoring of the cellular impedance (cell index) which reflects cell viability and proliferation, cell morphology and cell membrane integrity. Cell index was measured continuously for 24 h after exposure of liver HepG2 (A) and Hep3B (B), renal Caki-1 (C) and HEK293 (D), colon HT29 (E) and lung A549 (F) cells after exposure with carboxylated high pressure high temperature (HPHT) ND-20 and -100 nm, columns 1 and 2, respectively. Cell index was normalized at time 0 to ensure there was no inter-wall variability prior to the addition of nanoparticles. Control cells were not exposed to NDs. None of the tested cell lines presents significant change in cellular index in comparison with unexposed control cells[19]

None of the OH-, $NH_2$-, and COOH- HPHT NDs or detonation NDs was cytotoxic for Hek-293 cells for doses lower than 50 μg/ml. At higher doses, the ranking of ND according of toxicity is $-NH_2 \gg -OH > -COOH$, the aminated ND being highly cytotoxic.[23] However, in another study, comparing the cytotoxicity of unmodified detonation ND of 4–5 nm, $NH_2$-functionalized detonation ND and fluorescent (NV centers) ND of 45 nm on Hela (cervical adenocarcinoma) and HepG2 (hepatocarcinoma) cells, the authors did not observe any modification in cell death, survival, or on gene expression even at the very high dose of 1 mg/ml.[24] It has been also shown in vitro that detonation 2–10 nm ND, with and without surface modifications by acid or base, i.e. ND-raw, ND-COOH and ND-$SO_3Na$, are non-cytotoxic, do not generate ROS, do not modify adenosine triphosphate ATP (the molecule that fuels life) production, and do not induce any modification in cell morphology in a large variety of human cell lines, including neuroblastoma, macrophage, keratinocyte, and PC-12 cells.[25] The same team further compared the cytotoxicity of 2–10 nm detonation ND with carbon black (CB), multi-walled carbon nanotubes (MWNT), and single-walled carbon nanotubes (SWNT) on neuroblastoma cells and macrophages, and found that the toxicity follows the order of SWNT > MWNT > CB > ND.[26,27] Similarly, Liu et al. measured the toxicity of ND in comparison with carbon nanotubes (CNTs) on human lung A549 carcinoma cells and HFL-1 normal fibroblasts, and their results indicated that at the concentration of 0.1–100 μg/ml, two types of carboxyl-modified ND with a particle size of 5 and 100 nm, respectively, did not reduce cell viability or alter the protein expression profiles. In contrast, carboxyl-modified CNTs exposed under the same conditions induced significant cell toxicity.[20] Finally, another comparison of the biological effects induced after exposure of Hela cells with different carbon nanomaterials, namely, multi-walled carbon nanotubes, graphene oxide, and ND, concluded that ND exhibited the highest cell uptake ratio but the least cytotoxicity.[28]

Only a few studies reported cytotoxicity of ND after in vitro exposure. For example, both detonation ND and OH-ND showed time- and concentration-dependent cytotoxicity on immortalized human cord endothelial cells and on A549.[29,30] However, the purity and the size of NDs used in these studies were not characterized. Additionally, a strong size- and concentration-dependent cytotoxicity was observed after exposure of macrophages to synthetic ND (6, 60, 100, 250, and 500 nm);[31] however, it was observed that macrophage could be a particularly nanomaterial-sensitive cell type.[26]

### 12.2.2 GENOTOXICITY

Genotoxicity is one of the most adverse cellular effects since it means that a drug or substance alters genome integrity by damaging DNA. The alteration of the DNA molecule could be the consequence of a direct interaction between the stressor and the DNA molecule (for example, the energy deposited on DNA when exposed to ionizing radiation) or of an indirect interaction through the generation of reactive oxygen species that can secondarily attack the DNA molecule. Once DNA is damaged, cells could (i) repair the damage without any further consequence, (ii) repair the damage with errors that indeed could be hazardous especially in term of risk of cancer, or (iii) ultimately cells could die if the number of damages is too high. The analysis of the genotoxicity of a compound/nanomaterial could be done by directly measuring the alteration of the DNA molecule or by indirectly measuring the expression of genes or activity of proteins known to be induced in case of DNA alterations. For example, studies of DNA fragmentation in macrophages, HT-29 human colorectal adenocarcinoma cells, A549 lung cancer cells, and 3T3-L1 embryonic fibroblasts exposed to ND indicate the absence of any DNA breaks.[32,33] By analyzing the location of a protein that specifically recognizes DNA double strand breaks (phospho-H2AX), it was shown that COOH-NDs of different sizes do not induce any genotoxicity on lung, liver, and renal cell lines even at high doses of ND exposure.[19] However, the study of ND by Xing et al. gave a different result, since the authors showed that incubation of embryonic stem cells with ND led to a slightly increased expression of DNA repair proteins (p53 and OGG-1), COOH-ND inducing a higher number of damages than the raw ND. Detonation NDs were found to be genotoxic on glioblastomas cell lines; however, the ND purity was not high (superior to 95%) and thus genotoxicity could be related to ND impurities (see Chapter 5).[34] Up to now, no sign of nuclear entry of ND was described in the literature; therefore, the possible DNA damages observed in a few works are likely caused by indirect mechanisms (i.e. through the generation of ROS). In a majority of studies, NDs are found to be non-genotoxic and, in case of genotoxicity, the DNA damages caused by either the COOH-ND or the raw ND are much less severe compared with MWNT.[35] In agreement with the absence of genotoxicity, the lack of ROS generation after cell exposure with detonation ND, raw-ND, COOH-ND, and $SO_3Na$-ND- of different sizes was shown.[19,25,26]

### 12.2.3 BIOLOGICAL EFFECT OF ND ON STEM CELLS

Stem cells are a unique cell population with the ability to undergo both self-renewal and differentiation with the potential to give rise to all cell lineages of an entire organism (if embryonal stem cells) or to give lineage in given tissues.[36] It has been shown that embryonal stem cells are highly sensitive to DNA damages, notably after exposure to UV irradiation[37] or multi-walled carbon nanotubes.[38] Regarding NDs, pristine detonation NDs were found to be non-cytotoxic on embryonic stem cells.[35] It was also published that COOH-ND (100–250 nm) (100 μg/ml for 4 h) do not interfere with the gene expression or with cell proliferation, and do not induce any cytotoxicity or DNA damage even after a long term culture of embryonic fibroblasts,[33] mesenchymal stem cells,[39] embryonal carcinoma stem cells,[40] and osteo-progenitor.[21] However, one paper reported that COOH-ND (of unknown purity) induced in embryonic stem cells an increase of the p53 protein expression (the protein guardian of the genome integrity) and of other proteins reflecting specifically the occurrence of DNA double strand breaks. Also indicative of a cellular stress, embryonic stem cells die or start to differentiate.[35] Again, it appears that even if DNA damages occurred after

COOH-ND exposure of stem cells, DNA damages were quite less severe than those caused by multi-walled carbon nanotubes.[38]

Nanomaterials emerge as a potential platform for tracking, targeting, understanding, and, if necessary, treating normal and cancer stem cells.[41]

## 12.3 INTERACTION OF ND WITH BLOOD CELLS OR BLOOD PROTEINS

For medical applications, one of the major routes of ND administration consists in intravenous injection. Consequently, it is of main importance to establish if ND could modify blood homeostasis or the function of blood proteins. Recently, many investigations of ND interactions with blood and blood components have been performed ex-vivo by mixing a given blood protein with ND and by measuring adsorption, structural modifications of proteins, and eventually by testing if the biological function of the given protein is altered or not. For example, using the small stable enzyme lysozyme as a test protein, it was shown by observing variation in the Fourier Transform Infrared spectra that the protein adsorption leads to observable changes in protein conformation, together with a significant decrease in the protein functional activity. Using this method, the adsorption of albumin, which accounts for over 60% of all plasma proteins, and of gamma-globin (immunoglobin) was observed; the adsorption being more important on 5-nm ND as compared to a 100-nm ND, but it occurred to a lesser extent as compared to lysozyme.[42] Using ND of average size of 75 nm, it was shown that the adsorption of gamma-globulin and of albumin is very low; however, in this work the consequence of this interaction on blood homeostasis was not studied.[43] Interestingly, Lin et al. studied the influence of 5 and 100 nm COOH-ND exposure on the oxygenation and de-oxygenation states of hemoglobin and on the micro-rheological properties of human red blood cells in vitro. It was found that cell viability was not altered and hemolysis did not exceed 2 to 4% as compared with hemolysis induced by Triton X-100. However, after interaction with ND, deformation and aggregation of red blood cells were observed.[44] Another important parameter for studying haematotoxicity concerns the coagulation. Effectively, when the blood is exposed to exogenous factors (like, for example, tissue damage), blood coagulation is activated and leads to generation of thrombin and fibrin clot that stops the bleeding. Results of a coagulation test, done after ex-vivo incubation of 5–10 nm ND with blood samples of healthy volunteers, indicate that ND do not induce any delay or modification of coagulation efficacy.[45]

It should be mentioned that Kumari et al. analyzed the acute effects of COOH-ND on platelet function (blood cells that play a central role in hemostasy and arterial thrombosis). They found that when administered intravenously in mice, ND induced widespread pulmonary thromboembolism.[46] These data are contrary to what is described in the literature, since overall ND do not seem to induce blood toxicity, but when we looked in detail at the physical and chemical ND characteristics, we noticed that ND preparation was not pure and probably contained traces of metals, and more importantly their surface appeared graphitized.[46]

Finally, in comparison with NDs, carbon nanotubes but not fullerenes (nC60) induce the aggregation of human blood platelets.[47,48] Overall, in vitro data suggests that NDs can be used effectively as a bio-label and drug delivery tool in ambient conditions, without complicating the blood's physiological conditions; however, controlling the blood cell morphology in the course of treatment could be necessary.[44]

## 12.4 FATE IN THE BODY AND IN VIVO TOXICITY OF ND

All in vitro tests enable us to screen for the cytotoxicity and genotoxicity and to understand the mechanisms of action of numerous NDs differing in size, coating, and functionalization, and to test NDs at different doses and over different periods of time. Indeed, these tests cannot exactly mimic what could happen when the whole organism is exposed to nanoparticles, and thus in vivo studies on animal models are quite necessary. Contamination of animals could be performed by inhalation, ingestion of drinking water or food additives, dermal penetration by skin contact, and injection of engineered nanomaterials. After introducing nanomaterials into the body, whatever the route of administration, their fate in terms of translocation, biodistribution, elimination, accumulation, and mechanisms of their interaction with living systems should be analyzed and understood in short and long term for estimating hazards and risks of NDs. For these in vivo studies, the critical point is that they require an exact quantification of ND for obtaining a balance of 100% between ND entrance, ND retention, and ND elimination. Scanning electron microscopy on a tissue extract and in vivo fluorescence imaging[15] are useful for qualitative ND detection, but are not adapted for quantitative measurements. Radionuclide tracer technique can be useful for the research in this field due to its easy detection, high sensitivity, good reliability, and the possibility to quantify. NDs were successfully labeled by chemical grafting of rhenium-188 ($^{188}$R),[49] iodide-125 ($^{125}$I),[50] and fludeoxyglucose ($^{18}$F).[51] However, in case of grafting, markers could be separated from the ND during the transit in animals and thus could be at the origin of many artefacts and of contradictory results. Girard et al. proposed a promising alternative by describing a method for synthetizing core tritium labeled ND,[52] opening the way for quantitative biodistribution and pharmacokinetics studies.

Up to now, no studies have focused on the biodistribution and toxicity of ND after inhalation because of the difficulty of controlling the physical and chemical properties of ND under conditions of inhalation. As an alternative, intratracheal instillation has become a widely applied technique for the investigation of the consequences of respiration of nanoparticles due to its simplicity and reproducibility. Using $^{188}$Re ND, an analysis of biodistribution after intratracheal instillation showed that ND distributed mainly in the spleen, liver, bones, and heart, in addition to the high retention in the lungs.[49] However, by electron microscopy observation, which is a less sensitive method, no ND were found in the mice's liver, spleen, and lymphatic node, even when the mice were instilled with a very high dose (20 mg/kg).[53] Yuan et al. hypothesize that most of the ND entering lungs were rapidly engulfed by macrophages as previously shown[54] and eliminated by mucociliary transport since it becomes difficult to discern ND at 90 days post-exposure.[53] Although the data are convincing, the absence of ND translocation from the lungs to the whole organism via the systemic circulation needs to be based on quantitative measurement.

The distribution of ND was also studied after intravenous injection $^{125}I$-labeled raw ND of 50 nm in mice.[50] ND predominantly accumulated in the liver, spleen, and lungs. About 60% of the injected ND was found in the liver 30 min after injection and the level remained constant over one month. Data were confirmed by Raman spectroscopy, NDs were entrapped in the lungs and liver for a long time, and NDs were barely detected in urine and feces.[50] Rojas et al. labeled NH2-ND of 350 nm with another radionuclide $^{18}F$ and, by positron emission tomography (PET), they observed the distribution of the intravenous injected ND in mice. Their results showed that NDs were mainly distributed in the lungs, spleen, and liver, and excreted into the urinary tract. Their research further indicated that the addition of surfactant agents did not significantly change this distribution pattern.[51] Up to now, because of the absence of a possible quantification of ND, studies on the behavior and biokinetics of NDs in the whole organism were very fragmented. However, these studies are essential since understanding, according to the route of entrance, in which organs NDs are located, how long they stay, and how they are eliminated will help us to identify the organs in which one should look for potentially hazardous biological effects.

Regarding no more the location of NDs but their toxicity, a significant pulmonary toxicity was found after intratracheal instillation of 4-nm detonation ND and 50-nm HPHT-ND in mice that with no evidence of oxidative stress.[53] The acute toxicity of detonation ND administrated by intratracheal instillation (0.8, 4, and 20 mg/kg) showed a dose-dependent toxicity of ND in the lungs, liver, kidneys, and blood.[49] Specifically in the lungs, inflammatory responses characterized by neutrophils and foamy alveolar macrophage accumulation, with progressive apparition of fibrosis, were observed.[55] It should be mentioned that in this work ND purity was over 95% and traces of metals such as Fe (0.1%), Al (0.1%), and Ba, Mn, and Ni (0.001% for each), known to be cytotoxic through ROS generation, were present. However, again by making a comparison with carbon nanotubes and other carbon related nanoparticles, the respiratory toxicity of NDs, even for the highest dose, is much smaller.[56–58]

Yuan et al. used $^{125}I$ labeled ND 50 nm for intravenous injection in mice at 30 mg/kg and observed no sign of abnormality in terms of survival, weight, and lethargy in mice 28 days after injection.[50] Intravenously injected COOH-ND had either an excellent hemocompatibility with negligible hemolytic and thrombogenic activities and with no sign of inflammation[59] or induced huge toxicity through activation of blood platelets and thromboembolism.[46] However, looking precisely at the treatment of NDs and their physical and chemical characterization, Kumari et al. analyzed not NDs but graphitized particles.[60]

After intraperitoneal injection of ND, for instance, Bakowicz et al. showed in rats that no immune responses were observed and very few inflammatory cells were found in liver, kidneys, and peritoneum.[61] After intraperitoneal injection of 100-nm NV fluorescent in mice over 5 months, fluorescent ND particles can be drained from the injection sites by macrophages and selectively accumulated in the axillary lymph nodes of the treated mice. Observation of animals and tissue analysis reveal that NDs are non-toxic even when a large quantity, up to 75 mg/kg body weight, of the particles was administered to the living animals.[62]

In another study, Puzyr et al. further observed no inflammatory symptoms in mice after subcutaneous exposure to ND for three months,[63] the particles were localized in the subcutaneous space at the injection point. However, when the same NDs were intravenously injected into a rabbit, a number of blood biochemical parameters were affected, suggesting that exposure routes could influence the toxicity effects of ND observed in animals, perhaps via the first proteins that entered into contact with ND.

Oral administration of ND hydrosols in mice food for a period of 6 months did not cause death or affected the growth or the internal organs (liver, lungs, heart, kidneys, and pancreas), weight dynamics, and neither did it affect the mouse's reproductive ability.[1]

Finally, concerning intravenous injection, the main route of administration for medical application, intravenously injected COOH-ND had either an excellent hemocompatibility with negligible hemolytic and thrombogenic activities and with no sign of inflammation[59] or induced huge toxicity through activation of blood platelets and thromboembolism; however, as mentioned above, Kumari et al. used uncharacterized and probably graphitized NDs.[46]

## 12.5 ND ON BACTERIA AND FUNGI

Depending on their surface composition, NDs kill gram-positive and -negative bacteria rapidly and efficiently. The presence of partially oxidized and negatively charged surfaces, specifically those containing anhydric acid, yields higher antibacterial activity.[64] Visualization of the morphologic interaction between the nanoparticles and *Staphylococcus aureus* (bacteria) and *Candida albicans* (fungi) showed that NDs, which are dielectrics and exhibit a positive zeta potential, uniformly surrounded the microorganisms, without causing visible damage and destruction of cells. On the contrary, all metal nanoparticles (Ag, Au) with negative zeta potential have cell damaging properties.[65] Lin et al. have shown that both 5 and 100 nm NDs decrease the viability without altering the capacity of division of the protozoa *Thermus thermophile* and *Paramecium caudatum*. The 5 nm NDs are more toxic than the 100 nm; and non-COOH-NDs are more toxic than COOH-NDs.[66] Ultrafine detonation 5 nm NDs attached to gram-negative *Escherichia coli* bacterial cell wall surface and provoked its destruction.[67] The ND interaction with bacteria might be physical in nature, or due to highly reactive surface that could generate ROS and thus alter cell wall surface.[67] However, the bactericidal properties of NDs could be controlled by covering their surface with proteins.[67] Overall, it seems that NDs with negative zeta potential are more toxic for microorganisms compared to animal/human cells; however, as for human cells, COOH-NDs, which have a proper surface, seem to be quite less toxic.

## 12.6 CONCLUSIONS AND CHALLENGING ISSUES

The absence of obvious acute adverse effects of ND on in vitro and on experimental animals indeed should be confirmed, but published data suggest that nanoparticles can form a basis for new nanotechnologies and methods for medicine. However, NDs will be used in the medical fields, of course, if the medical benefit is proven, but also, and this is more complicated, if they do not present short and long term risks for human health. For this, it is first of all essential to work with labeled NDs, especially tritium core-labeled NDs,[52] for targeting toxicological research on organs accumulating ND that necessitates robust quantitative analysis of ND behavior in the whole organism. These studies should be conducted as a function of the predicted medical applications since depending on the mode of ND administration (inhalation, intravenous injection, etc.). The comportment of a given ND is probably greatly dependent on the proteins interacting with it. Once the retention organs have been identified, comprehensive analyzes of hazards in terms of cytotoxicity, genotoxicity, immunotoxicity, etc., should be addressed on in vitro models and then in vivo. However, quantification of ND uptake for in vitro experiments is also very important for comparing the ND sensitivity of different cell lines and types or the toxicity of different ND, and indeed for concluding that a given ND is not toxic. Until the availability of radiolabeled ND, flow cytometry (a technique available in numerous biology laboratories) was well adapted for studying labeled free ND cellular uptake. Effectively, it was shown that the internalization of 100 nm ND, of 20, 50, and 100 nm COOH-ND, and of detonation ND markedly change the light scattering characteristics of the cells. It results in a dramatic increase in the side scatter (SSC) that reflects cell granulometry with a concomitant decrease in the forward scatter that reflects cell size. An example of the usefulness of flow cytometry for measuring ND entrance into the cells is given in Fig. 12.2. The enhancement in SSC makes light scattering a highly competitive parameter for the identification and recovery of cells that contain ND. It could help to compare uptake in different cell lines, to discriminate living and dead cells with and without ND.[5,19,21]

For easy comparison between cell lines' responses and different ND, the metric for qualifying ND exposure is crucial. As for chemical toxicity, people often used the mass of ND per milliliter of medium for qualifying ND/nanomaterial exposure but, in fact, it is not sufficient. The exposure can be compared only if the dose is also given in mass per surface occupied by the adherent cells ($\mu g/cm^2$). Another major challenge concerns the choice of ND concentrations for conducting in vitro and in vivo experiments, knowing that in vivo doses cannot be converted directly to in vitro concentrations. For example, for $TiO_2$ nanoparticles, overload effects in rats were reported at 10 $mg/m^3$, and this corresponds to lung burden amounts of approximately 3 $\mu g$ $NMs/cm^2$ of lung surface. Hence, a range of testing doses from 0.1 to 10 $\mu g/cm^2$

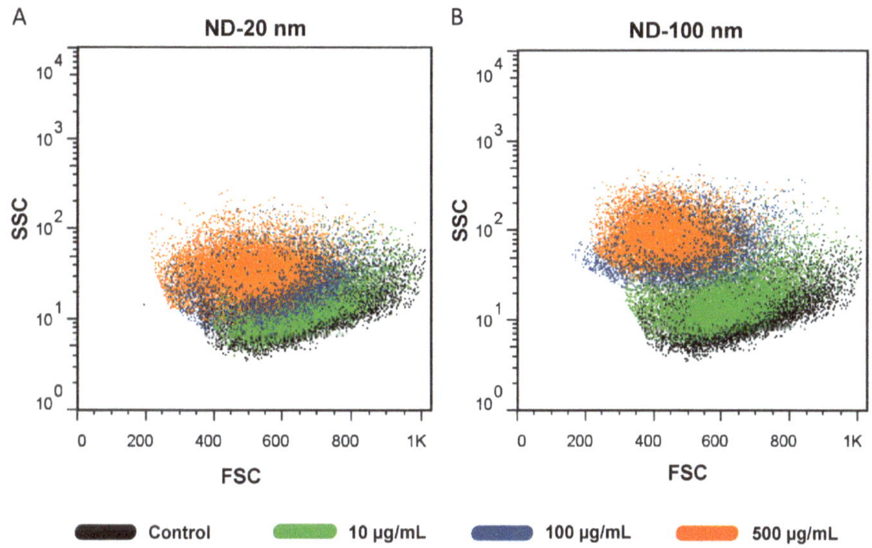


**FIGURE 12.2**

Flow cytometry for estimating COOH-ND-cell interactions. Forward scatter (FSC) (cell size) versus side scatter (SSC) (cell granulometry) analysis (a) is presented for a human renal cancer cell line (Caki-1) exposed to carboxylated HPHT ND-20 nm (A) and to ND-100 nm (B). Flow cytometry sensitivity allowed detection of Caki-1 cell-ND interactions from 10 μg/ml (green) compared with non-exposed cells (black). The SCC signal increased with increasing concentration (100 μg/ml in blue and 500 μg/ml in orange)[19]

might provide valuable information regarding the dose dependency of particle toxicity.[18]

Another important point concerns the physical and chemical characterization of ND, their purity, graphitization, traces of metals, and their size and zeta potential not only in water but also in cell culture medium.[5,19] As an illustration, it was shown that the most toxic NDs were detonation NDs with the smallest size and slight content of non-diamond carbon, while NDs of larger size and free from impurities had no significant influence on cell proliferation and morphology, however, with great differences according to the cell lines, the mesenchymal stem cells being sensitive, while the osteosarcoma tumoral cells being quite insensitive.[68] The different cell lines are cultivated in different culture media and consequently differences in protein adsorption on ND could influence the nanoparticles entrance and biological activity as it was previously shown for single-walled carbon nanotubes and amorphous silica.[69]

The current recommendation in the field of toxicology of nanoparticles is that each nanoparticle should be considered as singular. It means that no extrapolation of toxic effects of given nanomaterial cannot be made to another nanomaterial, even if it differs from the first one by only one parameter, such as size. Having in mind the huge potential on ND for medical applications (see Chapter 16), and thus the huge number of different NDs that will be candidates for medical applications, owing to the differences in synthesis, size, charge, purification, coating, functionalization, grafting (see Chapters 2, 8, and 14), and knowing that NDs could be alone or associated with other nanomaterials or partners of complex architectures, the work for hazard assessment of NDs is tremendous. Fortunately, with the growing number of nanomaterials, rapid and robust approaches for high throughput testing of nanomaterial safety have recently emerged, permitting us to manage this challenge.[70–74]

## REFERENCES


1. Schrand, A.; Ciftan Hens, S.; Shenderova, O. Nanodiamond Particles: Properties and Perspectives for Bioapplications. *Crit. Rev. Solid State Mater. Sci.* **2009**, *34* (1–2), 18–74.
2. Moore, L.; Chow, E. K.; Osawa, E.; Bishop, J. M.; Ho, D. Diamond-Lipid Hybrids Enhance Chemotherapeutic Tolerance and Mediate Tumor Regression. *Adv. Mater.* **2013**, *25* (26), 3532–3541.
3. Chow, E. K.; Zhang, X. Q.; Chen, M.; Lam, R.; Robinson, E.; Huang, H.; Schaffer, D.; Osawa, E.; Goga, A.; Ho, D. Nanodiamond Therapeutic Delivery Agents Mediate Enhanced Chemoresistant Tumor Treatment. *Sci. Transl. Med.* **2011**, *3* (73), 73ra21.
4. Lam, R.; Ho, D. Nanodiamonds as Vehicles for Systemic and Localized Drug Delivery. *Expert Opinion Drug Deliv.* **2009**, *6* (9), 883–895.
5. Grall, R.; Girard, H.; Saad, L.; Petit, T.; Gesset, C.; Combis-Schlumberger, M.; Paget, V.; Delic, J.; Arnault, J. C.; Chevillard, S. Impairing the Radioresistance of Cancer Cells by Hydrogenated Nanodiamonds. *Biomaterials* **2015**, *61*, 290–298.
6. Yu, S. J.; Kang, M. W.; Chang, H. C.; Chen, K. M.; Yu, Y. C. Bright Fluorescent Nanodiamonds: No Photobleaching and Low Cytotoxicity. *J. Am. Chem. Soc.* **2005**, *127* (50), 17604–17605.
7. Hui, Y.; Chang, H. Recent Developments and Applications of Nanodiamonds as Versatile Bioimaging Agents. *J. Chin. Chem. Soc.* **2014**, *61*, 67–76.
8. Puzyr, A.; Pozdnyakova, I.; Bondar, V. Design of a Luminescent Biochip With Nanodiamonds and Bacterial Luciferase. *Phys. Solid State* **2004**, *46* (4), 761–763.
9. Fu, C. C.; Lee, H. Y.; Chen, K.; Lim, T. S.; Wu, H. Y.; Lin, P. K.; Wei, P. K.; Tsao, P. H.; Chang, H. C.; Fann, W. Characterization and Application of Single Fluorescent Nanodiamonds as Cellular Biomarkers. *Proc. Natl. Acad. Sci. USA* **2007**, *104* (3), 727–732.
10. Chao, J. I.; Perevedentseva, E.; Chung, P. H.; Liu, K. K.; Cheng, C. Y.; Chang, C. C.; Cheng, C. L. Nanometer-Sized Diamond Particle as a Probe for Biolabeling. *Biophys. J.* **2007**, *93* (6), 2199–2208.
11. Cheng, C.; Perevedentseva, E.; Tu, J.; Chung, H.; Chenga, C. Direct and In Vitro Observation of Growth Hormone Receptor Molecules in A549 Human Lung Epithelial Cells by Nanodiamond Labeling. *Appl. Phys. Lett.* **2007**, *90*, 163903.
12. Gruber, A.; Drabenstedt, A.; Tietz, C.; Fleury, L.; Wrachtrup, J.; von Borczyskowski, C. Scanning Confocal Optical Microscopy and Magnetic Resonance on Single Defect Centers. *Science* **1997**, *276* (5321), 2012–2014.

13. Barnard, A. S. Diamond Standard in Diagnostics: Nanodiamond Biolabels Make Their Mark. *Analyst* **2009**, *134* (9), 1751–1764.
14. Eidi, H.; David, M. O.; Crepeaux, G.; Henry, L.; Joshi, V.; Berger, M. H.; Sennour, M.; Cadusseau, J.; Gherardi, R. K.; Curmi, P. A. Fluorescent Nanodiamonds as a Relevant Tag for the Assessment of Alum Adjuvant Particle Biodisposition. *BMC Med.* **2015**, *13*, 144.
15. Igarashi, R.; Yoshinari, Y.; Yokota, H.; Sugi, T.; Sugihara, F.; Ikeda, K.; Sumiya, H.; Tsuji, S.; Mori, I.; Tochio, H.; Harada, Y.; Shirakawa, M. Real-Time Background-Free Selective Imaging of Fluorescent Nanodiamonds In Vivo. *Nano Lett.* **2012**, *12* (11), 5726–5732.
16. Huang, L. C.; Chang, H. C. Adsorption and Immobilization of Cytochrome c on Nanodiamonds. *Langmuir* **2004**, *20* (14), 5879–5884.
17. Krueger, A. New Carbon Materials: Biological Applications of Functionalized Nanodiamond Materials. *Chemistry* **2008**, *14* (5), 1382–1390.
18. Kroll, A.; Dierker, C.; Rommel, C.; Hahn, D.; Wohlleben, W.; Schulze-Isfort, C.; Gobbert, C.; Voetz, M.; Hardinghaus, F.; Schnekenburger, J. Cytotoxicity Screening of 23 Engineered Nanomaterials Using a Test Matrix of Ten Cell Lines and Three Different Assays. *Part. Fibre Toxicol.* **2011**, *8*, 9.
19. Paget, V.; Sergent, J. A.; Grall, R.; Altmeyer-Morel, S.; Girard, H. A.; Petit, T.; Gesset, C.; Mermoux, M.; Bergonzo, P.; Arnault, J. C.; Chevillard, S. Carboxylated Nanodiamonds Are Neither Cytotoxic Nor Genotoxic on Liver, Kidney, Intestine and Lung Human Cell Lines. *Nanotoxicology* **2014**, *8* (Suppl 1), 46–56.
20. Liu, K.; Cheng, C.; Chang, C.; Chao, J. Biocompatible and Detectable Carboxylated Nanodiamond on Human Cell. *Nanotechnology* **2007**, *18*, 325102.
21. Fang, C. Y.; Vaijayanthimala, V.; Cheng, C. A.; Yeh, S. H.; Chang, C. F.; Li, C. L.; Chang, H. C. The Exocytosis of Fluorescent Nanodiamond and Its Use as a Long-Term Cell Tracker. *Small* **2011**, *7* (23), 3363–3370.
22. Zhu, Y.; Li, J.; Li, W.; Zhang, Y.; Yang, X.; Chen, N.; Sun, Y.; Zhao, Y.; Fan, C.; Huang, Q. The Biocompatibility of Nanodiamonds and Their Application in Drug Delivery Systems. *Theranostics* **2012**, *2* (3), 302–312.
23. Marcon, L.; Riquet, F.; Vicogne, D.; Szunerits, S.; Bodartb, J.; Boukherrouba, R. Cellular and In Vivo Toxicity of Functionalized Nanodiamond in Xenopus Embryos. *J. Mater. Chem.* **2010**, *20*, 8064–8069.
24. Moore, L.; Grobarova, V.; Shen, H.; Man, H. B.; Micova, J.; Ledvina, M.; Stursa, J.; Nesladek, M.; Fiserova, A.; Ho, D. Comprehensive Interrogation of the Cellular Response to Fluorescent, Detonation and Functionalized Nanodiamonds. *Nanoscale* **2014**, *6* (20), 11712–11721.
25. Schrand, A. M.; Huang, H.; Carlson, C.; Schlager, J. J.; Omacr Sawa, E.; Hussain, S. M.; Dai, L. Are Diamond Nanoparticles Cytotoxic? *J. Phys. Chem. B* **2007**, *111* (1), 2–7.
26. Schrand, A.; Dai, L.; Schlager, J.; Hussain, S.; Osawa, E. Differential Biocompatibility of Carbon Nanotubes and Nanodiamonds. *Diam. Relat. Mater.* **2007**, *16*, 2118–2123.
27. Zhu, Y.; Li, J.; Li, W.; Zhang, Y.; Yang, X.; Chen, N.; Sun, Y.; Zhao, Y.; Fan, C.; Huang, Q. The Biocompatibility of Nanodiamonds and Their Application in Drug Delivery Systems. *Theranostics* **2012**, *2* (3), 302–312.
28. Zhang, X.; Hu, W.; Li, J.; Taob, L.; Wei, Y. A Comparative Study of Cellular Uptake and Cytotoxicity of Multi-walled Carbon Nanotubes, Graphene Oxide, and Nanodiamond. *Toxicol. Res.* **2012**, *1*, 62.
29. Solarska-Sciuk, K.; Gajewska, A.; Skolimowski, J.; Gajek, A.; Bartosz, G. Cellular Redox Homeostasis in Endothelial Cells Treated With Nonmodified and Fenton-Modified Nanodiamond Powders. *Biotechnol. Appl. Biochem.* **2014**, *61* (5), 593–602.

30. Solarska-Sciuk, K.; Gajewska, A.; Glinska, S.; Michlewska, S.; Balcerzak, L.; Jamrozik, A.; Skolimowski, J.; Burda, K.; Bartosz, G. Effect of Functionalized and Nonfunctionalized Nanodiamond on the Morphology and Activities of Antioxidant Enzymes of Lung Epithelial Cells (A549). *Chem.-Biol. Interact.* **2014**, *222C*, 135–147.
31. Thomas, V.; Halloran, B. A.; Ambalavanan, N.; Catledge, S. A.; Vohra, Y. K. In Vitro Studies on the Effect of Particle Size on Macrophage Responses to Nanodiamond Wear Debris. *Acta Biomater.* **2012**, *8* (5), 1939–1947.
32. Huang, H.; Pierstorff, E.; Osawa, E.; Ho, D. Active Nanodiamond Hydrogels for Chemotherapeutic Delivery. *Nano Lett.* **2007**, *7* (11), 3305–3314.
33. Liu, K. K.; Wang, C. C.; Cheng, C. L.; Chao, J. I. Endocytic Carboxylated Nanodiamond for the Labeling and Tracking of Cell Division and Differentiation in Cancer and Stem Cells. *Biomaterials* **2009**, *30* (26), 4249–4259.
34. Hinzmann, M.; Jaworski, S.; Kutwin, M.; Jagiełło, J.; Koziński, R.; Wierzbicki, M.; Grodzik, M.; Lipińska, L.; Sawosz, E.; Chwalibog, A. Nanoparticles Containing Allotropes of Carbon Have Genotoxic Effects on Glioblastoma Multiforme Cells. *Int. J. Nanomed.* **2014**, *9*, 2409–2417.
35. Xing, Y.; Xiong, W.; Zhu, L.; Osawa, E.; Hussin, S.; Dai, L. DNA Damage in Embryonic Stem Cells Caused by Nanodiamonds. *ACS Nano* **2011**, *5* (3), 2376–2384.
36. Odorico, J. S.; Kaufman, D. S.; Thomson, J. A. Multilineage Differentiation From Human Embryonic Stem Cell Lines. *Stem Cells* **2001**, *19* (3), 193–204.
37. Lin, T.; Chao, C.; Saito, S.; Mazur, S. J.; Murphy, M. E.; Appella, E.; Xu, Y. p53 Induces Differentiation of Mouse Embryonic Stem Cells by Suppressing Nanog Expression. *Nat. Cell Biol.* **2005**, *7* (2), 165–171.
38. Zhu, L.; Chang, D. W.; Dai, L.; Hong, Y. DNA Damage Induced by Multiwalled Carbon Nanotubes in Mouse Embryonic Stem Cells. *Nano Lett.* **2007**, *7* (12), 3592–3597.
39. Blaber, S. P.; Hill, C. J.; Webster, R. A.; Say, J. M.; Brown, L. J.; Wang, S. C.; Vesey, G.; Herbert, B. R. Effect of Labeling With Iron Oxide Particles or Nanodiamonds on the Functionality of Adipose-Derived Mesenchymal Stem Cells. *PLoS ONE* **2013**, *8* (1), e52997.
40. Hsu, T. C.; Liu, K. K.; Chang, H. C.; Hwang, E.; Chao, J. I. Labeling of Neuronal Differentiation and Neuron Cells With Biocompatible Fluorescent Nanodiamonds. *Sci. Rep.* **2014**, *4*, 5004.
41. Ferreira, L.; Karp, J. M.; Nobre, L.; Langer, R. New Opportunities: The Use of Nanotechnologies to Manipulate and Track Stem Cells. *Cell Stem Cell* **2008**, *3* (2), 136–146.
42. Perevedentseva, E.; Su, F.; Su, T.; Lin, Y.; Cheng, C.; Karmenyan, A.; Priezzhev, A.; Lugovtsov, A. Laser-Optical Investigation of the Effect of Diamond Nanoparticles on the Structure and Functional Properties of Proteins. *Quantum Electron.* **2010**, *40* (12), 1089–1093.
43. Wasdo, S.; Barber, D.; Denslow, N.; Powers, K. W.; Palazuelos, M.; Stevens, S.; Moudgil, B.; Roberts, S. Differential Binding of Serum Proteins to Nanoparticles. *Int. J. Nanotechnol.* **2008**, *5* (1), 92–115.
44. Lin, Y. C.; Tsai, L. W.; Perevedentseva, E.; Chang, H. H.; Lin, C. H.; Sun, D. S.; Lugovtsov, A. E.; Priezzhev, A.; Mona, J.; Cheng, C. L. The Influence of Nanodiamond on the Oxygenation States and Micro Rheological Properties of Human Red Blood Cells In Vitro. *J. Biomed. Opt.* **2012**, *17* (10), 101512.
45. Mona, J.; Kuo, C.; Priezzhev, A.; Cheng, C. Adsorption of Human Blood Plasma on Nanodiamond and Its Influence on Activated Partial Thromboplastin Time. *Diam. Relat. Mater.* **2013**, *39*, 73–77.

46. Kumari, S.; Singh, M. K.; Singh, S. K.; Gracio, J. J.; Dash, D. Nanodiamonds Activate Blood Platelets and Induce Thromboembolism. *Nanomedicine* **2014**, *9* (3), 427–440.
47. Lacerda, S. H.; Semberova, J.; Holada, K.; Simakova, O.; Hudson, S. D.; Simak, J. Carbon Nanotubes Activate Store-Operated Calcium Entry in Human Blood Platelets. *ACS Nano* **2011**, *5* (7), 5808–5813.
48. Semberova, J.; De Paoli Lacerda, S. H.; Simakova, O.; Holada, K.; Gelderman, M. P.; Simak, J. Carbon Nanotubes Activate Blood Platelets by Inducing Extracellular $Ca^{2+}$ Influx Sensitive to Calcium Entry Inhibitors. *Nano Lett.* **2009**, *9* (9), 3312–3317.
49. Zhang, X.; Yin, J.; Kang, C.; Li, J.; Zhu, Y.; Li, W.; Huang, Q.; Zhu, Z. Biodistribution and Toxicity of Nanodiamonds in Mice After Intratracheal Instillation. *Toxicol. Lett.* **2010**, *198* (2), 237–243.
50. Yuan, Y.; Yuanwei, C.; Jia-Hui, L.; Haifang, W.; Yuanfang, L. Biodistribution and Fate of Nanodiamonds In Vivo. *Diam. Relat. Mater.* **2009**, *18* (1), 95–100.
51. Rojas, S.; Gispert, J. D.; Martin, R.; Abad, S.; Menchon, C.; Pareto, D.; Victor, V. M.; Alvaro, M.; Garcia, H.; Herance, J. R. Biodistribution of Amino-Functionalized Diamond Nanoparticles. In Vivo Studies Based on 18F Radionuclide Emission. *ACS Nano* **2011**, *5* (7), 5552–5559.
52. Girard, H. A.; El-Kharbachi, A.; Garcia-Argote, S.; Petit, T.; Bergonzo, P.; Rousseau, B.; Arnault, J. C. Tritium Labeling of Detonation Nanodiamonds. *Chem. Commun.* **2014**, *50* (22), 2916–2918.
53. Yuan, Y.; Wang, X.; Jia, G.; Jia-Hui, L.; Wang, T.; Gud, Y.; Yanga, S.; Zhenb, S.; Wang, H.; Liua, Y. Pulmonary Toxicity and Translocation of Nanodiamonds in Mice. *Diam. Relat. Mater.* **2010**, *19* (4), 291–299.
54. Geiser, M. Update on Macrophage Clearance of Inhaled Micro- and Nanoparticles. *J. Aerosol Med. Pulm. Drug Deliv.* **2010**, *23* (4), 207–217.
55. Li, J.; Zhu, Y.; Li, W.; Zhang, X.; Peng, Y.; Huang, Q. Nanodiamonds as Intracellular Transporters of Chemotherapeutic Drug. *Biomaterials* **2010**, *31* (32), 8410–8418.
56. Lam, C. W.; James, J. T.; McCluskey, R.; Hunter, R. L. Pulmonary Toxicity of Single-Wall Carbon Nanotubes in Mice 7 and 90 Days After Intratracheal Instillation. *Toxicol. Sci.* **2004**, *77* (1), 126–134.
57. Warheit, D. B.; Laurence, B. R.; Reed, K. L.; Roach, D. H.; Reynolds, G. A.; Webb, T. R. Comparative Pulmonary Toxicity Assessment of Single-Wall Carbon Nanotubes in Rats. *Toxicol. Sci.* **2004**, *77* (1), 117–125.
58. Muller, J.; Huaux, F.; Moreau, N.; Misson, P.; Heilier, J. F.; Delos, M.; Arras, M.; Fonseca, A.; Nagy, J. B.; Lison, D. Respiratory Toxicity of Multi-wall Carbon Nanotubes. *Toxicol. Appl. Pharmacol.* **2005**, *207* (3), 221–231.
59. Li, H.-C.; Hsieh, F.-J.; Chen, C.-P.; Chang, M.-Y.; Hsieh, P. C. H.; Chen, C.-C.; Hung, S.-U.; Wu, C.-C.; Chang, H.-C. The Hemocompatibility of Oxidized Diamond Nanocrystals for Biomedical Applications. *Diam. Relat. Mater.* **2007**, *16*, 2124.
60. Petit, T.; Arnault, J.; Girard, H.; Sennour, M.; Bergonzo, P. Early Stages of Surface Graphitization on Nanodiamond Probed by X-Ray Photoelectron Spectrosco. *Phys. Rev. B* **2011**, *84*, 233407.
61. Bąkowicz, K.; Mitura, S. Biocompatibility of NCD. *J. Wide Bandgap Mater.* **2002**, *9* (9), 261–272.
62. Vaijayanthimala, V.; Cheng, P. Y.; Yeh, S. H.; Liu, K. K.; Hsiao, C. H.; Chao, J. I.; Chang, H. C. The Long-Term Stability and Biocompatibility of Fluorescent Nanodiamond as an In Vivo Contrast Agent. *Biomaterials* **2012**, *33* (31), 7794–7802.

63. Puzyr, A.; Baron, A.; Purtov, K.; Bortnikov, E.; Skobelev, N.; Mogilnay, O.; Bondara, V. Nanodiamonds With Novel Properties: A Biological Study. *Diam. Relat. Mater.* **2007**, *16*, 2124–2128.
64. Wehling, J.; Dringen, R.; Zare, R. N.; Maas, M.; Rezwan, K. Bactericidal Activity of Partially Oxidized Nanodiamonds. *ACS Nano* **2014**, *8* (6), 6475–6483.
65. Chwalibog, A.; Sawosz, E.; Hotowy, A.; Szeliga, J.; Mitura, S.; Mitura, K.; Grodzik, M.; Orlowski, P.; Sokolowska, A. Visualization of Interaction Between Inorganic Nanoparticles and Bacteria or Fungi. *Int. J. Nanomed.* **2010**, *5*, 1085–1094.
66. Lin, Y. C.; Perevedentseva, E.; Tsai, L. W.; Wu, K. T.; Cheng, C. L. Nanodiamond for Intracellular Imaging in the Microorganisms In Vivo. *J. Biophotonics* **2012**, *5* (11–12), 838–847.
67. Chatterjee, A.; Perevedentseva, E.; Jani, M.; Cheng, C. Y.; Ye, Y. S.; Chung, P. H.; Cheng, C. L. Antibacterial Effect of Ultrafine Nanodiamond Against Gram-Negative Bacteria Escherichia coli. *J. Biomed. Opt.* **2015**, *20* (5), 051014.
68. Keremidarska, M.; Ganeva, A.; Mitev, D.; Hikov, T.; Presker, R.; Pramatarova, L.; Krasteva, N. Comparative Study of Cytotoxicity of Detonation Nanodiamond Particles With an Osteosarcoma Cell Line and Primary Mesenchymal Stem Cells. *Biotechnol. Biotechnol. Equip.* **2014**, *28* (4), 733–739.
69. Dutta, D.; Sundaram, S. K.; Teeguarden, J. G.; Riley, B. J.; Fifield, L. S.; Jacobs, J. M.; Addleman, S. R.; Kaysen, G. A.; Moudgil, B. M.; Weber, T. J. Adsorbed Proteins Influence the Biological Activity and Molecular Targeting of Nanomaterials. *Toxicol. Sci.* **2007**, *100* (1), 303–315.
70. Lin, S.; Zhao, Y.; Xia, T.; Meng, H.; Ji, Z.; Liu, R.; George, S.; Xiong, S.; Wang, X.; Zhang, H.; Pokhrel, S.; Madler, L.; Damoiseaux, R.; Nel, A. E. High Content Screening in Zebrafish Speeds up Hazard Ranking of Transition Metal Oxide Nanoparticles. *ACS Nano* **2011**, *5* (9), 7284–7295.
71. Anguissola, S.; Garry, D.; Salvati, A.; O'Brien, P. J.; Dawson, K. A. High Content Analysis Provides Mechanistic Insights on the Pathways of Toxicity Induced by Amine-Modified Polystyrene Nanoparticles. *PLoS ONE* **2014**, *9* (9), e108025.
72. Harris, G.; Palosaari, T.; Magdolenova, Z.; Mennecozzi, M.; Gineste, J. M.; Saavedra, L.; Milcamps, A.; Huk, A.; Collins, A. R.; Dusinska, M.; Whelan, M. Iron Oxide Nanoparticle Toxicity Testing Using High-Throughput Analysis and High-Content Imaging. *Nanotoxicology* **2015**, *9* (Suppl 1), 87–94.
73. Jan, E.; Byrne, S. J.; Cuddihy, M.; Davies, A. M.; Volkov, Y.; Gun'ko, Y. K.; Kotov, N. A. High-Content Screening as a Universal Tool for Fingerprinting of Cytotoxicity of Nanoparticles. *ACS Nano* **2008**, *2* (5), 928–938.
74. Collins, A. R.; Annangi, B.; Rubio, L.; Marcos, R.; Dorn, M.; Merker, C.; Estrela-Lopis, I.; Cimpan, M. R.; Ibrahim, M.; Cimpan, E.; Ostermann, M.; Sauter, A.; Yamani, N. E.; Shaposhnikov, S.; Chevillard, S.; Paget, V.; Grall, R.; Delic, J.; de-Cerio, F. G.; Suarez-Merino, B.; Fessard, V.; Hogeveen, K. N.; Fjellsbø, L. M.; Pran, E. R.; Brzicova, T.; Topinka, J.; Silva, M. J.; Leite, P. E.; Ribeiro, A. R.; Granjeiro, J. M.; Grafström, R.; Prina-Mello, A.; Dusinska, M. High throughput toxicity screening and intracellular detection of nanomaterials. *Wiley Interdiscip. Rev. Nanomed. Nanobiotechnol.* **2016 Jun 7**. http://dx.doi.org/10.1002/wnan.1413 [Epub ahead of print]. PMID: 27273980.

CHAPTER

# Nanodiamonds embedded in shells

# 13

**Jitka Neburkova*,†, Jan Vavra*, Helena Raabova*,‡, Goutam Pramanik*, Jan Havlik*,§, Petr Cigler***

*Institute of Organic Chemistry and Biochemistry, v.v.i., Academy of Sciences of the Czech Republic, Prague, Czech Republic* First Faculty of Medicine, Charles University, Prague, Czech Republic† Department of Inorganic Chemistry, University of Chemistry and Technology Prague, Prague, Czech Republic‡ Faculty of Science, Charles University, Prague, Czech Republic§*

## Contents



## 13.1 INTRODUCTION

Nanodiamonds (NDs) are a diverse group of carbon nanomaterials with a diamond core that contains various lattice impurities (see Chapter 5) and can be terminated with a vast number of chemical groups (see Chapter 8). Since the late 1990s, NDs have been increasingly studied as a unique material for applications ranging from materials science and physics to biology and medicine. Other chapters of this book focus on summarizing the main features of NDs, and a number of recent review articles have covered topics including ND properties and applications[1–3]; photophysics[4]; surface chemistry[5–7]; and the use of NDs in bioimaging,[8–10] drug delivery,[11] and nanoscale medicine.[12]

Here, we focus on a specific area of ND modification based on the creation of compact shell structures on the ND surface. The general theme of these approaches is the creation of a new ND interface that converts the diamond cores into fundamentally different, hybrid nanoparticles (NPs). The shells are thick enough (typically more than 1 nm) to avoid the characteristic ND atomic/molecular interactions at the interface. The new NPs thus outwardly appear to be chemical structures different from NDs, while the physical and spectroscopic properties of the diamond core [such as fluorescence from color point defects – nitrogen vacancy (NV) and silicon vacancy (SiV) centers] can either be modulated by the presence of the shell or kept intact.

Coating of NDs using the shells emerged from a general chemical approach that has been studied extensively over the past two decades: pathways leading to various ND chemical functionalizations via surface chemical transformation have been described for a large number of chemical groups and ND types.[5–7] Nevertheless, the surface chemistry of NDs is typically not uniform, and due to low reactivity and poor steric accessibility, monofunctional terminations of NDs are extremely difficult to achieve. Both unmodified and chemically modified NDs also show undesirable attributes typical for other colloids of inorganic nanocrystals, such as a tendency to aggregate in the presence of salts, poor solubility in most organic solvents, and strong adsorption of biomolecules.[7] Moreover, NDs are heterogeneous in shape,[13] which can be problematic because shape can have a strong effect on their biological behavior and performance.[14,15]

Creation of ND particles capped with various shell structures can overcome some of these issues and limitations. This chapter aims to present an overview of current research on NDs embedded in shells, show the available synthetic approaches to these core–shell hybrid NPs, and assess their properties and emerging applications. Composite materials in which NDs are embedded in continuous macroscopic or microscopic phases (see Chapter 15), as well as NDs directly terminated with various chemical groups, are beyond the scope of this chapter.

## 13.2 NANODIAMONDS EMBEDDED IN SHELLS

### 13.2.1 SILICA COATINGS ON NANODIAMONDS

Chemical modification of the ND surface is necessary for almost any advanced application of these particles. Silica can be conveniently used as a linker between the ND core and a functional coating of polymers[16,17] or biomolecules.[18] Silica-based coatings also can be used to modify the roundness,[14,15] size, and polydispersity[17,19]

**Hydrolysis:** $$Si(OC_2H_5)_4 + 4\,H_2O \xrightarrow{NH_3} Si(OH)_4 + 4\,C_2H_5OH$$

**Polycondensation:** $$\mathbf{ND}\text{–}OH + Si(OH)_4 \longrightarrow \mathbf{ND}\text{–}O\text{–}Si(OH)_3 + H_2O$$

$$\mathbf{ND}\text{–}O\text{–}Si(OH)_3 + Si(OH)_4 \longrightarrow \mathbf{ND}\text{–}O\text{–}Si(OH)_2\text{–}O\text{–}Si(OH)_3 + H_2O$$

**FIGURE 13.1**

Representation of Stöber process, where silane is hydrolyzed under basic catalysis mediated by ammonia. Subsequent polycondensation of silanol creates weakly crosslinked amorphous silica from the surface of hydroxyl terminated ND. Similar mechanism is employed when functionalized silane, such as APTES or TMSPMA is used instead of TEOS

of NDs. Although methods to directly chemically modify the ND surface exist, very few of these provide quantitative conversion of surface groups. Most methods have been developed for detonation NDs (DNDs), which are considered the most reactive NDs, and are often not transferable to NDs of a different origin. Silica can be used to normalize the surface of any type of ND, installing free silanol groups on the surface. This enables ND particles to take advantage of the well-investigated silane chemistry toolbox, while retaining their unique fluorescent properties.

The typical silica-coating procedure utilizes a modified Stöber process[20] involving base-catalyzed hydrolysis of a silyl ether, often tetraethyl orthosilicate (TEOS), forming silanol. This is followed by a polycondensation reaction of silanols to form amorphous silica. The silica-coating procedure targets the hydroxyl groups on the ND surface as represented in Fig. 13.1. The surface of NDs after oxidative pretreatment generally contains a mixture of functional groups dominated by hydroxyl groups followed by a minor percentage of carboxylic groups.[21] The surface content of carboxylic groups has been reported to be between 2–7% after oxidative treatment.[22,23] The carboxylic groups provide ND particles with negative zeta potential, promoting stabilization of colloids in aqueous solution.

### *Silication methods*

Silication by (3-aminopropyl)trimethoxysilane (APTES) was the first silica-coating procedure to be used on NDs. It covers ND particles with a layer of positively charged amino-functionalized silica. First described in 2006 by A. Krüger and collaborators,[18] this silication protocol enables covalent attachment of small peptides to the surface of DNDs, utilizing amino-functionalized silica as a linker. The surface carboxyl groups were reduced to hydroxyls prior to coating with APTES. Although the particles partially aggregated during the silication step, they were able to undergo grafting of peptides and attachment of biotin.[24] The effect of the alkoxy substituent of silane on surface loading was further investigated, revealing that methoxysilane shows higher reactivity and interparticle condensation than ethoxysilane.[25] The APTES coating on NDs was further utilized to attach a spacer between the DND and glycoprotein binding molecule.[26] A slightly different approach utilizing a non-ionic surfactant to stabilize the dispersion of high-pressure high-temperature (HPHT) NDs during silication with APTES has been used to graft peptides onto NDs.[27] Carbon nanotubes have also been successfully attached to NDs using APTES as a linker. The effect of amino-functionalization with APTES on the toxicity of NDs has been assessed.[28,29] The roundness of NPs has a profound effect on their cellular fate, which has been assessed through studies of NDs with silica coatings of different thicknesses.[14] APTES coating has been used to stabilize the dispersion of NDs under Stöber reaction conditions. Amino-functionalized NDs also have been studied as a potential platform for gene delivery.[15]

A different approach to silication of NDs utilizes adsorbed polyvinylpyrrolidone (PVP) as an initial coating.[30] It stabilizes dispersion of ND particles in a basic water–ethanol mixture, enabling a layer of silica to be grown on each individual particle by Stöber reaction with TEOS as the silica precursor. This procedure yields thicker silica shells, typically 10–20 nm, which in turn normalizes the irregular shape of NDs. Silica-encapsulated NDs have been applied in the field of vitrified ceramics research.[31,32]. Recently, researchers described an elaborate silicate architecture on NDs,[17] combining PVP-stabilized growth of 20-nm-thick silica from TEOS, followed by a layer of APTES silica with 1,2-bis triethoxysilyl ethane (BTSE) as a crosslinking agent to prevent hydrolysis of APTES silica. Targeted ND-based particles bearing a plasmonic gold shell for selective thermal ablation of cancer cells have also been produced utilizing a PVP-mediated silica layer.[33] Functionalized silanes are even more prone to hydrolysis than TEOS-based silica.[34] Complete hydrolysis of APTES silica can be achieved in 0.1 M HCl.[18]

An interesting approach to deagglomerate DNDs by beads-assisted sonication[35] simultaneously stabilizes the newly formed ND particles by silanization with an acrylate-modified silane. 3-(Trimethoxysilyl)propyl methacrylate (TMSPMA) was used to form a thin (less than 1 nm) silica layer as a platform to graft the highly biocompatible hydrophilic polymer poly[N-(2-hydroxypropyl) methacrylamide] (PHPMA) from the surface of fluorescent NDs.[16] Selective targeting of these particles to cancer cells has been demonstrated.[36]

Liposome-based silica encapsulation is yet another method to coat NDs, and produces monodisperse stable composite particles[19] that can be used for in vivo imaging. By a combination of magnetic modulation of fluorescent ND emission with computational phase-sensitive signal recovery, up to 100-fold increases in the signal-to-noise ratio have been achieved.[37] See Chapter 17 for more information on spin properties of NV centers.

Porous silica has been investigated as a potential drug delivery platform owing to its large surface-to-volume ratio. Multifunctional nanocomposite particles have been created, synergistically combining the high loading capacity of mesoporous silica with the bioimaging properties of NDs.[38,39] The synthetic strategy presented here takes advantage of surfactant molecules serving as the pore-generating template as well as the phase transfer agent, thus making this method feasible even for inherently hydrophobic NPs.[40]

The basic Stöber silication protocol with TEOS as a precursor provides NDs with unifom silica layer with thickness that can be controlled by adjustment of reaction

**Table 13.1** Comparison of different silication protocols for ND

| Silica precursor | Usual thickness of silica layer | Reactive moiety for further modifications | Reference |
|---|---|---|---|
| APTES | 2–4 nm | Amino group | 14,15,18,24–28 |
| TEOS | 10–20 nm | Silanol group | 14,15,17,19,31,33,37 |
| TMSPMA | 1–2 nm | Methacrylate group | 16,35,36 |

conditions.[14] Silication with functionalized silane tends to create thin layers, where non-silanol functionality sterically hinders further polycondensation. Silication protocols are summarized in Table 13.1. Silane chemistry is highly versatile: different silanes can be used in a single reaction creating multifunctionalized layer, or in a multistage reaction for synthesis of multilayer composite coating. When increased hydrolytic stability of silica coating is needed, it is advisable to add a crosslinker containing multiple silane groups (such as BTSE), that promotes formation of more stable silica layers.[17]

### *Analytical methods for characterization of silica shells on NDs*

The presence of a silica coating on NDs can be assessed conveniently by infrared spectroscopy (IR). Intensive peaks at around 1110 $cm^{-1}$ correspond to the asymmetric stretching vibrations of Si–O–Si.[17] Zeta potential measurements and dynamic light scattering can be used to examine the colloidal properties of coated NDs. The zeta potential indicates the degree of electrostatic repulsion between similarly charged particles in a solution. See Chapter 11 regarding the interaction of ND with the solvents. The colloidal stability of a system is generally considered to be good when the absolute value of the zeta potential is above 30 mV. The zeta potential of particles coated with unmodified silica in solution at neutral pH is around −40 mV. A similar zeta potential is observed for silica-coated NDs,[19] while the zeta potential for particles with APTES silica coating is −27 mV.[27] Dynamic light scattering (DLS) is a non-destructive and very sensitive method. The measured intensity of light scattering, which originates from NDs randomly moving in solution, fluctuates over time depending on the size of the NDs. It is a useful indication of the aggregation state of the NP system.

Thicker shells (5–20 nm) can be visualized using bright field imaging on a transmission electron microscope (TEM). The irregular spiky shape of unmodified NDs (Fig. 13.2(A))[13] contrasts with micrographs of rounded silica-coated NDs (Fig. 13.2(B)).[17] The tightly arranged crystalline diamond core has a higher average electron density than the well-hydrated, weakly crosslinked amorphous silica. The overlapping of individual particles is often an artifact caused by chaotic particle adsorption onto the TEM grid, rather than particles being interconnected by silica layer. Thus, it is always beneficial to assess the colloidal distribution of a system by a different method such as dynamic light scattering. Thinner shells (APTES or methacrylate-modified silica) are visible only using high-resolution TEM (HRTEM). Electron energy loss spectroscopy (EELS) integrated with HRTEM in scanning TEM

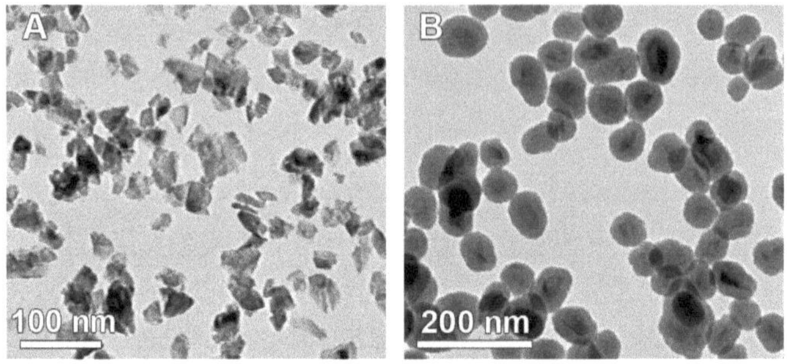


**FIGURE 13.2**

Bright field TEM micrograph of unmodified spiky NDs (A) and NDs rounded by a silica coating (B). Adapted from: Rehor, I., et al. Small 2014, 10, 1106–1115[17]

(STEM) mode is a powerful tool to spatially resolve the different elemental composition of the core ND and silica coating,[17] see Chapter 3 for more details. The overall surface loading of NDs is usually determined from thermogravimetric and elemental analyses.

### 13.2.2 POLYMER COATINGS

For biomedical applications (see Chapter 16), NDs must form stable colloids in aqueous media, including buffers and solutions of high ionic strength. Stability can be ensured by electrostatic (charged particles repel each other) or steric repulsion by a surface layer of hydrophilic polymers preventing close contact between NDs and subsequent aggregation caused by van der Waals interactions.[41] For this purpose, NDs can be covered with different types of polymers (Fig. 13.3) that meet certain criteria for biocompatibility and hydrophilicity. Polymers should be neutral and able to prevent non-specific interactions of NDs with proteins or cells, the so-called antifouling effect.[42]

Usually, polyethylene glycol (PEG) or its derivatives are used. However, hydroxyl groups can be enzymatically oxidized in vivo to aldehydes, which further react with proteins.[42] This drawback can be overcome by using a methoxy-functionalized polymer such as poly[PEG methyl ether methacrylate] (PPEGMA).[53] Immune reaction to PEGylated NPs and other disadvantages of PEG, known as the "PEG dilemma," were reported recently.[66] Therefore, other suitable polymers, such as methacrylate derivatives or polyether polyols, are often employed. PHPMA is a linear, biodegradable, non-immunogenic polymer with functionalizable side chains.[66] Hyperbranched polyglycerol (PG) can cover NDs more densely than linear chains of other polymers. Its hydroxyl groups are more hydrophilic than PEG ether groups and less susceptible to oxidation or thermal stress than PEG hydroxyl groups.[58,66]

Some polymers are chosen for NDs modification to provide a certain biological function. For example, polyethylene-imine (PEI),[44,45] polyallylamine hydrochloride

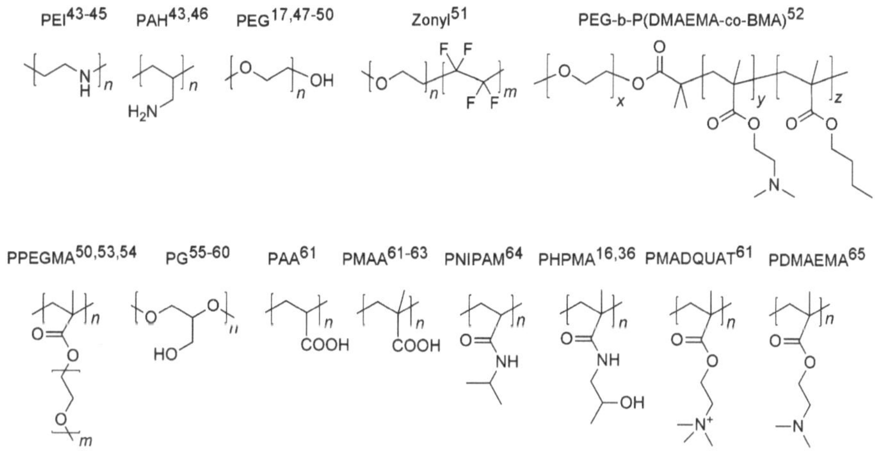


**FIGURE 13.3**

Structure of polymers that have been used to coat NDs by "grafting to" (first row) and "grafting from" approaches (second row). PMADQUAT: poly[2-(methacryloyloxy)ethyl trimethylammonium chloride]

(PAH),[43] and poly[2-(dimethylamino)ethyl methacrylate] (PDMAEMA)[65] are used for preparation of gene delivery vectors. These polymers form complexes with negatively charged nucleic acids and improve their transfection efficiency compared to conventionally used transfection agents. Poly(*N*-isopropylacrylamide) (PNIPAM) or PAH, which are pH- and thermo-responsive, on NDs are useful for sensing applications.[46,64]

There are two approaches to coat NDs with a polymer shell: (i) non-covalent adsorption through electrostatic or hydrophobic interaction and (ii) covalent grafting, which can be divided to the "**grafting to**" and "**grafting from**" approach (Fig. 13.4(A)). In "grafting to" methods, the polymer is first synthesized in solution and then attached to the ND surface via covalent bond formation. In "grafting from" methods, polymerization occurs directly on the ND surface.

## *Non-covalent approach*

Polymers can adsorb onto NDs spontaneously from solution if the interaction of NDs with the polymer is more favorable than with solvent. Polymer connections to NDs occur through various mechanisms, dependent on the internal structure of the polymer.[67] The whole chain may possess the same affinity to the ND surface, such as the positively charged polymers PEI or PAH, which adsorb onto NDs due to an electrostatic interaction with the negatively charged NDs (Fig. 13.4(A)).[43–46] The colloidal stability of such conjugates can be negatively influenced by possible adherence of one polymer chain to many NDs, which can cause aggregation. NDs need to be mixed

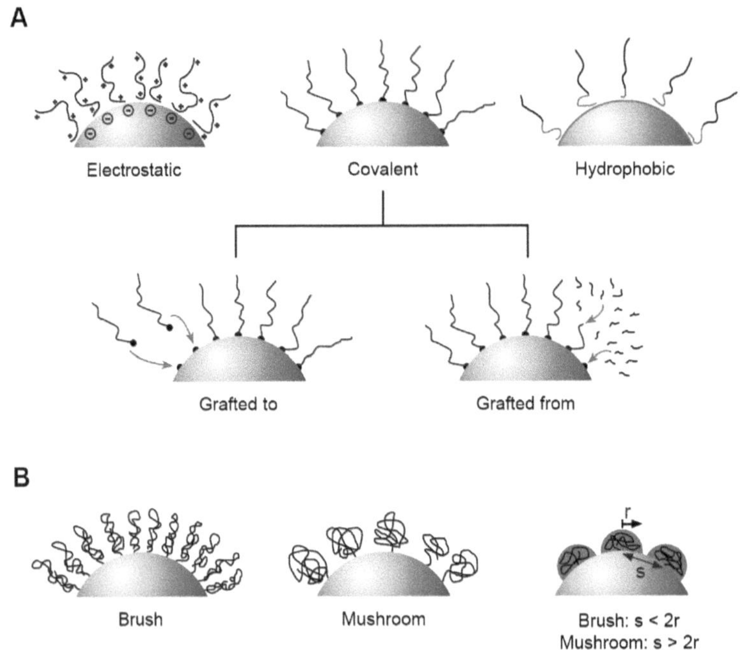


**FIGURE 13.4**

(A) Methods of polymers' attachment to NDs by various approaches: using adsorption of positively charged polymer by electrostatic interaction with the negatively charged ND surface; covalent attachment (and the difference between "grafting to" and "grafting from" method); attachment of a block copolymer through hydrophobic interaction of the anchor attracted to ND surface. (B) Two conformations of polymers ("brush" and "mushroom") on ND surface regarding to their surface grafting density

with positively charged polymers by slowly adding one component to another, usually during sonication.

Block copolymers typically include one block that is attracted to NDs and serves as an anchor, and a second, hydrophilic block that does not have significant affinity to NDs (Fig. 13.4(A)). Block copolymers containing both positively charged PDMAEMA and hydrophobic butyl methacrylate (BMA) segments, for example, interact with negatively charged and hydrophobic moieties on the ND surface. The PEG block forms a hydrophilic shell, ensuring the colloidal stability.[52]

### *"Grafting to" approach*

In most cases, end-modified polymers are employed. One end binds to the ND surface via a chemical bond (usually an amide or ester bond), and the rest of the polymer is exposed to solution and can be further modified with biomolecules to generate NDs that can be used for targeting or drug delivery.[17,48,49,68,69] Marcon et al. immobilized

Zonyl polymer on NDs by a multistep reaction that included preparation of NDs with azido groups that subsequently reacted with alkynyl–Zonyl polymer. Zonyl polymer is a block copolymer composed of PEG and perfluoroalkyl chains. Attachment of the polymer improves the stability of NDs in phosphate buffer saline (PBS) and has an antifouling effect, as shown by a decrease in bovine serum albumin adsorption onto NDs.[51]

The density of the surface polymer coating is a crucial parameter determining the stability of NDs in various solutions. Polymer coils that are well-separated on the surface form so-called "mushroom" conformations. Overlapping polymer coils on surfaces with a high polymer concentration are known as "brushes" (Fig. 13.4(B)).[67] "Grafting to" methods usually yield less dense and more poorly protective coatings than "grafting from" methods. The "grafting to" approach suffers from several limitations, such as steric hindrance. Polymer chains have to diffuse through an existing layer of polymer shell to reach complementary groups on the surface.

### *"Grafting from" approach*

Polymers can be introduced to the ND surface in a "grafting from" approach using different polymerization methods. One common strategy is radical living polymerization, in which brush thickness, polydispersity, and composition are controllable. Atom transfer radical polymerization (ATRP) and reversible addition-fragmentation chain transfer (RAFT) polymerization are directly initiated from initiator-functionalized surfaces.[42] ATRP includes halide-terminated NDs (ND–Br or ND–Cl) and metal halide as a catalyst. Li et al. were the first to introduce hydrophobic poly(tert-butyl methacrylate) brushes hydrolyzable by acids to hydrophilic poly(methacrylic acid) (PMAA) brushes,[63] which can covalently bind protein.[62] Positively charged polymer-coated NDs, prepared by a similar procedure with DMAEMA as a monomer, were able to complex plasmid DNA, resulting in more efficient expression than using PEI (25 kDa).[65]

In RAFT polymerization, an RAFT agent needs to be introduced to the ND surface first. Huynh et al. polymerized NDs with PPEGMA monomer using the radical initiator azobisisobutyronitrile (AIBN). Although NDs were covered with polymer, a cisplatin complex was still able to adsorb onto the ND surface, yielding an ND-polymer-cisplatin conjugate, which was more toxic to cancer cells than free cisplatin or other conjugates.[53]

In addition to living polymerization, conventional radical polymerization has been implemented to prepare stable polymer-modified NDs. Chang et al. prepared surface-graphitized NDs prone to reacting with various monomers (acrylic or methyl methacrylic acid) using the radical initiator benzoyl peroxide, sonication, and microwave irradiation.[61] A significant improvement in ND colloidal stability was achieved by use of PHPMA polymer, grown on the terminal methacrylate groups of thin silica coating, with the thermolabile radical initiator AIBN. Propargylacrylamide or 3-(azidopropyl)methacrylamide, which copolymerizes with PHPMA, enables further modification of NDs with biorthogonal reactions. This polymerization process was carefully optimized using different concentrations, solvents, times, and temperatures. The higher viscosity of the solvent and lower temperature postpones the termination and prolongs the reaction time. The optimization process succeeded in providing NDs stable even in 1 M NaCl. Because non-specific binding of proteins and cells was almost completely reduced, ND-PHPMA modified with cyclic peptide containing tripeptide Arginine–Glycine–Aspartic acid (RGD) motif showed a high ability to specifically target cancer cells.[16,36]

Non-radical ring-opening polymerization of glycidol (initiated at the oxygen-containing functionalities of NDs) modifies NDs with a hyperbranched PG layer.[58] This dense and thick shell ensures great stability in physiological media and buffers.[55] Hydroxyl groups of ND–PG can be substituted with azido groups, which can be used for attachment of biomolecules such as RGD peptide. In combination with the PG shielding from non-specific interactions with cells, RGD enables specific targeting of cancer cells. ND–PG–RGD particles selectively kill cancer cells when modified with a cisplatin complex[60] or the drug doxorubicin.[59] ND–PG particles have also been utilized in gene delivery and as a magnetic resonance contrast agent.[56,57]

#### *Analytical methods for characterization of polymers on NDs*

Basic methods for evaluation of ND–polymer conjugate preparation include thermogravimetric analysis, dynamic light scattering, and zeta potential measurement. Thermogravimetric analysis determines the weight ratio of polymer and ND from weight loss during temperature increases. The polymer itself degrades rapidly above 200 °C, while NDs oxidize around 600 °C.[58] The molecular structure of polymers can be examined using various spectroscopic methods, such as nuclear magnetic resonance (NMR) and IR spectroscopy.[16,60] The structure of the polymeric shell can be analyzed using TEM.[16]

Interestingly, some NDs with neutral polymers attached still exhibited negative zeta potential, limited dispersibility, and the ability to non-covalently bind the drug doxorubicin or proteins, which can indicate an insufficient coating of the surface.[47,48,50,51,68,69]

### 13.2.3 PROTEIN COATINGS

When nanomaterials are introduced into a biological system, the proteins present in the biofluid coat the NP surface to form a bio–nano interface, the so-called "protein corona".[70] The protein corona is composed of an inner layer of proteins with a lifetime of several hours in slow exchange with the environment (the hard corona) and an outer layer of weakly bound proteins characterized by a faster exchange rate with the free proteins (the soft corona).[71] Due to the hard corona's long lifetime, it interacts with cellular receptors rather than the pristine NP surface and defines the fate of NPs in a biological environment.[72] The interaction between proteins and the surface of a nanomaterial plays a major role in determining the biocompatibility of the nanomaterial.[72] The surface properties of the NP (charge, physical and chemical structure, hydrophobicity/hydrophilic-

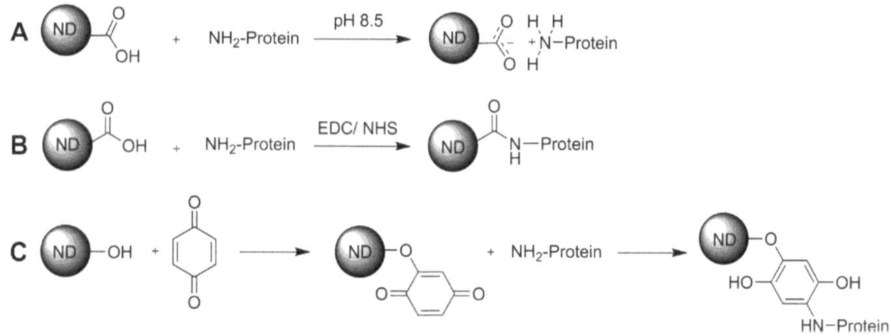


**FIGURE 13.5**

Schematic representation of an ND surface coated with proteins. (A) Non-covalent electrostatic interaction, (B) covalent coating with activation of the surface carboxyl groups on the ND with EDC and NHS, followed by conjugation with the primary amino groups of protein through amide bond formation, (C) covalent coating on a benzoquinone-activated ND

ity, size, curvature, etc.) determine the strength of adsorption, number of adsorbed biomolecules, conformations, orientations, ordering, and so forth, and can influence the structure, properties, and correspondingly, the functional activities of adsorbed biomolecules.[72–74]

### *Non-covalent coatings*

The protein coating serves as an effective stabilizing agent and greatly suppresses NP agglomeration.[75,76] Carboxylated/oxidized NDs exhibit exceptionally high affinity for biomolecules, including proteins,[77–80] enzymes,[23,81] antibodies,[82] and peptides.[83] The affinity is so high that proteins in dilute solution can be easily captured by NDs[78] and remain attached to the ND surfaces even after thoroughly washing the ND particles with deionized water.[77] Several proteins, such as albumin,[75,84] $\gamma$-globulin,[85] $\alpha$-lactalbumin,[75] $\alpha$-bungarotoxin ($\alpha$-BTX),[86] bone morphogenetic protein-2 (BMP-2),[80] cytochrome c,[77,78] lysozyme,[74,84,23,87], myoglobin,[78] ferritin,[88] silk fibroin spheres,[89] and albumin-derived copolymer,[90] have been used as non-covalent coatings on ND surfaces to study protein interaction with the ND surface and for biomedical applications, which are discussed in detail in Chapter 16.

The adsorption of protein on carboxylated NDs is established primarily through electrostatic attraction between the surface-terminated anionic groups ($-COO^-$) and the positively charged amino groups ($-NH_3^+$) in the protein (Fig. 13.5(A)).[77,91] In addition to the electrostatic interactions, hydrogen bonds can also form between $-NH_3^+$ and CO-containing surface groups.[92] Small rigid proteins, such as lysozyme, form multilayers with significant conformational changes, while larger soft proteins (albumin) form a monolayer with minor conformational changes.[84]

### Covalent coatings

Non-covalent coating of NDs with proteins is often relatively easy to achieve; however, it is nonspecific and unstable in blood serum.[93] In contrast, covalent bonding between the biomolecule and the ND is more stable and controllable.[94] A variety of surface functional groups, including esters, hydroxyls, and carboxylic acids, enables a diverse range of biomolecular coatings.[95] One of the most reliable grafting methods for immobilization of large proteins is the coupling of amino groups from their side chains onto carboxylated NDs by the formation of an amide bond (Fig. 13.5(B)).[5] Using this method, researchers have coated carboxylated NDs with bovine serum albumin,[96] streptavidin,[97] trypsin,[98] transferrin,[99] growth hormone,[100] peptide nucleic acids (PNA),[101] and antibodies.[102] The coated NDs have been applied for *in-vivo* imaging, proteolysis of model proteins, and targeted labeling of mammalian cells. Purtov et al. demonstrated immobilization of rabbit anti-mouse antibody and bovine serum albumin on hydroxyl-terminated NDs via benzoquinone chemistry (Fig. 13.5(C)) to specifically target an antigen (mouse IgG).[93]

Chemical or physical modifications of ND surfaces with biomolecules via diverse chemical routes provide NDs with specific functionalities. With further development, bioconjugated NDs will offer exciting new opportunities for a wide range of cell labeling and targeting applications.

### Analytical methods for characterization of proteins on NDs

Fourier transform infrared spectroscopy (FTIR) measurement provides information about the conformational changes in adsorbed or immobilized proteins on NDs. The most characteristic IR bands of proteins arise from the peptide backbone vibrations at 1600–1700 $cm^{-1}$ from C=O, C–N stretching (amide I) and 1510–1580 $cm^{-1}$ from N–H deformation and C–N stretching (amide II). In addition, a maximum near 3250 $cm^{-1}$ can be attributed to N–H stretching.[103,104] The amide I region is more sensitive than the amide II region, and used to estimate the protein structural changes, via the frequency shift.[74,105]

UV–vis absorbance spectroscopy in the 230–300 nm range, which is mainly attributable to aromatic residues, including tryptophan, tyrosine, and phenylalanine, can provide valuable information regarding whether a major structural change occurred in proteins upon ND coating. A blue-shift of absorbance of the ND–protein complex compared to free protein indicates exposure of an aromatic residue to an environment with more polarity, suggesting that a major protein unfolding occurred upon coating of the ND surface. Comparison of the characteristic Soret band for a surface-bound heme-containing protein, such as cytochrome c or myoglobin, with that of the free protein also provides information about the conformational changes upon immobilization on the ND surface.[77,106] Circular dichroism (CD) spectroscopy can be used for quantitative analysis to determine the extent of conformational changes of proteins upon binding to the ND surface. These findings are often related to retention of the protein's biological functionality.[107]

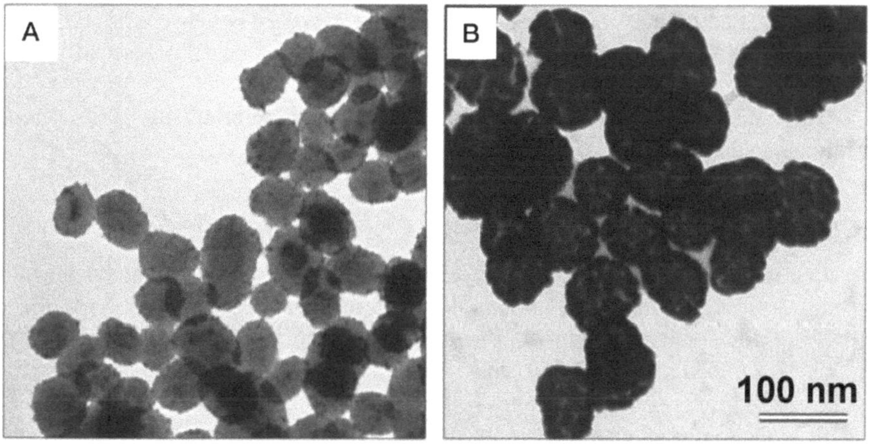


**FIGURE 13.6**

TEM images showing (A) silica-coated NDs with gold seeds, (B) GNSs grown from these precursors. Adapted from: Rehor, I., et al. Adv. Healthc. Mater. 2015, 4, 460–468[33]

## 13.2.4 METALLIC SHELLS ON NANODIAMONDS

Metal-coated NDs forming a core–shell structure are still a very uncommon type of NP. Various systems based on a combination of NDs and metal NPs have been reported; however, these are generally nanocomposite mixtures[108] or structures in which one particle type is attached to the other.[109–111] The various types of NDs prepared this way serve mainly in two fields – as a platform for observation of optical phenomena (predominantly ND-gold NP systems)[112–114] and as diamond-supported metal catalysts for redox reactions containing mostly platinum and its mixtures.[115–117]

To date, two studies have described a metallic core–shell structure on NDs. Rehor et al.[33] prepared particles in which a gold nanoshell (GNS) grows from gold seeds attached to silica-coated NDs. The formation of the GNS caused a plasmonic band with a maximum at 675 nm to emerge in absorption spectra (dark blue solution). The structure of the particles was confirmed by TEM (Fig. 13.6), and the GNSs were characterized in detail by high angle annular dark field scanning transmission electron microscopy (HAADF-STEM) electron tomography; the STEM mode is explained in more details in Chapter 3. The PEG coating and attachment of the protein transferrin enabled targeting of the particles to cancer cells. The plasmonic properties of the gold shell were then utilized as opto-thermal convertors for highly effective *in-vitro* thermal ablation of cancer cells.

Minanti et al.[118] described preparation of ND-gold core–shell particles in which the gold shell is grown directly on 100–200 nm diamond particles containing fluorescent silicon-vacancy centers. The existence of gold nanostructures larger than gold seeds was confirmed by absorption spectroscopy and scanning electron microscopy, which revealed a nano-rough surface. The presence of NDs was confirmed by diamond Raman spectroscopy and by observing the maintained fluorescence of the silicon-vacancy color centers.

These materials have high therapeutic potential and are promising in the field of bioimaging applications.[33,118]

### 13.2.5 NANODIAMOND CORE–SHELL SP²/SP³ HYBRIDS

In contrast to $sp^2$-hybridized graphite, $sp^3$-hybridized diamond is thermodynamically metastable under ambient conditions. Formation of diamond from graphite therefore requires conditions with high pressure and high temperature.[1] For NPs, the graphite-diamond phase balance is also influenced by particle surface energy, which is inversely proportional to particle size. NDs smaller than 5 nm are therefore more stable than their graphitic counterparts.[119]

The conversion of a diamond lattice to graphite is inhibited by high activation energy, and for bulk diamonds, it requires heating above 1700 °C. However, graphitization can be catalyzed at much lower temperatures by trace amounts of oxygen and hydrogen or by transition metals.[120] Above 900 °C, significant graphitization of NDs starts from the particle surface by creation of ND–graphite core–shell structures. This complicates the work-up of NDs after annealing, which is used as final step in formation of fluorescent centers.[121,122] The transition is accompanied by the decomposition of various surface groups.[123,124] An analysis by Raman spectroscopy and HRTEM showed loss of oxygen-containing functionalities in NDs annealed in a vacuum at 550–850 °C, followed by a reconstruction of the surface. When NDs were annealed in a vacuum at 1100 °C, a continuous onion-like shell wrapped around a diamond core was produced.[125]

The dependence of surface graphitization on temperature has also been studied by X-ray photoelectron spectroscopy (XPS). This surface analytical technique provides information about both elemental composition and binding energy and can therefore distinguish $sp^2$- from $sp^3$-hybridized carbon atoms (for more details see Chapter 11). ND samples annealed in an inert atmosphere for 1 hour at 900 °C or 1500 °C were compared with untreated sample. The relative atomic content of carbon significantly increased and the relative atomic content of oxygen and nitrogen decreased for the sample annealed at 1500 °C. While the ratio of $sp^2/sp^3$ content in the sample annealed at 900 °C increased only slightly from 0.44 to 0.55, the sample annealed at 1500 °C had a ratio of 6.08, indicating that almost complete graphitization had taken place.[124] Petit et al. obtained similar results for samples annealed under a vacuum.[126] For this study the different graphitization stages were also characterized using HRTEM.[127]

Different $sp^2/sp^3$ hybrids comprising a diamond core and a graphitic shell were tested as a possible catalyst for oxidative dehydrogenation of propane at 550 °C.[128,129] Optimal catalytic performance and selectivity was achieved for the hybrid prepared from NDs annealed at 1100 °C for 4 hours. According to HRTEM characterization (Fig. 13.7), the surface of these particles was reconstructed into a curved, distorted graphene outer shell. A diamond core covered with an amorphous carbon

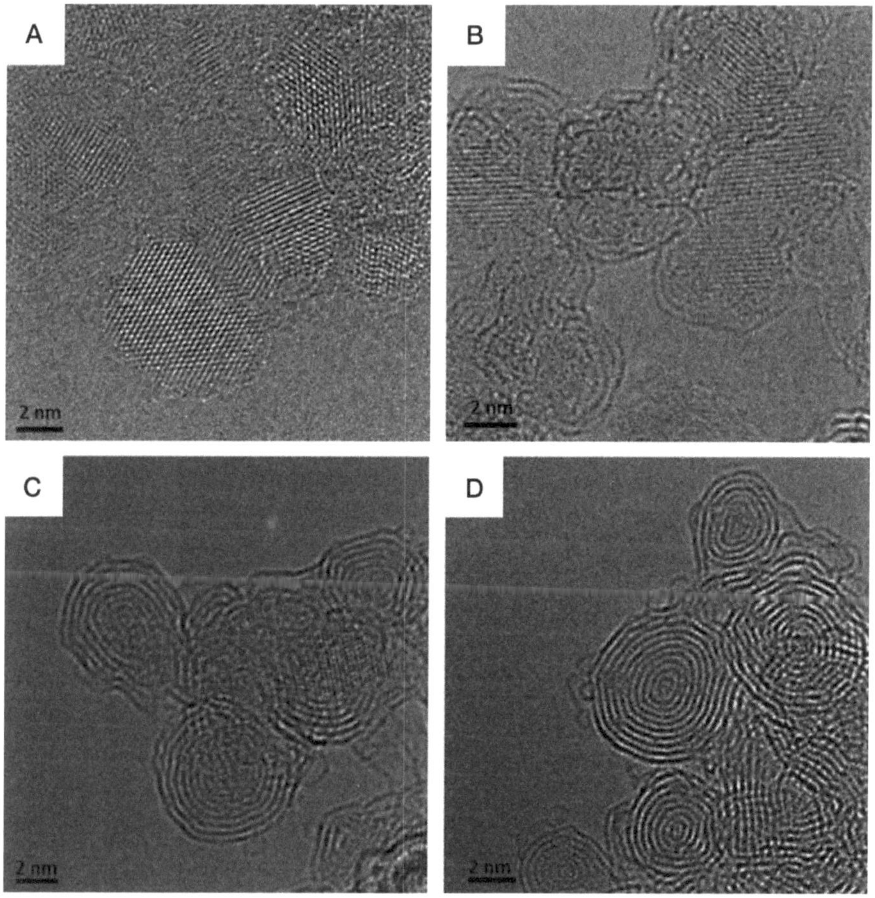


**FIGURE 13.7**

HRTEM images of NDs annealed at the following temperatures: (A) room temperature, (B) 1000 °C, (C) 1300 °C, and (D) 1500 °C. Adapted from: Sun, X., et al. ChemCatChem 2014, 6, 2270–2275[129]

layer with low catalytic activity and selectivity was present on particles annealed below 800 °C. On particles annealed at 1300 °C, the surface was fully transformed into an onion-like carbon shell multilayer, resulting in a sharp drop in catalytic activity. However, the selectivity remained similar. Upon further annealing at 1500 °C, all particles were transformed into spherical carbon onions.

Similar ND hybrids have been used for electrocatalytic applications. Zhang et al. used NDs with $sp^2$ shells as a high conductive support of Pt electrocatalyst for methanol oxidation.[130] The Pt NPs were deposited on the ND surface by a microwave-assisted reduction method. Dong et al. prepared nitrogen-doped hybrids by mixing ND hybrids annealed at 1400 °C with melamine. The mixture was then annealed at 1000 °C. The obtained material has shown promising properties as a catalyst for metal-free oxygen reduction reactions.[131]

## 13.3 CONCLUSIONS

Core–shell hybrids on NDs are a novel group of nanomaterials that can significantly alter the colloidal, biological, spectral, and functional properties of ND cores.[132] For certain applications, they provide a powerful alternative to direct chemical functionalization of the ND surface, bridging the known chemical, physico-chemical, and biological properties of the shells with the unique behavior of the diamond cores.[133] The shells also serve as a spacing interlayer between the neighboring nanoenvironment and ND, enabling fine tuning of core–shell particles' interactions, which are barely achievable directly on the diamond lattice. Numerous protocols leading to the synthesis of shell coatings on NDs have been established, followed by an increasing number of advanced applications in bioimaging, catalysis, sensing, drug delivery, and other disciplines.

We envision the core–shell hybrids on NDs will enable substantial extension and better control of biological applications of NDs. However, for biosensor applications, very thin and compact antifouling biocompatible surface coatings must be developed, as the interfacial volume influencing individual NV centers is spatially limited. We believe this is currently a very important direction in design and preparation of core–shell hybrids on NDs.

## REFERENCES

1. Mochalin, V. N.; Shenderova, O.; Ho, D.; Gogotsi, Y. The Properties and Applications of Nanodiamonds. *Nat. Nanotechnol.* **2012**, *7*, 11–23.
2. Krueger, A. Beyond the Shine: Recent Progress in Applications of Nanodiamond. *J. Mater. Chem.* **2011**, *21*, 12571–12578.
3. Rehor, I.; Slegerova, J.; Havlik, J.; Raabova, H.; Hyvl, J.; Muchova, E.; Cigler, P. Nanodiamonds: Behavior in Biological Systems and Emerging Bioapplications. In *Carbon Nanomaterials for Biomedical Applications, Vol. 5;* Zhang, M., Naik, R. R., Dai, L., Eds.; Springer International Publishing: Cham, 2016, pp. 319–361.
4. Schirhagl, R.; Chang, K.; Loretz, M.; Degen, C. L. Nitrogen-Vacancy Centers in Diamond: Nanoscale Sensors for Physics and Biology. *Annu. Rev. Phys. Chem.* **2014**, *65*, 83–105.
5. Krueger, A.; Lang, D. Functionality Is Key: Recent Progress in the Surface Modification of Nanodiamond. *Adv. Funct. Mater.* **2012**, *22*, 890–906.
6. Arnault, J. C. Surface Modifications of Nanodiamonds and Current Issues for Their Biomedical Applications. In *Novel Aspects of Diamond, Vol. 121;* Yang, N., Ed.; Springer International Publishing: Cham, 2015, pp. 85–122.
7. Shenderova, O. A.; McGuire, G. E. Science and Engineering of Nanodiamond Particle Surfaces for Biological Applications (Review). *Biointerphases* **2015**, *10*, 30802.

8. Perevedentseva, E.; Lin, Y.-C.; Jani, M.; Cheng, C.-L. Biomedical Applications of Nanodiamonds in Imaging and Therapy. *Nanomed.* **2013**, *8*, 2041–2060.
9. Hui, Y. Y.; Chang, H.-C. Recent Developments and Applications of Nanodiamonds as Versatile Bioimaging Agents. *J. Chin. Chem. Soc.* **2014**, *61*, 67–76.
10. Slegerova, J.; Rehor, I.; Havlik, J.; Raabova, H.; Muchova, E.; Cigler, P. Nanodiamonds as Intracellular Probes for Imaging in Biology and Medicine. In *Intracellular Delivery II;* Prokop, A., Iwasaki, Y., Harada, A., Eds., Fundamental Biomedical Technologies, Vol. 7; Springer Netherlands: Dordrecht, 2014, pp. 363–401.
11. Zhu, Y. The Biocompatibility of Nanodiamonds and Their Application in Drug Delivery Systems. *Theranostics* **2012**, *2*, 302–312.
12. Vaijayanthimala, V.; Lee, D. K.; Kim, S. V.; Yen, A.; Tsai, N.; Ho, D.; Chang, H.-C.; Shenderova, O. Nanodiamond-Mediated Drug Delivery and Imaging: Challenges and Opportunities. *Expert Opin. Drug Deliv.* **2015**, *12*, 735–749.
13. Rehor, I.; Cigler, P.; Precise Estimation of HPHT Nanodiamond Size Distribution Based on Transmission Electron Microscopy Image Analysis. *Diam. Relat. Mater.* **2014**, *46*, 21–24.
14. Chu, Z.; Zhang, S.; Zhang, B.; Zhang, C.; Fang, C.-Y.; Rehor, I.; Cigler, P.; Chang, H.-C.; Lin, G.; Liu, R.; et al. Unambiguous Observation of Shape Effects on Cellular Fate of Nanoparticles. *Sci. Rep.* **2014**, *4*, 4495.
15. Chu, Z.; Miu, K.; Lung, P.; Zhang, S.; Zhao, S.; Chang, H.-C.; Lin, G.; Li, Q. Rapid Endosomal Escape of Prickly Nanodiamonds: Implications for Gene Delivery. *Sci. Rep.* **2015**, *5*, 11661.
16. Rehor, I.; Mackova, H.; Filippov, S. K.; Kucka, J.; Proks, V.; Slegerova, J.; Turner, S.; Van Tendeloo, G.; Ledvina, M.; Hruby, M.; et al. Fluorescent Nanodiamonds with Bioorthogonally Reactive Protein-Resistant Polymeric Coatings. *ChemPlusChem* **2014**, *79*, 21–24.
17. Rehor, I.; Slegerova, J.; Kucka, J.; Proks, V.; Petrakova, V.; Adam, M.-P.; Treussart, F.; Turner, S.; Bals, S.; Sacha, P.; et al. Fluorescent Nanodiamonds Embedded in Biocompatible Translucent Shells. *Small* **2014**, *10*, 1106–1115.
18. Krüger, A.; Liang, Y.; Jarre, G.; Stegk, J. Surface Functionalisation of Detonation Diamond Suitable for Biological Applications. *J. Mater. Chem.* **2006**, *16*, 2322–2328.
19. Bumb, A.; Sarkar, S. K.; Billington, N.; Brechbiel, M. W.; Neuman, K. C. Silica Encapsulation of Fluorescent Nanodiamonds for Colloidal Stability and Facile Surface Functionalization. *J. Am. Chem. Soc.* **2013**, *135*, 7815–7818.
20. Stöber, W.; Fink, A.; Bohn, E. Controlled Growth of Monodisperse Silica Spheres in the Micron Size Range. *J. Colloid Interface Sci.* **1968**, *26*, 62–69.
21. Wolcott, A.; Schiros, T.; Trusheim, M. E.; Chen, E. H.; Nordlund, D.; Diaz, R. E.; Gaathon, O.; Englund, D.; Owen, J. S. Surface Structure of Aerobically Oxidized Diamond Nanocrystals. *J. Phys. Chem. C* **2014**, *118*, 26695–26702.
22. Wang, X.; Ruslinda, A. R.; Ishiyama, Y.; Ishii, Y.; Kawarada, H. Higher Coverage of Carboxylic Acid Groups on Oxidized Single Crystal Diamond (001). *Diam. Relat. Mater.* **2011**, *20*, 1319–1324.
23. Nguyen, T.-T.-B.; Chang, H.-C.; Wu, V. W.-K. Adsorption and Hydrolytic Activity of Lysozyme on Diamond Nanocrystallites. *Diam. Relat. Mater.* **2007**, *16*, 872–876.
24. Krueger, A.; Stegk, J.; Liang, Y.; Lu, L.; Jarre, G. Biotinylated Nanodiamond: Simple and Efficient Functionalization of Detonation Diamond. *Langmuir* **2008**, *24*, 4200–4204.
25. Krueger, A.; Ozawa, M.; Jarre, G.; Liang, Y.; Stegk, J.; Lu, L. Deagglomeration and Functionalisation of Detonation Diamond. *Phys. Status Solidi A* **2007**, *204*, 2881–2887.

26. Yeap, W. S.; Tan, Y. Y.; Loh, K. P. Using Detonation Nanodiamond for the Specific Capture of Glycoproteins. *Anal. Chem.* **2008**, *80*, 4659–4665.
27. Vial, S.; Mansuy, C.; Sagan, S.; Irinopoulou, T.; Burlina, F.; Boudou, J.-P.; Chassaing, G.; Lavielle, S. Peptide-Grafted Nanodiamonds: Preparation, Cytotoxicity and Uptake in Cells. *ChemBioChem* **2008**, *9*, 2113–2119.
28. Weng, M.-F.; Chang, B.-J.; Chiang, S.-Y.; Wang, N.-S.; Niu, H. Cellular Uptake and Phototoxicity of Surface-Modified Fluorescent Nanodiamonds. *Diam. Relat. Mater.* **2012**, *22*, 96–104.
29. Moore, L.; Grobárová, V.; Shen, H.; Man, H. B.; Míčová, J.; Ledvina, M.; Štursa, J.; Nesladek, M.; Fišerová, A.; Ho, D. Comprehensive Interrogation of the Cellular Response to Fluorescent, Detonation and Functionalized Nanodiamonds. *Nanoscale* **2014**, *2014*, 11712–11721.
30. Graf, C.; Vossen, D. L. J.; Imhof, A.; van Blaaderen, A. A General Method To Coat Colloidal Particles with Silica. *Langmuir* **2003**, *19*, 6693–6700.
31. Zhao, D.; Wang, Z.; Xi, Y.; Zou, Q.; Li, X.; Wang, B.; Guo, X.; Liang, M.; Li, W.; Wang, M.; et al. Preparation of Silica-Coated Ultrafine Diamond and Dispersion in Ceramic Matrix. *Mater. Lett.* **2013**, *113*, 134–137.
32. Yan, N.; Zhao, D. P.; Wang, L.; Zou, Q.; Xi, Y. Y.; Guo, X. P.; Wang, B.; Wang, Z. L.; Wang, L. P.; Dai, W. J.; et al. Preparation and Sintering of Silica-Coated Ultrafine Diamonds–vitrified Bond Composite Powders. *Int. J. Refract. Met. Hard Mater.* **2014**, *43*, 212–215.
33. Rehor, I.; Lee, K. L.; Chen, K.; Hajek, M.; Havlik, J.; Lokajova, J.; Masat, M.; Slegerova, J.; Shukla, S.; Heidari, H.; et al. Plasmonic Nanodiamonds: Targeted Core–Shell Type Nanoparticles for Cancer Cell Thermoablation. *Adv. Healthc. Mater.* **2015**, *4*, 460–468.
34. Dove, P. M.; Han, N.; Wallace, A. F.; Yoreo, J. J. D. Kinetics of Amorphous Silica Dissolution and the Paradox of the Silica Polymorphs. *Proc. Natl. Acad. Sci.* **2008**, *105*, 9903–9908.
35. Liang, Y.; Ozawa, M.; Krueger, A. A General Procedure to Functionalize Agglomerating Nanoparticles Demonstrated on Nanodiamond. *ACS Nano* **2009**, *3*, 2288–2296.
36. Slegerova, J.; Hajek, M.; Rehor, I.; Sedlak, F.; Stursa, J.; Hruby, M.; Cigler, P. Designing the Nanobiointerface of Fluorescent Nanodiamonds: Highly Selective Targeting of Glioma Cancer Cells. *Nanoscale* **2015**, *7*, 415–420.
37. Sarkar, S. K.; Bumb, A.; Wu, X.; Sochacki, K. A.; Kellman, P.; Brechbiel, M. W.; Neuman, K. C. Wide-Field in Vivo Background Free Imaging by Selective Magnetic Modulation of Nanodiamond Fluorescence. *Biomed. Opt. Express* **2014**, *5*, 1190–1202.
38. von Haartman, E.; Jiang, H.; Khomich, A. A.; Zhang, J.; Burikov, S. A.; Dolenko, T. A.; Ruokolainen, J.; Gu, H.; Shenderova, O. A.; Vlasov, I. I.; et al. Core–shell Designs of Photoluminescent Nanodiamonds with Porous Silica Coatings for Bioimaging and Drug Delivery I: Fabrication. *J. Mater. Chem. B* **2013**, *1*, 2358–2366.
39. Prabhakar, N.; Näreoja, T.; Haartman, E.; von Karaman, D. Ş; Jiang, H.; Koho, S.; Dolenko, T. A.; Hänninen, P. E.; Vlasov, D. I.; Ralchenko, V. G.; et al. Core–Shell Designs of Photoluminescent Nanodiamonds with Porous Silica Coatings for Bioimaging and Drug Delivery II: Application. *Nanoscale* **2013**, *5*, 3713–3722.
40. Kim, J.; Lee, J. E.; Lee, J.; Yu, J. H.; Kim, B. C.; An, K.; Hwang, Y.; Shin, C.-H.; Park, J.-G.; Kim, J.; et al. Magnetic Fluorescent Delivery Vehicle Using Uniform Mesoporous Silica Spheres Embedded with Monodisperse Magnetic and Semiconductor Nanocrystals. *J. Am. Chem. Soc.* **2006**, *128*, 688–689.

41. de Gennes, P. G. Polymers at an Interface; A Simplified View. *Adv. Colloid Interface Sci.* **1987**, *27*, 189–209.
42. Barbey, R.; Lavanant, L.; Paripovic, D.; Schüwer, N.; Sugnaux, C.; Tugulu, S.; Klok, H.-A. Polymer Brushes via Surface-Initiated Controlled Radical Polymerization: Synthesis Characterization, Properties, and Applications. *Chem. Rev.* **2009**, *109*, 5437–5527.
43. Alhaddad, A.; Durieu, C.; Dantelle, G.; Le Cam, E.; Malvy, C.; Treussart, F.; Bertrand, J.-R. Influence of the Internalization Pathway on the Efficacy of siRNA Delivery by Cationic Fluorescent Nanodiamonds in the Ewing Sarcoma Cell Model. *PLoS ONE* **2012**, *7*, e52207.
44. Chen, M.; Zhang, X.-Q.; Man, H. B.; Lam, R.; Chow, E. K.; Ho, D. Nanodiamond Vectors Functionalized with Polyethylenimine for siRNA Delivery. *J. Phys. Chem. Lett.* **2010**, *1*, 3167–3171.
45. Zhang, X.-Q.; Chen, M.; Lam, R.; Xu, X.; Osawa, E.; Ho, D. Polymer-Functionalized Nanodiamond Platforms as Vehicles for Gene Delivery. *ACS Nano* **2009**, *3*, 2609–2616.
46. Petrakova, V.; Rehor, I.; Stursa, J.; Ledvina, M.; Nesladek, M.; Cigler, P. Charge-Sensitive Fluorescent Nanosensors Created from Nanodiamonds. *Nanoscale* **2015**, *7*, 12307–12311.
47. Takimoto, T.; Chano, T.; Shimizu, S.; Okabe, H.; Ito, M.; Morita, M.; Kimura, T.; Inubushi, T.; Komatsu, N. Preparation of Fluorescent Diamond Nanoparticles Stably Dispersed Under a Physiological Environment Through Multistep Organic Transformations. *Chem. Mater.* **2010**, *22*, 3462–3471.
48. Wang, D.; Li, Y.; Tian, Z.; Cao, R.; Yang, B. Transferrin-Conjugated Nanodiamond as an Intracellular Transporter of Chemotherapeutic Drug and Targeting Therapy for Cancer Cells. *Ther. Deliv.* **2014**, *5*, 511–524.
49. Zhang, B.; Li, Y.; Fang, C.-Y.; Chang, C.-C.; Chen, C.-S.; Chen, Y.-Y.; Chang, H.-C. Receptor-Mediated Cellular Uptake of Folate-Conjugated Fluorescent Nanodiamonds: A Combined Ensemble and Single-Particle Study. *Small* **2009**, *5*, 2716–2721.
50. Zhang, X.; Fu, C.; Feng, L.; Ji, Y.; Tao, L.; Huang, Q.; Li, S.; Wei, Y. PEGylation and polyPEGylation of Nanodiamond. *Polymer* **2012**, *53*, 3178–3184.
51. Marcon, L.; Kherrouche, Z.; Lyskawa, J.; Fournier, D.; Tulasne, D.; Woisel, P.; Boukherroub, R. Preparation and Characterization of Zonyl-Coated Nanodiamonds with Antifouling Properties. *Chem. Commun.* **2011**, *47*, 5178.
52. Lee, J. W.; Lee, S.; Jang, S.; Han, K. Y.; Kim, Y.; Hyun, J.; Kim, S. K.; Lee, Y. Preparation of Non-Aggregated Fluorescent Nanodiamonds (FNDs) by Non-Covalent Coating with a Block Copolymer and Proteins for Enhancement of Intracellular Uptake. *Mol. BioSyst.* **2013**, *9*, 1004–1011.
53. Huynh, V. T.; Pearson, S.; Noy, J.-M.; Abboud, A.; Utama, R. H.; Lu, H.; Stenzel, M. H. Nanodiamonds with Surface Grafted Polymer Chains as Vehicles for Cell Imaging and Cisplatin Delivery: Enhancement of Cell Toxicity by POEGMEMA Coating. *ACS Macro Lett.* **2013**, *2*, 246–250.
54. Shi, Y.; Liu, M.; Wang, K.; Huang, H.; Wan, Q.; Tao, L.; Fu, L.; Zhang, X.; Wei, Y. Direct Surface PEGylation of Nanodiamond via RAFT Polymerization. *Appl. Surf. Sci. Part B* **2015**, *357*, 2147–2153.
55. Boudou, J.-P.; David, M.-O.; Joshi, V.; Eidi, H.; Curmi, P. A. Hyperbranched Polyglycerol Modified Fluorescent Nanodiamond for Biomedical Research. *Diam. Relat. Mater.* **2013**, *38*, 131–138.
56. Zhao, L.; Nakae, Y.; Qin, H.; Ito, T.; Kimura, T.; Kojima, H.; Chan, L.; Komatsu, N. Polyglycerol-Functionalized Nanodiamond as a Platform for Gene Delivery: Derivatiza-

tion, Characterization, and Hybridization with DNA. *Beilstein J. Org. Chem.* **2014**, *10*, 707–713.

57. Zhao, L.; Shiino, A.; Qin, H.; Kimura, T.; Komatsu, N.; Synthesis Characterization, and Magnetic Resonance Evaluation of Polyglycerol-Functionalized Detonation Nanodiamond Conjugated with Gadolinium(III) Complex. *J. Nanosci. Nanotechnol.* **2015**, *15*, 1076–1082.
58. Zhao, L.; Takimoto, T.; Ito, M.; Kitagawa, N.; Kimura, T.; Komatsu, N. Chromatographic Separation of Highly Soluble Diamond Nanoparticles Prepared by Polyglycerol Grafting. *Angew. Chem. Int. Ed.* **2011**, *50*, 1388–1392.
59. Zhao, L.; Xu, Y.-H.; Akasaka, T.; Abe, S.; Komatsu, N.; Watari, F.; Chen, X. Polyglycerol-Coated Nanodiamond as a Macrophage-Evading Platform for Selective Drug Delivery in Cancer Cells. *Biomaterials* **2014**, *35*, 5393–5406.
60. Zhao, L.; Xu, Y.-H.; Qin, H.; Abe, S.; Akasaka, T.; Chano, T.; Watari, F.; Kimura, T.; Komatsu, N.; Chen, X. Platinum on Nanodiamond: A Promising Prodrug Conjugated with Stealth Polyglycerol, Targeting Peptide and Acid-Responsive Antitumor Drug. *Adv. Funct. Mater.* **2014**, *24*, 5348–5357.
61. Chang, I. P.; Hwang, K. C.; Ho, J. A.; Lin, C.-C.; Hwu, R. J.-R.; Horng, J.-C. Facile Surface Functionalization of Nanodiamonds. *Langmuir* **2010**, *26*, 3685–3689.
62. Dahoumane, S. A.; Nguyen, M. N.; Thorel, A.; Boudou, J.-P.; Chehimi, M. M.; Mangeney, C. Protein-Functionalized Hairy Diamond Nanoparticles. *Langmuir* **2009**, *25*, 9633–9638.
63. Li, L.; Davidson, J. L.; Lukehart, C. M. Surface Functionalization of Nanodiamond Particles via Atom Transfer Radical Polymerization. *Carbon* **2006**, *44*, 2308–2315.
64. Su, S.; Wei, J.; Zhang, K.; Qiu, J.; Wang, S. Thermo- and pH-Responsive Fluorescence Behaviors of Sulfur-Functionalized Detonation Nanodiamond-poly(N-Isopropylacrylamide). *Colloid Polym. Sci.* **2015**, *293*, 1299–1305.
65. Zhang, P.; Yang, J.; Li, W.; Wang, W.; Liu, C.; Griffith, M.; Liu, W. Cationic Polymer Brush Grafted-Nanodiamond via Atom Transfer Radical Polymerization for Enhanced Gene Delivery and Bioimaging. *J. Mater. Chem.* **2011**, *21*, 7755.
66. Amoozgar, Z.; Yeo, Y. Recent Advances in Stealth Coating of Nanoparticle Drug Delivery Systems. *Wiley Interdiscip. Rev. Nanomed. Nanobiotechnol.* **2012**, *4*, 219–233.
67. Netz, R. R.; Andelman, D. Neutral and Charged Polymers at Interfaces. *Phys. Rep.* **2003**, *380*, 1–95.
68. Dong, Y.; Cao, R.; Li, Y.; Wang, Z.; Li, L.; Tian, L. Folate-Conjugated Nanodiamond for Tumor-Targeted Drug Delivery. *RSC Adv.* **2015**, *5*, 82711–82716.
69. Zhang, T.; Cui, H.; Fang, C.-Y.; Cheng, K.; Yang, X.; Chang, H.-C.; Forrest, M. L. Targeted Nanodiamonds as Phenotype-Specific Photoacoustic Contrast Agents for Breast Cancer. *Nanomed.* **2015**, *10*, 573–587.
70. Lynch, I.; Dawson, K. A. Protein–Nanoparticle Interactions. *Nano Today* **2008**, *3*, 40–47.
71. Mahmoudi, M.; Simchi, A.; Imani, M.; Shokrgozar, M. A.; Milani, A. S.; Häfeli, U. O.; Stroeve, P. A New Approach for the in Vitro Identification of the Cytotoxicity of Superparamagnetic Iron Oxide Nanoparticles. *Colloids Surf. B Biointerfaces* **2010**, *75*, 300–309.
72. Mahmoudi, M.; Lynch, I.; Ejtehadi, M. R.; Monopoli, M. P.; Bombelli, F. B.; Laurent, S. Protein–Nanoparticle Interactions: Opportunities and Challenges. *Chem. Rev.* **2011**, *111*, 5610–5637.

73. Mu, Q.; Jiang, G.; Chen, L.; Zhou, H.; Fourches, D.; Tropsha, A.; Yan, B. Chemical Basis of Interactions Between Engineered Nanoparticles and Biological Systems. *Chem. Rev.* **2014**, *114*, 7740–7781.
74. Perevedentseva, E.; Cai, P.-J.; Chiu, Y.-C.; Cheng, C.-L. Characterizing Protein Activities on the Lysozyme and Nanodiamond Complex Prepared for Bio Applications. *Langmuir* **2011**, *27*, 1085–1091.
75. Tzeng, Y.-K.; Faklaris, O.; Chang, B.-M.; Kuo, Y.; Hsu, J.-H.; Chang, H.-C. Superresolution Imaging of Albumin-Conjugated Fluorescent Nanodiamonds in Cells by Stimulated Emission Depletion. *Angew. Chem. Int. Ed.* **2011**, *50*, 2262–2265.
76. Kato, H.; Nakamura, A.; Horie, M.; Endoh, S.; Fujita, K.; Iwahashi, H.; Kinugasa, S. Preparation and Characterization of Stable Dispersions of Carbon Black and Nanodiamond in Culture Medium for in Vitro Toxicity Assessment. *Carbon* **2011**, *49*, 3989–3997.
77. Huang, L.-C. L.; Chang, H.-C. Adsorption and Immobilization of Cytochrome c on Nanodiamonds. *Langmuir* **2004**, *20*, 5879–5884.
78. Kong, X. L.; Huang, L. C. L.; Hsu, C.-M.; Chen, W.-H.; Han, C.-C.; Chang, H.-C. High-Affinity Capture of Proteins by Diamond Nanoparticles for Mass Spectrometric Analysis. *Anal. Chem.* **2005**, *77*, 259–265.
79. Huang, H.; Pierstorff, E.; Osawa, E.; Ho, D. Protein-Mediated Assembly of Nanodiamond Hydrogels into a Biocompatible and Biofunctional Multilayer Nanofilm. *ACS Nano* **2008**, *2*, 203–212.
80. Steinmüller-Nethl, D.; Kloss, F. R.; Najam-Ul-Haq, M.; Rainer, M.; Larsson, K.; Linsmeier, C.; Köhler, G.; Fehrer, C.; Lepperdinger, G.; Liu, X.; et al. Strong Binding of Bioactive BMP-2 to Nanocrystalline Diamond by Physisorption. *Biomaterials* **2006**, *27*, 4547–4556.
81. Chung, P.-H.; Perevedentseva, E.; Tu, J.-S.; Chang, C. C.; Cheng, C.-L. Spectroscopic Study of Bio-Functionalized Nanodiamonds. *Diam. Relat. Mater.* **2006**, *15*, 622–625.
82. Huang, T. S.; Tzeng, Y.; Liu, Y. K.; Chen, Y. C.; Walker, K. R.; Guntupalli, R.; Liu, C. Immobilization of Antibodies and Bacterial Binding on Nanodiamond and Carbon Nanotubes for Biosensor Applications. *Diam. Relat. Mater.* **2004**, *13*, 1098–1102.
83. Kong, X.; Huang, L. C. L.; Liau, S.-C. V.; Han, C.-C.; Chang, H.-C. Polylysine-Coated Diamond Nanocrystals for MALDI-TOF Mass Analysis of DNA Oligonucleotides. *Anal. Chem.* **2005**, *77*, 4273–4277.
84. Aramesh, M.; Shimoni, O.; Ostrikov, K.; Prawer, S.; Cervenka, J. Surface Charge Effects in Protein Adsorption on Nanodiamonds. *Nanoscale* **2015**, *7*, 5726–5736.
85. Perevedentseva, E. V.; Su, F. Y.; Su, T. H.; Lin, Y. C.; Cheng, C. L.; Karmenyan, A. V.; Priezzhev, A. V.; Lugovtsov, A. E. Laser-Optical Investigation of the Effect of Diamond Nanoparticles on the Structure and Functional Properties of Proteins. *Quantum Electron.* **2011**, *40*, 1089–1093.
86. Liu, K.-K.; Chen, M.-F.; Chen, P.-Y.; Lee, T. J. F.; Cheng, C.-L.; Chang, C.-C.; Ho, Y.-P.; Chao, J.-I. Alpha-Bungarotoxin Binding to Target Cell in a Developing Visual System by Carboxylated Nanodiamond. *Nanotechnology* **2008**, *19*, 205102.
87. Chao, J.-I.; Perevedentseva, E.; Chung, P.-H.; Liu, K.-K.; Cheng, C.-Y.; Chang, C.-C.; Cheng, C.-L. Nanometer-Sized Diamond Particle as a Probe for Biolabeling. *Biophys. J.* **2007**, *93*, 2199–2208.
88. Ermakova, A.; Pramanik, G.; Cai, J.-M.; Algara-Siller, G.; Kaiser, U.; Weil, T.; Tzeng, Y.-K.; Chang, H. C.; McGuinness, L. P.; Plenio, M. B.; et al. Detection of a Few Metallo-Protein Molecules Using Color Centers in Nanodiamonds. *Nano Lett.* **2013**, *13*, 3305–3309.

89. Khalid, A.; Mitropoulos, A. N.; Marelli, B.; Simpson, D. A.; Tran, P. A.; Omenetto, F. G.; Tomljenovic-Hanic, S. Fluorescent Nanodiamond Silk Fibroin Spheres: Advanced Nanoscale Bioimaging Tool. *ACS Biomater. Sci. Eng.* **2015**, *1*, 1104–1113.
90. Wu, Y.; Ermakova, A.; Liu, W.; Pramanik, G.; Vu, T. M.; Kurz, A.; McGuinness, L.; Naydenov, B.; Hafner, S.; Reuter, R.; et al. Programmable Biopolymers for Advancing Biomedical Applications of Fluorescent Nanodiamonds. *Adv. Funct. Mater.* **2015**, *25*, 6576–6585.
91. John, P.; Polwart, N.; Troupe, C. E.; Wilson, J. I. B. The Oxidation of Diamond: The Geometry and Stretching Frequency of Carbonyl on the (100) Surface. *J. Am. Chem. Soc.* **2003**, *125*, 6600–6601.
92. Tu, J.-S.; Perevedentseva, E.; Chung, P.-H.; Cheng, C.-L. Size-Dependent Surface CO Stretching Frequency Investigations on Nanodiamond Particles. *J. Chem. Phys.* **2006**, *125*, 174713.
93. Purtov, K. V.; Petunin, A. I.; Burov, A. E.; Puzyr, A. P.; Bondar, V. S. Nanodiamonds as Carriers for Address Delivery of Biologically Active Substances. *Nanoscale Res. Lett.* **2010**, *5*, 631–636.
94. Say, J. M.; van Vreden, C.; Reilly, D. J.; Brown, L. J.; Rabeau, J. R.; King, N. J. C. Luminescent Nanodiamonds for Biomedical Applications. *Biophys. Rev.* **2011**, *3*, 171–184.
95. Schrand, A. M.; Hens, S. A. C.; Shenderova, O. A. Nanodiamond Particles: Properties and Perspectives for Bioapplications. *Crit. Rev. Solid State Mater. Sci.* **2009**, *34*, 18–74.
96. Mohan, N.; Chen, C.-S.; Hsieh, H.-H.; Wu, Y.-C.; Chang, H.-C. In Vivo Imaging and Toxicity Assessments of Fluorescent Nanodiamonds in Caenorhabditis Elegans. *Nano Lett.* **2010**, *10*, 3692–3699.
97. Chang, B.-M.; Lin, H.-H.; Su, L.-J.; Lin, W.-D.; Lin, R.-J.; Tzeng, Y.-K.; Lee, R. T.; Lee, Y. C.; Yu, A. L.; Chang, H.-C. Highly Fluorescent Nanodiamonds Protein-Functionalized for Cell Labeling and Targeting. *Adv. Funct. Mater.* **2013**, *23*, 5737–5745.
98. Wei, L.; Zhang, W.; Lu, H.; Yang, P. Immobilization of Enzyme on Detonation Nanodiamond for Highly Efficient Proteolysis. *Talanta* **2010**, *80*, 1298–1304.
99. Weng, M.-F.; Chiang, S.-Y.; Wang, N.-S.; Niu, H. Fluorescent Nanodiamonds for Specifically Targeted Bioimaging: Application to the Interaction of Transferrin with Transferrin Receptor. *Diam. Relat. Mater.* **2009**, *18*, 587–591.
100. Cheng, C.-Y.; Perevedentseva, E.; Tu, J.-S.; Chung, P.-H.; Cheng, C.-L.; Liu, K.-K.; Chao, J.-I.; Chen, P.-H.; Chang, C.-C. Direct and in Vitro Observation of Growth Hormone Receptor Molecules in A549 Human Lung Epithelial Cells by Nanodiamond Labeling. *Appl. Phys. Lett.* **2007**, *90*, 163903.
101. Gaillard, C.; Girard, H. A.; Falck, C.; Paget, V.; Simic, V.; Ugolin, N.; Bergonzo, P.; Chevillard, S.; Arnault, J. C. Peptide Nucleic Acid–nanodiamonds: Covalent and Stable Conjugates for DNA Targeting. *RSC Adv.* **2013**, *4*, 3566–3572.
102. Mkandawire, M.; Pohl, A.; Gubarevich, T.; Lapina, V.; Appelhans, D.; Rödel, G.; Pompe, W.; Schreiber, J.; Opitz, J. Selective Targeting of Green Fluorescent Nanodiamond Conjugates to Mitochondria in HeLa Cells. *J. Biophotonics* **2009**, *2*, 596–606.
103. Cantor, C. R.; Schimmel, P. R. *Biophysical Chemistry;* Freeman: San Francisco, 1980.
104. Tamm, L. K.; Tatulian, S. A. Infrared Spectroscopy of Proteins and Peptides in Lipid Bilayers. *Q. Rev. Biophys.* **1997**, *30*, 365–429.
105. Militello, V.; Casarino, C.; Emanuele, A.; Giostra, A.; Pullara, F.; Leone, M. Aggregation Kinetics of Bovine Serum Albumin Studied by FTIR Spectroscopy and Light Scattering. *Biophys. Chem.* **2004**, *107*, 175–187.

106. Lin, C.-L.; Lin, C.-H.; Chang, H.-C.; Su, M.-C. Protein Attachment on Nanodiamonds. *J. Phys. Chem. A* **2015**, *119*, 7704–7711.
107. Wang, H.-D.; Niu, C. H.; Yang, Q.; Badea, I. Study on Protein Conformation and Adsorption Behaviors in Nanodiamond Particle–Protein Complexes. *Nanotechnology* **2011**, *22*, 145703.
108. Habibi, B.; Jahanbakhshi, M. Sensitive Determination of Hydrogen Peroxide Based on a Novel Nonenzymatic Electrochemical Sensor: Silver Nanoparticles Decorated on Nanodiamonds. *J. Iran. Chem. Soc.* **2015**, *12*, 1431–1438.
109. Zhang, B.; Fang, C.-Y.; Chang, C.-C.; Peterson, R.; Maswadi, S.; Glickman, R. D.; Chang, H.-C.; Ye, J. Y. Photoacoustic Emission from Fluorescent Nanodiamonds Enhanced with Gold Nanoparticles. *Biomed. Opt. Express* **2012**, *3*, 1662–1729.
110. Cheng, L.-C.; Chen, H. M.; Lai, T.-C.; Chan, Y.-C.; Liu, R.-S.; Sung, J. C.; Hsiao, M.; Chen, C.-H.; Her, L.-J.; Tsai, D. P. Targeting Polymeric Fluorescent Nanodiamond–Gold/silver Multi-Functional Nanoparticles as a Light-Transforming Hyperthermia Reagent for Cancer Cells. *Nanoscale* **2013**, *5*, 3931–3940.
111. Martín, R.; Menchón, C.; Apostolova, N.; Victor, V. M.; Álvaro, M.; Herance, J. R.; García, H. Nano-Jewels in Biology. Gold and Platinum on Diamond Nanoparticles as Antioxidant Systems Against Cellular Oxidative Stress. *ACS Nano* **2010**, *4*, 6957–6965.
112. Liu, Y. L.; Sun, K. W. Plasmon-Enhanced Photoluminescence from Bioconjugated Gold Nanoparticle and Nanodiamond Assembly. *Appl. Phys. Lett.* **2011**, *98*, 153702.
113. Huck, A.; Kumar, S.; Shakoor, A.; Andersen, U. L. Controlled Coupling of a Single Nitrogen-Vacancy Center to a Silver Nanowire. *Phys. Rev. Lett.* **2011**, *106*, 96801.
114. Potamianos, D.; Papadakis, I.; Kakkava, E.; Bourlinos, A. B.; Trivizas, G.; Zboril, R.; Couris, S. Nonlinear Optical Response of Gold-Decorated Nanodiamond Hybrids. *J. Phys. Chem. C* **2015**, *119*, 24614–24620.
115. Wang, Y.; Zang, J.; Dong, L.; Pan, H.; Yuan, Y.; Wang, Y. Graphitized Nanodiamond Supporting PtNi Alloy as Stable Anodic and Cathodic Electrocatalysts for Direct Methanol Fuel Cell. *Electrochimica Acta* **2013**, *113*, 583–590.
116. Lu, R.; Zang, J. B.; Wang, Y. H.; Zhao, Y. L. Microwave Synthesis and Properties of Nanodiamond Supported PtRu Electrocatalyst for Methanol Oxidation. *Electrochimica Acta* **2012**, *60*, 329–333.
117. Magdalinova, N. A.; Kalmykov, P. A.; Klyuev, M. V. Hydrogenation Catalysts Based on Platinum- and Palladium-Containing Nanodiamonds. *Russ. J. Gen. Chem.* **2014**, *84*, 33–39.
118. Minati, L.; Cheng, C. L.; Lin, Y. C.; Hees, J.; Lewes-Malandrakis, G.; Nebel, C. E.; Benetti, F.; Migliaresi, C.; Speranza, G. Synthesis of Novel Nanodiamonds–Gold Core Shell Nanoparticles. *Diam. Relat. Mater.* **2015**, *53*, 23–28.
119. Wang, C.; Chen, J.; Yang, G.; Xu, N. Thermodynamic Stability and Ultrasmall-Size Effect of Nanodiamonds. *Angew. Chem. Int. Ed.* **2005**, *44*, 7414–7418.
120. Butenko, Y. V.; Kuznetsov, V. L.; Chuvilin, A. L.; Kolomiichuk, V. N.; Stankus, S. V.; Khairulin, R. A.; Segall, B. Kinetics of the Graphitization of Dispersed Diamonds at "Low" Temperatures. *J. Appl. Phys.* **2000**, *88*, 4380.
121. Havlik, J.; Petrakova, V.; Rehor, I.; Petrak, V.; Gulka, M.; Stursa, J.; Kucka, J.; Ralis, J.; Rendler, T.; Lee, S.-Y.; et al. Boosting Nanodiamond Fluorescence: Towards Development of Brighter Probes. *Nanoscale* **2013**, *5*, 3208–3211.
122. Stursa, J.; Havlik, J.; Petrakova, V.; Gulka, M.; Ralis, J.; Zach, V.; Pulec, Z.; Stepan, V.; Zargaleh, S. A.; Ledvina, M.; et al. Mass Production of Fluorescent Nanodiamonds with a Narrow Emission Intensity Distribution. *Carbon* **2016**, *96*, 812–818.

123. Danilenko, V. V. Assessment of the Influence of Particle Size and Degree of Cohesion of Detonation Nanodiamonds on Their Thermal Stability. *J. Superhard Mater.* **2009**, *31*, 218–225.
124. Xie, F. Y.; Xie, W. G.; Gong, L.; Zhang, W. H.; Chen, S. H.; Zhang, Q. Z.; Chen, J. Surface Characterization on Graphitization of Nanodiamond Powder Annealed in Nitrogen Ambient. *Surf. Interface Anal.* **2010**, *42*, 1514–1518.
125. Zang, J.; Wang, Y.; Bian, L.; Zhang, J.; Meng, F.; Zhao, Y.; Ren, S.; Qu, X. Surface Modification and Electrochemical Behaviour of Undoped Nanodiamonds. *Electrochimica Acta* **2012**, *72*, 68–73.
126. Petit, T.; Arnault, J.-C.; Girard, H. A.; Sennour, M.; Bergonzo, P. Early Stages of Surface Graphitization on Nanodiamond Probed by X-Ray Photoelectron Spectroscopy. *Phys. Rev. B* **2011**, *84*.
127. Petit, T.; Arnault, J.-C.; Girard, H. A.; Sennour, M.; Kang, T.-Y.; Cheng, C.-L.; Bergonzo, P. Oxygen Hole Doping of Nanodiamond. *Nanoscale* **2012**, *4*, 6792–6799.
128. Wang, R.; Sun, X.; Zhang, B.; Sun, X.; Su, D. Hybrid Nanocarbon as a Catalyst for Direct Dehydrogenation of Propane: Formation of an Active and Selective Core–Shell $sp^2/sp^3$ Nanocomposite Structure. *Chem. – Eur. J.* **2014**, *20*, 6324–6331.
129. Sun, X.; Wang, R.; Zhang, B.; Huang, R.; Huang, X.; Su, D. S.; Chen, T.; Miao, C.; Yang, W. Evolution and Reactivity of Active Oxygen Species on sp2@sp3 Core–Shell Carbon for the Oxidative Dehydrogenation Reaction. *ChemCatChem* **2014**, *6*, 2270–2275.
130. Zang, J.; Wang, Y.; Bian, L.; Zhang, J.; Meng, F.; Zhao, Y.; Qu, X.; Ren, S. Bucky Diamond Produced by Annealing Nanodiamond as a Support of Pt Electrocatalyst for Methanol Electrooxidation. *Int. J. Hydrog. Energy* **2012**, *37*, 6349–6355.
131. Dong, L.; Zang, J.; Su, J.; Jia, Y.; Wang, Y.; Lu, J.; Xu, X. Nanodiamond/nitrogen-Doped Graphene (Core/Shell) as an Effective and Stable Metal-Free Electrocatalyst for Oxygen Reduction Reaction. *Electrochimica Acta* **2015**, *174*, 1017–1022.
132. Neburkova, J.; Vavra, J.; Cigler, P. Coating nanodiamonds with biocompatible shells for applications in biology and medicine. *Curr. Opin. Solid State Mater. Sci.* **2016**. http://dx.doi.org/10.1016/j.cossms.2016.05.008.
133. Kozak, O.; Sudolska, M.; Pramanik, G.; Cigler, P.; Otyepka, M.; Zboril, R. Photoluminescent Carbon Nanostructures. *Chem. Mater.* **2016**, *28*, 4085–4128.

## ABBREVIATIONS

| | |
|---|---|
| **NDs** | nanodiamonds |
| **NPs** | nanoparticles |
| **NV** | nitrogen vacancy centers |
| **SiV** | silicon vacancy centers |
| **DNDs** | detonation nanodiamonds |
| **TEOS** | tetraethyl orthosilicate |
| **APTES** | (3-aminopropyl)trimethoxysilane |
| **HPHT** | high-pressure high-temperature |
| **PVP** | polyvinylpyrrolidone |
| **BTSE** | 1,2-bis triethoxysilyl ethane |
| **TMSPMA** | 3-(trimethoxysilyl)propyl methacrylate |
| **PHPMA** | poly[N-(2-hydroxypropyl)methacrylamide] |

| | |
|---|---|
| **IR** | infrared spectroscopy |
| **DLS** | dynamic light scattering |
| **TEM** | transmission electron microscope |
| **HRTEM** | high-resolution transmission electron microscope |
| **EELS** | electron energy loss spectroscopy |
| **STEM** | scanning transmission electron microscope |
| **PMADQUAT** | poly[2-(methacryloyloxy)ethyl trimethylammonium chloride] |
| **PEG** | polyethylene glycol |
| **PPEGMA** | poly[PEG methyl ether methacrylate] |
| **PG** | polyglycerol |
| **PEI** | polyethylene-imine |
| **PAH** | polyallylamine hydrochloride |
| **PDMAEMA** | poly[2-(dimethylamino)ethyl methacrylate] |
| **PNIPAM** | poly(*N*-isopropylacrylamide) |
| **BMA** | butyl methacrylate |
| **PBS** | phosphate buffer saline |
| **ATRP** | atom transfer radical polymerization |
| **RAFT** | reversible addition-fragmentation chain transfer polymerization |
| **PMAA** | poly(methacrylic acid) |
| **AIBN** | azobisisobutyronitrile |
| **RGD** | tripeptide arginine glycine aspartic acid |
| **NMR** | nuclear magnetic resonance |
| **$\alpha$-BTX** | $\alpha$-bungarotoxin |
| **BMP-2** | bone morphogenetic protein-2 |
| **PNA** | peptide nucleic acids |
| **FTIR** | Fourier transform infrared spectroscopy |
| **CD** | circular dichroism |
| **GNS** | gold nanoshell |
| **HAADF-STEM** | high angle annular dark field scanning transmission electron microscopy |
| **XPS** | X-ray photoelectron spectroscopy |

## ACKNOWLEDGEMENTS

This work was supported by the Grant Agency of the Czech Republic Project No. 16-16336S, Ministry of Health of the Czech Republic, grant nr. 15e33094A and by Ministry of Education, Youth and Sports project MOBILITY, grant nr. 7AMB16FR029.

CHAPTER

# Nanodiamonds in composites: polymer chemistry and tribology

# 14

**Ioannis Neitzel***, **Vadym N. Mochalin**†, **Yury Gogotsi***

*A.J. Drexel Nanomaterials Institute and Department of Materials Science & Engineering, Drexel University, Philadelphia, PA, United States* Department of Chemistry, Missouri University of Science & Technology, Rolla, MO, United States*†

**Contents**



## 14.1 INTRODUCTION TO POLYMER NANOCOMPOSITES

Polymer composite materials are widely used as structural components due to their high strength and low weight. Common examples include continuous fiber reinforced polymers such as carbon fiber–epoxy composites, which have Young's moduli of up to ~150 GPa.[1] Polymer–particulate composites represent non-continuous reinforced matrices, where, for example, the stiffness and fracture toughness are improved.[2] Polymer nanocomposites (PNCs) belong to the subgroup of particulate composites. These PNCs can in turn be used as a new matrix material for continuous fiber reinforced composites, possibly resulting in composites with further improved properties.

By definition,[3] a PNC is a multiphase material, in which, in contrast to microcomposites, one of the phases has one, two, or three dimensions of less than 100 nm, or the average distances between different phases of the composite are less than 100 nm. Another difference of PNCs and traditional polymer composites is the large amount of "interphase" material. The term "interphase" is used to designate the polymer affected by the presence of fillers to such extent that it behaves as a new phase, often

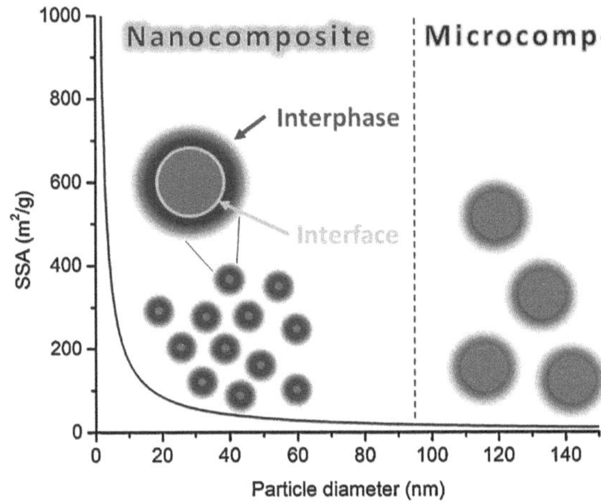


**FIGURE 14.1**

Specific surface area (SSA) as a function of particle diameter. As the particle diameter decreases, the amount of interfacial material increases. For optimal mechanical properties, a strong interface, which transfers load from the filler to the interphase and polymer matrix is mandatory. Composites containing particles smaller than 100 nm fall into the subclass of nanocomposites

having a structure and characteristics significantly different from those of the neat polymer. The interphase properties depend on many factors, including the size and surface chemistry of the filler. Due to the small size of nanofillers resulting in a large surface area, the amount of polymer in contact with the filler, and therefore affected by the filler, dramatically increases. When the volume content of the nanofiller is large enough, the interphase becomes the dominant phase in the composite (Fig. 14.1). Controlling the properties of the interphase is the key to tailoring the properties of the nanocomposites for applications. Resulting PNCs can have manifold properties, tailored for the application by selecting an appropriate polymer, the material of the filler and its surface chemistry, and are of interest for various applications, ranging from mechanical to electrical or optical ones.

Various PNCs with different polymers and fillers were studied in the past. For example, nanosilica was added for mechanical reinforcement,[4] carbon nanotubes (CNTs) improved electrical conductivity,[5] and gold nanoparticles enhanced the dielectric contrast[6] in polymers. These studies show that various properties can be improved due to the addition of nanofillers. However, in general the reported improvements in properties are below expectations.[7]

One major challenge common to all nanoparticles is agglomeration. Although this feature can sometimes be beneficial, for most applications, the agglomeration of nanofillers is undesirable, since the properties of the agglomerates are significantly inferior to the nanoparticle properties itself.[8] Thus, reducing agglomerate size and improving the dispersion (optimal are random or uniform dispersions) of

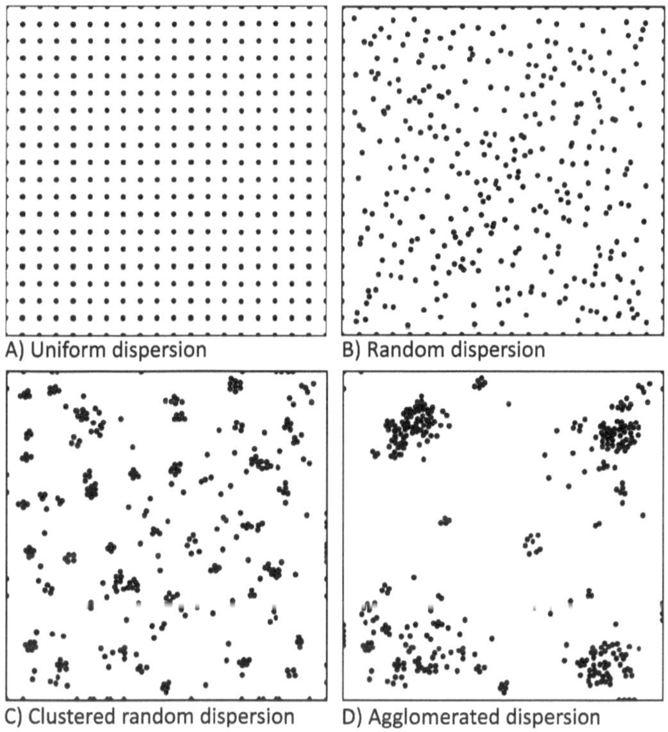


**FIGURE 14.2**

Possible dispersion and agglomeration states in polymer nanocomposites. (A) and (B) Uniform and random dispersions are desired since this state allows taking advantage of the large surface area of nanofillers. (C) and (D) Clustered and agglomerated dispersions are undesired since the properties of agglomerates and clusters can be significantly different from the filler properties. From Khare, H. S.; Burris, D. L. A Quantitative Method for Measuring Nanocomposite Dispersion. *Polymer* **2010**, *51* (3), 719–729[9]

nanoparticles is of major importance (Fig. 14.2).[9] Models for nanoparticle–polymer composites which take the filler shape, filler–filler, and filler–polymer interactions into account can provide guidance on how to select nanofillers and change the surface properties of the fillers to decrease agglomerate size and improve dispersion in the polymer.[10] In practice, the optimal interaction parameters can be achieved by changing the surface chemistry of the filler, selecting polymer molecules with appropriate functional groups, and choosing applicable processing methods.

In summary, to further improve the properties of PNCs it is important to consider the following: first, a nanofiller with properties beneficial for the application should be selected, e.g. good mechanical properties for load bearing composites, high (or low) electrical conductivity for conducting (or insulating) materials, transparency or high index of refraction for optical applications. Second, the nanofiller should have

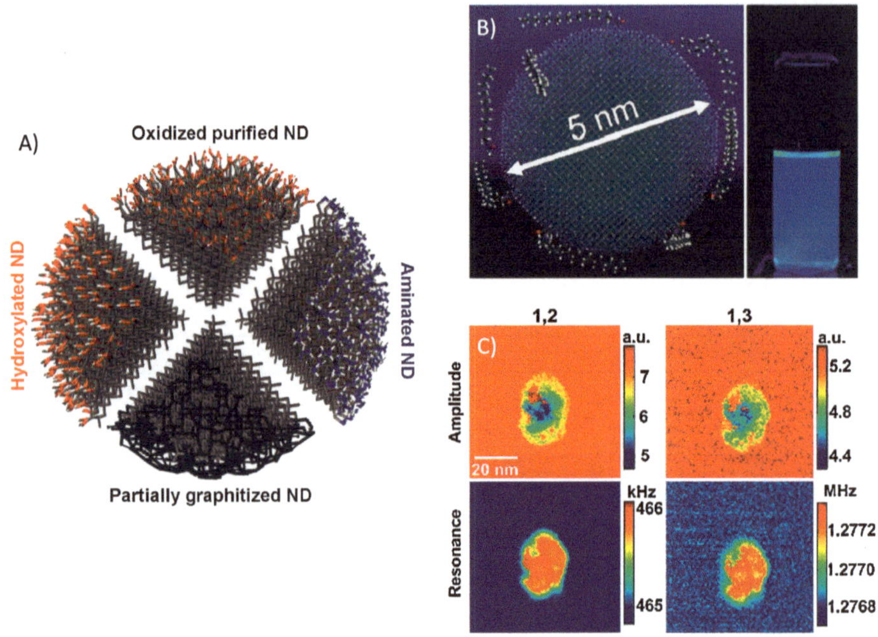


**FIGURE 14.3**

(A) Computer models showing different surface chemistries of ND[12] and (B) a model of ND functionalized with octadecylamine (ODA).[28] Bright fluorescence of this material was observed when dispersed in different hydrophobic solvents. (C) AFM scans of ND–ODA particles using generalized phase imaging scans (resonance and amplitude signals are shown), based on broad band detection and multi-eigenmode operations at different combinations of topographic and BE eigenmodes.[91] Due to the high spatial resolution, the grafted octadecylamine molecules to ND become traceable

a large surface area (small particle size). Finally, an optimal nanofiller should have a tailorable surface chemistry allowing for control of the nanofiller dispersion, polymer interface and the interphase.

Nanodiamond (ND) produced by detonation synthesis (Chapter 2: Production and purification of NDs) in large commercial quantities[11,12] is an optimal nanofiller for polymer composites. It has many of the unique properties of bulk diamond, including superior mechanical properties (hardness of 10 on Mohs scale and Young's modulus of ~1220 GPa), excellent tribological properties,[13] a high thermal conductivity (one of the highest among all materials ~500 $WmK^{-1}$ [14]), and a high refraction index to name a few. ND is composed of small nearly spherical ~5 nm diameter diamond particles with a specific surface area of ~350 $m^2/g$, a narrow particle size distribution, and a rich and tailorable surface chemistry[15–19] (Fig. 14.3). Advantages of ND particles for nanocomposites are multiple and stem from its unique properties: (i) diamond structure that provides superior Young's modulus, hardness, high thermal conductivity and electrical resistivity, low coefficient of friction, chemical stability, and biocompatibility; (ii) small and uniform size, eliminating the need for costly fractionation procedures required for many other nanoparticles (Chapter 2: Production and purification of NDs); (iii) nearly spherical shape as compared to other carbon nanomaterials such as nanotubes or graphene; (iv) large and accessible external surface, maximizing interactions with the polymer, i.e., the interphase volume; and (v) rich and tailorable surface chemistry (Chapter 8: Current issues and challenges in surface chemistry of nanodiamonds), providing great flexibility for the rational design of the polymer–ND interface. The role of these properties is discussed in detail in the literature.[20] In general, ND fulfills all requirements to a perfect nanofiller. Various gas and wet chemistry treatments can be applied to optimize NDs properties for a specific application, as discussed in the above chapters.

In this chapter, we summarize the latest developments in the field of polymer–ND composites and focus on how mechanical, thermal and tribological properties are improved due to the addition of ND and address the role of the interphase.

## 14.2 POLYMER–ND COMPOSITES: THE INFLUENCE OF ND AND ITS SURFACE CHEMISTRY

In this section, we describe recent developments in polymer–ND design, focusing on the mechanical and thermal properties of the composites and give examples on how dispersion can be improved. We will provide representative examples of how ND can improve mechanical properties and how its surface chemistry and interphase can be tailored to achieve optimal properties.

In the presented studies, various mechanical characterization techniques are used to measure properties that are important for potential applications and to understand reinforcing mechanisms, briefly introduced in the following. The performance of bulk composites is commonly measured using tensile and toughness tests. To gain information at various temperatures and loading conditions, dynamic mechanical analysis (DMA) is suitable. Nanoindentation is a powerful indentation technique for measuring the visco-elastic properties of materials with high spatial resolution, while only requiring small samples. It enables researchers to investigate reinforcing mechanisms on the micron- and nanoscale. A review on the characterization of PNCs using nanoindentation is available.[21] Further mechanical information about the properties of polymer composites on the nanoscale can be obtained using AFM. An AFM system uses sharp tips that can be used to measure and manipulate matter at surfaces with atomic resolution.[22]

This section is split into three subsections following the classic subdivision of polymer matrices into thermoplastic, elastomeric, and thermosetting. Chemical structures of most of the presented polymer matrices are shown in Fig. 14.4. In addition to the presented examples, the reader can also refer to recent review papers, addressing the properties of polymer–ND composites.[20,23,24]

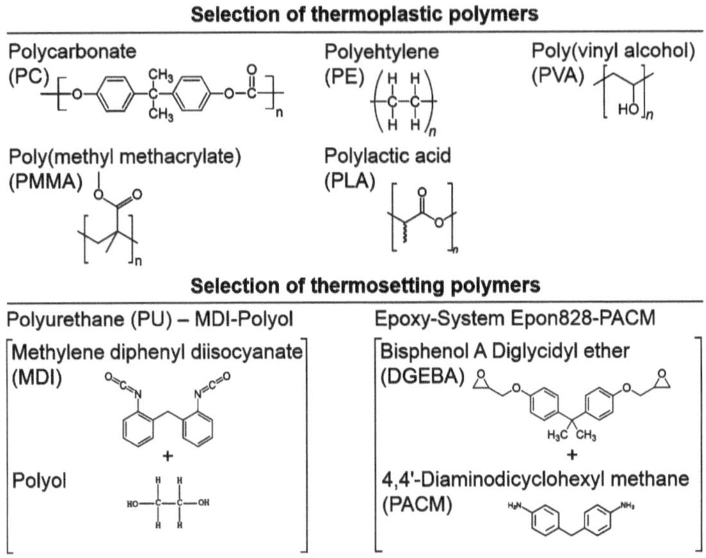


**FIGURE 14.4**

Chemical structures of some thermoplastic and thermosetting polymer matrices investigated in the presented studies[92–94]

## 14.2.1 THERMOPLASTIC–ND COMPOSITES

In the first presented study, the differences in the reinforcing effect of ND depending on its surface chemistry are investigated, where two different thermoplastic polymers (polycarbonate (PC, Fig. 14.4) and Poly(methyl methacrylate) (PMMA, Fig. 14.4)) were reinforced with as-received and different functionalized NDs.[25] Hardness and Young's moduli of these nanocomposites were increased by up to 100%, when the appropriate surface functionality was selected. Young's moduli of PC and PMMA with same contents of different functionalized NDs are compared in Fig. 14.5. Highest improvements in mechanical properties are obtained when using amide functionalized ND for PMMA and amino functionalized ND for PC.[25]

In the above study, the main reason for improved properties is ascribed to the selection of an appropriate surface functionalization. In contrast, a mismatching surface functionalization can lead to significant decreases in mechanical properties. For example, the Young's moduli and hardness of a Kevlar–ND composite measured by nanoindentation were reduced by more than 50%. In this case, according to the authors, the attractive interactions between NDs COOH groups and Kevlar molecules played a detrimental role, destroying the supramolecular structure of the aramide.[26] In another example, the scratch resistance of a polyacrylic polymer could either be increased when incorporating hydrophobic ND (functionalization with oleylamine (OLA)) or decreased when using oxidized ND.[27]

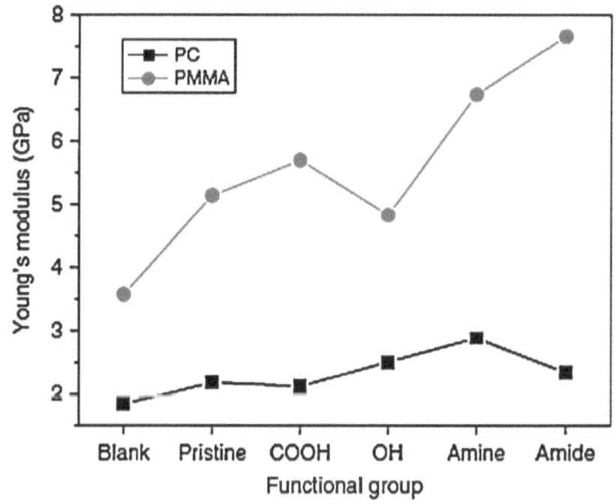


**FIGURE 14.5**

Young's modulus of polycarbonate (PC) and poly-(methyl methacrylate) (PMMA) composites with different functionalized NDs. From Khare, H. S.; Burris, D. L. A Quantitative Method for Measuring Nanocomposite Dispersion. *Polymer* **2010**, *51* (3), 719–729[25]

These examples emphasize the importance of selecting a ND with a surface chemistry that matches to the polymer matrix and provides properties required in the application.

Besides changing ND surface functional groups, another approach is the covalent attachment of polymers (Fig. 14.3B, Fig. 14.9A).[28,29] Polyethylene (PE, Fig. 14.4) is a widely used polymer consisting of a $-CH_2-$ backbone, with no side chain functionalities. To improve the affinity of ND to this specific polymer matrix, alkyl chains of variable length were grafted to ND. This approach allowed to gain new insights into interphase formation. Differential Scanning Calorimetry (DSC) was performed to characterize the polyethylene–ND composites. In a DSC experiment the response of a material to heat is measured and allows to characterize various properties.[30] Increased crystallinity, crystallization, and melting temperature of the nanocomposites with increasing ND content and alkyl chain length were found. Following this trend, Young's moduli and hardness values measured by AFM were increased by a factor of 2.5 and 4.5, respectively; the increase was larger with longer grafted alkyl chains (Fig. 14.6A). These improvements show the impact of ND surface chemistry on the properties of the interphase.[31] For a cross-linked polyethylene (PEX) improved mechanical properties were reported as well, albeit with a maximum improvement in yield stress of only 30% with 0.5 wt.% addition of as-received ND (Fig. 14.6B). Larger additions resulted in smaller improvements.[32] Comparison of these two studies shows that better mechanical properties can be achieved with specifically optimized ND surface chemistry, matched to the properties of the polymer matrix. This finding is also confirmed for other polymer matrices as discussed below.

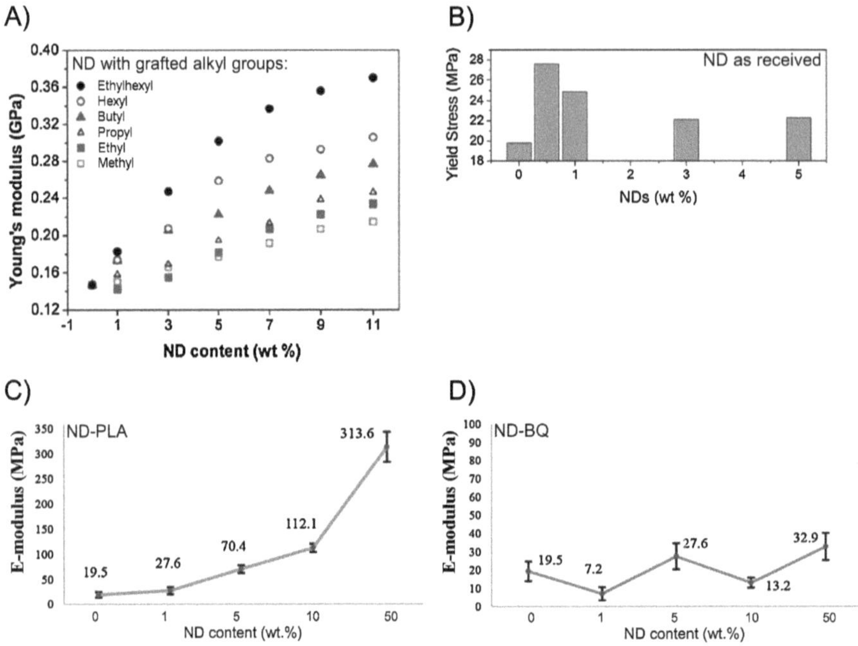


**FIGURE 14.6**

(A) Young's modulus of polyethylene–ND composites as functions of ND content and grafted alkyl chain length to ND. From Jee, A.-Y.; Lee, M. Thermal and Mechanical Properties of Alkyl-Functionalized Nanodiamond Composites. *Curr. Appl. Phys.* **2011**, *11* (5), 1183–1187.[31] (B) Yield stress of PEX–ND composites as a function of content, using as-received ND. From Roumeli et al., Factors Controlling the Enhanced Mechanical and Thermal Properties of Nanodiamond-Reinforced Cross-Linked High Density Polyethylene. *J. Phys. Chem. B* **2014**, *118* (38), 11341–11352.[32] (C) Summary of E-modulus values determined by tensile testing of composites made with ND–PLA, (D) of composites made with ND–BQ. Composites made with ND–PLA have a six times increased E-Modulus at a rather low addition of 10 wt.%, while composites made with ND–BQ show no improvement in E-Modulus, underlining the importance of selecting an appropriate surface functionalization. From Sun et al., Reinforced Degradable Biocomposite by Homogenously Distributed Functionalized Nanodiamond Particles. *Macromol. Mater. Eng.* **2015**, *300* (4), 436–447[38]

Poly-Lactic Acid (PLA, Fig. 14.4) is a common biodegradable polymer, used in tissue engineering. Storage modulus, tensile modulus, tensile strength, and the thermal stability of PLA were improved due to the addition of 5 wt.% as-received ND. An increased crystallinity of the matrix (measured by DSC), as well as the attractive molecular interactions between ND and PLA molecules (deduced from Thermogravimetric analysis (TGA) data, also discussed in more detail in Chapter 11: Interactions with solvent), are possible reasons for the improvements.[33]

An optical isomer (enantiomer) of PLA, poly-L-(Lactic-Acid) (PLLA), which is used in manufacturing fixation devices for bone fracture surgery, was reinforced by a specifically modified ND for this application. In order to increase the affinity of ND towards this polymer and improve the dispersion of the nanofiller, a hydrophobic amine-, octadecylamine (ODA), was covalently grafted[28] to NDs surface (Figs. 14.3B, 14.3C, and 14.7A), resulting in a decrease of nanoparticle–nanoparticle attractive interactions[34] and a better affinity of hydrophobic ND–ODA towards hydrophobic PLLA.[35] In contrast to pristine ND, the resulting ND–ODA material is hydrophobic, which is beneficial for dispersion in hydrophobic polymers. TEM images of PLLA–ND–ODA films (Figs. 14.7B and 14.7C) show single ND–ODA particles and loosely connected ND–ODA agglomerates dispersed in the matrix. The elongated shape of these agglomerates is essential. According to polymer–nanofiller theory,[36] the high aspect ratio ND–ODA agglomerates in the composites containing 3 wt.% and 10 wt.% ND–ODA (Figs. 14.7B and 14.7C) corresponds to the penultimate step before obtaining single dispersed particles. The covalent modification of ND with ODA translates into very good mechanical properties: PLLA–ND–ODA composites containing up to 10 wt.% ND–ODA showed an up to 2 times increase in Young's modulus and up to 8 times higher hardness when compared to neat PLLA, measured using nanoindentation.[35] As expected, unmodified ND reinforced the PLLA to a lesser extent, being in agreement with the above study.[33] The bulk compression modulus of a PLLA–ND–ODA composite containing 10 wt.% ND–ODA was increased by 22%, and a 316% increase in fracture energy was observed at the same time. It is speculated that ND–ODA induced crazing, observed by light microscopy and TEM, is responsible for the large increase in fracture energy (Figs. 14.7D–F).[37] Besides mechanical reinforcement, ND–ODA is fluorescent under UV light,[28] potentially allowing monitoring of the performance and bioresorption of bone fixation devices made of PLLA–ND–ODA.

In the context of effects of the nanofillers surface chemistry, the results of another study are striking. Also focusing on the reinforcing mechanism of ND in composites for biomedical applications, poly(L-lactide-co-ε-caprolactone) P(LLA-co-CL) was reinforced by the addition of ND functionalized with polylactide (ND-PLA), synthesized by a "grafting from" technique and benzoquinone-modified ND (ND–BQ). Composites made with ND–PLA have a six times increased E-Modulus (tensile test) at a rather low ND content of 10 wt.%, while composites made with ND–BQ show no improvement in E-Modulus (Figs. 14.6C and 14.6D). It is speculated that the interactions between the grafted polylactide and P(LLA-co-CL), physical crosslinks and an improved dispersion in the polymer matrix lead to these improvements.[38]

For other polymer matrices, the importance of ND dispersion has been highlighted as well. Poly-(methyl methacrylate) (PMMA, Fig. 14.4) is used for various applications in the automotive (light systems), electronic (screens), and biomedical (implants) industries. The impact strength and fracture toughness of PMMA–ND composites were increased due to the addition of only 0.1 wt.% of acid-purified ND. At ND contents of 0.8 wt.%, Young's modulus was 80% higher and the glass transition temperature ($T_g$) increased by 20°C. The improved properties are ex-

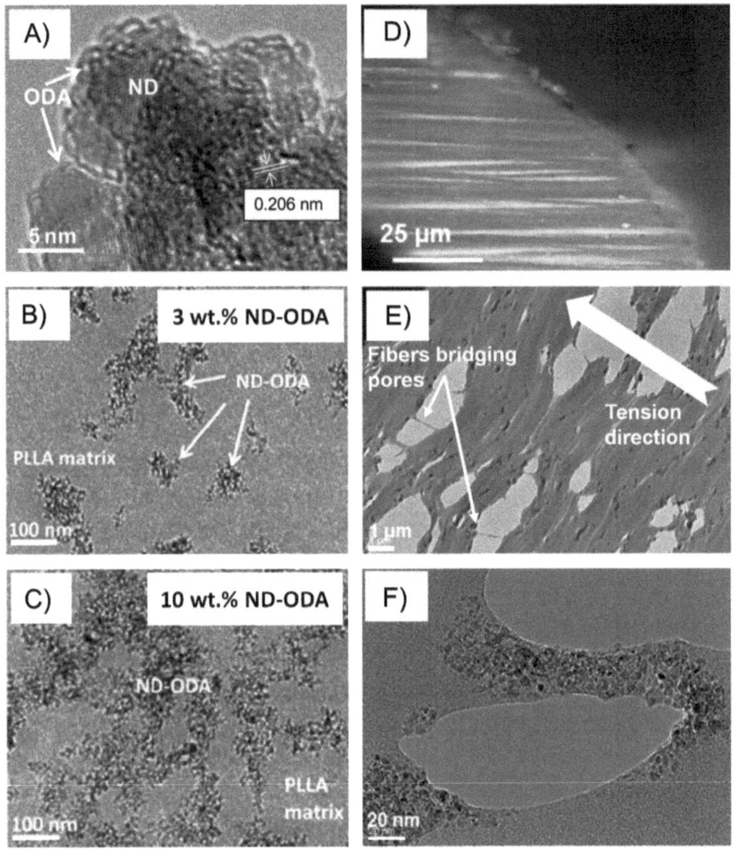


**FIGURE 14.7**

TEM images of ND–ODA and PLLA–ND–ODA composites[35]: (A) At high resolution the grafted ODA chains are seen on ND particles. Low resolution TEM images of PLLA–ND–ODA nanocomposites show good dispersion and reduced agglomeration of ND–ODA particles: composites containing (B) 3 wt.% and (C) 10 wt.% ND–ODA show high aspect ratio agglomerates, being the penultimate step before obtaining single dispersed particles according to polymer–nanofiller theory.[36] Light and TEM micrographs of fractured tensile PLLA–ND–ODA specimen containing 10 wt.% ND–ODA[37]: (D) Crazing observed using light microscopy, (E) TEM micrograph showing crazing, and (F) High magnification TEM micrograph showing typical pores and fibers bridging a crazing gap in ND–ODA reinforced PLLA

plained by hydrogen bonding between the polymer carbonyl groups and ND surface hydroxyl groups.[39] However, TEM micrographs of the produced nanocomposites showed significant agglomeration of ND, suggesting that further improvements are possible when single dispersed ND particles become available. In this context it is

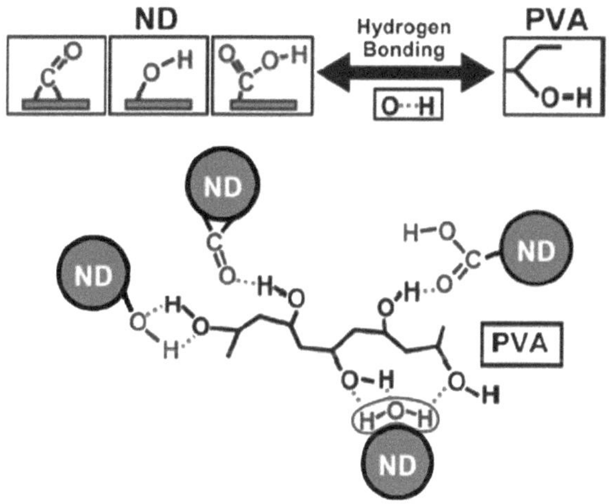


**FIGURE 14.8**

Illustration of the possible interactions between PVA and ND functional groups. Hydrogen bonding between oxygen containing functional groups on ND and hydroxyl-groups of PVA can explain the significant improvement in the mechanical properties of PVA–ND composites. From Morimune et al., Poly(Vinyl Alcohol) Nanocomposites With Nanodiamond. *Macromolecules* **2011**, *44* (11), 4415–4421[42]

also suggested that the addition of other nanoparticles can lead to synergetic effects (Section 14.2.4), improving dispersion.

Polyvinyl alcohol (PVA, Fig. 14.4) has numerous biomedical applications including soft tissue replacements, artificial cartilage, catheters, artificial skin, artificial pancreas, and hemodialysis membranes.[40] The use of ND to reinforce PVA was investigated in several studies. For example, a 40% improved tensile modulus and a 70% larger fracture energy were measured due to the addition of 1 vol.% of ND purified by a treatment with *o*-xylene to remove any adsorbed organics. The observed improved properties were explained by favorable interactions between ND–COOH groups and the hydroxyl groups of PVA.[41]

In a similar study, using "Gohsenol NH-18" PVA, a significantly larger improvement in tensile modulus (150%), increased yield strength and fracture energy were measured. The improvements were explained by an 8% increased crystallinity of PVA and hydrogen bonding between PVA hydroxyl groups and ND oxygen containing surface functional groups and adsorbed water (Fig. 14.8).[42]

In the above studies, the composites were manufactured by solution mixing and casting. Another approach to manufacture nanocomposites is to start from monomers and perform in situ polymerization directly on the ND surface. Resulting composites are expected to have an improved dispersion since ND agglomerates can be broken by the growing polymer chains during the reaction. At the same time, when the polymers are covalently attached to the ND surface, the strongest interface between the polymer

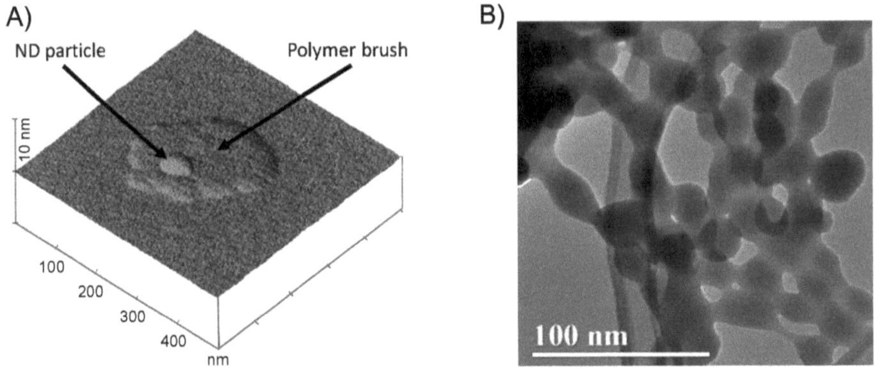


**FIGURE 14.9**

(A) AFM scan of a ND–PiBMA-1 polymer brush. The ND–polymer brush has a diameter of 300 nm, ~100 times larger than the pristine ND particle. From Li et al., Surface Functionalization of Nanodiamond Particles via Atom Transfer Radical Polymerization. *Carbon* **2006**, *44* (11), 2308–2315.[43] (B) High-resolution helium ion microscopy (HIM) image of the PEDOT–ND composite. From Tamburri et al., Detonation Nanodiamonds Tailor the Structural Order of PEDOT Chains in Conductive Coating Layers of Hybrid Nanoparticles. *J. Mater. Chem. C* **2014**, *2* (19), 3703–3716[45]

and ND is formed, which is mandatory for an effective load and heat transfer from the matrix to the nanofiller.

ND with covalently linked polymer chains (polymer–ND brushes) were synthesized using the atom transfer radical polymerization of poly(iso-butyl methacrylate) on the ND surface. The brushes were characterized by Fourier transform infrared spectroscopy (FTIR), nuclear magnetic resonance spectroscopy (NMR), TGA and AFM. Single ND–polymer brushes were visualized using AFM as shown in Fig. 14.9A. According to this data, the single ND–polymer brushes have diameters of up to 300 nm.[43]

Using a similar approach, polyimides were grown on ND particles. It was found that the XRD peak of polyimide at $2\Theta = 4.9°$ disappeared in ND composites, providing evidence that long range interactions between polyimide chains are disrupted due to the introduction of ND. The produced nanocomposite had a 25% higher Young's modulus and a 15% higher hardness at a ND content of 5 wt.%.[29]

For biomedical applications, biodegradable poly($\varepsilon$-caprolactone) was polymerized from the ND surface.[44] Successful grafting was confirmed by FTIR and TGA measurements. The synthesized ND is hydrophobic, dispersing well in the organic solvents, such as tetrahydrofuran and chloroform.

Also, in a recent study on an electrically conducting polymer, poly(3,4-ethylenedioxythiophene) (PEDOT), the polymer–ND composite was obtained using a grafting from approach applied to ND agglomerates. ND agglomerates significantly reduced the activation time of the polymerization process. The resulting composites

(Fig. 14.9B) have an up to 3 times higher modulus and hardness when compared to the neat polymer, measured by AFM.[45]

In summary, as-received ND can be used to reinforce various thermoplastic matrices, but to take full advantage of ND as an additive and create more consistent and ultimately strong composites, tailoring the surface chemistry of ND for each specific application is highly important.

### 14.2.2 ELASTOMER–ND COMPOSITES

An elastomer is defined as a cross-linked, amorphous polymer above its $T_g$.[46] Elastomers were probably the first type of polymers used for commercial composites (rubbers) and are heavily used in the automotive industry for tires, braking systems, chassis, interior parts or haptic applications, to name a few.[47,48] Their thermo-mechanical properties can be tailored by adjusting their chemical structure and composition.[49] For future applications, an additive is required that allows further improving properties of elastomers and broadening their application.

Improved mechanical properties, such as cohesive strength, rupture and wear resistance due to the addition of ND, were reported for a variety of elastomers including fluorinated elastomers and rubbers.[50] The effect of the surface functionalization of ND on the mechanical properties of polysiloxane films was studied as well, where mechanical properties such as engineering stress and tensile strength were increased due to the addition of silanized ND. The improvement in mechanical properties was attributed to a reduction in ND agglomerate size during a silylation reaction, which also helps to remove adsorbed water from the ND surface, rendering the material hydrophobic.[51]

For a thermoplastic elastomer (TPE) based on polypropylene (PP) and ethylene-propylene-diene rubber (EPDM), the storage modulus was increased over a broad temperature range from −80 to 150°C due to the addition of 5 wt.% graphitized ND. Nanoindentation experiments revealed increases in hardness and reduced modulus of 100% and ~140%, respectively. At the same time the thermal decomposition temperature was increased by ~40°C.[52]

The reinforcement of rubbers with ND was addressed in several studies and is summarized in a recent review.[53]

Still, as of now, only a few studies on elastomer–ND composites exist. Further studies on the reinforcing mechanisms of ND in elastomeric polymer matrices are required and are of great interest for various applications. The covalent incorporation of ND into elastomers is a promising approach that could yield materials with an improved chemical resistance, increased mechanical properties, and higher thermal stability.

### 14.2.3 THERMOSET–ND COMPOSITES

A cured thermosetting polymer is called a thermoset,[54] where the resin–hardener system cures into an infusible, insoluble polymer network. Thermosets find numerous

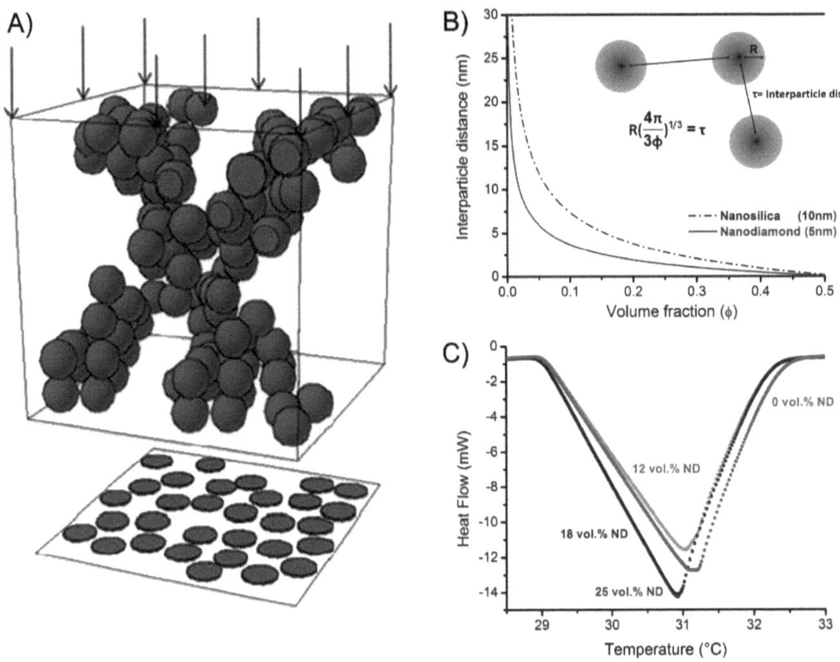


**FIGURE 14.10**

(A) Model of a high concentration ND composite. (B) Relationship between the volume fraction of the filler and the interparticle distance. In comparison to larger particles, the interparticle separation in nanodiamond composites is shorter at lower concentrations and disappears earlier. (C) DSC curves for epoxy–ND composites with 0–25 vol.% ND. A more negative slope of the recorded DSC curves relates to an increased thermal conductivity[56]

applications in industry as adhesives, structural components, polymer concretes, and matrix materials. These applications require superior mechanical properties. Similar to thermoplastics, ND can be incorporated into thermosets to improve their mechanical, thermal and tribological properties.

A common thermosetting polymer is epoxy, widely used as a matrix material for carbon fiber reinforced composites.[55] To further improve the mechanical properties of the matrix material (Epon828-PACM, Fig. 14.4) and to study the reinforcing potential of ND, epoxy–ND composites with ND contents of up to 35 vol.% were produced using as-received ND. Hardness and Young's moduli of these composites measured by nanoindentation were improved by 300% and 700%, respectively, reaching modulus values of up to 20 GPa and resulted in an increased scratch resistance. At such high ND contents, the composite material can be thought of as an ND network infiltrated by a polymer acting as a binder (Fig. 14.10A).[56] NDs small particle size results in a much shorter average interparticle distance at lower filler loadings, making ND advantageous compared to other nanofillers, such as nanosilica (Fig. 14.10B). The high loadings of ND resulted in a 30% improved thermal conductivity when compared to neat epoxy, measured by DSC (Fig. 14.10C).[56] Further improvements in thermal conductivity might be possible when purified or aminated ND is used to create a stronger polymer–ND interface.

While high loadings of ND result in remarkably high hardness and Young's moduli of the epoxy–ND composites, lower concentrations can increase the mechanical properties as well. The bulk Young's modulus of a Bisphenol A diglycidyl ether (DGEBA, Fig. 14.4) Jeffamine–D2000 epoxy system measured in tensile tests was 25% higher due to the addition of 0.5 wt.% as-received ND, which also increased the polymer decomposition temperature. However, due to poor dispersion, the storage modulus of the epoxy composite was significantly reduced with addition of ND,[57] emphasizing the importance of a good dispersion to optimize the mechanical properties of ND composite materials.

The mode I (tensile stress on a pre-crack) and II (shear stress on a pre-crack) fracture toughness of polymer composites are important characteristics for structural applications, where it was shown that the mode II fracture toughness can be improved by applying CNTs.[58] For epoxy–ND composites it was confirmed that, besides an improved Young's modulus and hardness, the mode II fracture toughness of epoxy–ND composites with ND loadings of 0.1 wt.% can be increased. This is because ND is thought to hinder shear deformation, improving fracture toughness.[59]

To strengthen the epoxy–nanofiller interface, functionalization of the nanofillers was explored. Following the analogy with molecular curing agents, which for epoxy resins are typically amines, nanoparticles were modified in a way to form surface amino groups that are expected to react with epoxy groups of the resin. CNTs functionalized with amino groups significantly improved the flexural modulus and strength of an epoxy matrix when compared to unmodified CNTs.[60] A route to aminated ND (ND–$NH_2$) was developed by covalent linking ethylenediamine to ND surface, making it possible to form a covalently bonded network of ND–$NH_2$ and epoxy molecules (Fig. 14.11A).[61] To confirm covalent bonding between the diamine molecules and ND particles, FTIR, TGA, and DSC studies were performed on ND–$NH_2$. Then ND–$NH_2$ reacted with the epoxy resin and a strong covalent epoxy–ND interface was formed. The resulting Young's modulus of a composite containing 3.5 vol.% ND–$NH_2$ was improved by 60% (Fig. 14.11B).[62] The higher increase in Young's modulus compared to composite prepared with as-received ND can be explained by covalent bond formation due to reaction between amino groups of ND–$NH_2$ and the epoxide groups of the resin. Also, it was found that to manufacture epoxy–ND–$NH_2$ composites with uniformly dispersed NDs it is important to use ND–$NH_2$ that was kept in an inert solvent without drying (Figs. 14.12E and 14.12F). This study also discussed the interference of amino groups of ND–$NH_2$ with the amino groups of curing agent used for this epoxy system: special attention was paid to compensate for the excess of amino groups introduced by ND–$NH_2$ in order to keep the right ratio of the amino-to-epoxy groups in the system and to maximize the Young's modulus of the composite. Young's moduli of neat epoxy samples containing an excess of amino groups, equivalent to the number of amino groups introduced by

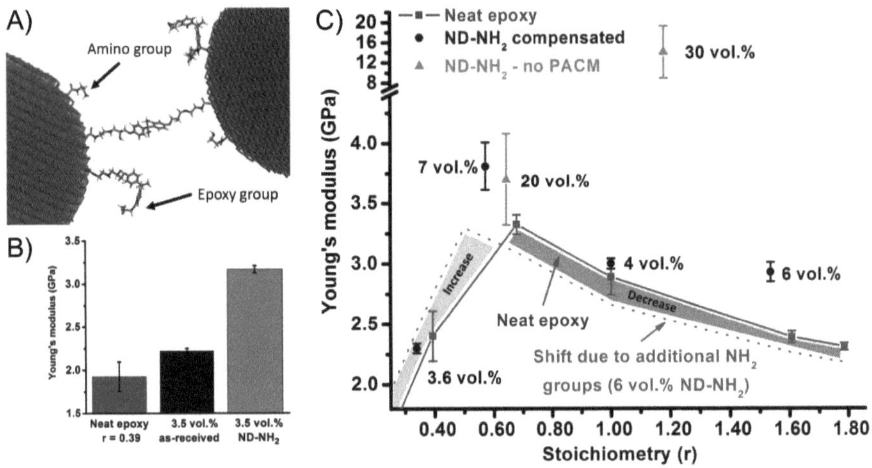


**FIGURE 14.11**

(A) Computer model showing the covalent incorporation of aminated ND into an epoxy matrix. (B) The addition of as-received ND results in an increase of Young's modulus. For aminated ND (ND–$NH_2$), a much higher increase in Young's modulus (~60%) was measured at the same ND content of 3.5 vol.%, due to covalent bond formation between amino groups of ND–$NH_2$ and the epoxide groups of the resin. (C) Young's moduli of neat epoxy, compensated epoxy–ND–$NH_2$ composites (spheres) and composites cured with ND–$NH_2$ only (triangles). The effect of the additional amino groups on Young's modulus introduced by 6 vol.% ND–$NH_2$ at different stoichiometries is shown as well (dotted grey line)[62]

6 vol.% ND–$NH_2$ have been calculated and are plotted in Fig. 14.11C (dotted line). To avoid an undesired reduction in Young's moduli of the epoxy–ND–$NH_2$ composites, it is critical to compensate for the additionally introduced amino groups. With proper compensation, the Young's moduli of composites are always higher compared to non-compensated epoxy–ND–$NH_2$ composites (Fig. 14.11C, circles).[62] Remarkably, due to the large number of amino groups formed on NDs surface, it is possible to cure epoxy solely with ND–$NH_2$, yielding composites with Young's moduli of up to 20 GPa (Fig. 14.11C, triangles).

In another study, ND was aminated in a similar manner with a long-chain amine. Improved fracture toughness and stiffness were observed,[63] confirming that ND with reactive amino groups is an excellent material to improve the properties of epoxies.

## 14.2.4 SYNERGETIC EFFECTS OF ND IN REINFORCING POLYMERS

Recent studies on polymer–ND composites suggest that the reinforcing effect of two fillers or more combined can be larger than the reinforcing effect achieved by the individual fillers, due to interactions in-between them. Such "synergetic" effects, when

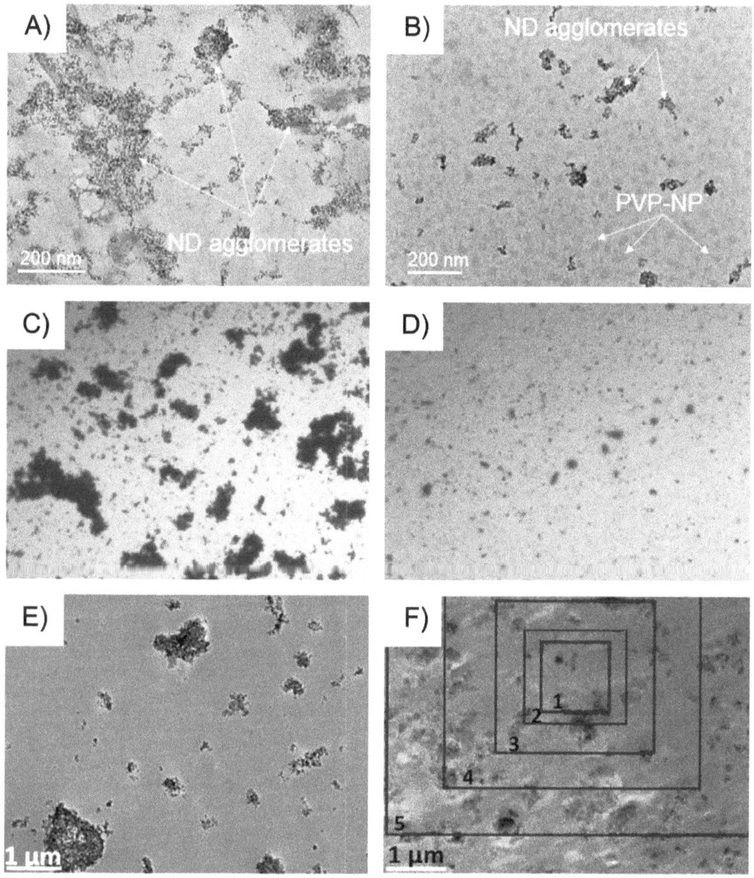


**FIGURE 14.12**

(A) TEM image of ND dispersion in poly(vinylpyrrolidone), (B) improved ND dispersion in poly(vinylpyrrolidone) with PVP-nanoparticles. From Attia et al., Nanodiamond–Polymer Nanoparticle Composites and Their Thin Films. *J. Nanopart. Res.* **2014**, *16* (4).[66]
(C) Microscope images of polymer–urethan films containing SWCNTs alone and (D) with NDs, magnified with a $40\times$ objective.[68] ND reduces agglomerate size and improves dispersion of SWCNTs. (E) TEM image of an epoxy–ND–$NH_2$ composite manufactured using dried ND–$NH_2$ powder, showing large ND–$NH_2$ agglomerates of up to 1 µm in size. (F) TEM images of epoxy–ND–$NH_2$ composites manufactured with ND–$NH_2$ stored in THF suspension after the synthesis prior to composite manufacturing, showing improved dispersion due to this adjusted manufacturing process. Different magnifications, distinguished by five frames, were used to evaluate the state of ND dispersion and agglomeration[62]

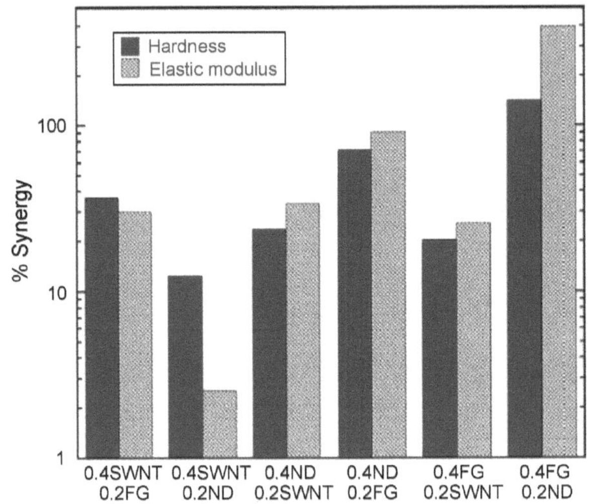


**FIGURE 14.13**

Synergetic effects of different carbon nanofillers in PVA. The highest synergetic effect was found when few layered graphene (FG) and ND were combined in a PVA composite. From Prasad et al., Extraordinary Synergy in the Mechanical Properties of Polymer Matrix Composites Reinforced With 2 Nanocarbons. *Proc. Natl. Acad. Sci. USA* **2009**, *106* (32), 13186–13189[67]

the total effect exceeds the sum of the individual contributions, were observed when combining a micrometer-sized filler with ND. For example, polyurethane foam containing a mixture of 10 μm diameter glass spheres and ND showed a 30% improved strain resistance.[64]

A reduced mechanical stress, as defined by the Mooney–Rivlin equation, was observed due to the addition of glass beads in combination with only 0.1 wt.% of as-received ND, without any surface functionalization, to polydimethylsiloxane. Viscosity measurements of the melts suggest that ND changes the conformation of polydimethylsiloxane molecules, resulting in improved mechanical properties.[65]

For a poly(vinylpryrrolidone) (PVP) matrix it was recently reported that nanoparticles synthesized from the same polymer (PVP-NP) can significantly improve dispersion of as-received ND. It was found that low concentrations of the polymer nanoparticles lead to improved ND dispersion, where the polymer nanoparticles are probably breaking ND agglomerates during ultrasonication (Figs. 14.12A and 14.12B). The obtained composites have an increased UV absorbance, while remaining transparent in the visible range of the spectrum until ND concentrations of up to 20 wt.%.[66]

In the case of a PVA-based nanocomposite, the strongest synergetic effects were observed when two nanocarbons, few layered graphene and ND, were combined (Fig. 14.13). The highest Young's modulus – 9.3 GPa, 14 times higher compared to the neat PVA matrix – was achieved when CNTs were used with ND at a total nanofiller loading of only 0.6 wt.%. The mechanism of the synergism is still unclear.[67] Other studies report the positive effect of ND addition on CNT dispersion in a urethane thin film,[68] which might explain high increases in mechanical properties due to improved dispersion at small additions of ND (Figs. 14.12C and 14.12D).

Also, for an epoxy–polyurethane (EP–UP) polymer blend synergetic properties are reported, where the EP–UP ratio significantly effects ND dispersion and the degradation temperature of the compounds.[69]

For a copolymer–ND composite, ethylene-1-octene–ND, the amount of crystalline orthorhombic phase and its crystallization rate were increased due to the addition of 5 wt.% of as-received ND. This resulted in an improved Young's modulus and larger strain-to-failure values.[70] V-shaped ND–copolymer brushes were produced in a different study and might be optimal for further improvements of copolymers.[71]

Further studies are required to reveal synergetic effects when applying ND, going beyond the extent of the presented studies. It is expected that dispersion and mechanical properties are enhanced when combining various fillers. Composites for tribological applications already benefit from synergetic effects, often containing various fillers, the kind and concentration of the fillers being critical for the tribological performance.[72] The following section (Section 14.2.5) presents some insights into the tribological performance of polymer–ND composites and may help to initiate future studies investigating synergetic effects of these materials.

### 14.2.5 TRIBOLOGICAL PROPERTIES OF POLYMER–ND COMPOSITES

Polymers are commonly used in tribological applications to reduce friction between two surfaces, e.g., as bearings between steel counterparts or as coatings in joint prostheses.[73–76] To improve their tribological performance and produce solid, self-lubricating materials, polymer composites with additives such as graphite, bronze, and alumina were proven to be effective even to the point of replacing metals commonly used in high performance wear applications.[77] Further improvements in tribological properties are possible when macroscopic and nanoscopic fillers are combined,[78] eventually leading to "synergetic" effects, as described in Section 14.2.4.

Since it is well known that diamond has excellent tribological properties,[79] it is expected that the tribological properties of polymers can be improved by the addition of ND. In the following, various studies are presented that address the improved tribological performance of polymer–ND composites.

Polytetrafluoroethylene (PTFE) is a well-known and widely used low-friction polymer with a high chemical and thermal stability, often used in tribological applications.[80] To further improve its tribological properties, as-received ND was introduced into this matrix. Reduced friction coefficients and an improved wear resistance with a significant reduction in the wear track width, at 23 and 150°C were measured.[81] The reduced wear was explained not by the low friction coefficient of ND, but rather by a change in the microstructure of the nanocomposite observed by SEM.

Up to 5 times reduction in the wear rate of PTFE-ND composites with 2 wt.% of ND, as well as decreased wear of polyacrylamide–, polymethylmethacrylate–, and polyethylene–ND composites were observed in another study.[82] Studies on polyurethane–ND composites with 0.5 wt.% of ND confirmed that tribological properties of different thermoplastics can be improved due to the addition of ND.[83]

Another study reported a relationship between ND agglomerate size and the wear performance of PTFE–ND composites. As previously reported, polymer–ND composites contain ND agglomerates, the sizes of which can be divided into the following three groups: 1st >0.6 μm; 2nd 0.2–0.5 μm; and 3rd ~100 nm.[84] In this study, a decrease in wear was observed for small agglomerates, while larger agglomerates resulted in increased wear.[85] Friction coefficients follow the same trend, indicating that small agglomerate size leads to better tribological properties.

For an epoxy–ND composite containing as-received ND, an alumina counterbody was damaged by the epoxy–ND agglomerates contained within these composites, suggesting high hardness of the agglomerates, which can eventually replace micron-sized diamond particles in cutting tools. In this study wear also decreased when agglomerate size decreased. Minimal friction coefficients measured for epoxy composites containing 7.5 vol.% ND, were reduced by a factor 4 reaching values of 0.1.[86]

Recently, it was shown that functionalized nanofillers can further improve the wear resistance of thermosetting polymer matrices by up to 800%.[87] For the study on epoxy–ND composites these findings were confirmed: a composite with a high content of aminated ND–$NH_2$ (33 vol.%) has a friction coefficients of $0.06 \pm 0.02$, being 54% less when compared to composites produced with up to 50 vol.% of as-received ND.[86] This friction coefficient value is close to the one reported for carbide-derived[88] or diamond-like carbon films,[89] showing the potential of this composites for tribological applications. To investigate mechanisms leading to the improved tribological properties, AFM scans were performed. AFM line scans on epoxy–ND composites made with as-received ND showed a sharp change in friction force when the AFM tip moved from the matrix to ND clusters (Fig. 14.14). Composites made with aminated ND–$NH_2$ have a rather different structure: transitions between low and high friction force areas are smoother, indicating the formation of an interphase around ND–$NH_2$ particles that might be responsible for the better performance of these composites (Fig. 14.14).[86]

For a different epoxy system (Epilox T 19-36/700 – Epilox Cross linker H 10-31), friction coefficients decreased with increasing ND addition, confirming the positive effect of ND on the tribological performance of various polymer matrices.[90]

## 14.3 FUTURE WORK

This review of literature illustrates that ND is an excellent material to improve many important properties of various polymer matrices. However, future advances in ND dispersion and optimization of the ND–polymer interface and interphase hold an even greater and largely unexplored potential for further improvements in properties. Also, synergetic effects when combining several fillers might result in unexpected improve-

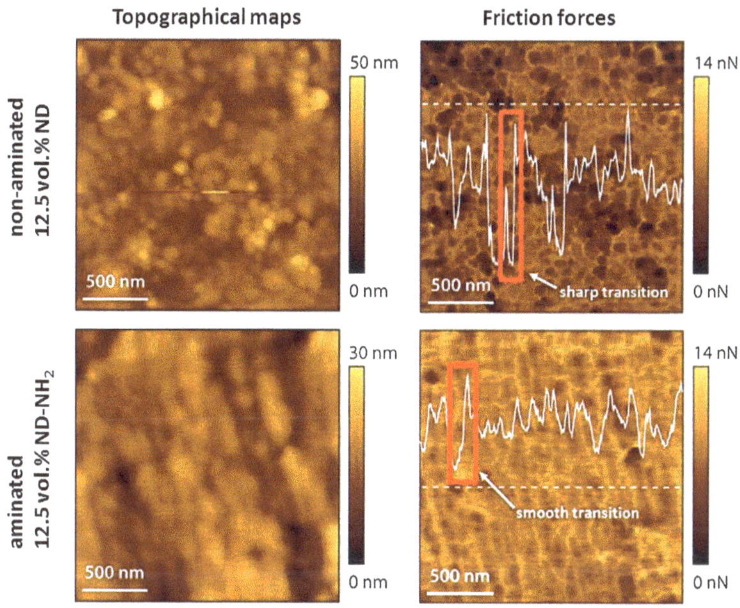


**FIGURE 14.14**

Topographical and friction force maps recorded with AFM on epoxy–ND composites containing 12.5 vol.% ND and ND–$NH_2$. AFM line scans show sharp transitions in frictional forces for composites containing as-received ND, while composites containing ND–$NH_2$ have a smoother transition in frictional forces[86]

ments that require further studies. To fully explore the opportunities provided by ND in polymer matrix composites, new processing techniques and routes of chemical modification of ND surfaces as well as new, smarter ways to control the ND–polymer interface and interphase need to be developed.

# REFERENCES


1. Selzer, R.; Friedrich, K. Mechanical Properties and Failure Behaviour of Carbon Fibre-Reinforced Polymer Composites Under the Influence of Moisture. *Composites Part A: Appl. Sci. Manuf.* **1997**, *28* (6), 595–604.
2. Ahmed, S.; Jones, F. R. A Review of Particulate Reinforcement Theories for Polymer Composites. *J. Mater. Sci.* **1990**, *25* (12), 4933–4942.
3. Ajayan, P. M.; Schadler, L. S.; Braun, P. V. *Nanocomposite Science and Technology;* Wiley–VCH, 2003.
4. Chen, C.; Justice, R. S.; Schaefer, D. W.; Baur, J. W. Highly Dispersed Nanosilica-Epoxy Resins With Enhanced Mechanical Properties. *Polymer* **2008**, *49* (17), 3805–3815.

5. Bauhofer, W.; Kovacs, J. Z. A Review and Analysis of Electrical Percolation in Carbon Nanotube Polymer Composites. *Compos. Sci. Technol.* **2009**, *69* (10), 1486–1498.
6. Bockstaller, M. R.; Thomas, E. L. Optical Properties of Polymer-Based Photonic Nanocomposite Materials. *J. Phys. Chem. B* **2003**, *107* (37), 10017–10024.
7. Seidel, G. D.; Lagoudas, D. C. Micromechanical Analysis of the Effective Elastic Properties of Carbon Nanotube Reinforced Composites. *Mech. Mater.* **2006**, *38* (8–10), 884–907.
8. Bika, D. G.; Gentzler, M.; Michaels, J. N. Mechanical Properties of Agglomerates. *Powder Technol.* **2001**, *117* (1–2), 98–112.
9. Khare, H. S.; Burris, D. L. A Quantitative Method for Measuring Nanocomposite Dispersion. *Polymer* **2010**, *51* (3), 719–729.
10. Li, Y.; Kröger, M.; Liu, W. K. Nanoparticle Geometrical Effect on Structure, Dynamics and Anisotropic Viscosity of Polyethylene Nanocomposites. *Macromolecules* **2012**, *45* (4), 2099–2112.
11. Shenderova, O. A.; McGuire, G. Nanocrystalline Diamond. In *Nanomaterials Handbook;* Gogotsi, Y., Ed.; CRC Taylor and Francis Group: Boca Raton, 2006, pp. 203–237.
12. Mochalin, V. N.; Shenderova, O.; Ho, D.; Gogotsi, Y. The Properties and Applications of Nanodiamonds. *Nat. Nanotechnol.* **2012**, *7* (1), 11–23.
13. Grill, A. Tribology of Diamondlike Carbon and Related Materials: An Updated Review. *Surf. Coat. Technol.* **1997**, *94–95*, 507–513.
14. Balandin, A. A. Thermal Properties of Graphene and Nanostructured Carbon Materials. *Nat. Mater.* **2011**, *10* (8), 569–581.
15. Osswald, S.; Yushin, G.; Mochalin, V.; Kucheyev, S. O.; Gogotsi, Y. Control of $sp^2/sp^3$ Carbon Ratio and Surface Chemistry of Nanodiamond Powders by Selective Oxidation in Air. *J. Am. Chem. Soc.* **2006**, *128* (35), 11635–11642.
16. Mochalin, V.; Osswald, S.; Gogotsi, Y. Contribution of Functional Groups to the Raman Spectrum of Nanodiamond Powders. *Chem. Mater.* **2009**, *21* (2), 273–279.
17. Krueger, A.; Lang, D. Functionality Is Key: Recent Progress in the Surface Modification of Nanodiamond. *Adv. Funct. Mater.* **2012**, *22* (5), 890–906.
18. Krueger, A. The Chemistry of Nanodiamond. In *Nanodiamond;* Williams, O., Ed.; RSC Nanoscience and Nanotechnology; 2014, pp. 49–88.
19. Etzold, B. J. M.; Neitzel, I.; Kett, M.; Strobl, F.; Mochalin, V. N.; Gogotsi, Y. Layer-by-Layer Oxidation for Decreasing the Size of Detonation Nanodiamond. *Chem. Mater.* **2014**, *26* (11), 3479–3484.
20. Mochalin, V. N.; Gogotsi, Y. Nanodiamond–Polymer Composites. *Diam. Relat. Mater.* **2015**, *58*, 161–171.
21. Díez-Pascual, A. M.; Gómez-Fatou, M. A.; Ania, F.; Flores, A. Nanoindentation in Polymer Nanocomposites. *Prog. Mater. Sci.* **2015**, *67*, 1–94.
22. Sugimoto, Y.; Pou, P.; Abe, M.; Jelinek, P.; Perez, R.; Morita, S.; Custance, O. Chemical Identification of Individual Surface Atoms by Atomic Force Microscopy. *Nature* **2007**, *446* (7131), 64–67.
23. Dolmatov, V. Y. Polymer–Diamond Composites Based on Detonation Nanodiamonds. Part 3. *J. Superhard Mater.* **2007**, *29* (4), 199–205.
24. Gerasin, V. A.; Evgenii, M. A.; Karbushev, V. V.; Valerii, G. K.; Galina, P. K.; Raisa, V. T.; Kudryavtsev, Y. V. New Approaches to the Development of Hybrid Nanocomposites: From Structural Materials to High-Tech Applications. *Russ. Chem. Rev.* **2013**, *82* (4), 303.
25. Jee, A.-Y.; Lee, M. Mechanical Properties of Polycarbonate and Poly(Methyl Methacrylate) Films Reinforced With Surface-Functionalized Nanodiamonds. *J. Nanosci. Nanotechnol.* **2011**, *11* (1), 533–536.

26. Comet, M.; Pichot, V.; Siegert, B.; Britz, F.; Spitzer, D. Detonation Nanodiamonds for Doping Kevlar. *J. Nanosci. Nanotechnol.* **2010**, *10* (7), 4286–4292.
27. Khalilnezhad, P.; Sajjadi, S. A.; Zebarjad, S. M. Effect of Nanodiamond Surface Functionalization Using Oleylamine on the Scratch Behavior of Polyacrylic/Nanodiamond Nanocomposite. *Diam. Relat. Mater.* **2014**, *45*, 7–11.
28. Mochalin, V. N.; Gogotsi, Y. Wet Chemistry Route to Hydrophobic Blue Fluorescent Nanodiamond. *J. Am. Chem. Soc.* **2009**, *131* (13), 4594–4595.
29. Zhang, Q.; Naito, K.; Tanaka, Y.; Kagawa, Y. Grafting Polyimides From Nanodiamonds. *Macromolecules* **2008**, *41* (3), 536–538.
30. Menczel, J. D.; Prime, R. B. *Thermal Analysis of Polymers: Fundamentals and Applications;* Wiley–VCH, 2008, pp. 1–688.
31. Jee, A.-Y.; Lee, M. Thermal and Mechanical Properties of Alkyl-Functionalized Nanodiamond Composites. *Curr. Appl. Phys.* **2011**, *11* (5), 1183–1187.
32. Roumeli, E.; Pavlidou, E.; Avgeropoulos, A.; Vourlias, G.; Bikiaris, D. N.; Chrissafis, K. Factors Controlling the Enhanced Mechanical and Thermal Properties of Nanodiamond-Reinforced Cross-Linked High Density Polyethylene. *J. Phys. Chem. B* **2014**, *118* (38), 11341–11352.
33. Zhao, Y.-Q.; Lau, K.-T.; Kim, J-k.; Xu, C.-L.; Zhao, D.-D.; Li, H.-L. Nanodiamond/Poly (Lactic Acid) Nanocomposites: Effect of Nanodiamond on Structure and Properties of Poly (Lactic Acid). *Composites Part B: Eng.* **2010**, *41* (8), 646–653.
34. Hooper, J. D.; Schweizer, K. S. Theory of Phase Separation in Polymer Nanocomposites. *Macromolecules* **2006**, *39* (15), 5133–5142.
35. Zhang, Q.; Mochalin, V. N.; Neitzel, I.; Knoke, I. Y.; Han, J.; Klug, C. A.; Zhou, J. G.; Lelkes, P. I.; Gogotsi, Y. Fluorescent PLLA-Nanodiamond Composites for Bone Tissue Engineering. *Biomaterials* **2011**, *32* (1), 87–94.
36. Akcora, P.; Liu, H.; Kumar, S. K.; Moll, J.; Li, Y.; Benicewicz, B. C.; Schadler, L. S.; Acehan, D.; Panagiotopoulos, A. Z.; Pryamitsyn, V.; Ganesan, V.; Ilavsky, J.; Thiyagarajan, P.; Colby, R. H.; Douglas, J. F. Anisotropic Self-Assembly of Spherical Polymer-Grafted Nanoparticles. *Nat. Mater.* **2009**, *8* (4), 354–359.
37. Zhang, Q.; Mochalin, V. N.; Neitzel, I.; Hazeli, K.; Niu, J.; Kontsos, A.; Zhou, J. G.; Lelkes, P. I.; Gogotsi, Y. Mechanical Properties and Biomineralization of Multifunctional Nanodiamond–PLLA Composites for Bone Tissue Engineering. *Biomaterials* **2012**, *33* (20), 5067–5075.
38. Sun, Y.; Finne-Wistrand, A.; Waag, T.; Xing, Z.; Yassin, M.; Yamamoto, A.; Mustafa, K.; Steinmüller-Nethl, D.; Krueger, A.; Albertsson, A. C. Reinforced Degradable Biocomposite by Homogenously Distributed Functionalized Nanodiamond Particles. *Macromol. Mater. Eng.* **2015**, *300* (4), 436–447.
39. Protopapa, P.; Kontonasaki, E.; Bikiaris, D.; Parasekevopoulos, K.; Koidis, P. Reinforcement of a PMMA Resin for Fixed Interim Protheses With Nanodiamonds. *Dent. Mater. J.* **2010**, *30* (2), 222–231.
40. Hassan, C. M.; Peppas, N. A. Structure and Applications of Poly(Vinyl Alcohol) Hydrogels Produced by Conventional Crosslinking or by Freezing/Thawing Methods. In Biopolymers/Pva Hydrogels/Anionic Polymerisation Nanocomposites, Vol. 153, Abe, A., Ed.; Springer-Verlag Berlin: Berlin, 2000, pp. 37–65.
41. Kurkin, T. S.; Ozerin, A. N.; Kechek'yan, A. S.; Gritsenko, O. T.; Ozerina, L. A.; Alkhanishvili, G. G.; Sushchev, V. G.; Dolmatov, V. Y. The Structure and Properties of Polymer Composite Fibers Based on Poly(Vinyl Alcohol) and Nanodiamond of Detonation Synthesis. *Nanotechnol. Russ.* **2010**, *5*, 340–351 (Compendex).

42. Morimune, S.; Kotera, M.; Nishino, T.; Goto, K.; Hata, K. Poly(Vinyl Alcohol) Nanocomposites With Nanodiamond. *Macromolecules* **2011**, *44* (11), 4415–4421.
43. Li, L.; Davidson, J.; Lukehart, C. Surface Functionalization of Nanodiamond Particles via Atom Transfer Radical Polymerization. *Carbon* **2006**, *44* (11), 2308–2315.
44. Ma, W.; Yu, X.; Qu, X.; Zhang, Q. Functionalization of Agglomerating Nanodiamonds With Biodegradable Poly(ε-Caprolactone) Through Surface-Initiated Polymerization. *Diam. Relat. Mater.* **2016**, *62*, 14–21.
45. Tamburri, E.; Guglielmotti, V.; Matassa, R.; Orlanducci, S.; Gay, S.; Reina, G.; Terranova, M. L.; Passeri, D.; Rossi, M. Detonation Nanodiamonds Tailor the Structural Order of PEDOT Chains in Conductive Coating Layers of Hybrid Nanoparticles. *J. Mater. Chem. C* **2014**, *2* (19), 3703–3716.
46. Angelopoulos, A. P. Introduction to Physical Polymer Science, by L.H. Sperling. *Mol. Cryst. Liq. Cryst. Sci. Technol., Sect. A* **2002**, *381*, 183–184.
47. Harper, C. A., Ed., *Handbook of Plastics, Elastomers, and Composites;* 4th Edition; McGraw–Hill, 2002, p. 884.
48. Schwartz, C. J.; Darden, M. A. The Role of Friction in Controlling Haptic Characteristics of Textured Polymer Surfaces. In *International Polyolefins Conference 2011: Evolving Technology for the Global Economy;* 2011.
49. Sheth, J. P.; Xu, J.; Wilkes, G. L. Solid State Structure–Property Behavior of Semicrystalline Poly(Ether-Block-Amide) PEBAX® Thermoplastic Elastomers. *Polymer* **2003**, *44* (3), 743–756.
50. Dolmatov, V. Y. Composition Materials Based on Elastomer and Polymer Matrices Filled with Nanodiamonds of Detonation Synthesis. *Nanotechnol. Russ.* **2009**, *4*, 556–575 (Compendex).
51. Dolmatov, V. Y. Polymer–Diamond Composites Based on Detonation Nanodiamonds. Part 2. *J. Superhard Mater.* **2007**, *29* (2), 65–75.
52. Jahromi, F. H.; Katbab, A. A. Nanodiamond-Based PP/EPDM Thermoplastic Elastomer Composites: Microstructure, Tribo-dynamic, and Thermal Properties. *J. Appl. Polym. Sci.* **2012**, *125* (3), 1942–1950.
53. Shakun, A.; Vuorinen, J.; Hoikkanen, M.; Poikelispää, M.; Das, A. Hard Nanodiamonds in Soft Rubbers: Past, Present and Future – A Review. *Composites Part A: Appl. Sci. Manuf.* **2014**, *64*, 49–69.
54. IUPAC. Compendium of Chemical Terminology. In Gold Book, Vol. 2.3.3; ICT Press, 2014, p. 1533.
55. Soutis, C. Carbon Fiber Reinforced Plastics in Aircraft Construction. *Mater. Sci. Eng. A* **2005**, *412* (1–2), 171–176.
56. Neitzel, I.; Mochalin, V.; Knoke, I.; Palmese, G. R.; Gogotsi, Y. Mechanical Properties of Epoxy Composites With High Contents of Nanodiamond. *Compos. Sci. Technol.* **2011**, *71* (5), 710–716.
57. Spitalsky, Z.; Kromka, A.; Matejka, L.; Cernoch, P.; Kovarova, J.; Kotek, J.; Slouf, M. Effect of Nanodiamond Particles on Properties of Epoxy Composites. *Adv. Compos. Lett.* **2008**, *17* (1), 29–34.
58. Tugrul Seyhan, A.; Tanoglu, M.; Schulte, K. Mode I and Mode II Fracture Toughness of E-Glass Non-crimp Fabric/Carbon Nanotube (CNT) Modified Polymer Based Composites. *Eng. Fract. Mech.* **2008**, *75* (18), 5151–5162.
59. Ayatollahi, M. R.; Alishahi, E.; Shadlou, S. Mechanical Behavior of Nanodiamond/Epoxy Nanocomposites. *Int. J. Fract.* **2011**, *170* (1), 95–100.

60. Ma, P.-C.; Mo, S.-Y.; Tang, B.-Z.; Kim, J.-K. Dispersion, Interfacial Interaction and Re-Agglomeration of Functionalized Carbon Nanotubes in Epoxy Composites. *Carbon* **2010**, *48* (6), 1824–1834.
61. Mochalin, V. N.; Neitzel, I.; Etzold, B. J. M.; Peterson, A.; Palmese, G.; Gogotsi, Y. Covalent Incorporation of Aminated Nanodiamond Into an Epoxy Polymer Network. *ACS Nano* **2011**, *5* (9), 7494–7502.
62. Neitzel, I.; Mochalin, V. N.; Niu, J.; Cuadra, J.; Kontsos, A.; Palmese, G. R.; Gogotsi, Y. Maximizing Young's Modulus of Aminated Nanodiamond–Epoxy Composites Measured in Compression. *Polymer* **2012**, *53* (25), 5965–5971.
63. Guo, H.; Sheng, H.; Peng, X.; Yu, X.; Naito, K.; Qu, X.; Zhang, Q. Preparation and Mechanical Properties of Epoxy/Diamond Nanocomposites. *Polym. Compos.* **2014**, *35* (11), 2144–2149.
64. Sirotinkin, N.; Voznyakovskii, A.; Ershova, A. Model of Formation of Three-Dimensional Polyurethane Films Modified by Detonation Nanodiamonds. *Phys. Solid State* **2004**, *46* (4), 746–747.
65. Gavrilov, A.; Voznyakovskii, A. Rheological Characteristics and Relaxation Properties of Polymer–Nanodiamond Composites. *Russ. J. Appl. Chem.* **2009**, *82* (6), 1041–1045.
66. Attia, N. F.; Rao, J. P.; Geckeler, K. E. Nanodiamond–Polymer Nanoparticle Composites and Their Thin Films. *J. Nanopart. Res.* **2014**, *16* (4).
67. Prasad, K. E.; Das, B.; Maitra, U.; Ramamurty, U.; Rao, C. N. R. Extraordinary Synergy in the Mechanical Properties of Polymer Matrix Composites Reinforced With 2 Nanocarbons. *Proc. Natl. Acad. Sci. USA* **2009**, *106* (32), 13186–13189.
68. Hens, S. C.; Cunningham, G.; McGuire, G.; Shenderova, O. Nanodiamond-Assisted Dispersion of Carbon Nanotubes and Hybrid Nanocarbon-Based Composites. *Nanosci. Nanotechnol. Lett.* **2011**, *3* (1), 75–82.
69. Xia, S.; Liu, Y.; Zhang, L.; Wang, D.; Zou, W.; Peng, J.; Cao, S. Tribological Mechanism Improving the Wear Resistance of Polyurethane/Epoxy Interpenetrating Polymer Network via Nanodiamond Hybridization. *J. Appl. Polym. Sci.* **2014**, *131* (10).
70. Korobko, A. P.; Bessonova, N. P.; Krasheninnikov, S. V.; Konyukhova, E. V.; Drozd, S. N.; Chvalun, S. N. Nanodiamonds as Modifier of Ethylene-1-Octene Copolymer Structure and Properties. *Diam. Relat. Mater.* **2007**, *16* (12), 2141–2144.
71. Cheng, J. L.; He, J. P.; Li, C. X.; Yang, Y. L. Facile Approach to Functionalize Nanodiamond Particles With V-Shaped Polymer Brushes. *Chem. Mater.* **2008**, *20* (13), 4224–4230.
72. Zhang, Z.; Breidt, C.; Chang, L.; Haupert, F.; Friedrich, K. Enhancement of the Wear Resistance of Epoxy: Short Carbon Fibre, Graphite, PTFE and Nano-$TiO_2$. *Composites Part A: Appl. Sci. Manuf.* **2004**, *35* (12), 1385–1392.
73. Klein, J. Shear, Friction, and Lubrication Forces Between Polymer-Bearing Surfaces. *Annu. Rev. Mater. Sci.* **1996**, *26* (1), 581–612.
74. Lancaster, J. K. Polymer-Based Bearing Materials: The Role of Fillers and Fibre Reinforcement. *Tribology* **1972**, *5* (6), 249–255.
75. Rehbein, P.; Wallaschek, J. Friction and Wear Behaviour of Polymer/Steel and Alumina/Alumina Under High-Frequency Fretting Conditions. *Wear* **1998**, *216* (2), 97–105.
76. Widmer, M. R.; Heuberger, M.; Vörös, J.; Spencer, N. D. Influence of Polymer Surface Chemistry on Frictional Properties Under Protein-Lubrication Conditions: Implications for Hip-Implant Design. *Tribol. Lett.* **2001**, *10* (1), 111–116.
77. (a) Friedrich, K.; Zhang, Z.; Schlarb, A. K. Effects of Various Fillers on the Sliding Wear of Polymer Composites. *Compos. Sci. Technol.* **2005**, *65* (15–16), 2329–2343;

(b) Donnet, C.; Erdemir, A. Historical Developments and New Trends in Tribological and Solid Lubricant Coatings. *Surf. Coat. Technol.* **2004**, *180–181*, 76–84.
78. Chang, L.; Zhang, Z.; Ye, L.; Friedrich, K. Tribological Properties of High Temperature Resistant Polymer Composites With Fine Particles. *Tribol. Int.* **2007**, *40* (7), 1170–1178.
79. Robertson, J. Properties of Diamond-Like Carbon. *Surf. Coat. Technol.* **1992**, *50* (3), 185–203.
80. Khedkar, J.; Negulescu, I.; Meletis, E. I. Sliding Wear Behavior of PTFE Composites. *Wear* **2002**, *252* (5–6), 361–369.
81. Lee, J.-Y.; Lim, D.-S. Tribological Behavior of PTFE Film With Nanodiamond. *Surf. Coat. Technol.* **2004**, *188–189*, 534–538.
82. Bobrovnitchii, G. S.; Skury, A. L. D.; Monteiro, S. N.; Tardim, R. C. Effect of Nanodiamond Addition on the Mechanical Properties of Polycrystalline Metallic and Polymeric Composites. *Mater. Sci. Forum* **2010**, *660–661*, 848–853.
83. Voznyakovskii, A.; Ginzburg, B.; Rashidov, D.; Tochil'nikov, D.; Tuichiev, S. Structure, Mechanical, and Tribological Characteristics of Polyurethane Modified With Nanodiamonds. *Polym. Sci., Ser. A* **2010**, *52* (10), 1044–1050.
84. Bershtein, V.; Karabanova, L.; Sukhanova, T.; Yakushev, P.; Egorova, L.; Lutsyk, E.; Svyatyna, A.; Vylegzhanina, M. Peculiar Dynamics and Elastic Properties of Hybrid Semi-interpenetrating Polymer Network–3-D Diamond Nanocomposites. *Polymer* **2008**, *49* (4), 836–842.
85. Lim, D. P.; Lee, J. Y.; Lim, D. S.; Ahn, S. G.; Lyo, I. W. Effect of Reinforcement Particle Size on the Tribological Properties of Nano-diamond Filled Polytetrafluoroethylene Based Coating. *J. Nanosci. Nanotechnol.* **2009**, *9* (7), 4197–4201.
86. Neitzel, I.; Mochalin, V.; Bares, J. A.; Carpick, R. W.; Erdemir, A.; Gogotsi, Y. Tribological Properties of Nanodiamond–Epoxy Composites. *Tribol. Lett.* **2012**, *47* (2), 195–202.
87. Burris, D. L.; Zhao, S.; Duncan, R.; Lowitz, J.; Perry, S. S.; Schadler, L. S.; Sawyer, W. G. A Route to Wear Resistant PTFE via Trace Loadings of Functionalized Nanofillers. *Wear* **2009**, *267* (1–4), 653–660.
88. Carroll, B.; Gogotsi, Y.; Kovalchenko, A.; Erdemir, A.; McNallan, M. J. Effect of Humidity on the Tribological Properties of Carbide-Derived Carbon (CDC) Films on Silicon Carbide. *Tribol. Lett.* **2003**, *15* (1), 51–55.
89. Liu, Y.; Erdemir, A.; Meletis, E. I. A Study of the Wear Mechanism of Diamond-Like Carbon Films. *Surf. Coat. Technol.* **1996**, *82* (1–2), 48–56.
90. Koumoulos, E. P.; Jagadale, P.; Lorenzi, A.; Tagliaferro, A.; Charitidis, C. A. Evaluation of Surface Properties of Epoxy–Nanodiamonds Composites. *Composites Part B: Eng.* **2015**, *80*, 27–36.
91. Guo, S.; Solares, S. D.; Mochalin, V.; Neitzel, I.; Gogotsi, Y.; Kalinin, S. V.; Jesse, S. Multifrequency Imaging in the Intermittent Contact Mode of Atomic Force Microscopy: Beyond Phase Imaging. *Small* **2012**, *8* (8), 1264–1269.
92. Klein, R. A. SFPE Handbook of Fire Protection Engineering (1995): 2nd Edition. *Fire Saf. J.* **1997**, *29* (1), 61–63.
93. Palmese, G. R.; McCullough, R. L. Effect of Epoxy–Amine Stoichiometry on Cured Resin Material Properties. *J. Appl. Polym. Sci.* **1992**, *46* (10), 1863–1873.
94. Madbouly, S. A.; Otaigbe, J. U. Recent Advances in Synthesis, Characterization and Rheological Properties of Polyurethanes and POSS/Polyurethane Nanocomposites Dispersions and Films. *Prog. Polym. Sci.* **2009**, *34* (12), 1283–1332.

CHAPTER

# Detonation nanodiamonds in oils and lubricants

# 15

**Valerii Y. Dolmatov**

*(FSUE SCTB) "Technolog", Saint-Petersburg, Russia*

**Contents**



## 15.1 LUBRICATING COMPOSITION WITH A SOLID FRICTION MODIFIER

There are widely known lubricant compositions, in which solid materials are used as antifriction additives, in particular metal powders, graphite, molybdenum compounds, and classical diamond. Widely used metal powders are those of copper, tin, lead, silver, and bronze. Additive powders of graphite, molybdenum, and tungsten disulfides are also often used. An analysis of published data provides a basis for the following conclusions: the best anti-friction properties are powders of copper, silver, and molybdenum disulfide. The use of these additives allows realizing friction poly-boundary mode that reduces the coefficient of friction and wear, as well as increases the maximum loads in friction units. However, the use of well-known compositions is limited due to low rheological and anti-corrosive properties and sedimentation stability.[5,9]

## 15.2 THE SYNTHESIS OF DND

The first industrial manufacture of detonation diamonds (DNDs) was set up at Altai Research-and-Production Association (Biisk, Russia) in 1984.[15,17] The first area of research and application of DNDs and their half-product, diamond-containing stock (DCS), was to examine the possibility of using the DCS and DNDs in oil formulations for various purposes.[14,1]

## 15.3 THE TRIBOLOGICAL PROPERTIES OF DND

The main indicators of the quality of lubricating compositions include the limiting load on the friction assembly, the coefficient of friction, the bulk temperature in the friction area, oil viscosity, oil film denseness, and the scuffing load capacity of oils. The presence of DCS nanoparticles in the lubricant contributes to the formation of a dense lubricant layer, reliably separating the contacting surfaces and improving the quality of all of the above indicators of oils and lubricants. The nanometer size of DND particles and their rounded shape (see Chapter 2) make them rather attractive for various antifriction and antiwear formulations.[18,11,2] The advantages of DNDs as friction modifiers over other substances of this kind consist in the following:

- Efficiency at very low concentrations in a base oil;
- Compatibility with various synthetic and mineral oils;
- Ecological safety of the carbon additive to oil compared with metallic and fluorocarbon particles and substances.

This creates necessary prerequisites for a wide introduction of DND-containing additives.

The presence of a large amount of ultrasmall graphite-like particles and nanodiamonds in the DCS changes the properties of a lubricating film and the nature of the interaction between friction surfaces. The concentration of particles may be as high as $10^{14}$ cm$^{-3}$ in the fluid and $10^8$–$10^9$ in a monolayer with area of 1 cm$^2$; the distance between separate fractal aggregates (DNDs or DCS) will be 0.1–0.5 μm.[20] The viscosity of the fluid increases, and so does the strength of the lubricating film, and, as a consequence, the carrying capacity of the tribocoupling is improved.

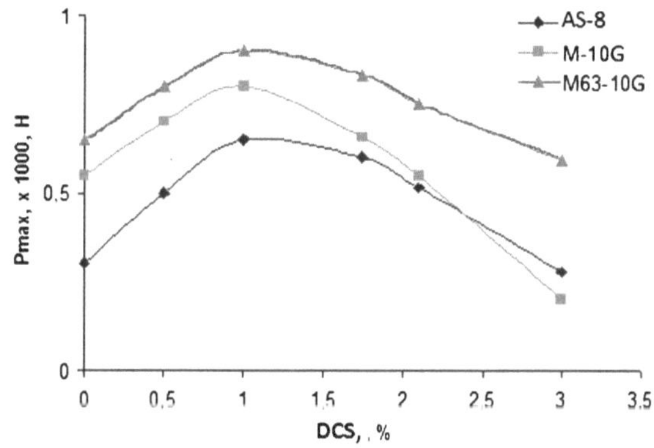


**FIGURE 15.1**

Dependence of the carrying capacity of motor oils on the concentration of DCS (for abbreviations see Glossary)

A portion of the DCS powder corresponding to the percentage content is thoroughly mixed with the oil and passed through a number 005 sieve (sieve size). After this the magnetic cleaning of oil is done, and then it is dispersed for 15 min at a frequency of 22 kHz.

Before the test, samples are thoroughly wiped, washed with petrol and dried, followed by breaking-in carried out on the test oil until the disappearance of abrasion traces.

Oils with DCS test under conditions of sliding friction were held in a "disk-block" and under "mechanical friction."

DNDs constituting ~50 wt.% in the DCS and possessing a certain cutting capacity in the submicrocutting mode play an active part in the running-in of mated surfaces and "healing" of their microscopic defects, including those newly formed. This process is nearly complete when the actual contact area is reached, in which the lubrication occurs in the semi-fluid or fluid mode (Fig. 15.1). In fluid friction, rubbing surfaces are separated by a layer of oil, whose thickness is greater than the sum of the heights of the surface roughness of a conjugate pair. In half-liquid friction, there is mixed friction, both fluid and the boundary (friction surfaces are covered with the finest oil film formed as a result of the molecular forces and chemical reactions of active oil molecules with the friction surface).

The goal of Sakovich et al.[14] was to create a lubricating formulation with a solid friction modifier that would possess good antifriction, antiwear, and antiwelding properties at a low content of the solid phase in a formulation. The authors suggested a lubricating formulation composed of an oil base and DCS, with the DCS composition specified as follows (wt.%): nonabrasive diamonds (DNDs) 2.0–99.0 and graphite 1.0–98. The formulation contained 0.01–1.0 wt.% DCS and oil base being the rest.

## 15.4 DCS ACTION MECHANISM IN THE LUBRICATING COMPOSITION

The role of the base of a lubricating formulation with the DCS can be played by mineral or synthetic lubricating oils and coolant-cutting fluids containing necessary functional additives that provide oxidation resistance, foaming stability, and washing-detergent, anticorrosion, and other properties.[12]

The coagulation stability in an oil base is provided by the fact that the size of fractal carbon clusters in the DCS in energetic equilibrium does not exceed 30–60 nm. This circumstance rules out stratification of lubricating formulations during a long time (more than six months).

The mechanisms by which the DCS acts in a lubricating formulation are manifested in several aspects providing its efficiency.[1]

- Having small size, DCS particles saturate the friction surfaces by filling their imperfections, with the boundary friction and wear decreasing, especially under high loads and at a deficiency of the lubricating material; also, seizure and formation of scores on the friction surfaces are ruled out.
- The DCS makes the lubricating formulation more viscous in thin films due to the dispersed structuring, with the dynamic strength of the film increasing and its "brittle" cracking at high velocities ruled out due to the termination of cracks on clusters.
- The DCS diminishes the viscosity of a lubricating formulation at low temperatures by lowering the glass-transition temperature of the dispersion-filled medium. As a result, the temperature range in which lubricating formulations can be used becomes 5–10°C wider.
- The DCS provides a residual effect (for more than 60 h) after the lubricating formulation is replaced with the base oil. This effect is due to the firm mechanical adsorption and diffusive fixation of carbon clusters on friction surfaces.
- Under heavy loads and maximum displacement of the fluid phase from between the friction surfaces, nonabrasive diamonds in the DCS operate as rolling microbearings, which makes the ultimate load sustained by a friction pair higher without scuffing; for example, the load of a steel–bronze pair is increased from 16 to 72 MPa.

## 15.5 OPTIMAL DCS CONTENT IN THE LUBRICATING COMPOSITION

The content of the solid friction modifier (DCS) in a lubricating formulation depends on the nature and mode of operation of a device (machine, mechanism, unit) in which it is used. For example, for heavily loaded units with mostly large sliding surfaces, it is 0.3–1.0 wt.%.[14] For medium-loaded mechanisms, e.g., internal combustion engines (of carburetor or diesel types), the DCS concentration in the lubricating formulation is 0.01–0.3 wt.%. This concentration provides the optimal effect by reducing the friction and fuel consumption and by improving the compression and dynamic characteristics.

For lightly loaded mechanisms, e.g., metal-working machinery with circulatory lubricating systems, the content of DCS in lubricating formulations is 0.01–0.15 wt.%. At DCS contents below 0.01 wt.%, no significant changes in the service properties of lubricating formulations are observed.

## 15.6 THE METHOD OF PRODUCING DIAMOND-CONTAINING COMPOSITION

A lubricating formulation with a solid friction modifier is obtained in a technologically simple way as follows. A DCS produced by detonation synthesis is dehydrated by vacuum drying and is charged in required amounts to a preliminary-mixing vessel, into which the oil base, e.g., mineral oil is also introduced. Being thoroughly mixed, the mixture is forced by pumps through mesh filters and magnetic separators and, finally, through finishing-dispersion devices, e.g., disintegrators, acoustic (ultrasonic) mixers, or mills, with the subsequent filtration through fine filters to remove coarse inclusions and admixtures.

## 15.7 MOTOR AND TRANSMISSION OILS WITH DCS

Tests of motor oils (AS-8, M-10G, M63-10G manufactured in Russia) demonstrated that the smallest frictional torque is observed for all oils at a DCS content of 0.5–1 wt.%.[9] The sustaining counter pressure of motor oils with a DCS additive grows as the powder concentration in oil is raised to 1 wt.%. Further increase in the concentration reduces the sustaining counter pressure. For AS-8 and M-10G, this parameter falls significantly. A possible explanation of this behavior of the oils is that at substantial DCS concentrations the continuity of the oil layer is deteriorated when the oil is heated to high temperatures.

For TSP-10, TAD-17, and TAP-15V gear-box oils manufactured in Russia, the minimum friction is also observed in the range from 0.5 to 1.0 wt.% DCS, with the temperature in the friction zone of the samples noticeably decreasing (by 10–20°) as the DCS concentration increases to 0.5%. At higher DCS concentrations, the temperature again increases, which is accounted for by the decrease in the sustaining counter pressure of the oils at powder concentrations exceeding 1 wt.%.

Carbon contained in the DCS is inert in its common state and does not enter into reactions with a metal. Temperature, pressure, and presence of chemically active substances are necessary for initiating the adhesive behavior of carbon. High temperatures arise at surface-contact places in the course friction. The presence of carbon and chemically active substances results in that surface layers are saturated with carbon, i.e., secondary structures are formed (austenite, martensite, white zones). The specific surface energy decreases, and so does the adhesion component of the friction, with the friction coefficient decreasing as a result.

The run-in time of internal combustion engines (ICEs) with a DCS-based lubricating formulation can be reduced from 60 to 5.5 h; the wear of the cylinder–piston-group articles becomes 1.5–2.5 times smaller, with the internal friction in the engine decreasing by 25–30%; and the compression in the engine cylinders grows by 10–17%. The optimal DCS content in the lubricating formulation is 0.02–0.04 wt.%.

It was found that the DCS additive has no adverse effect on the service properties of motor oils, the thermo-oxidative stability is not impaired, the corrosion capacity for nonferrous metals and alloys does not increase, and antifriction and antiwear properties of the oils are improved.[20]

With the DCS used to modify industrial and gear-box oils, the friction coefficient in gears and mechanisms falls by 20–30%, the wear of friction surfaces becomes 2.0–2.5 times lower, the smoothness of unit motion is improved, and the gear-box noise decreases.

## 15.8 DND MODIFICATION WITH LANTHANIDES, MG-SULFONATE, AND POLYTETRAFLUOROETHYLENE

In [6], the conditions of modification of purified DNDs with complex lanthanide sulfonates to obtain stable dispersions in oils were studied, and the possibility of using modified DNDs as ecologically safe corrosion inhibitors and friction modifiers was assessed. The author developed a method for synthesis of combined sulfonates of rare-earth elements (REEs) for oleophilic modification of DNDs to obtain stable DND suspensions in oils. It was found that modified DNDs are effective friction modifiers. Formulations containing La–Mg systems with DNDs and ultradispersed polytetrafluoroethylene (UPTFE) were suggested as oil additives and substance to be used in machining the cylinder–piston group of two-stroke engines.

Because the high surface activity of DNDs leads to their aggregation even in liquid polar media, it is possible to stabilize the degree of dispersity and structure of DND suspensions in nonpolar media via modification of their surface (see Chapter 8). It was assumed that functional groups on the DND surface are involved in the coordination with metal ions taking part in the synthesis of sulfonates (Ln, Mg, Zn).

It was found that DNDs modified in the stage of synthesis of REE complexes form in industrial oil suspensions stable during up to a year. Under the same conditions, complexes with the DCS precipitate during a month (it should be noted that the DCS composition was not determined in [6]).

Tribological tests of the formulations synthesized in the study were performed in I-12A or I-20A industrial oil (Russia). The content of additives in the oil was 2% in all cases.

The highest scuffing load (5.0 kN) was recorded when a complex of oil-soluble lanthanum–magnesium sulfonates was added to the oil. An additional introduction of

DNDs (0.026–0.1 wt.%) into the system resulted in that the anti-scuffing properties of the formulation were impaired by 30% (3.5 kN).

Modification with nanodiamonds of an UPTFE-containing lubricating formulation strongly changed the nature of the friction interaction in a rubbing pair. For example, for the I-12A + 0.33% UPTFE + 0.026% DND formulation, the friction coefficient remained invariant during the whole test on reaching the steady-state mode (0.011), in contrast to the friction coefficient (0.042) of the I-12A + UPTFE formulation. In this case, the friction-zone temperature increased stepwise on passing the minimum friction coefficient for the I-12A + UPTFE formulation and, by the end of tests, was twice that for the I-12A + UPTFE + DND (0.026%) formulation.

In the author's opinion,[6] the occurring decrease in the friction coefficient of the UPTFE-DND formulation, compared with UPTFE, may be due to the change in the scheme of the stress–strain state in the metal–polymer contact. The introduction of DNDs into the UPTFE matrix probably reduces the shear resistance between matrix layers in the direction perpendicular to the relative displacement velocity and leads to a decrease in the deformation component of the friction force. In addition, the heat removal from the tribo-zone is improved in the presence of DNDs.

Full-scale tests demonstrated that application of lubricants containing nanodiamonds together with UPTFE results in that the engine temperature becomes lower and the maximum engine rpm speed increases.

Thus, it was found that, in contrast to the DCS, DNDs form suspensions stable during up to a year. La–Mg systems are the best in the scuffing load, and formulations including La–Mg systems with DNDs and UPTFE, the best in the friction coefficient.

## 15.9 TECHNOLOGY FOR PRODUCING THE LUBRICANT COMPOSITION WITH DCS AND SURFACTANTS

Dolmatov and Pimenov[3] suggested a method for obtaining a lubricating formulation by mechanically mixing oil in which a surfactant is dissolved with 5–10 wt.% DCS preliminarily dehydrated by vacuum drying at 80–110°C. A cationic nitrogen-containing substance taken in an amount of 2–10% relative to the DCS mass is used as a surfactant. After being formed, the mixture is kept for no less than 60 min. The time of treatment on a vibrocavitation homogenizer is 20–60 min.

The vibrocavitation treatment improves the surface activity of dispersed particles and, accordingly, accelerates processes that yield interfaces with the required properties providing a high homogeneity and stability of the formulation. The maximum size of solid particles does not exceed 1 μm. No visible changes were observed in the formulation after three months of storage, with good service properties preserved during the whole storage time.

**Table 15.1** Characteristics of diamond-containing lubricating formulations and the effect of their application

| No. | Parameter, measurement units | Value of parameter |
|---|---|---|
| 1. | DND content, wt.% | 0.01–0.3 |
| 2. | Particle concentration in 1 $cm^3$ | $\sim 10^{14}$ |
| 3. | Decrease in temperature in the contact zone, % | 16–20 |
| 4. | Decrease in the friction coefficient, % | 20–30 |
| 5. | Increase in the limiting load on a friction pair (steel-bronze), MPa | from 16 to 72 (by a factor of ~4) |
| 6. | Decrease in the wear of mated articles by a factor of $n$ | $n = 1.5–3.0$ |
| **Internal combustion engines** | | |
| 7. | Decrease in the run-in time by a factor of $m$ | $m = 10–12$ |
| 8. | Decrease in the fuel consumption, % | 3–6 |
| 9. | Increase in power, % | 4–8 |
| 10. | Increase in the cylinder compression, % | 10–17 |
| **Coolant-cutting fluids** | | |
| 11. | Decrease in the cutting torque, % | 20–30 |
| 12. | Increase in the tool life by a factor of $p$ | $p = 1.5–4.0$ |

## 15.10 THE MAIN RESULTS OF THE USE OF DIAMOND LUBRICANT COMPOSITIONS

Table 15.1 lists selected characteristics of diamond-containing lubricating formulations and the effect of their application.

The DCS strongly affects the majority of structural-mechanical and service characteristics of all kinds of lubricant materials. In storage compounds, the chemical and electrochemical corrosion of metals decreases when lowering the amount of additives, which are, as a rule, corrosive substances, or when abandoning these at all. In sealing pastes and thread lubricants, the air-tightness of joints is improved, the corrosion of contacting surfaces decreases, and the shear stresses and torques are stabilized and become lower.

## 15.11 COOLANTS WITH DCS FOR CUTTING METALS

Coolant-cutting fluids (CCFs) have been developed for mechanical processing by cutting on the basis of mineral oils into which DCS, chlorosulfonated fat, and sulfur-containing polyisobutylene are introduced.[13,12,4] These CCFs proved to be effective in manual and machined thread cutting, hole drilling, grinding of ferrites, etc. Application of a CCF with the DCS makes the tool life 1.5–4.0 times longer, improves the cutting processes by diminishing cutting forces, and makes it possible to avoid using some CCFs (e.g., sulfo-frezol and oleic acid) that contaminate the environment and are harmful for workers' health.

The correct choice of the content of CCF components and DCS makes it possible to raise by a factor of ~8 the running-in ability of friction surfaces (compared with that in pure oil).

A CCF for mechanical processing of metals has been developed,[13] which is based on mineral oil and additionally contains the following components (wt.%): sulfurized polyisobutylene 5–15, chlorosulfonated fat 2–5, DCS 1–5.

In the authors' opinion,[13,12,4] the developed CCFs will replace sulfo-frezol and oleic acid. The efficiency is the highest in manual and machined thread cutting on difficult-to-machine steels and alloys.

Application of the developed CCFs enables the following:

1. The cutting process can be stabilized, and the cutting torque can be reduced by 20–30%.
2. The CCFs can be used for thread cutting and hole drilling in difficult-to-machine materials.
3. The formulations can be used in automated CCF supply systems because of the small size of DCS particles.
4. The tool life can be made 1.5–4.0 times longer as compared with the best CCFs.
5. The working conditions can be improved because the CCFs contain no components dangerous to human health.

## 15.12 PLASTIC LUBRICANTS WITH DCS AND DNDS

In plastic lubricants, DCS or DNDs are introduced into the base oil directly or via an intermediate medium and, depending on the technology used, can be involved in the formation of a structural skeleton. In lubricants of this kind, the DCS also reduces the friction coefficient by 20–30%, with the wear of the friction pairs becoming 1.3–3.0 times smaller.[14]

It was noted in [12] that the introduction of 1–5 wt.% DCS into a lubricant used in noncutting shaping of metals makes it possible to substantially reduce the friction coefficient and improve the surface quality of the products being obtained.

In the presence of the DCS, the boundary layer can assume a structure and an elasticity of shape, similarly to a solid substance, under a high specific load exceeding in many cases the flow point of a metal being processed. Under these conditions, ultramicroscopic surface defects of a material being processed are subjected to an adsorption influence. This provides the effect in which the strain is relieved. With the closeness of contact increasing further, charging of the working surfaces of the tribocontact by diamond particles becomes possible. Apparently, a noticeable effect can be obtained in molding, dragging, sizing, and deep drawing.

Compared with the presently used ordinary graphite (particle size 1–5 μm), the DCS has a number of advantages:

1. DCS particles easily penetrate into the deformation zone, being retained in surface microheterogeneities of an unworked part and a tool.

2. Having a large specific surface area, the DCS forms a continuous lubricating layer between the rubbing surfaces and thereby rules out their direct contact.

## 15.13 SOLID LUBRICANTS

A method for obtaining solid lubricating materials on the basis of petroleum oil with the use of dry DND powders was suggested in the patent.[7] These oil formulations are used to wear-protect the surface of friction units. To raise the sedimentation stability of a formulation, the DND powder was hydrophobized by its impregnation with solutions of fluorine-containing oligomers and octyltriethoxysilane (e.g., PER-240 perfluorpropyl ether or perfluoropolyester acid) dissolved in trifluorotrichloroethane. On being impregnated for 24 h, modified DNDs were dried at room temperature and then heated to 120°C for 3 h. Then, the resulting powder was ground in butanol in a planetary mill with zirconium dioxide ($ZrO_2$) balls. The reactive ethoxy groups at the silicon atom make octyltriethoxysilane capable of binding with the surface functional groups (–OH, –COOH) of DNDs, thereby imparting oleophilic properties to the surface. The ratio between octyltriethoxysilane and the fluorine-containing oligomer was 50:1 to 1:50 at their total content in the nanodiamond additive of 5.1–40.8 wt.%, with the amount of DNDs being 59.2–94.9 wt.%, respectively.

To prepare lubricating materials, a suspension of modified deagglomerated DNDs in butanol is mixed in an amount equivalent to 0.015 wt.% in terms of DNDs (relative to oil) with oil, and the mixture is placed in a rotary evaporator in which the solvent is evaporated and the final product is formed.

The resulting composition 6–10-fold surpasses in sedimentation stability and tribotechnical characteristics the prototype composition[7] at the same mass content of DNDs.

## 15.14 THE ADVANTAGES OF PLASTIC AND SOLID LUBRICANTS

It was shown in [8] that, compared with base lubricants, such as solid oil, Tsiatim, lithol, etc., paste lubricants with DCS provide a 1.2–3.0 times better wear resistance of friction units, 6–20% higher antifriction properties, 1.2–3.0 times shallower surface breaks, and 16–19% lower temperature in the contact zone.

A significant effect is obtained when DNDs and DCS are used in a solid hydrocarbon-based lubricant deposited on abrasive wheels in grinding. The composition developed in the study makes it possible raising the grinding-burn resistance by 20–50% and reducing the roughness of a surface being processed by 1–2 classes.

In addition, industrial tests demonstrated that the output capacity of grinding procedures grows by 20–30%, the smearing of the abrasive surface decreases, and the temperature of abrasive grains and of the surface of an article being processed becomes lower.

The properties and compositions of numerous oil formulations were also described in [10,16,19,20].

## 15.15 CONCLUSIONS

1. Detonation nanodiamonds and diamond-containing stock are effective friction modifiers in oil formulations and are compatible with mineral and synthetic oils.
2. Addition of DCS and DNDs provides the following effects:
   - The boundary friction and wear are made lower;
   - A protective layer is formed on friction surfaces, with this layer hindering the adhesion scuffing and surface chipping;
   - The friction coefficient decreases by 20–30%;
   - The tool life becomes 1.5–4.0 times longer, and the wear of the friction pair decreases by a factor of 1.3–3.0;
   - The limiting load on a friction pair becomes 4.0 times higher.
3. The optimal amount of the DCS in oils is 0.05–0.50 wt.%, depending on the working conditions of a diamond–oil formulation.
4. Modification of DNDs with lanthanide sulfonates yields suspensions in oil that are stable up to a year and provide the minimum friction coefficient (0.011).
5. Oil compositions with DSC and DND offer new promising prospects for the creation and use of safe lubricants for all fields of application.

## REFERENCES

1. Badaev, F. Z.; Privalko, Yu. N.; Gubarevich, T. M. Method for Obtaining a Carbon Dispersion in Oil, C09G1/02 Pat. U.S.S.R 1658641, Publ. May 1988.
2. Dolmatov, V. Yu. Detonation Synthesis Ultradispersed Diamonds: Properties and Applications. *Russ. Chem. Rev.* **2001**, *70* (7), 607–626.
3. Dolmatov, V. Yu.; Pimenov, Yu. A. Method for Obtaining a Diamond-Containing Formulation, C10N30/06 Pat. RU 2567058, Publ. Oct. 2015.
4. Dolmatov, V. Yu. Detonation Nanodiamonds Production, Properties and Application; Professional: Saint-Petersburg, 2011.
5. Goryunov, V. M. Metal Plaque Lubricant in Heavily Loaded Hinges. In *Thermal Dynamics and Simulation of the External Friction*; Nauka: M., 1975, pp. 102–105 (in Russian).
6. Ivanov, D. M. Technology of Oil-Soluble Lanthanide and Mg Sulfonates as Corrosion Inhibitors and Friction Modifiers, Abstract of a Dr. Sci. Thesis, Yekaterinburg, 2006 (in Russian).
7. Ivanov, D. M. Additive to Petroleum Lubricant Oils, Based on Ultradispersed Diamonds, and Method for Its Synthesis, C10M161/00 Pat. RU 2554002, Publ. Apr. 2014.
8. Karpov, I. V.; Red'kin, V. E.; Krushenko, G. G. Ultrasonic Impregnation of Slippery Contact With Lubricants, Modified With Nanopowder. *Russ. Metall. (Met.)* **2010**, *11*, 17–19.
9. Kolisnichenko, L. F. Effect of Molybdenum Chalcogenides on Tribotechnical Properties of Lubricating Materials. *Powder Metall.* **1987**, *11*, 65–71.

10. Kompan, V.; Slobodsky, V. Anti-friction Additive and Method for Using Same, C10M125/02 Pat. US 5614477 A, Publ. March 1997.
11. Martonik, M. V.; Primacheva, L. G.; Red'kin, V. E.; Staver, A. M. Grinding-and-Polishing Formulation, C09G1/02 Pat. RU 2005758, Publ. Jan. 1994.
12. Red'kin, V. E. Lubricants With Ultradispersed Diamond Ultradispersed Diamond-Graphite Powder. *Chem. Technol. Fuels Oils* **2004**, *40* (3), 32–35.
13. Red'kin, V. E.; Staver, A. M.; Shangin, A. P. Coolant-Cutting Fluid for Mechanical Processing of Metals, C10M141/08 Pat. RU 2009186, Publ. March 1994.
14. Sakovich, G. H.; Zharkova, A. S.; Petrov, E. A. Results of Research Into the Physicochemical Process of Detonation Synthesis and Nanodiamond Applications. *Nanotechnol. Russ.* **2013**, *8* (9–10), 581–591.
15. Trefilov, V. I.; Savvakin, G. I.; Skorohod, V. V. Specific Features of the Structure of Ultradispersed Diamonds Produced by High-Temperature Synthesis Under Explosion Conditions. *Dokl. Akad. Nauk SSSR* **1978**, *239* (4), 839–841 (in Russian).
16. Shenderova, O. A., Gruen, D. M., Eds. *Ultracristalline Diamond, Synthesis, Properties and Application*; Williams Andrew Publishing: Norwich, New York, USA, 2006, 600 p.
17. Volkov, K. V.; Danilenko, V. V.; Elin, V. I. Synthesis of Diamond From the Carbon in the Detonation Products of Explosives. *Combust. Explos. Shock Waves* **1990**, *26* (3), 366–368.
18. Xu, T.; Zhao, J.; Xue, Q. Study on the Properties of Ultradispersed Diamonds Soot as an Oil Additive. *Tribol. Trans.* **1997**, *40*, 178–184.
19. Yashchenko, N. K.; Ogorodnik, V. V. Diamond-Based Antifriction Material, C10M2201/05 Pat. US 5158695 A, Publ. Oct. 1992.
20. Zaharov, A. A.; Redkin, V. E.; Staver, A. M.; Lyamkin, A. I.; Kuchma, E. I. Lubricant Formulation, Pat. Application WO 93/01261 (PCT/RU 91/00134), 1993.

## GLOSSARY

**AS-8 (Russia)** multigrade engine oil for the lubrication of gasoline engines of trucks; Analogue of Mobil Delvac 1320, Mobil Delvac MX 15w/40 and 10w/30, ROTELLA X 20/20w ("Shell")

**M-10G (Russia)** a universal engine oil for highly augmented gasoline and diesel engines

**M63-10 G** multigrade motor oil (viscosity at 100°C 100.5 $mm^2/s$, viscosity at −18°C 2600–10400 $mm^2/s$)

**TSP-10 (Russia)** gear oil for heavily loaded cylindrical, bevel, spiral bevel gears of trucks

**TAD 17 (Russia)** multigrade mineral oil for transmissions of cars and trucks

**TAP-15 (Russia)** gear oil for gear boxes trucks

**I-12A (Russia)** a general purpose industrial oil, with no additives, for assemblies and mechanisms, hydraulic fluids, kinematic viscosity at 40°C – $mm^2/s$ – 13–17.

**I-20A (Russia)** industrial oil, similar to I-12A, but the kinematic viscosity at 40°C – $mm^2/s$ – 29–35.

CHAPTER

# Biomarkers and drug delivery applications

16

**Be-Ming Chang***, **Dean Ho**†, **Huan-Cheng Chang***

*Institute of Atomic and Molecular Sciences, Academia Sinica, Taipei, Taiwan** *Division of Oral Biology and Medicine and Department of Bioengineering, UCLA, Los Angeles, CA, United States*†

**Contents**



## 16.1 INTRODUCTION

The applications of nanocarbon particles in life sciences have emerged as an expanding research area since the discovery of fullerenes in 1985 and carbon nanotubes in 1991.[1] Nanodiamond (ND) is a member of this nanocarbon family. In contrast to other forms of carbon, ND constitutes entirely $sp^3$ bonds and carries several extraordinary physical, chemical, and optical properties (see previous chapters of this book). Two common methods of making NDs involve (a) detonation of carbon-containing explosives and (b) grinding microdiamond powders synthesized under static high-pressure high-temperature (HPHT) conditions[2] (see Chapter 2). These two types of ND have large specific surface areas and are amenable to tailoring of diverse functional groups, such as hydroxyl, halide, amine, and thiol groups, for covalent and noncovalent linkage with bioactive molecules. A key breakthrough that places ND in the forefront of biological applications is the creation of photostable color centers in HPHT diamond powders (see Chapter 7). These NDs are monocrystalline and contain abundant nitrogen impurities for conversion into vacancy-related color centers, forming fluorescent nanodiamonds (FNDs). Of particular interest is the nitrogen-vacancy center, which possesses a number of remarkable photophysical features that enable FNDs to play an important role in interdisciplinary fields encompassing biology, chemistry, physics, and materials sciences, leading to new discoveries.

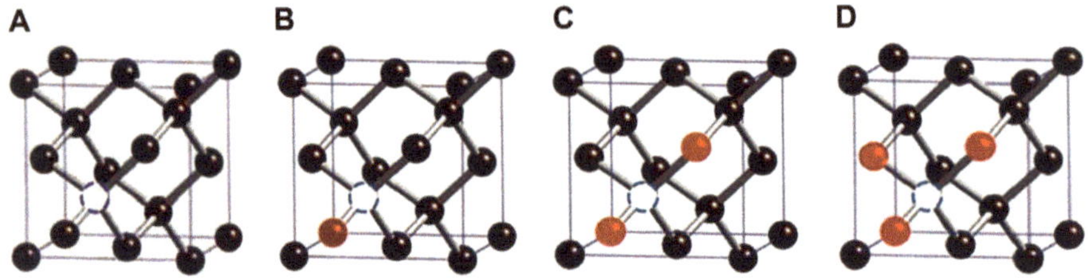


**FIGURE 16.1**

Structures of vacancy-related centers in diamond: (A) GR1, (B) NV, (C) H3, (D) N3. The carbon atoms, nitrogen atoms, and vacancies are denoted by black spheres, dark red spheres, and blue dashed circles, respectively.[7] From Hui, Y. Y.; Chang, H.-C. Recent Developments and Applications of Nanodiamonds as Versatile Bioimaging Agents. *J. Chin. Chem. Soc.* **2014**, *61*, 67–76

In this book chapter, we present a representative case where FNDs are applied as fluorescent biomarkers for long-term tracking of living cells (including stem cells) both in vitro and in vivo. Additionally, detonation nanodiamonds (DNDs) are also explored as drug delivery and imaging agents, and the ND-enabled application of powerful combination therapy optimization technologies that have been validated at the in vitro, in vivo, and clinical stages will be described in detail.

## 16.2 FLUORESCENT NANODIAMOND

Pure diamond crystals are optically transparent to visible light. Colors in diamond originate from impurities, such as nitrogen or boron, within the crystal lattice.[3] More than a hundred types of luminescent color center have been discovered and the nitrogen-incorporated defects are the most common.[4] To distinguish from pure and boron-doped diamonds, the nitrogen-containing diamonds are classified as type I diamonds. Depending on the nitrogen concentration, type I diamonds are further sorted into two groups, Ia and Ib. Type Ia has a nitrogen concentration in the range of $500 \sim 3000$ ppm, typically found in natural diamonds, whereas type Ib contains less than 500 ppm nitrogen, typically found in HPHT synthetic diamonds. The nitrogen atoms are structurally isolated in type Ib diamond, giving the material a yellowish color, but they form aggregates in type Ia diamond, keeping the material colorless. None of these NDs are fluorescent.

To make NDs brightly fluorescent, carbon vacancies in the crystal lattice have to be introduced. These vacancies can be created through radiation damage with high-energy particles such as 2 MeV electrons.[5] Subsequent thermal annealing at high temperatures (e.g. 800°C) promotes the migration of the vacancies to form stable complexes with nitrogen atoms. Fig. 16.1 illustrates the structures of four different types of vacancy-related color centers in type I diamond: GR1, $NV^-$, H3, and N3.[6,7] These centers are capable of emitting near-infrared, red, green, and blue fluorescence

**Table 16.1** Properties of vacancy-related color centers in diamond[a]

| Centers | ZPL/nm | $\lambda_{em}$/nm | Q.Y. |
|---|---|---|---|
| $V^0$ (GR1) | 741.2 | 898 | 0.014 |
| $NV^-$ | 637.6 | 685 | 0.99 |
| N-V-N (H3) | 503.5 | 531 | 0.95 |
| $N_3$+V (N3) | 415.4 | 445 | 0.29 |

[a] *Listed are zero-phonon lines (ZPL), emission maxima ($\lambda_{em}$), fluorescence quantum yields (Q.Y.). From Hui, Y. Y.; Chang, H.-C. Recent Developments and Applications of Nanodiamonds as Versatile Bioimaging Agents. J. Chin. Chem. Soc.* ***2014****, 61, 67–76*

(Table 16.1), respectively, depending on the number of nitrogen atoms next to the vacancies at the nearest neighbor sites. Of these four types of color center, the $NV^-$ is most noteworthy. It is a point defect comprising a substitutional nitrogen atom adjacent to a vacancy site with $C_{3v}$ symmetry. The center exhibits a zero-phonon line (ZPL) at 638 nm, corresponding to the electronic transition $^3A_2 \rightarrow {}^3E$,[8] and a broad phonon sideband peaking at ~560 nm. When excited by green yellow light, it emits far-red fluorescence at ~700 nm with a quantum yield close to 100%. It was first reported in 2005 that the far-red fluorescence from $NV^-$-containing FNDs can be readily detected in live cells[9] since it is well separated in wavelength from the autofluorescence (400–600 nm) of biological samples.[10] It is this important characteristic that makes FNDs well suited as biomarkers even at the single particle level.

Another outstanding feature of the $NV^-$ center is its photostability. Fig. 16.2A displays three typical time traces of the fluorescence intensities of single $NV^-$-containing FND particles (35 and 100 nm in diameter) and Alex Fluor 546 molecules.[11] The fluorescence intensities of both FNDs stay nearly the same over 300 s, while the Alex Fluor 546 photobleaches very rapidly in 12 s under the same excitation conditions. The fluorescence of the FNDs is so stable that no sign of blinking and photobleaching has been observed (Fig. 16.2A and Fig. 16.2B) even under intense laser excitation (~1 $MW/cm^2$ in Fig. 16.2C). The photostability is preserved even when the particle size is reduced to 35 nm because the centers are embedded deep in the crystal matrix. Further studies of 5 nm NDs containing single $NV^-$ centers showed no fading of the fluorescence.[12] Harnessing the remarkable optical properties of this center, researchers in the field have exploited $NV^-$ as a powerful tool for bioimaging and quantum sensing applications.[13]

The advent of nanoscience and nanotechnology has brought about diverse applications of surface-functionalized nanoparticles in various branches of life sciences. Compared with metal, semiconductor, and oxide nanoparticles, NDs have the advantages of being carbon-based. Researchers in the field can therefore utilize well-established chemical methods for conjugation of NDs with various bioactive ligands or biomolecules for technological and medical applications.[14] For pristine NDs, their surfaces are always contaminated with $sp^2$ carbon atoms. To remove these graphitic structures as well as residual chemical compounds, chemical treatments with strong oxidative acids,[15] air oxidation,[16] or ozone[17] are often applied. Following the treat-

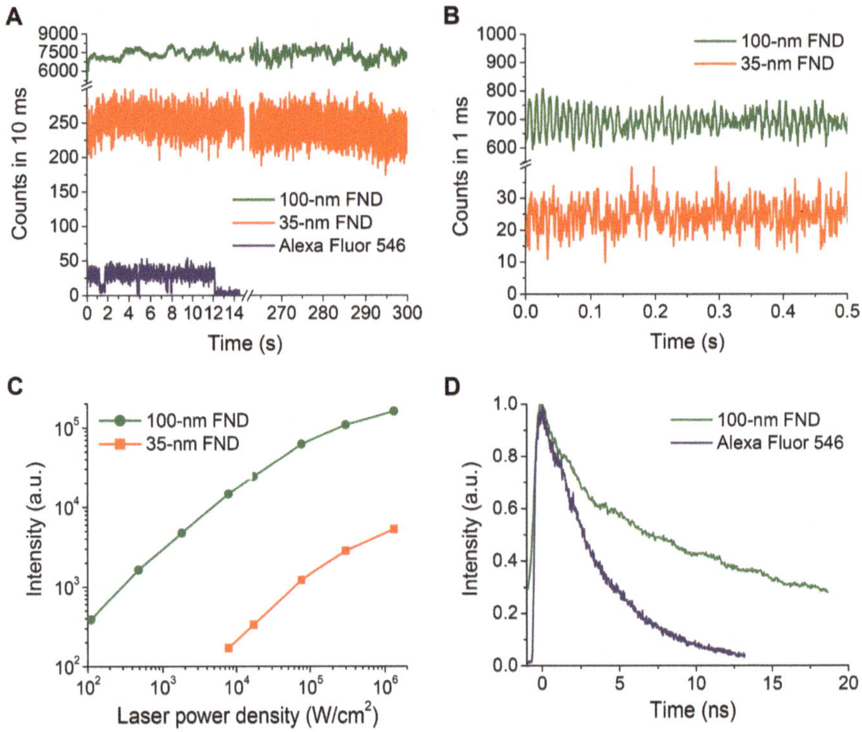


**FIGURE 16.2**

(A) Typical time traces of the fluorescence from a single 100 nm FND (green), a single 35-nm FND (red), and a single Alexa Fluor 546 dye molecule (blue). (B) Time traces of the fluorescence from a single 100 nm FND (green) and a single 35 nm FND (red) acquired with a time resolution of 1 ms. (C) Plot of fluorescence intensity as a function of laser power density over the range of $1 \times 10^2$–$1 \times 10^6$ W/cm$^2$ for 100 nm and 35 nm FNDs. (D) Fluorescence lifetime measurements of 100 nm FNDs (green) and Alexa Fluor 546 dye molecules (blue). The observed lifetimes are ~17 ns and ~4 ns for FND and Alexa Fluor 546, respectively.[11] From Fu et al., Characterization and Application of Single Fluorescent Nanodiamonds as Cellular Biomarkers. *Proc. Natl. Acad. Sci. USA* **2007**, *104*, 727–732

ments, particles are terminated with a variety of functional groups, including carboxyl (–$COO^-$) and carbonyl (–C=O) groups. The presence of carboxyl groups makes it convenient for subsequent bioconjugation of FNDs using carbodiimide chemistry.

A variety of organic and biological molecules have been conjugated onto NDs either covalently or non-covalently. Fig. 16.3 shows two examples of the bioconjugation for FNDs: (a) protein conjugation by physical adsorption, and (b) covalent conjugation of proteins through a polymeric buffer layer. In the latter approach, grafting of the polymers on NDs is established by chemical reactions between the end-functionalized polymer chains and the complementary functional groups on the solid substrate.[18] Poly-L-lysine is one of the first polymers covalently grafted on the

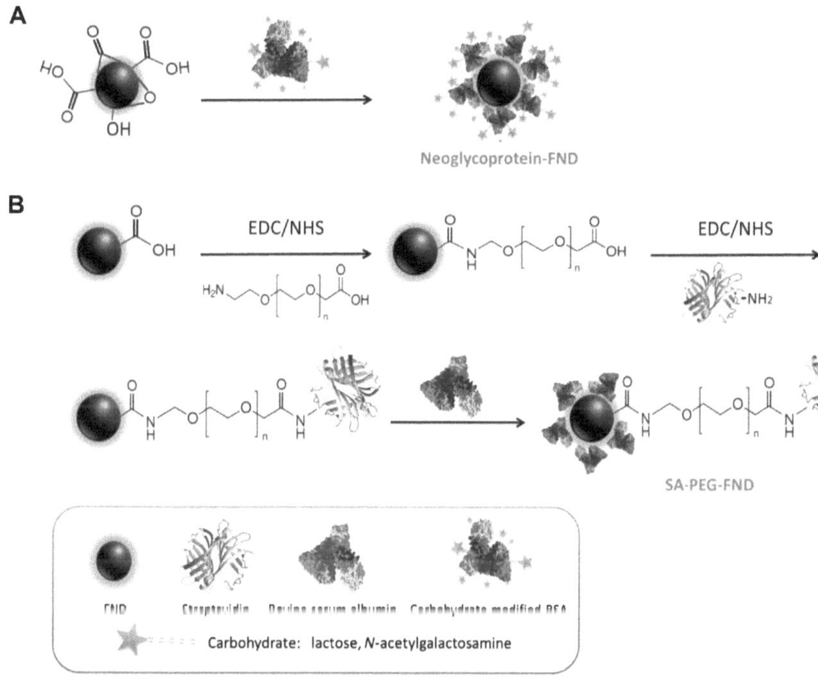


**FIGURE 16.3**

Work flows of the bioconjugation of FNDs by (A) physical adsorption and (B) covalent linking through activation of the surface carboxyl groups. Abbreviations: FND (fluorescent nanodiamond), EDC (N-(3-dimethylaminopropyl)-N'-ethylcarbodiimide hydrochloride), NHS (N-hydroxysuccinimide), and PEG (polyethylene glycol).[18] From Chang et al., Highly Fluorescent Nanodiamonds Protein-Functionalized for Cell Labeling and Targeting. *Adv. Funct. Mater.* **2013**, *23*, 5737–5745

carboxylated surface of ND through amide bond formation.[15] Researchers later used polyelectrolytes (such as polyallylamine)[19] or organosilanes[20] for the surface functionalization of NDs with amino groups. Experiments showed that coating of NDs with hyperbranched polyglycerol by ring-opening polymerization can significantly improve their colloidal stability in water and buffers.[21] For bioconjugation, in order to reduce steric constraints and retain the adsorbate's activities, methods have been developed to insert inert spacers between the biomolecules of interest and NDs. This step is crucial for immobilization of enzymes since their active sites may be sterically hindered after attachment to the ND surface. Another function of the spacers is that it helps suppress non-specific interactions and protein conformational changes caused by strong biomolecule-particle interactions.[22,23] One of the most commonly used spacers is polyethylene glycol (PEG), which has low toxicity and a low degree of specific interactions with biomolecules. They can be covalently conjugated with carboxylated NDs via the amino groups on their termini. These spacers are available in various discrete-lengths and also provide carboxyl groups on the other termini for carbodiimide crosslinking with $NH_2$-containing bioactive ligands or protein molecules.

A concern for practical use of NDs in biology is that they aggregate easily in physiological medium. Without proper surface modification, the colloidal stability of NDs decreases with the increasing ionic strengths of the solution. This poses a serious obstacle for the biolabeling application of FNDs since physiological buffers and culture media all contain high concentrations of salts. Tzeng et al. have found a simple solution for this problem.[24] They showed that through noncovalent conjugation with serum albumin, the FND colloids are stable in phosphate-buffered saline (PBS) over weeks and suitable for superresolution imaging. The noncovalent conjugation is established by the interplay of electrostatic forces, hydrogen bonding, and hydrophobic interactions between the nanoparticles and the protein molecules.[25] The approach is simple, straightforward and, more importantly, compatible with the methods commonly used by biologists.

## 16.3 CELL LABELING AND TRACKING

With the methods depicted in Fig. 16.3, Chang et al. conjugated FNDs with proteins of different types for ligand-based targeting and antibody-based labeling of mammalian cells.[18] They used nitrogen-rich type Ib diamond powders as the starting material to fabricate high-brightness FNDs. Additionally, the BSA coating is applied to resist flocculation of the particles in PBS and reduce their nonspecific binding with cells. Liver cells were employed as the specific labeling and targeting models. In the first example of this study, acid-treated FNDs (~100 nm in diameter) were noncovalently conjugated with glycoproteins such as asialofetuin (AF) or neoglycoproteins, i.e., BSA chemically modified with multiple sugar residues such as N-acetylgalactosamine (GalNAc) and lactose (Lac). These FND bioconjugates were then incubated with HepG2 cells for uptake by endocytosis. Confocal fluorescence imaging confirmed the presence of Lac–BSA–FNDs, TriGalNAc–BSA–FNDs, and AF–FNDs inside the HepG2 cells, but negligible for BSA–FNDs. The result is in agreement with flow cytometric analysis which showed that the amount of the cellular uptake follows the trend of Lac–BSA–FND > AF–FND > TriGalNAc–BSA–FND > BSA–FND. In the second example of this work, FNDs were conjugated with antibodies for specific labeling of CD44 antigens on cell surfaces. Two human breast carcinoma cell lines (MCF-7 and ASB145-1R) were used and they expressed different levels of CD44, presenting an interesting case for comparison. The researchers first synthesized SA–PEG–FNDs, prepared by PEGylation of FNDs followed by covalent conjugation with streptavidin (SA), and then used the nanoparticle bioconjugates to target biotinylated antibodies that bind specifically with the extracellular epitopes of CD44. It was found that the FND labeling method performed comparably well as the conventional approach using organic dye (such as DyLight488) as the fluorescent tag, demonstrating that the protein-functionalized FNDs are useful for targeted labeling of mammalian cells. An additional advantage of this method is that the events before and after endocytosis of the FND particles in the individual cells can be repeatedly examined by confocal fluorescence microscopy, thanks to the excellent photostability of the $NV^-$ centers.

The long-term labeling capability suggests a promising application of FNDs, that is, to find slow-proliferating and quiescent cancer stem cells (CSCs) by fluorescence imaging. CSCs have long been considered to be a source of tumor initiation but still yearn for effective tools for their identification and isolation. This is mainly because the present methods of finding these cells often use molecular markers, which suffer from limited chemical and/or photophysical stability, disallowing tracking of the cells over a long period of time. Lin et al. were the first to apply FNDs to track CSCs in vitro.[26] They first carried out genotoxicity tests with comet and micronucleus assays for human fibroblasts and breast cancer cells to confirm that the FNDs neither caused DNA damage nor impaired cell growth. After that, the ASB145-1R cells were used as the model cell line for CSCs and compared the performance of FNDs with the widely used cell trackers, carboxyfluorescein diacetate succinimidyl ester (CFSE) and 5-ethynyl-2′-deoxyuridine (EdU). Their results showed that the technique not only enabled quantitative analysis of the FND-labeled cells by flow cytometry but also outperformed CFSE and EdU in regards to its long-term tracking capability (up to 20 days). Further development and application of the FND platform are expected to enhance our understanding of the biochemical properties of CSCs in vitro as well as in vivo.

Among various exciting applications of nanoparticles in stem cell research, one of the most challenging is how to manipulate and track these cells in vivo. For this particular application, a robust and reliable cell labeling technique is required in order to closely interrogate their development, fate, and contribution to regenerating tissues in animals. While quantum dots have been considered as a promising material, the potential toxicity of this cadmium-based compound impedes its real-world biomedical applications. The biocompatible, non-toxic, and brightly fluorescent diamond nanoparticles serve well the purpose. Fang et al. pioneered the study by examining in detail the exocytosis of FND particles (size ~100 nm) from three cell lines, HeLa cervical cancer cells, 3T3-L1 pre-adipocytes, and 489-2.1 multipotent stromal cells.[27] Both fluorescence imaging and flow cytometric analysis revealed that there was little excretion of the endocytosed FND particles from the HeLa cells and the $NV^-$ fluorescence could still be detected in the individual cells up to 8 generations. It demonstrates that FNDs are useful as long-term cell trackers in vitro.

To conduct in vivo studies, the first hurdle to overcome is the high background derived from the autofluorescence of biological tissues during imaging. While the far-red emission of FND is well separate from the autofluorescence in wavelength, the identification of single FND-labeled cells in tissues is still challenging. To address this issue, the distinct fluorescence lifetime of FND comes into play a role. For endogenous fluorophores such as flavin and exogenous fluorophores such as fluorescein in biological samples, their emission lifetimes are typically in the range of 1–5 ns,[28] which are considerably shorter than the emission decays of ~10 ns and ~20 ns for $NV^-$ centers in bulk and nanoscale diamonds (Fig. 16.2D), respectively. Hence, various time-gating techniques can be applied to recover the fluorescence signals of FNDs in high background and enhance the image contrast. A good demonstration of this concept is that of Faklaris et al. who have previously acquired time-gated fluorescence images of FNDs in HeLa cells.[29] During the imaging, photons were collected within a proper time window to avoid interference from autofluorescence. The method allowed dramatic improvement of the image contrast as well as precise localization of the FNDs inside the cells, even though each of these particles hosted only a single $NV^-$ center in its matrix.

Stem cells are a group of undifferentiated biological cells that can self-renew and differentiate into various specialized cells. Understanding the role of stem cells in vivo is important in life science research but many questions remain unanswered. The use of techniques such as fluorescent labeling and gene transfection to trace their fates in vivo has raised critical concerns about how they affect the cell growth and differentiation capacity. The FND labeling method, in contrast, is free of these issues and thus holds great potential and promise for such applications. Wu et al. have employed FNDs as biolabels for tracking the specific engraftment of transplanted lung stem cells (LSCs) in a mouse model.[30] In this study, in vitro experiments first verified that the FND labeling did not eliminate the cells' properties of self-renewal and differentiation into type I and type II pneumocytes. The FND-labeled LSCs were then injected into lung-injured mice via tail veins. The mice were finally sacrificed at different time points and their organs were collected to search for the injected LSCs by fluorescence microscopy. As commonly encountered in tissue section imaging, the high autofluorescence background prevented clear identification of the cells of interest in tissues. Yet, thanks to the distinct fluorescence lifetime of the $NV^-$ centers, the contrast of images can be greatly enhanced by using fluorescence lifetime imaging microscopy (FLIM). As shown in Fig. 16.4, the location of the transplanted FND-labeled LSCs can be precisely determined in the lung tissues with single cell resolution. The LSCs were clearly discerned even after the cells were stained with hematoxylin and eosin (H&E) to aid histological examination. The method is suitable for rare stem cell studies. It also serves well a novel platform for monitoring the homing of different types of stem cells (such as mesenchymal stem cells), not only in small animal models (such as mice and rats) but also in large animal models (such as miniature pigs) for preclinical experimentation. It has important implications for future stem cell therapy and cancer treatment.

## 16.4 OPTIMIZING COMBINATORIAL NANOMEDICINE

Both FNDs and DNDs have been widely explored as drug delivery and imaging agents. This is due to their aforementioned beneficial properties that include facile processing parameters that render them safe for biomedical applications, unique surface chemical properties, and ability to markedly enhance the efficacy and safety of treatment outcomes. Prior studies have demonstrated that DND-doxorubicin (NDX)

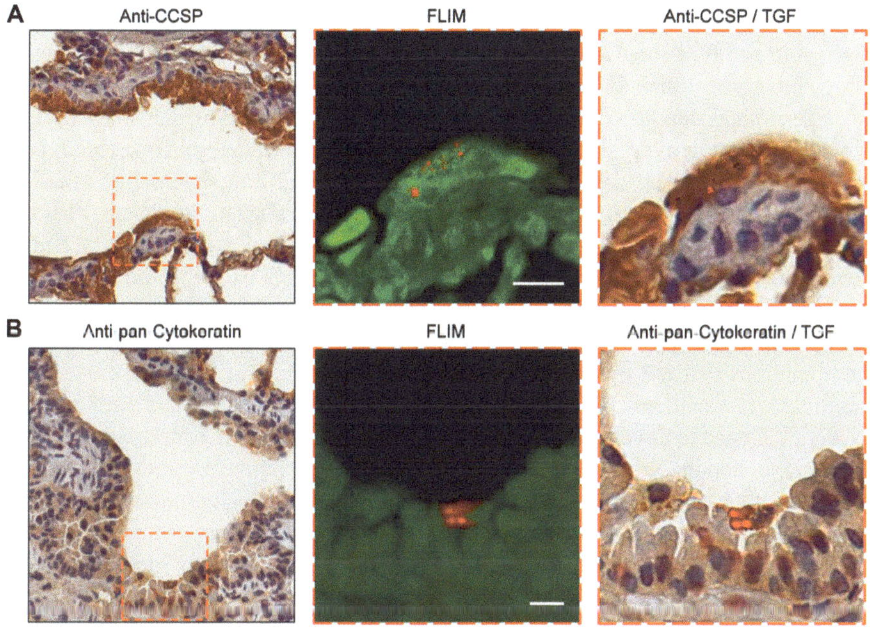


**FIGURE 16.4**

Identification of transplanted FND-labeled LSCs in bronchiolar epithelia of a mouse by FLIM and time-gated fluorescence (TGF) imaging of FNDs (red) as well as immunostaining of the lung tissue with two cell markers, (A) CCSP (brown) and (B) pan-cytokeratin (brown). Scale bars: 10 μm.[30] From Wu et al., Tracking the Engraftment and Regenerative Capabilities of Transplanted Lung Stem Cells Using Fluorescent Nanodiamonds. *Nat. Nanotechnol.* **2013**, *8*, 682–689

complexes can combat cancer drug resistance, resulting in substantially reduced preclinical 4T1 breast and LT2M liver tumor sizes while these tumor models are virtually not responsive to the administration of drug alone. When the drug dosages were doubled to lethal levels, early mortality was observed. However, lethal dosages delivered as NDX resulted in complete animal survival and the smallest tumors observed in the study.[31] In the area of gadolinium-based contrast imaging, DNDs have mediated among the highest ever reported per-gadolinium relaxivity values.[32]

A broad spectrum of nanomaterials is being explored as drug delivery and imaging agents. Many of these nanomaterials exhibit unique properties that can accelerate their translation into the clinic, including biocompatibility, scalable synthesis parameters, ability to improve drug retention or pharmacokinetic properties, as well as reduction in drug toxicity, among others. As specific nanomaterials begin to emerge as best-in-class platforms for the delivery of specific drugs for specific indications, it is important to evaluate their roles in combination therapy with other nanomaterial–drug complexes and/or unmodified drugs. DNDs have been particularly effective as preclinical anthracycline delivery vehicles and magnetic resonance imaging compounds. As they continue to be evaluated for potential clinical validation, it is important to combine their role as monotherapeutic delivery agents with other drugs or drug delivery materials towards optimal combination therapy strategies.

The clinical standard of monotherapy is often characterized by delivering drugs at the maximum tolerated dose (MTD). When combination therapy is employed, however, MTD cannot be used due to issues such as drug–drug interactions that can substantially affect the efficacy and safety of treatment. Current approaches to combination therapy design cannot systematically identify the specific drug-dose ratios that result in the best possible therapeutic outcomes. This is due to the fact that the drug-dose parameter space in combination therapy is virtually infinite, making it virtually impossible for high throughout screening and other similar methods to identify combinations that are optimized for multiple criteria such as multiple efficacy and multiple safety markers, particularly when primary samples are used. Therefore, there is a critical need to develop technology platforms that can identify optimal drug-dose ratios in an actionable fashion without the need for predictive algorithms or modeling, but rather, utilize experimental data to rationally develop combinations that are optimized based on composition and drug-dose ratio. To address this pressing need, a powerful technology platform named Phenotypic Personalized Medicine (PPM) was recently developed.[33,34] PPM does not rely on genetic or mechanistic information to design combination therapies. Instead, it correlates drug input dosing with phenotypic response (with phenotype defined as efficacy such as tumor size reduction, minimizing toxicity, or a combination of multiple phenotypic readouts) using a mapping process that definitively identifies a list of experimentally backed and optimally ranked combinations.[35] PPM is also disease indication-agnostic, meaning it can be applied to every disease. In the context of DND drug delivery, PPM was recently utilized to design optimized combinations containing 3 DND-drug complexes and unmodified paclitaxel. In this study, multiple breast cancer cell lines were used as testing platforms, and multiple control cell lines were used to optimize therapeutic safety. PPM-optimized combinations with defined drug-dose ratios outperformed unmodified drug combinations, single-drug DND-modified and unmodified agents, as well as randomly tested DND-drug combinations using a therapeutic window of maximal cancer cell apoptosis and maximal control cell viability as the phenotypic readout (Fig. 16.5). This was achieved without the use of search algorithms or modeling. In addition, the optimization process can be achieved using multiple orders of reduction in the number of required tests. PPM has also been extended towards other indications ranging from additional types of cancer to infectious diseases and regenerative medicine, where drugs that are completely ineffective on their own can result in marked enhancements in therapeutic efficacy when delivered as part of a PPM-optimized drug combination.

In addition to in vitro and preclinical demonstration of PPM optimization, PPM was also recently harnessed in a landmark clinical study that was able to dynamically and actionably optimize patient-specific immunosuppression during the entire course of treatment. For this first in-human pilot trial, PPM was applied towards a liver transplant immunosuppression indication to individualize tacrolimus administration

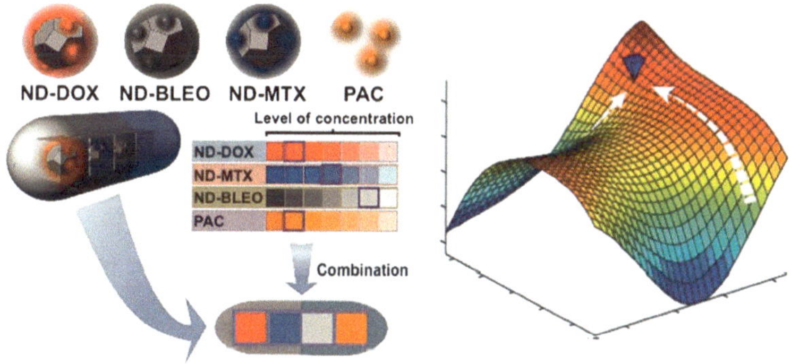


**FIGURE 16.5**

Phenotypic personalized medicine (PPM) was utilized to optimize nanodiamond-based combination therapy. Optimal drug-dose ratios were identified without using predictive algorithms or modeling.[34] From Wang et al., Mechanism-Independent Optimization of Combinatorial Nanodiamond and Unmodified Drug Delivery Using a Phenotypically Driven Platform Technology. *ACS Nano* **2015**, *9*, 3332–3344

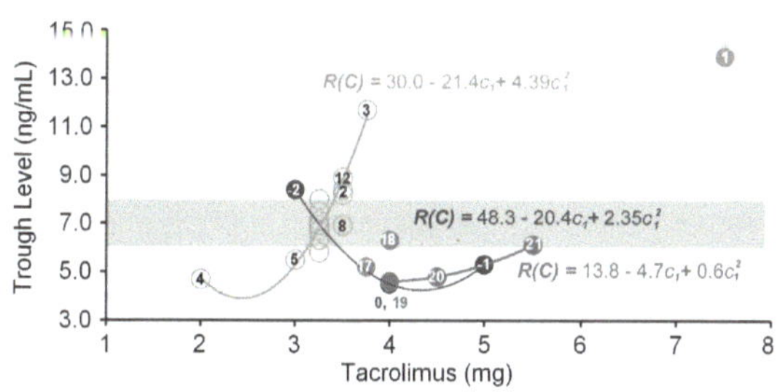


**FIGURE 16.6**

Phenotypic personalized medicine (PPM) was clinically validated in a first in-human pilot clinical trial to personalized post-transplant immunosuppression. PPM is disease indication-agnostic. Therefore, it will play a powerful role in optimizing combination therapy design, including for those involving nanomedicine-based multi-drug therapy[36]

towards the prevention of graft rejection.[36] The input was defined as the immunosuppressant dosage while the output was defined as the serum tacrolimus level, with each patient being prescribed a unique target tacrolimus range that also changed during the course of treatment for several patients. In this trial, PPM implementation demonstrated that a parabola represents the patient response to therapeutic treatment. This major finding made it possible to identify the specific drug dosages that mediate optimal tacrolimus trough levels being measured. Patients also frequently had their regimens modified by the introduction of new drugs, changes in drug dosages, or additional procedures such as hemodialysis. PPM was able to dynamically optimize patient treatment by recalibrating patient response to construct a new parabola, demonstrating the actionable nature of this approach (Fig. 16.6).

Importantly, PPM optimizes drug development at each stage of testing. It is important to note that while in vitro PPM optimization results in the identification of experimentally verified lead combinations with defined drug-dose ratios, the in vitro optimization stage is primarily used to identify drug candidates and combinations using multi-parameter optimization that will be subsequently tested in vivo. In vivo testing re-optimizes the combinatorial design process using substantially fewer animals compared to conventional brute force drug development. Lead combinations identified during in vivo PPM implementation can subsequently be re-optimized during clinical testing utilizing individualized/patient-specific testing as demonstrated by the recent immunosuppression PPM clinical pilot trial. In summary, the optimized administration of nanomedicine in the clinic will likely involve its use in combination therapy. Therefore, the realization of PPM, a clinically-validated genomics/mechanism-independent and indication agnostic technology is expected to develop the best possible nanomedicines and unmodified combination therapies that are simultaneously optimized for efficacy and safety.

## 16.5 CONCLUSIONS AND OUTLOOKS

In the past decade, NDs have rapidly developed into an invaluable tool for life science studies. With the combined properties of excellent photostability and inherent biocompatibility, FNDs afford scientists a powerful means for optical imaging both in vitro and in vivo. In particular, the successful applications of FNDs as long-term cell trackers have opened up new avenues for advanced scientific investigation in stem cell research. While the FND-enabled technology has a number of superior features to the conventional fluorescence imaging using organic dyes and fluorescent proteins as biomarkers, it has several limitations. First, the size of the presently used FNDs (typically 30–100 nm in diameter) is too large to allow labeling of intracellular organelles and monitoring of cell trafficking. Reduction of the particle size to the sub-20 nm range is required. Second, the penetration depth of the fluorescence light is too low to allow deep tissue imaging, which is of importance for preclinical development and clinical translation of the techniques. A solution to latter problem is to conjugate FNDs and DNDs with Gd(III) ions or iron oxide nanoparticles for magnetic resonance imaging (MRI).[32,37] Alternatively, radioisotope labeling of FNDs and DNDs with tracers such as $^{18}F$ (lifetime = 109.8 min) can be applied for positron emission tomography (PET).[38] The integration of the FND labeling technology into the existing modalities (MRI, PET, and others) will offer not only higher sensitivity in whole-animal imaging but also higher resolution in tissue-section imaging. Combining the powerful capabilities uniquely enabled by FNDs and DNDs with the use of novel optimization platforms, which will ensure that the administration of these agents in clinical settings yield the best possible patient therapeutic and imaging outcomes, will allow scientists to see more, explore deeper, and mediate maximal efficacy and safety, providing a new paradigm for both fundamental and clinical research.

## REFERENCES


1. Hong, G.; Diao, S.; Antaris, A. L.; Dai, H. Carbon Nanomaterials for Biological Imaging and Nanomedicinal Therapy. *Chem. Rev.* **2015**, *115*, 10816–10906.
2. Mochalin, V. N.; Shenderova, O.; Ho, D.; Gogotsi, Y. The Properties and Applications of Nanodiamonds. *Nat. Nanotechnol.* **2011**, *7*, 11–23.
3. Zaitsev, A. M. *Optical Properties of Diamond: A Data Handbook;* Springer-Verlag Berlin Heidelberg GmbH: Morwell, 2001.
4. Davies, G., Ed. *Properties and Growth of Diamond;* EMIS Datareviews Series, Vol. 9; INSPEC, The Institute of Electrical Engineers: London, UK, 1994.
5. Campbell, B.; Choudhury, W.; Mainwood, A.; Newton, M.; Davies, G. Lattice Damage Caused by the Irradiation of Diamond. *Nucl. Instrum. Methods A* **2002**, *476*, 680–685.
6. Davies, G.; Lawson, S. C.; Collins, A. T.; Mainwood, A.; Sharp, S. Vacancy-Related Centers in Diamond. *Phys. Rev. B* **1992**, *46*, 13157–13170.
7. Hui, Y. Y.; Chang, H.-C. Recent Developments and Applications of Nanodiamonds as Versatile Bioimaging Agents. *J. Chin. Chem. Soc.* **2014**, *61*, 67–76.
8. Doherty, M. W.; Manson, N. B.; Delaney, P.; Jelezko, F.; Wrachtrup, J.; Hollenberg, L. C. L. The Nitrogen-Vacancy Colour Centre in Diamond. *Phys. Rep.* **2013**, *528*, 1–45.
9. Yu, S.-J.; Kang, M.-W.; Chang, H.-C.; Chen, K.-M.; Yu, Y.-C. Bright Fluorescent Nanodiamonds: No Photobleaching and Low Cytotoxicity. *J. Am. Chem. Soc.* **2005**, *127*, 17604–17605.
10. Billinton, N.; Knight, A. W. Seeing the Wood Through the Trees: A Review of Techniques for Distinguishing Green Fluorescent Protein From Endogenous Autofluorescence. *Anal. Biochem.* **2001**, *291*, 175–197.
11. Fu, C.-C.; Lee, H.-Y.; Chen, K.; Lim, T.-S.; Wu, H.-Y.; Lin, P.-K.; Wei, P.-K.; Tsao, P.-H.; Chang, H.-C.; Fann, W. Characterization and Application of Single Fluorescent Nanodiamonds as Cellular Biomarkers. *Proc. Natl. Acad. Sci. USA* **2007**, *104*, 727–732.
12. Tisler, J.; Balasubramanian, G.; Naydenov, B.; Kolesov, R.; Grotz, B.; Reuter, R.; Boudou, J.-P.; Curmi, P. A.; Sennour, M.; Thorel, A.; Börsch, M.; Aulenbacher, K.; Erdmann, R.; Hemmer, P. R.; Jelezko, F.; Wrachtrup, J. Fluorescence and Spin Properties of Defects in Single Digit Nanodiamonds. *ACS Nano* **2009**, *3*, 1959–1965.
13. Schirhagl, R.; Chang, K.; Loretz, M.; Degen, C. L. Nitrogen-Vacancy Centers in Diamond: Nanoscale Sensors for Physics and Biology. *Annu. Rev. Phys. Chem.* **2014**, *65*, 83–105.
14. Krueger, A.; Lang, D. Functionality Is Key: Recent Progress in the Surface Functionalization of Nanodiamond. *Adv. Funct. Mater.* **2012**, *22*, 890–906.
15. Huang, L. C. L.; Chang, H.-C. Adsorption and Immobilization of Cytochrome *c* on Nanodiamonds. *Langmuir* **2004**, *20*, 5879–5884.
16. Osswald, S.; Yushin, G.; Mochalin, V.; Kucheyev, S. O.; Gogotsi, Y. Control of $sp^2/sp^3$ Carbon Ratio and Surface Chemistry of Nanodiamond Powders by Selective Oxidation in Air. *J. Am. Chem. Soc.* **2006**, *128*, 11635–11642.
17. Shenderova, O.; Koscheev, A.; Zaripov, N.; Petrov, I.; Skryabin, Y.; Detkov, P.; Turner, S.; Tendeloo, G. V. Surface Chemistry and Properties of Ozone-Purified Detonation Nanodiamonds. *J. Phys. Chem. C* **2011**, *115*, 9827–9837.
18. Chang, B.-M.; Lin, S.-H.; Su, L.-J.; Lin, W.-D.; Lin, R.-J.; Tzeng, T.-K.; Lee, R. T.; Lee, Y. C.; Yu, A. L.; Chang, H.-C. Highly Fluorescent Nanodiamonds Protein-Functionalized for Cell Labeling and Targeting. *Adv. Funct. Mater.* **2013**, *23*, 5737–5745.

19. Vial, S.; Mansuy, C.; Sagan, S.; Irinopoulou, T.; Burlina, F.; Boudou, J. P.; Chassaing, G.; Lavielle, S. Peptide-Grafted Nanodiamonds: Preparation, Cytotoxicity and Uptake in Cells. *ChemBioChem* **2008**, *9*, 2113–2119.
20. Bumb, A.; Sarkar, S. K.; Billington, N.; Brechbiel, M. W.; Neuman, K. C. Silica Encapsulation of Fluorescent Nanodiamonds for Colloidal Stability and Facile Surface Functionalization. *J. Am. Chem. Soc.* **2013**, *135*, 7815–7818.
21. Zhao, L.; Takimoto, T.; Ito, M.; Kitagawa, N.; Kimura, T.; Komatsu, N. Chromatographic Separation of Highly Soluble Diamond Nanoparticles Prepared by Polyglycerol Grafting. *Angew. Chem. Int. Ed.* **2011**, *50*, 1388–1392.
22. Lee, J. W.; Lee, S.; Jang, S.; Han, K. Y.; Kim, Y.; Hyun, J.; Kim, S. K.; Lee, Y. Preparation of Non-aggregated Fluorescent Nanodiamonds (FNDs) by Non-covalent Coating With a Block Copolymer and Proteins for Enhancement of Intracellular Uptake. *Mol. BioSyst.* **2013**, *9*, 1004–1011.
23. Rehor, I.; Mackova, H.; Filippov, S. K.; Kucka, J.; Proks, V.; Slegerova, J.; Turner, S.; Van Tendeloo, G.; Ledvina, M.; Hruby, M.; Cigler, P. Fluorescent Nanodiamonds With Bioorthogonally Reactive Protein-Resistant Polymeric Coatings. *Chempluschem* **2014**, *79*, 21–24.
24. Tzeng, Y.-K.; Faklaris, O.; Chang, B.-M.; Kuo, Y.; Hsu, J.-H.; Chang, H.-C. Superresolution Imaging of Albumin-Conjugated Fluorescent Nanodiamonds in Cells by Stimulated Emission Depletion. *Angew. Chem. Int. Ed.* **2011**, *50*, 2262–2265.
25. Kong, X. L.; Huang, L. C. L.; Hsu, C.-M.; Chen, W.-H.; Han, C.-C.; Chang, H.-C. High-Affinity Capture of Proteins by Diamond Nanoparticles for Mass Spectrometric Analysis. *Anal. Chem.* **2005**, *77*, 259–265.
26. Lin, H.-H.; Lee, H.-W.; Lin, R.-J.; Huang, C.-W.; Liao, Y.-C.; Fang, J.-M.; Chen, Y.-T.; Lee, T.-C.; Yu, A. L.; Chang, H.-C. Tracking and Finding Slow-Proliferating/Quiescent Cancer Stem Cells With Fluorescent Nanodiamonds. *Small* **2015**, *11*, 4394–4402.
27. Fang, C.-Y.; Vaijayanthimala, V.; Cheng, C.-A.; Yeh, S.-H.; Chang, C.-F.; Li, C.-L.; Chang, H.-C. The Exocytosis of Fluorescent Nanodiamond and Its Use as a Long-term Cell Tracker. *Small* **2011**, *7*, 3363–3370.
28. Chang, C. W.; Sud, D.; Mycek, M. A. Fluorescence Lifetime Imaging Microscopy. *Methods Cell Biol.* **2007**, *81*, 495–524.
29. Faklaris, O.; Garrot, D.; Joshi, V.; Druon, F.; Boudou, J. P.; Sauvage, T.; Georges, P.; Curmi, P. A.; Treussart, F. Detection of Single Photoluminescent Diamond Nanoparticles in Cells and Study of the Internalization Pathway. *Small* **2008**, *4*, 2236–2239.
30. Wu, T.-J.; Tzeng, Y.-K.; Chang, W.-W.; Cheng, C.-A.; Kuo, Y.; Chien, C.-H.; Chang, H.-C.; Yu, J. Tracking the Engraftment and Regenerative Capabilities of Transplanted Lung Stem Cells Using Fluorescent Nanodiamonds. *Nat. Nanotechnol.* **2013**, *8*, 682–689.
31. Chow, E. K.; Zhang, X.-Q.; Chen, M.; Lam, R.; Robinson, E.; Huang, H.; Schaffer, D.; Osawa, E.; Goga, A.; Ho, D. Nanodiamond Therapeutic Delivery Agents Mediate Enhanced Chemoresistant Tumor Treatment. *Sci. Transl. Med.* **2011**, *3*, 73ra21.
32. Manus, L. M.; Mastarone, D. J.; Waters, E. A.; Zhang, X. Q.; Schultz-Sikma, E. A.; Macrenaris, K. W.; Ho, D.; Meade, T. J. Gd(III)-Nanodiamond Conjugates for MRI Contrast Enhancement. *Nano Lett.* **2010**, *10*, 484–489.
33. Weiss, A.; Ding, X.; van Beijnum, J.; Wong, I.; Wong, T.; Berndsen, R.; Dormond, O.; Dallinga, M.; Shen, L.; Schlingemann, R.; Pili, R.; Ho, C.-M.; Dyson, P.; van den Bergh, H.; Griffioen, A.; Nowak-Sliwinska, P. Rapid Optimization of Drug Combinations for the Optimal Angiostatic Treatment of Cancer. *Angiogenesis* **2015**, *18*, 233–244.

34. Wang, H.; Lee, D.-K.; Chen, K.-Y.; Chen, J.-Y.; Zhang, K.; Silva, A.; Ho, C.-M.; Ho, D. Mechanism-Independent Optimization of Combinatorial Nanodiamond and Unmodified Drug Delivery Using a Phenotypically Driven Platform Technology. *ACS Nano* **2015**, *9*, 3332–3344.
35. Ho, D.; Wang, C.-H. K.; Chow, E. K.-H. Nanodiamonds: The Intersection of Nanotechnology, Drug Development, and Personalized Medicine. *Sci. Adv.* **2015**, *1*, e1500439.
36. Zarrinpar, A.; Lee, D.-K.; Silva, A.; Datta, N.; Kee, T.; Eriksen, C.; Weigle, K.; Agopian, V.; Kaldas, F.; Farmer, D.; Wang, S. E.; Busuttil, R.; Ho, C.-M.; Ho, D. Individualizing Liver Transplant Immunosuppression Using a Phenotypic Personalized Medicine Platform. *Sci. Transl. Med.* **2016**, *8*, 333ra49.
37. Nakamura, T.; Ohana, T.; Yabuno, H.; Kasai, R.; Suzuki, T.; Hasebe, T. Simple Fabrication of GD(III)-DTPA-Nanodiamond Particles by Chemical Modification for Use as Magnetic Resonance Imaging (MRI) Contrast Agent. *Appl. Phys. Express* **2013**, *6*, 015001.
38. Rojas, S.; Gispert, J. D.; Martin, R.; Abad, S.; Menchón, C.; Pareto, D.; Victor, V. M.; Alvaro, M.; Garcia, H.; Herance, J. R. Biodistribution of Amino-Functionalized Diamond Nanoparticles. *In vivo* Studies Based on $^{18}$F Radionuclide Emission. *ACS Nano* **2011**, *5*, 5552–5559.

CHAPTER

# Applications of color centers as nanoscopic sensors

17

**Elke Neu**

*Saarland University, Saarbrücken, Germany*

## COLOR CENTERS IN DIAMOND AS NANOSCOPIC SENSORS: INTRODUCTION

Color centers in diamond are versatile nanoscopic sensors: As point defects in the diamond lattice, their size is comparable to the lattice spacing in diamond (0.36 nm [1]). Thus, they can be considered as "artificial atoms" with truly atomic size, rendering them highly promising as nanoscopic sensors for sensing and imaging with potentially nanometer spatial resolution. More than 500 color centers in diamond have been reported,[2] and various of them might be useful in sensing applications. However, only the negatively charged nitrogen vacancy center (for simplicity termed NV center) has so far been extensively used as a sensor in nanoscopic imaging, and the following chapter will thus concentrate on the NV center. More general information on various color centers can be found in Chapter 7.

NV centers exhibit the necessary stability to be used as a reliable sensor: being incorporated into the chemically inert diamond lattice, they can be stable even very close to the surface of single crystal diamond (at $\approx$ 1 nm depth[3]) without suffering from light induced bleaching, i.e. permanent de-activation of the fluorescent color center. Color center stability also persists in nanodiamonds (NDs): silicon vacancy centers have been found to be stable in NDs as small as 1.6 nm [4] bringing the size of fluorescent NDs down to the level of single fluorescent dye molecules, furthermore allowing to bring the color center as a nanoscopic sensor very close to the sample under investigation.

In addition to being stable emitters of light, color centers in diamond, especially NV centers, host electronic spins and can be coupled to nuclear spins.[5] NV centers reveal outstanding spin properties owing to the unique properties of diamond as a host material: For many applications, it is crucial how long a spin can sustain its phase when being placed in a superposition state of spin-up and spin-down (coherence time of the spin). The lattice of tightly bound, low-mass carbon atoms shows a very weak spin-lattice interaction (phonon limited spin-relaxation times of $10^2$–$10^4$ s)[6] and does not limit the spin's phase coherence. Instead, the coherence times are limited by the presence of paramagnetic spins of impurities (e.g. nitrogen) and the nuclear spins in diamond.[6] However, the most abundant carbon isotope $^{12}C$ has no nuclear spin; the concentration of the paramagnetic $^{13}C$ isotope can be reduced to only 0.3% in isotopically engineered diamond.[6] As a result, in ultrapure diamond, a superposition state of an NV center's spin can be sustained for almost $T_2 = 2\,\text{ms}$ even at room temperature. Thus, NV spins located within an almost spin-free lattice can be used to sense magnetic fields external to the diamond with high sensitivity. Additionally, the NV's electron spin can be optically initialized into a defined state and its state can be optically read out allowing for so-called "optically detected magnetic resonance" (ODMR) of single NV spins used as a sensor.[7] For more information on electron paramagnetic resonance (also termed electron spin resonance, EPR/ESR) and nuclear magnetic resonance (NMR) techniques in diamond see Chapter 6.

The highly coherent, controllable spin states as well as the nature of the color center as stable light emitting dipole, form the basis of the sensing schemes introduced here. The chapter is outlined as follows:

The spins of electrons tightly bound to the NV defect are susceptible to magnetic and electric fields and can be used to detect even very weak fields as introduced in Sections 17.1.1 and 17.1.2. The temperature dependent spin properties of the NV center also render it highly promising as nanoscopic sensor for the local temperature (Section 17.1.2). Furthermore, the NV center as a stable light emitting dipole reveals the presence of other dipoles via near-field energy transfer processes and can thus be used as a probe for scanning near field optical imaging as introduced in Section 17.1.3.

The above mentioned imaging schemes typically reach spatial resolutions of $\leq 100\,\text{nm}$ when NV centers are used in scanning probe techniques, where the spatial resolution is mainly limited by the NV-to-sample distance and can thus in principle reach the nanometer regime (Sections 17.2.1 and 17.2.2). In these techniques, a scanning NV center is realized via attaching a ND to a commercial atomic force microscope (AFM) tip or by engineering a tip out of diamond. Scanning probe techniques are only sensitive to the surface of the investigated sample which can be advantageous but is not always desirable. As an alternative, NDs can be inserted into the sample under investigation e.g. into a (living) cell. Using this approach, NDs can be used as fluorescent labels for bio-applications (cf. Chapter 16) but also as quantum sensors for magnetic fields, temperature or for near field energy transfer sensing (see Section 17.2.1). However, using fluorescence microscopy, only objects that are spatially separated by more than roughly half the wavelength of the employed light can be spatially resolved (diffraction limit, typically 250 nm), thus limiting the spatial resolution of these sensing schemes. In recent years, several optical microscopy techniques have been developed that allow for nanometer resolution without the need for near-field or scanning probe techniques. These so-called far-field nanoscopy or superresolution techniques have also profited a lot from using NV centers in diamond and will be summarized in Section 17.3.

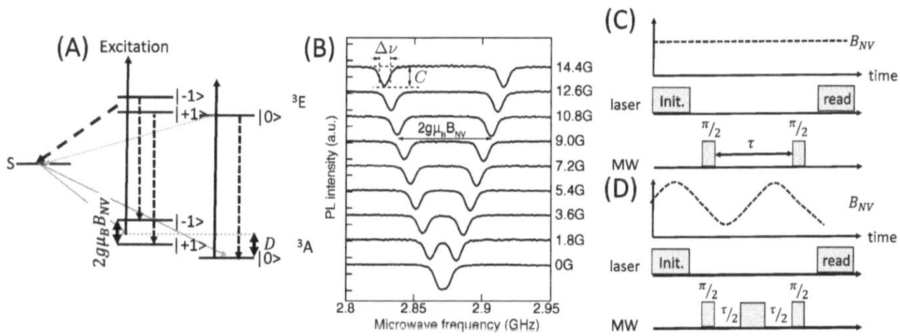


**FIGURE 17.1**

(A) Level scheme of the NV center. Optical transitions occur between triplet states ($^3$A, $^3$E); whereas singlet states (S) are responsible for the optical spin initialization and read out. Note that the transition (shelving) rate to the singlet states is strongly spin dependent, whereas the singlet states decay with similar probability to the $|0\rangle$, $|+1\rangle$, and $|-1\rangle$ states.[8] The Zeeman-splitting between the $|\pm1\rangle$ states is used for magnetometry. Already in zero-field, the states $|0\rangle$ and $|\pm1\rangle$ are split by $D = 2.87$ GHz. (B) ODMR spectrum of a single NV center in a magnetic field (reprinted with permission from L. Rondin et al., Rep. Prog. Phys. 77, 056503 (2014), © 2014 IOP publishing); (C) Ramsey-type pulse sequence for DC magnetometry, (D) Hahn-Echo-type pulse sequence for AC magnetometry

## 17.1 COLOR CENTERS IN DIAMOND AS SENSOR FOR MAGNETIC AND NEAR FIELDS AND TEMPERATURE

This section summarizes basic principles of sensing with color centers in diamond. It introduces basics of magnetic sensing with NV centers (17.1.1), electric field and temperature sensing (17.1.2) and optical sensing techniques using near-fields (17.1.3). Please note that this section concentrates only on the principles relevant for sensing, for a more general introduction to the properties of color centers please refer to Chapter 7, for more details on spin manipulation (ESR, NMR) please refer to Chapter 6.

### 17.1.1 MAGNETIC SENSING WITH NV CENTERS IN DIAMOND

#### *Principle*

NV centers in diamond are outstanding magnetic sensors providing an efficient mechanism for optical spin initialization and read out: As depicted in Fig. 17.1A, the NV ground state consists of three spin states: $m_s = 0 = |0\rangle$, $|+1\rangle$, and $|-1\rangle$. If the NV center resides in state $|0\rangle$, it has a high probability to decay back to $|0\rangle$ following optical excitation (cycling transition). If the NV center is in state $|1\rangle$ or $|-1\rangle$, it decays with enhanced probability to the singlets as depicted in Fig. 17.1A. This "intersystem-crossing" (ISC) is non-radiative and the singlet states have lifetimes exceeding 300 ns,[8] whereas the excited triplet state has a lifetime of around 10 ns.[8] Consequently, NV centers emit less fluorescence when being prepared in $|+1\rangle$ or $|-1\rangle$ states as a result of the "shelving" to the singlet states. Moreover, the spin-dependent shelving to the singlets induces also optical polarization of the spin: after undergoing optical excitation (typically with green laser light, 532 nm) for roughly 1 µs,[10] the NV center has been pumped (initialized) with high probability into its $|0\rangle$ state. The states $|+1\rangle$ and $|-1\rangle$ have identical energy (degenerate states) without external fields applied to the NV center. The splitting between $|0\rangle$ and $|\pm1\rangle$ amounts to $D = 2.87$ GHz in the absence of external fields and is thus termed the zero-field splitting. Magnetometry using NV centers is performed using the Zeeman-shift of the $|\pm1\rangle$ states: as depicted in Fig. 17.1A, an external magnetic field oriented along the NV axis $B_{NV}$ splits the $|\pm1\rangle$ states by $2g\mu_B B_{NV}$. Thus, NV magnetometry in general measures the magnetic field projection onto the NV axis $B_{NV}$. The spin resonances are detected via optically detected magnetic resonance (ODMR) as illustrated in Fig. 17.1B: The NV center is initialized in state $|0\rangle$ via optical pumping. Upon a frequency sweep of a microwave (MW) driving field, a decrease of the fluorescence indicates when the MW frequency is resonant with either the $|0\rangle \rightarrow |+1\rangle$ or the $|0\rangle \rightarrow |-1\rangle$ transition.

The ground state spin energies for the NV center are given by the Hamilton operator $H$ (neglecting hyperfine interaction with nuclear spins)[9]:

$$H = hDS_z + hE(S_x^2 + S_y^2) + \mu_B g(S_z B_z + B_y S_y + B_x S_x). \tag{17.1}$$

The coordinate system is chosen in a way that the $z$-axis is oriented along the high symmetry axis of the NV center which connects the nitrogen and the adjacent vacancy. $D$ is the zero-field-splitting introduced above, whereas the zero-field splitting parameter $E$ is usually much smaller and especially for NDs mostly determined by strain in the diamond lattice.[9] $S_x$, $S_y$, $S_z$ are the Pauli matrices, $g \approx 2$ is the Landé factor, and $\mu_B = 9.27 \times 10^{-24}\ \frac{\mathrm{J}}{\mathrm{T}}$. All terms in $H$ depending only on $S_z$ are so-called parallel terms, affecting energetic positions of the $|0\rangle$ and $|\pm1\rangle$ states (recall that $m_s$ is the eigenvalue of the $S_z$ operator), whereas terms including $S_x$ and $S_y$ lead to so-called spin-mixing. In essence, NV magnetometry measures the transition frequencies $|0\rangle \rightarrow |+1\rangle$ or $|0\rangle \rightarrow |-1\rangle$ ($\nu_\pm$). From Eq. (17.1), we find[9]:

$$\nu_\pm(B_{NV}) = D \pm \sqrt{\left(\frac{g\mu_B}{h} B_{NV}\right)^2 + E^2}. \tag{17.2}$$

Eq. (17.2) illustrates an important point: For external magnetic fields $g\mu_B B_{NV} \ll hE$, $\nu_\pm$ can be approximated as $\nu_\pm \approx D \pm E$. Thus, an NV center with non-negligible strain induced splitting $E$ is not susceptible to magnetic field changes in first-order.[9] To make the NV center sensitive for magnetic fields, a bias field $B_{bias} \gg \frac{hE}{g\mu_B}$ has to be applied. Generally, NDs exhibit a higher level of crystal strain introduced e.g. by mechanically crushing the diamonds.[11] For more information on how to analyze strain and defects in NDs via high resolution electron microscopy and Raman spectroscopy, see Chapters 3 and 4, respectively.

Applying a bias field is not always desirable thus avoiding highly strained NV centers is advantageous for magnetometry. Though being undesirable for magnetometry, strained NV center can be used as a sensor for external electric fields (see Section 17.1.2). Thus, for each sensing application and magnitude of external fields, careful consideration of the NV spin energies is necessary (see also Rondin et al.[9]).

### Sensitivity and techniques

As introduced above, the most straightforward way to use NV centers as magnetic sensors is to directly measure the ODMR spectrum and deduce $B_{NV}$ from the resonance positions. This technique is often employed when comparably strong magnetic fields e.g. from ferromagnetic structures are investigated.[12,13] The magnetic field sensitivity $\eta$ obtainable with this technique is[9]

$$\eta \approx \frac{h}{g\mu_B}\frac{\Delta\nu}{\sqrt{I_0 C}}. \tag{17.3}$$

Here $\Delta\nu$ is the ODMR linewidth, $I_0$ is the detected photon rate from the NV center and $C$ is the contrast in the ODMR measurement [fluorescence difference between $|0\rangle$ and $|\pm1\rangle$, for single NV centers typically 20%, see Fig. 17.1B]. Eq. (17.3) indicates that it is favorable to enhance the number of detected photons $I_0$ from a single NV center to enhance its sensitivity $\eta$ as a sensor, motivating research on diamond nanophotonic structures e.g. nanowires for efficient light extraction from diamond.[14–16] In Eq. (17.3), a fundamental limit for $\eta$ is given by the ODMR linewidth $\Delta\nu = \frac{1}{T_2^*}$, where $T_2^*$ is the inhomogeneous de-phasing time of the NV spin. In practice, this limit is not reached as continuously applied laser excitation and MW fields used to measure the ODMR spectrum, broaden the ODMR resonances ($\Delta\nu$ increases).[17] A way to overcome this limitation is to use a Ramsey-pulse sequence as depicted in Fig. 17.1C: After the NV is initialized to $|0\rangle$, a MW $\frac{\pi}{2}$ pulse creates a superposition $\Psi \sim |0\rangle + |1\rangle$. During the waiting time $\tau$, this superposition acquires a phase $\phi$ that depends on the constant (DC) magnetic field $B_{NV}$; the new state reads $\Psi \sim |0\rangle + e^{i\phi}|1\rangle$. A second $\frac{\pi}{2}$ pulse transforms this phase $\phi$ into a population difference between states $|0\rangle$ and $|1\rangle$ which is then optically read out via a second laser pulse. Thus, the Ramsey sequence separates spin-initialization, manipulation and magnetic field measurement (phase accumulation) in time and thus allows for enhanced sensitivity.[9]

With the above described technique, NV centers are sensitive to non-varying (DC) fields. Using an altered pulse sequence, oscillating (AC) fields can be measured with even higher sensitivity: The pulse sequence is altered to the so-called Hahn-Echo-type sequence in Fig. 17.1D: An additional $\pi$ pulse after half of the waiting time $\tau$ induces a swap of the phase $|0\rangle + e^{\frac{i\phi}{2}}|1\rangle \rightarrow e^{\frac{i\phi}{2}}|0\rangle + |1\rangle$. In a DC field, the $|1\rangle$ state again accumulates a phase of $\frac{\phi}{2}$ during the second half of the waiting time, thus canceling out the accumulated phase. However, the NV center is now sensitive to fields oscillating at the frequency of the pulse sequence. The spin coherence time $T_2$ is measured by sweeping the waiting time $\tau$ in the Hahn-Echo-sequence and fitting the resulting curve.[18] The NV's $T_2$ time is sensitive also to random magnetic fluctuations in the environment of the NV centers: For example, thermal fluctuations of nearby spins induce a fluctuating spin bath and reduce $T_2$. Such a "decoherence sensing" has been used to detect e.g. paramagnetic iron in the environment of NV centers.[19]

### 17.1.2 TEMPERATURE AND ELECTRIC FIELD SENSING WITH COLOR CENTERS IN DIAMOND

As discussed previously, Eq. (17.2) shows that a highly strained NV center without a bias magnetic field is not susceptible to magnetic fluctuations in first order. Though being undesirable for magnetometry, this situation allows using NV centers as sensors for external electric fields. When using NV centers as electric sensor, applying magnetic fields strictly perpendicular to the NV high symmetry axis can also be used to create eigenstates with a high sensitivity to electric instead of magnetic fields.[20] AC electric field sensitivities of 200 $\frac{\mathrm{V}}{\mathrm{cm}\sqrt{\mathrm{Hz}}}$ have been demonstrated, indicating that an elementary charge located at a distance of 150 nm from the NV sensor can be detected within 1 s.[20]

In addition to sensing electric fields, NV centers can also be used to measure the local temperature: The zero-field splitting $D$ of NV centers additionally depends on the local temperature $T$ as well as on crystal strain.[21] The dependence on the temperature can be illustrated as temperature changes induce crystal strain and thus influence the NV center. Under ambient conditions, $c_T = \frac{dD}{dT} = -74.2\,\frac{\mathrm{kHz}}{\mathrm{K}}$ gives the change of the zero field splitting parameter $D$ with temperature. As this change is very weak, great care has to be taken not to misinterpret e.g. residual magnetic field induced energy changes as temperature changes. In Neumann et al.,[21] a dedicated Ramsey sequence sensitive to temperature fluctuations that cancels axial magnetic sensitivity and eliminates resonance line broadening is introduced. In this sequence, the NV center resides for the same time in the $|+1\rangle$ and $|-1\rangle$ states so that phase changes due to magnetic fields have opposite signs and cancel. Changes on the order of mK have been measured. For color centers in bulk diamond, sensitivities of 5 $\frac{\mathrm{mK}}{\sqrt{\mathrm{Hz}}}$ were obtained.[21]

In this context, it should be noted that the NV center can also be used to directly sense mechanical strain e.g. in nanomechanical resonators (cantilevers).[22,23] Crystal strain induced changes in the NV center's level structure and can even be used to drive spin transitions in NV centers using the mechanical movement of a nanomechanical resonator instead of employing MW driving fields,[24] thus allowing for novel spin based hybrid systems.

### 17.1.3 NEAR FIELD IMAGING/SENSING USING COLOR CENTERS IN DIAMOND

While the approaches discussed above all rely on the electronic spin of NV centers, the following section summarizes how the fact that a single color center is an electric dipole can be used for sensing. Due to the strong dipole transitions in the visible/near-infrared spectral region, even single NV centers emit bright, stable photo-luminescence (PL). This characteristic as nanoscopic light source can be used for optical imaging with enhanced resolution: as detailed in Section 17.3, light microscopy has a limited spatial resolution of typically 200–300 nm. However, this only holds for the conventional case of far field microscopy: if a nanoscopic light source and the investigated sample are approached closer than typically the wavelength of light, imaging takes place via so-called optical near-fields and higher resolution is obtained [scanning near field optical microscopy (SNOM)]. This effect has been used to image samples via scanning an ND containing NV centers over a sample; using the NV PL to illuminate and image the sample.[25,26]

Moreover, very promising imaging schemes are feasible when near field energy transfer processes between NV centers and their environment are used: In the Förster or Fluorescence resonance energy transfer (FRET),[27] two dipoles (termed donor and acceptor) non-radiatively exchange energy. In detail, the donor dipole is excited by absorbing light (see Fig. 17.2) and transfers energy to the acceptor. The transferred energy excites the acceptor and it either emits luminescence or dissipates the energy non-radiatively. In both cases, the energy transfer constitutes an additional de-excitation pathway for the donor thus its excited state lifetime and its fluorescence rate is reduced. FRET-processes enable efficient energy transfer only for very small distances $r$ between the interacting dipoles: The efficiency $E$ for a FRET process is typically given by[28]

$$E = \frac{R_0^6}{R_0^6 + r^6} \qquad \text{with} \qquad R_0 = 0.211(\kappa^2 n^{-4} Q_D J(\lambda))^{\frac{1}{6}} \tag{17.4}$$

where $R_0$ is the Förster distance determined by the dipole orientation factor $\kappa$, the refractive index of the medium $n$, the donor's luminescence quantum efficiency $Q_D$, and the spectral overlap $J(\lambda)$ between the donor's emission and the acceptor's absorption band. $Q_D$ gives the probability for a radiative decay following an excitation of the donor in the absence of FRET processes. For NV centers, $R_0$ has been found to be 3.8 and 5.6 nm for the interaction between NV centers and the dye molecule DY781[28] or the IRDye800CW,[29] respectively. As it is discernible from Eq. (17.4), only for distances comparable to or smaller than $R_0$ efficient transfer occurs. Thus, FRET processes open the route towards high resolution optical imaging. One could envisage using NDs in cells to track proximity of e.g. dye molecules used as labels, whereas NDs stand out due to high bio-compatibility, stable fluorescence, and feasible functionalization (cf. Chapter 16). Moreover, FRET processes can also be used in scanning probe imaging to yield high resolution. This technique, termed FRET-SNOM, has been tested with semiconductor nanocrystals and single molecules as the FRET dipole in the scanning tip.[30,31] However, limited photostability of the emitter on the tip remained an issue. Color centers in diamond have been suggested as a stable alternative here.[30] First FRET imaging via scanning a ND attached to an AFM tip has been achieved recently: Tisler et al.[32] have used a ND scanning probe (cf. Section 17.2.1) to image graphene flakes on a substrate. FRET between a NV center

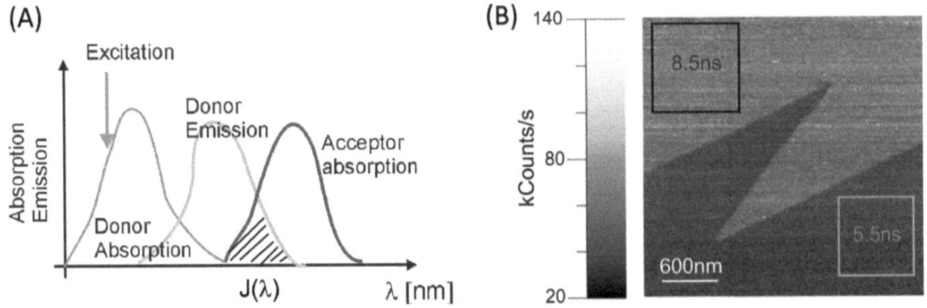


**FIGURE 17.2**

(A) Schematics of a typical FRET process: A donor fluorophore is excited via its absorption band using laser excitation (indicated by the arrow), if feasible in a region where the acceptor fluorophore does not absorb. Instead of emitting fluorescence, the donor fluorophore transfers its energy to an acceptor that starts to luminesce or dissipates the energy non-radiatively. In both cases, a decrease of the donor's PL intensity and lifetime will be observable. For the process efficiency, the spectral overlap $J(\lambda)$ between the donor's emission and the acceptor's absorption band is crucial. (B) Example of high resolution imaging of graphene flakes via an FRET process. Reprinted with permission from Nano Letters, 2013, 13, 3152–3156. © 2013 American Chemical Society. A scanning NV center serves as the donor, its fluorescence is quenched as result of near field transfer to graphene (note the differences in PL as well as the lifetime reduction from 8.5 to 5.5 ns). The edges of the graphene flakes have been imaged with a resolution of $\approx 20$ nm

on the tip and a graphene flake lowers the NV's PL and lifetime when it is located above the graphene flake (see Fig. 17.2). Edges of the graphene have been imaged with about 20 nm resolution (NV-to-graphene distance 16 nm). Graphene doesn't emit PL but dissipates the transferred energy. The nature of graphene as a 2 dimensional dipole modifies Eq. (17.4) to an $r^4$ dependence.[32] Scanning NV NDs have been also used to image silver nanowires on a substrate.[33] Here, the presence of the metal induces lifetime changes (local density of states, plasmonic modes, losses) the method is thus termed "fluorescence lifetime imaging." Similarly, scanning a nanosized metallic tip over NV centers can enhance PL when the tip is in close proximity and thus allows locating NV centers with 40 nm precision.[34]

## 17.2 APPLICATION OF DIAMOND COLOR CENTERS AS SENSORS

The following section illustrates examples of applications of color centers as sensors. The section is organized as follows: it first introduces sensing applications where NDs are directly used as sensors or following attachment to an AFM tip in scanning probe imaging (Section 17.2.1). Second, it introduces a novel concept to scanning probe

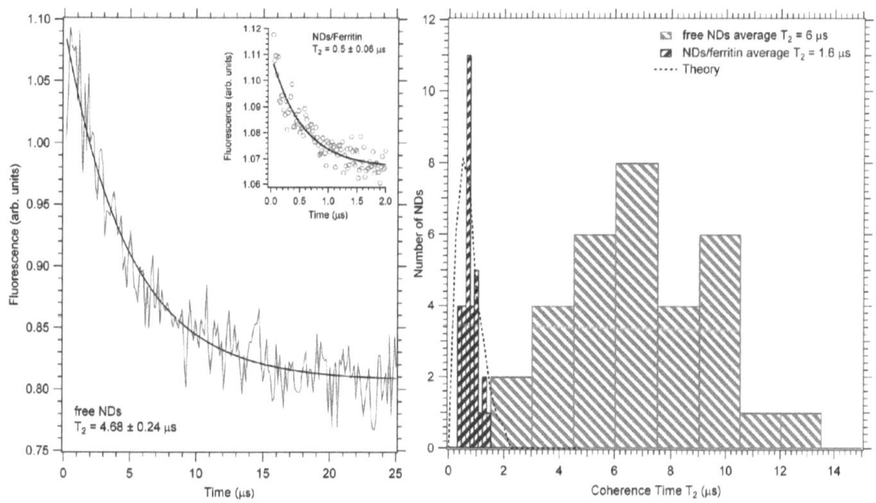


**FIGURE 17.3**

Sensing of Ferritin molecules on the surface of NDs using the change of coherence time $T_2$ induced by their magnetic noise. Reprinted with permission from Nano Letters, 2013, 13, 3305–3309. © 2013 American Chemical Society

imaging using color center in diamond nanostructures sculpted from single crystal diamond (Section 17.2.2).

## 17.2.1 SENSING USING NANODIAMONDS

### *NDs used directly as sensor*

NDs containing NV centers are versatile sensors in biological environments due to their bio-compatibility, possibility of cell uptake, bright fluorescence and high magnetic sensitivity even under physiological conditions (cf. [35] and Chapters 12 and 16). NDs were used to detect the presence of a few metallo-organic proteins, so-called ferritin molecules, that serve as iron storing proteins in many organisms. Fig. 17.3 summarizes some of the results presented in [19]: the magnetic noise induced by paramagnetic iron molecules (about 4500 stored in an individual ferritin molecule) reduces the $T_2$ spin coherence time of NV centers by an order of magnitude already in the presence of a few Ferritin molecules. The same holds for the $T_1$ time of the NV center. These results show the potential of relaxometry/decoherence measurements with NV centers in NDs. In principle, all kinds of processes inducing magnetic noise can be measured. This includes detecting the presence of paramagnetic ions like iron (in ferritin),[19,36] manganese[36] or gadolinium.[37,38] Moreover, NV centers have been proposed as probes for currents in biological systems, especially to detect the currents due to neuronal activity[39] and even the weak current of a single ion channel in a cell membrane,[40] the basic unit for communication of a cell with other cells. Please note that for all these applications, quality, purity, and size (distribution) of the NDs play a crucial role. For more information on the production and purification of NDs from high pressure high temperature diamond see Chapter 2.

As temperature sensors, NV centers in NDs witnessed the heating of a substrate close to a microwave conducting wire[21] or measured the local temperature in living cells.[41] The excellent heat conductivity of diamond ensures thermal equilibrium for all centers in NV ensembles in a ND. Additionally, the ND probes a small volume [on the order of $(100\,\mathrm{nm})^3$] and thus can finely detect temperature gradients e.g. in a cell heated via a laser-irradiated gold particle.[41]

### ND scanning probes

Magnetometers based on NV centers in scanning probes are highly appealing for many fields of research as they supply quantitative, nanoscale imaging of magnetic fields under ambient conditions. Furthermore, single NV centers do not magnetically perturb the sample allowing to perform measurements with a low back action onto the sample. The technique is thus especially promising to investigate nanoscale magnetism, providing insight into magnetic phenomena crucial e.g. for further developments in spintronic applications.

The technical challenge in scanning NV magnetometry is to supply a "scanning NV center". A straightforward approach to realize such as sensor is to attach a ND with a single NV to a commercial atomic force microscope (AFM) tip as illustrated in the upper row of Fig. 17.4. This approach enables scanning an NV within typically $\approx$100 nm above the sample surface.[12] The attachment to commercial AFM tips allows for a straightforward tip–sample-distance control via an AFM feedback loop. To set up a scanning NV magnetometer, a sample scanning AFM has to be combined with a confocal microscope for optical spin-read out of the NV center, cf. Fig. 17.4 upper row. Note that with a similar setup and the same type of scanning probes also near field imaging via energy transfer as described in Sect. 17.1.3 has been realized.[32]

Scanning NV magnetometry has been proposed[42,43] and realized [44] in 2008. Since then it has been used especially to exploit several open questions in nanomagnetism: First, the method has been used to map the core of a so-called magnetic vortex in a ferromagnetic square dot which was not possible before with other techniques due to its small size (10 nm).[12] Furthermore, via measuring the stray field of domain walls in ultra-thin ferromagnets, i.e. boarders between areas of opposite magnetization, diamond magnetometry managed to distinguish between different types of domain walls (Néel or Bloch walls) and thus to identify the presence or absence of certain interactions at interfaces.[13] NV magnetometry also witnessed the jumps of a domain wall between 2 positions (pinning sites) as close as 50 nm.[45] In a recent application, NV magnetometry is used to investigate ultrathin ferromagnetic films with perpendicular magnetic anisotropy (PMA) which are interesting for fundamental studies in nanomagnetism and to develop low-power spintronic devices.[46] In this context, the unknown surface density of magnetic moments $I_s$ and its local variation have to be determined. Fig. 17.4 shows the stray field of a PMA measured with a scanning NV magnetometer, from this $I_s$ can be inferred using line cuts of the stray field in a di-

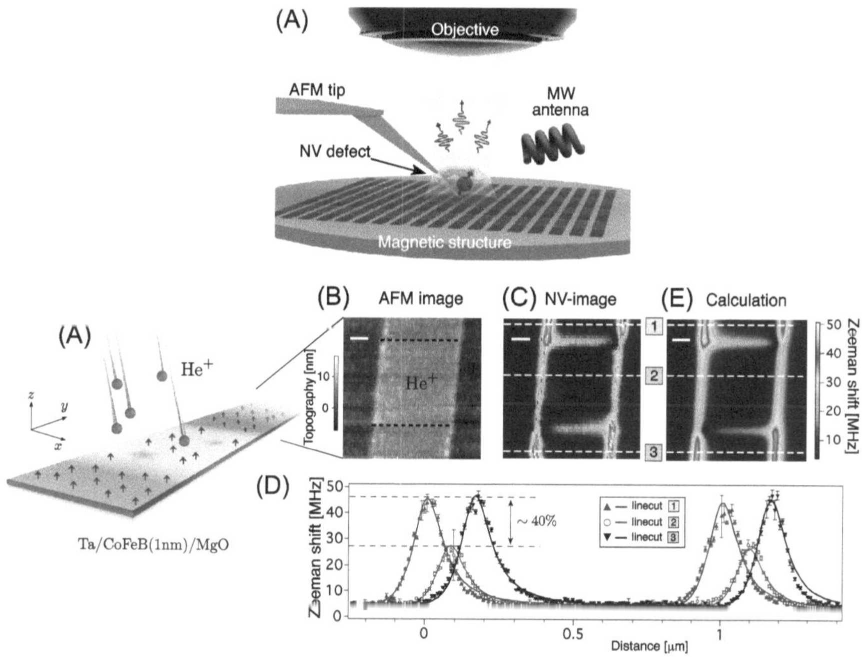


**FIGURE 17.4**

(Upper row) Schematic of a setup for scanning NV magnetometry using an ND based scanning probe. Reprinted figure with permission from Rondin et al., Appl. Phys. Lett., 2012, 100, 153118. © 2012 AIP publishing. A pre-selected nanodiamond with a single NV center is attached to an AFM tip e.g. by applying a glue to the AFM tip and picking up the ND. The ND is positioned in the focal point of a confocal microscope (indicated by the microscope objective) using a nanopositioning system. To enable keeping the NV center in the focus point of the confocal microscope during the measurement, the sample is scanned with respect to the tip using a second nanopositioning system. Additionally, a microwave antenna brought close to the tip enables coherent control of the NV center spin using microwaves. (Lower row) Direct measurement of the density of magnetic moment using scanning NV magnetometry in so-called ultrathin ferromagnetic films with perpendicular magnetic anisotropy (PMA). Reprinted figure with permission from T. Hingant et al., Phys. Rev. Appl. 4, 014003, 2015. © 2015 American Physical Society

rection perpendicular to the wire (see part D). To show that the method is capable of precisely determining local variations in $I_s$, a part of the magnetic wire has been irradiated with $He^+$ ions to induce intermixing at both Ta–CoFeB and CoFeB–MgO interfaces forming the wire (illustrated in lower row Fig. 17.4A). This reduces $I_s$ locally. Two important features are discernible in Fig. 17.4C. First, at the upper and lower end of the irradiated area, a stray field is generated indicating the expected abrupt change of $I_s$. Second, a lower stray field is found at the wire edges in the irradiated area (see line-cuts in D). From these phenomena, a decrease of $I_s$ by 40% can be inferred in the irradiated area. Thus, $I_s$ was determined with an uncertainty of only a few percent with a spatial resolution of about $(100\ \text{nm})^2$, an improvement by several orders of magnitude over existing methods.[46]

As mentioned above, the technical challenge here is to provide a scanning NV center. It should be noted that also "reversed experiments" have been performed where the sample under investigation e.g. a molecule is attached to a scanning probe tip and scanned over a stationary NV center e.g. Pelliccione et al.[47] However, these techniques are limited to investigations of substances that can be functionalized to an AFM tip.

### 17.2.2 ALL DIAMOND SCANNING PROBES

Grafting NDs to AFM tips to form NV scanning probes has been used in many applications. However, NV centers in NDs often suffer from short coherence times and high levels of crystal strain; both limiting their usability as magnetic sensors.[9] Furthermore, the grafted NDs do typically not reside at the apex of the AFM tip thus often limiting the minimal achievable NV-to-sample distance. Thus, a novel approach to provide scanning NV centers has been realized:[48–50] Instead of using NDs crafted to a tip, the complete scanning probe nanostructure is sculpted out of single crystal diamond using a dedicated nanofabrication process. Briefly, a diamond nanopillar on a thin (< 1 μm), transparent diamond platform is fabricated as illustrated in Fig. 17.5A. The cylindrical diamond nanopillar has a diameter of about 200 nm and a typical length of 1–2 μm. It forms a nanowaveguide for the PL emitted by the NV center[14,15] and has been found to enhance the rate of collected photons by almost an order of magnitude.[14] Single NV color centers are formed via ion implantation with low (keV) energy and annealing in close proximity (≈10 nm) to the end facet of the nanopillar. Thus, these devices have several advantages: First, the magnetic field sensitivity is high due to enhanced PL collection and potentially higher coherence times in high purity single crystal diamond. Second, as the color center can be located very close to the end facet of the pillar the NV-to-sample distance is reduced (with NVs implanted on average 10 nm deep, NV-to-sample distances of 25 nm have been realized[51]). To enable NV-sample distance control, all diamond scanning probe are attached via a holder to a tuning fork based AFM system.[48,49] Fig. 17.5 illustrates a recent application: An all-diamond scanning probe is scanned over a microwave stripline as a prototypical microwave circuit. A microwave current oscillating at GHz frequency is sent through the stripline. The NV spin transition frequency ($|0\rangle \rightarrow |-1\rangle$) is tuned into resonance with the microwave frequency using a static magnetic field. As a result, the right-handed circularly polarized magnetic field component of the microwave field $B'_{-,MW}$ drives Rabi oscillations on this transition as shown in Fig. 17.5B. The frequency of the Rabi oscillation depends on $B'_{-,MW}$ at the position of the NV center. To map $B'_{-,MW}$, the microwave current is applied for a fixed time period $\tau$. Depending on the magnitude of $B'_{-,MW}$, the scanning NV center accumulates a variable phase in the Rabi oscillation during the time $\tau$ and thus

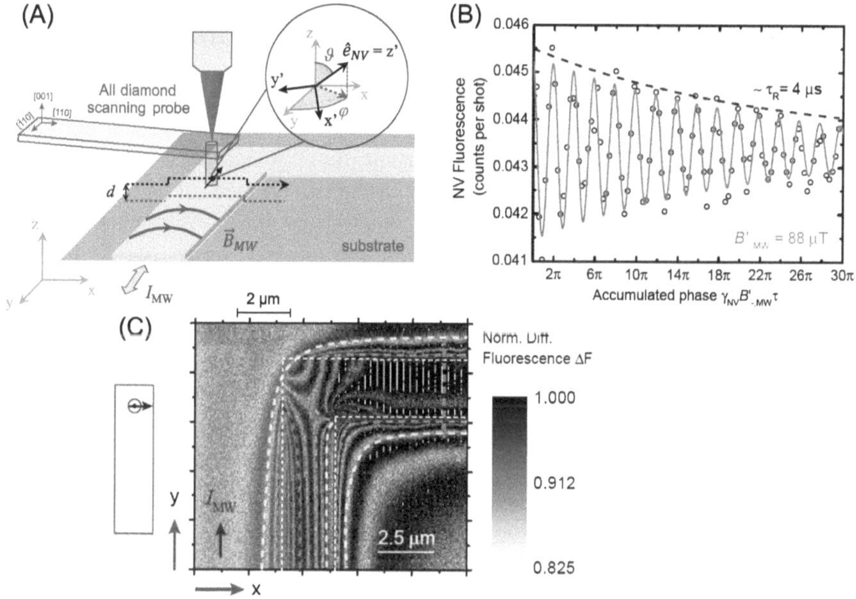


**FIGURE 17.5**

Scanning magnetometry with all diamond scanning probe (A) schematics of the measurement. An all diamond scanning probe with a single NV center is scanned over a microwave stripline to sense its magnetic stray field. Note that the optical access (excitation and PL collection) is established through the scanning device thus allowing for an opaque sample. (B) Rabi oscillation of the single NV spin measured for $B'_{-,MW} = 88$ μT. (C) Isofield lines of the microwave magnetic field as imaged with a single NV center. Images reprinted from P. Appel et al., New Journal of Physics, 17, 112001 (2015) under the creative commons license (CC BY 3.0)

appears brighter or darker upon spin readout. The result of the measurement is shown in Fig. 17.5C: Lines with equal brightness indicate a constant field $B'_{-,MW}$; measuring the Rabi frequency on these lines gives the value of $B'_{-,MW}$. Thus, scanning NV magnetometry here images GHz magnetic fields with nanoscale precision and a sensitivity of 15 $\frac{\mu T}{\sqrt{Hz}}$.[51] As a demonstration of the potential of the method, NV scanning magnetometry has imaged the magnetic field stemming from a single electron spin (here another NV center's electronic spin) at a distance of 50 nm corresponding to a detected field of 9 nT.[52] The same type of scanning probes has also been employed to measure surface spins on a proximal diamond surface using decoherence imaging.[53] Very recently, the applicability of scanning NV magnetometry has been extended to cryogenic temperature: All diamond NV scanning probes have been used to image vortices in superconducting materials,[54,55] thus opening up the route for NV magnetometry to investigate various phenomena at low temperature.

## 17.3 COLOR CENTERS IN FAR FIELD NANOSCOPY: SUPERRESOLUTION MICROSCOPY

### 17.3.1 SUPERRESOLUTION TECHNIQUES

Light microscopy is one of the most important imaging techniques in physics, material sciences, and life sciences. Especially in the life sciences, it enables long term observation of living cells and tissues. The spatial resolution of conventional light microscopy is limited by the so-called diffraction or Abbé limit: Two features can only be distinguished if they are separated by more than a distance $d$ given by

$$d = \frac{\lambda}{2n\sin(\alpha)}, \tag{17.5}$$

where $\lambda$ is the wavelength of the used light, and $\mathrm{NA} = n\sin(\alpha)$ the so-called numerical aperture of the microscope objective given by the refractive index $n$ of the medium in which the observation takes place and the half opening angle $\alpha$ under which the objective can collect light. Microscopes using visible light and optimized optics reach a resolution of about 200–300 nm. For example, in the life sciences, relevant structures in cells are smaller than this resolution. To image these structures, so-called superresolution techniques based on fluorescence microscopy have been developed. The following paragraph introduces one of these techniques, the so-called stimulated emission depletion microscopy (STED). To understand its working principle, we first look at the level scheme of a typical fluorophore used for STED: As depicted in Fig. 17.6C, the fluorophore is excited to a high lying excited state using laser light. From there it relaxes (wavy arrow) to a more stable excited state within typically picoseconds via phonon emission. Now the fluorophore could decay within typically nanoseconds to the ground state emitting fluorescence (dashed arrow). However, when using STED, a second laser (STED-laser) forces the fluorophore to undergo the so-called stimulated emission instead of the spontaneous fluorescence. Or in other words, the laser induces a fast stimulated decay of the system and thus empties or depletes the excited state before the spontaneous emission of fluorescence occurs. As the system has been forced to decay via the stimulated transition, a detection on the spontaneous fluorescence transition or after a STED laser pulse has been applied reveals no fluorescence; the fluorophore is switched off.

To use this depletion of fluorescence to reach an enhanced spatial resolution in a fluorescence microscope, excitation and STED laser have to be prepared with a distinct spatial intensity distribution: while the excitation laser resembles a Gaussian shape, the STED laser is modified to form a so-called doughnut mode with zero intensity in the center and a ring with high intensity as illustrated in Fig. 17.6B. The lasers are combined using dichroic mirrors and scanned over the sample simultaneously [see Fig. 17.6A]. Now all fluorophores in the region of high intensity of the STED laser will be efficiently switched off; only the ones residing in the center of the doughnut where a high excitation intensity and a low STED laser intensity is present

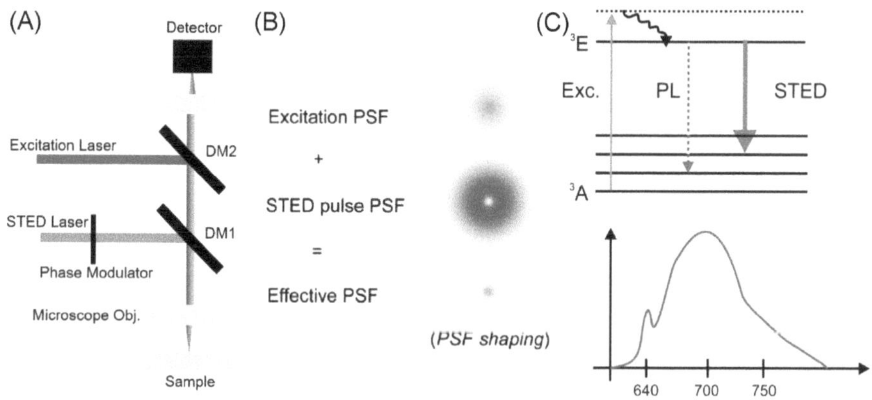


**FIGURE 17.6**

STED Schematics. (A) Basic building blocks of a STED-microscope: Excitation and STED laser are combined via dichroic mirrors (DM1 and DM2) and are focused onto the sample. The STED laser is spatially modulated using a phase plate (phase modulator) to resemble a doughnut profile. (B) Profiles of STED and excitation laser as well as effective PSF of the STED microscope. Image reprinted from L. Schermelleh et al., The Journal of Cell Biology, 190, 165–175 (2010) under the creative commons license (CC BY 3.0). (C) Level scheme of a typical fluorophore (for explanation see text). Below the level scheme, a schematic fluorescence spectrum of an NV center is drawn

are fluorescing. Thus, the microscope now has a high resolution (point spread function, PSF) below the diffraction limit and is thus called superresolution microscope or far-field nanoscope. Note that this also works for single fluorophores: only when the single emitter is located in the center, it can fluoresce; if it is moved out of the center, it is switched off. The resolution of the STED microscope is given by[56]

$$d_{STED}(I) = \frac{\lambda}{2n\sin(\alpha)\sqrt{\frac{1}{1+\frac{I}{I_s}}}}, \tag{17.6}$$

where $I$ is the intensity found at the doughnut crest and $I_s$ is the so-called threshold intensity, which is a fluorophore characteristic. As it is clear from Eq. (17.6), fluorophores that enable using high STED laser intensities are advantageous. This constitutes the main advantage of color centers in diamond as fluorophores for STED: They are highly photostable as they typically do not undergo laser-induced chemical reactions that often destroy the more commonly used fluorescent dye molecules and can thus withstand high laser power. In Rittweger et al.,[57] STED has been used to observe single NV centers with a resolution of 5.8 nm. Fig. 17.7A illustrates how the confocal point spread function is reduced below 10 nm when STED is applied. The high stability of the NV centers allowed for the variation of $I$ over a large range and to follow the scaling behavior in Eq. (17.6) as illustrated in Fig. 17.7B.

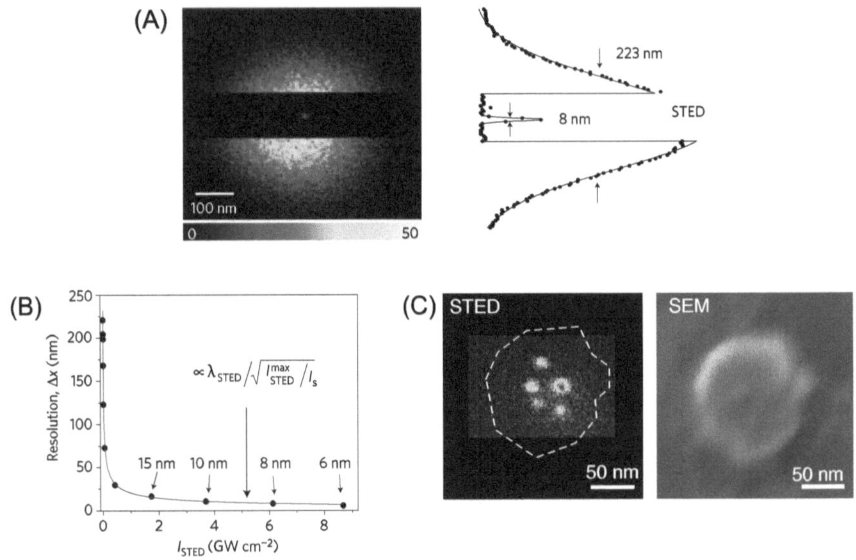


**FIGURE 17.7**

Results of STED imaging: (A) illustrates how the point spread function "shrinks" when STED is applied in comparison to confocal fluorescence microscopy. The fluorophore imaged here is a NV center in single crystal diamond. (B) Resolution scaling of STED with the applied STED laser intensity. Part (A)+(B): Reprinted by permission from Macmillan Publishers Ltd: E. Rittweger et al., Nature Photonics 3, 144–147 (2009), © 2009 Macmillan Publishers Ltd. (C) STED image revealing single NV centers in a 50 nm sized ND. Reprinted with permission from ACS Nano, 7, 10912–10919 (2013). © 2013 American Chemical Society

## 17.3.2 RECENT ACHIEVEMENTS USING COLOR CENTERS

The application of color centers in STED microscopy has led to several promising applications; examples are illustrated in the following. So far, STED has been realized with two types of color centers: mainly NV centers (e.g. Rittweger et al.[57]) and very recently also green fluorescent defects consisting of one vacancy next to two substitutional nitrogen atoms (NVN).[58] An important potential application for superresolution microscopy lies in applications of NV center spins for quantum technologies: When using NV electron spins as spin qubits, it is highly desirable to work with NV centers separated by only 10–30 nm as their spins can be coupled via magnetic dipole interactions and thus allow for high-fidelity two-qubit operations.[59] However, such NV centers cannot be addressed individually using conventional fluorescence microscopy, which would however be necessary to optically read out their spin-states. STED has been used to record the ODMR signal of individual NV centers separated by 50–110 nm.[60] To overcome the comparably high laser powers used in STED, a technique named spin-RESOLFT (reversible saturable optical linear fluorescence transitions) has been introduced. Similar to STED, a combination of a Gaussian laser beam, a doughnut-shaped beam, and microwaves is used to individually, coherently manipulate the spin of NV centers spaced by 150 nm.[59] It should be noted that superresolution techniques are also promising in the context of sensing with color centers, as they might open up routes towards magnetic and electric field, as well as temperature sensing, on the nanoscale without the need for scanning probe techniques. The high stability of NV centers together with the ability to fabricate photonic structures, here a so-called solid immersion lens (SIL), out of diamond have lead to a record high optical far field resolution of $2.4 \pm 0.3$ nm.[61] When observing color centers in bulk diamond, the lasers have to cross the diamond interface. Due to the high index of refraction of diamond, abberation leads to a non-optimal focusing. The SIL overcomes this problem opening the route to the observation of closely neighboring NV centers.[61] In NDs, STED can even resolve single NV centers in 40–250 nm sized NDs with a resolution of $\approx$10 nm as shown in Arroyo-Camejo et al.[62] and in Fig. 17.7C. For the sake of briefness, this chapter is restricted to the explanation of one superresolution technique. Additionally, there are so-called stochastic superresolution techniques: Single emitters can switch on and off (blinking) via internal processes. The localization of a single emitter can be determined with nanometer resolution by determining the maximum of the confocal PSF by a fit. Thus, if several NV centers in an ND do not luminesce in parallel but blink on and off, they can also be separated and imaged with superresolution.[63] Also active switching on and off by light has been shown.[64]

## CONCLUSION AND PERSPECTIVES

Single NV centers in diamond have shown the ability to sense magnetic fields, electric fields, temperature, crystal strain, and optical near fields with high precision, as well as demonstrated their usability in superresolution microscopy. Using fundamental effects of quantum mechanics such as entanglement or superpositions of quantum states, NV centers are sensors with unprecedentedly high sensitivity. Challenges in sensing with color centers are found in the non-standard nanofabrication procedures required for diamond photonic nanostructures[49] as well as in the optimization of the coherence times and stability for color centers close to surfaces.[65]

Quantum sensing using diamond is at present still highly focused on using NV centers. However, also other color centers have become potential candidates for quantum sensors in recent years: Single silicon vacancy centers have proven to be bright single photon emitters[66] which in contrast to NV centers have a narrow emission spectrum ($< 5$ nm) at room temperature. However, access to spin states seems to be limited to resonant optical techniques, and is thus only feasible at cryogenic temperature.[67] However, near field sensing with SiV centers might be feasible, as their narrow spectral features allow establishing a more defined overlap with other emitters. The short excited state lifetime of the SiV centers (1.7 ns [68]) rather suggests to use them as acceptors in the FRET process. During the last year, a novel color center based on germanium combined with a lattice vacancy has been discovered.[69] It can be fabricated reliably by various methods, rendering it a possible candidate for quantum technologies, however, it has not been shown to be usable as a quantum sensor so far.

## REFERENCES


1. Stoupin, S.; Shvyd'ko, Y. V. *Phys. Rev. B* **2011**, *83*, 104102.
2. Zaitsev, A. *Optical Properties of Diamond: A Data Handbook;* Springer, 2001.
3. Ofori-Okai, B. K.; Pezzagna, S.; Chang, K.; Loretz, M.; Schirhagl, R.; Tao, Y.; Moores, B. A.; Groot-Berning, K.; Meijer, J.; Degen, C. L. *Phys. Rev. B* **2012**, *86*, 081406.
4. Vlasov, I. I.; et al. *Nat. Nanotechnol.* **2014**, *9*, 54–58.
5. GurudevDutt, M.; Childress, L.; Jiang, L.; Togan, E.; Maze, J.; Jelezko, F.; Zibrov, A.; Hemmer, P.; Lukin, M. *Science* **2007**, *316*, 1312.
6. Balasubramanian, G.; Neumann, P.; Twitchen, D.; Markham, M.; Kolesov, R.; Mizuochi, N.; Isoya, J.; Achard, J.; Beck, J.; Tissler, J.; Jacques, V.; Hemmer, P.; Jelezko, F.; Wrachtrup, J. *Nat. Mater.* **2009**, *8*, 383–387.
7. Gruber, A.; Dräbenstedt, A.; Tietz, C.; Fleury, L.; Wrachtrup, J.; von Borczyskowski, C. *Science* **1997**, *276*, 2012–2014.
8. Robledo, L.; Bernien, H.; van der Sar, T.; Hanson, R. *New J. Phys.* **2011**, *13*, 025013.
9. Rondin, L.; Tetienne, J.-P.; Hingant, T.; Roch, J.-F.; Maletinsky, P.; Jacques, V. *Rep. Prog. Phys.* **2014**, *77*, 056503.
10. Manson, N. B.; Harrison, J. P.; Sellars, M. J. *Phys. Rev. B* **2006**, *74*, 104303.
11. Santori, C.; Barclay, P. E.; Fu, K.-M. C.; Beausoleil, R. G.; Spillane, S.; Fisch, M. *Nanotechnology* **2010**, *21*, 274008.
12. Rondin, L.; Tetienne, J.; Rohart, S.; Thiaville, A.; Hingant, T.; Spinicelli, P.; Roch, J.-F.; Jacques, V. *Nat. Commun.* **2013**, *4*, 2279.
13. Tetienne, J.-P.; et al. *Nat. Commun.* **2015**, *6*, 6733.
14. Babinec, T.; Hausmann, B.; Khan, M.; Zhang, Y.; Maze, J.; Hemmer, P.; Loncar, M. *Nat. Nanotechnol.* **2010**, *5*, 195–199.
15. Neu, E.; Appel, P.; Ganzhorn, M.; Miguel-Sanchez, J.; Lesik, M.; Mille, V.; Jacques, V.; Tallaire, A.; Achard, J.; Maletinsky, P. *Appl. Phys. Lett.* **2014**, *104*, 153108.
16. Momenzadeh, S. A.; Stöhr, R. J.; de Oliveira, F. F.; Brunner, A.; Denisenko, A.; Yang, S.; Reinhard, F.; Wrachtrup, J. *Nano Lett.* **2015**, *15*, 165–169, PMID: 25438091.
17. Dréau, A.; Lesik, M.; Rondin, L.; Spinicelli, P.; Arcizet, O.; Roch, J.-F.; Jacques, V. *Phys. Rev. B* **2011**, *84*, 195204.
18. Childress, L.; Taylor, J.; Sørensen, A.; Lukin, M. *Phys. Rev. Lett.* **2006**, *96*, 070504.
19. Ermakova, A.; Pramanik, G.; Cai, J.-M.; Algara-Siller, G.; Kaiser, U.; Weil, T.; Tzeng, Y.-K.; Chang, H. C.; McGuinness, L. P.; Plenio, M. B.; Naydenov, B.; Jelezko, F. *Nano Lett.* **2013**, *13*, 3305–3309.
20. Dolde, F.; Fedder, H.; Doherty, M. W.; Nöbauer, T.; Rempp, F.; Balasubramanian, G.; Wolf, T.; Reinhard, F.; Hollenberg, L.; Jelezko, F.; Wrachtrup, J. *Nat. Phys.* **2011**, *7*, 459–463.
21. Neumann, P.; Jakobi, I.; Dolde, F.; Burk, C.; Reuter, R.; Waldherr, G.; Honert, J.; Wolf, T.; Brunner, A.; Shim, J. H.; Suter, D.; Sumiya, H.; Isoya, J.; Wrachtrup, J. *Nano Lett.* **2013**, *13*, 2738–2742.
22. Teissier, J.; Barfuss, A.; Appel, P.; Neu, E.; Maletinsky, P. *Phys. Rev. Lett.* **2014**, *113*, 020503.

23. Ovartchaiyapong, P.; Lee, K. W.; Myers, B. A.; Jayich, A. C. B. *Nat. Commun.* **2014**, *5*, 4429.
24. Barfuss, A.; Teissier, J.; Neu, E.; Nunnenkamp, A.; Maletinsky, P. *Nat. Phys.* **2015**, *11*, 820–825.
25. Kühn, S.; Hettich, C.; Schmitt, C.; Poizat, J.-P.; Sandoghdar, V. *J. Microsc.* **2001**, *202*, 2–6.
26. Drezet, A.; Sonnefraud, Y.; Cuche, A.; Mollet, O.; Berthel, M.; Huant, S. *Micron* **2015**, *70*, 55–63.
27. Förster, T. *Ann. Phys.* **1948**, *437*, 55–75.
28. Tisler, J.; Reuter, R.; Lämmle, A.; Jelezko, F.; Balasubramanian, G.; Hemmer, P. R.; Reinhard, F.; Wrachtrup, J. *ACS Nano* **2011**, *5*, 7893–7898.
29. Mohan, N.; Chen, C.-S.; Hsieh, H.-H.; Wu, Y.-C.; Chang, H.-C. *Nano Lett.* **2010**, *10*, 3692–3699.
30. Sekatskii, S.; Dietler, G.; Letokhov, V. *Chem. Phys. Lett.* **2008**, *452*, 220–224.
31. Shubeita, G. T.; Sekatskii, S. K.; Dietler, G.; Potapova, I.; Mews, A.; Basché, T. *J. Microsc.* **2003**, *210*, 274–278.
32. Tisler, J.; Oeckinghaus, T.; Stöhr, R. J.; Kolesov, R.; Reuter, R.; Reinhard, F.; Wrachtrup, J. *Nano Lett.* **2013**, *13*, 3152–3156.
33. Schell, A. W.; Engel, P.; Werra, J. F. M.; Wolff, C.; Busch, K.; Benson, O. *Nano Lett.* **2014**, *14*, 2623–2627, PMID: 24694035.
34. Hui, Y. Y.; Lu, Y.-C.; Su, L.-J.; Fang, C.-Y.; Hsu, J.-H.; Chang, H.-C. *Appl. Phys. Lett.* **2013**, *102*, 013102.
35. Hall, L.; Simpson, D.; Hollenberg, L. *MRS Bull.* **2013**, *38*, 162–167.
36. Ziem, F. C.; Götz, N. S.; Zappe, A.; Steinert, S.; Wrachtrup, J. *Nano Lett.* **2013**, *13*, 4093–4098.
37. Kaufmann, S.; Simpson, D. A.; Hall, L. T.; Perunicic, V.; Senn, P.; Steinert, S.; McGuinness, L. P.; Johnson, B. C.; Ohshima, T.; Caruso, F.; Wrachtrup, J.; Scholten, R. E.; Mulvaney, P.; Hollenberg, L. *Proc. Natl. Acad. Sci.* **2013**, *110*, 10894–10898.
38. Tetienne, J.-P.; Hingant, T.; Rondin, L.; Cavaillès, A.; Mayer, L.; Dantelle, G.; Gacoin, T.; Wrachtrup, J.; Roch, J.-F.; Jacques, V. *Phys. Rev. B* **2013**, *87*, 235436.
39. Hall, L.; Beart, G.; Thomas, E.; Simpson, D.; McGuinness, L.; Cole, J.; Manton, J.; Scholten, R.; Jelezko, F.; Wrachtrup, J.; Petrou, S.; Hollenberg, L. C. L. *Sci. Rep.* **2012**, *2*, 401.
40. Hall, L. T.; Hill, C. D.; Cole, J. H.; Städler, B.; Caruso, F.; Mulvaney, P.; Wrachtrup, J.; Hollenberg, L. C. *Proc. Natl. Acad. Sci.* **2010**, *107*, 18777–18782.
41. Kucsko, G.; Maurer, P.; Yao, N.; Kubo, M.; Noh, H.; Lo, P.; Park, H.; Lukin, M. *Nature* **2013**, *500*, 54–58.
42. Taylor, J.; Cappellaro, P.; Childress, L.; Jiang, L.; Budker, D.; Hemmer, P.; Yacoby, A.; Walsworth, R.; Lukin, M. *Nat. Phys.* **2008**, *4*, 810–816.
43. Degen, C. L. *Appl. Phys. Lett.* **2008**, *92*, 243111.
44. Balasubramanian, G.; Chan, I.; Kolesov, R.; Al-Hmoud, M.; Tisler, J.; Shin, C.; Kim, C.; Wojcik, A.; Hemmer, P.; Krueger, A.; Hanke, T.; Leitenstorfer, A.; Bratschitsch, R.; Jelezko, F.; Wrachtrup, J. *Nature* **2008**, *455*, 648–651.
45. Tetienne, J.-P.; Hingant, T.; Kim, J.-V.; Diez, L. H.; Adam, J.-P.; Garcia, K.; Roch, J.-F.; Rohart, S.; Thiaville, A.; Ravelosona, D.; Jacques, V. *Science* **2014**, *344*, 1366–1369.
46. Hingant, T.; Tetienne, J.-P.; Martínez, L. J.; Garcia, K.; Ravelosona, D.; Roch, J.-F.; Jacques, V. *Phys. Rev. Appl.* **2015**, *4*, 014003.
47. Pelliccione, M.; Myers, B. A.; Pascal, L. M. A.; Das, A.; Bleszynski Jayich, A. C. *Phys. Rev. Appl.* **2014**, *2*, 054014.

48. Maletinsky, P.; Hong, S.; Grinolds, M.; Hausmann, B.; Lukin, M.; Walsworth, R.; Loncar, M.; Yacoby, A. *Nat. Nanotechnol.* **2012**, *7*, 320–324.
49. Appel, P.; Neu, E.; Ganzhorn, M.; Barfuss, A.; Batzer, M.; Gratz, M.; Tschöpe, A.; Maletinsky, P. arXiv:1604.00021, 2016.
50. Kleinlein, J.; Borzenko, T.; Münzhuber, F.; Brehm, J.; Kiessling, T.; Molenkamp, L. *Microelectron. Eng.* **2016**, *159*, 70–74.
51. Appel, P.; Ganzhorn, M.; Neu, E.; Maletinsky, P. *New J. Phys.* **2015**, *17*, 112001.
52. Grinolds, M.; Hong, S.; Maletinsky, P.; Luan, L.; Lukin, M.; Walsworth, R.; Yacoby, A. *Nat. Phys.* **2013**, *9*, 2145.
53. Luan, L.; Grinolds, M. S.; Hong, S.; Maletinsky, P.; Walsworth, R. L.; Yacoby, A. *Sci. Rep.* **2015**, *5*, 8119.
54. Thiel, L.; Rohner, D.; Ganzhorn, M.; Appel, P.; Neu, E.; Kleiner, R.; Koelle, D.; Maletinsky, P. arXiv:1511.02873, 2015.
55. Pelliccione, M.; Jenkins, A.; Ovartchaiyapong, P.; Reetz, C.; Emmanuelidu, E.; Ni, N.; Jayich, A. C. B. arXiv:1510.02780, 2015.
56. Hell, S. W. *Rev. Mod. Phys.* **2015**, *87*, 1169–1181.
57. Rittweger, E.; Han, K. Y.; Irvine, S. E.; Eggeling, C.; Hell, S. W. *Nat. Photonics* **2009**, *3*, 144–147.
58. Laporte, G.; Psaltis, D. *Biomed. Opt. Express* **2016**, *7*, 34–44.
59. Maurer, P.; Maze, J.; Stanwix, P.; Jiang, L.; Gorshkov, A. V.; Zibrov, A. A.; Harke, B.; Hodges, J.; Zibrov, A. S.; Yacoby, A.; Twitchen, D.; Hell, S.; Walworth, R.; Lukin, M. *Nat. Phys.* **2010**, *6*, 912–918.
60. Wildanger, D.; Maze, J. R.; Hell, S. W. *Phys. Rev. Lett.* **2011**, *107*, 017601.
61. Wildanger, D.; Patton, B. R.; Schill, H.; Marseglia, L.; Hadden, J.; Knauer, S.; Schönle, A.; Rarity, J. G.; O'Brien, J. L.; Hell, S. W.; Smith, J. M. *Adv. Mater.* **2012**, *24*, OP309–OP313.
62. Arroyo-Camejo, S.; Adam, M.-P.; Besbes, M.; Hugonin, J.-P.; Jacques, V.; Greffet, J.-J.; Roch, J.-F.; Hell, S. W.; Treussart, F. *ACS Nano* **2013**, *7*, 10912–10919.
63. Gu, M.; Cao, Y.; Castelletto, S.; Kouskousis, B.; Li, X. *Opt. Express* **2013**, *21*, 17639–17646.
64. Pfender, M.; Aslam, N.; Waldherr, G.; Neumann, P.; Wrachtrup, J. *Proc. Natl. Acad. Sci.* **2014**, *111*, 14669–14674.
65. Romach, Y.; Müller, C.; Unden, T.; Rogers, L. J.; Isoda, T.; Itoh, K. M.; Markham, M.; Stacey, A.; Meijer, J.; Pezzagna, S.; Naydenov, B.; McGuinness, L. P.; Bar-Gill, N.; Jelezko, F. *Phys. Rev. Lett.* **2015**, *114*, 017601.
66. Neu, E.; Steinmetz, D.; Riedrich-Möller, J.; Gsell, S.; Fischer, M.; Schreck, M.; Becher, C. *New J. Phys.* **2011**, *13*, 025012.
67. Müller, T.; Hepp, C.; Pingault, B.; Neu, E.; Gsell, S.; Schreck, M.; Sternschulte, H.; Steinmüller-Nethl, D.; Becher, C.; Atatüre, M. *Nat. Commun.* **2014**, *5*, 3328.
68. Sipahigil, A.; Jahnke, K. D.; Rogers, L. J.; Teraji, T.; Isoya, J.; Zibrov, A. S.; Jelezko, F.; Lukin, M. D. *Phys. Rev. Lett.* **2014**, *113*, 113602.
69. Iwasaki, T.; et al. *Sci. Rep.* **2015**, *5*, 12882.

CHAPTER

# Nanodiamonds for catalytic reactions

# 18

**Neeraj Gupta*, Qi Wang*, Guodong Wen*, Dangsheng Su*,†**

*Shenyang National Laboratory for Materials Science, Institute of Metal Research, Chinese Academy of Sciences, Shenyang, China* Dalian Institute of Chemical Physics, Chinese Academy of Sciences, Dalian, China†*

**Contents**



## 18.1 GENERAL ASPECTS OF NANODIAMONDS

Nanodiamond has been found in extraterrestrial materials such as interstellar dust, meteorites, and protoplanetary nebulae,[1] and was first produced by the detonation method[2] in the 1960s. But it remained unknown to the world until the end of the 1980s[3] and was able to gain scientific attention in the early 1990s. Research development in the field experienced a thrust only after commercial nanodiamonds became available. Initially, scientists used nanodiamonds for biomedical imaging[4] as a non-toxic alternative to semiconductor quantum dots; and applied them in nanoscale magnetic sensors.[5] The defects present in nanodiamonds can be used to measure and manipulate electrons so that they can lead to new quantum technology; and detailed account of defects in quantum physics research and applications in sensing is provided in Chapter 17. Their tunable properties and lower toxicity[6] in comparison to other carbon materials helped develop different composites, use them in applications in drug delivery, and promoted their use as "biomarkers" to measure the presence or severity of certain diseases. The reader is advised to see Chapter 16 for a detailed account.

### 18.1.1 PREPARATIVE METHODS

Nowadays, various methods are developed for the preparation of nanodiamond.[7] These include the detonation technique, high-energy ball milling of high-pressure high-temperature (HPHT) diamond microcrystals, laser ablation, autoclave synthesis from supercritical fluids, plasma-assisted chemical vapor deposition (CVD), chlorination of carbides, ion irradiation of graphite, electron irradiation of carbon "onions", and ultrasound cavitation. Currently, the most commonly used method for the preparation of nanodiamond is the detonation method, and all the details regarding preparation methods are provided in Chapter 2. Detonation nanodiamonds have average size between 2–5 nm, which is mostly found to be independent of the preparation conditions. They also have the tendency to form aggregates and de-aggregation may be required in some cases. Recently, stable nanodiamond–mineral oil dispersion was prepared by bead-milling of the oxidized ultradispersed diamond (UDD) in a dispersion medium to achieve efficient de-aggregation. Nanodiamonds of average size ~8 nm can be obtained by this method. Disintegration of the nanodiamond aggregates can also be performed with dry media assisted attrition milling.[8] This technique uses water-soluble nontoxic crystalline compounds, such as sodium chloride or sucrose, as milling media, which can be easily removed upon dissolution in water. During this process nanodiamond aggregates are milled with dry media and the size of the milling media can be reduced to the same scale during the process. Upon completion of the process, the media is easily removed by washing with water. The average particle size can be controlled by varying the milling time. Sometimes a chemical linkage such as the C–O–C ethereal linkage is also responsible for the aggregation,[9] which can be broken by a treatment with strong acids, such as hydroiodic acid (HI).

### 18.1.2 PROPERTIES, SURFACE STRUCTURE, AND MODIFICATION

UDD particles made by the detonation method are generally well-defined crystalline diamond core and are covered by a layer of unstructured amorphous carbon[6] including small fragments of graphitic carbon that are entrapped on the surface during preparation. The presence of $sp^2$-carbon on the surface sometimes can provide remarkable chemical and catalytic properties to nanodiamonds.[10] Various techniques have been used to determine the $sp^2/sp^3$ ratio of carbon materials. These include electron energy loss spectroscopy (EELS), Raman spectroscopy, X-ray photoelectron spectroscopy (XPS), and X-ray excited Auger electron spectroscopy (XAES). A microwave measurement method[11] is also developed recently to measure the degree of $sp^2$ hybridization in nanodiamond by using a cavity perturbation technique.

A detailed description of these methods can be found in Chapter 2 of this book. The unique surface of nanodiamond can also show special adsorption characteristics. For instance, during the study of adsorption behavior for the oxoacid anions,[12] it was shown that nanodiamond could selectively adsorb tungstate in weak acidic media. The adsorption behavior of nanodiamond was assigned to the functional groups on its surface. Therefore, it can be concluded that functionalizing nanodiamond can bring remarkable properties to the material (refer to Chapter 8). Modification and functionalization of its surface can be divided into three main categories:

(a) **Graphitization of the surface**
Annealing of the nanodiamond at different temperatures generally gives different degrees of graphitization on the surface. During the preparation process, it is unavoidable to keep the surface of these well-defined crystalline diamonds free from amorphous unstructured carbon, which is later reconstructed to well-ordered graphene layers with the increase of the temperature during the annealing process. This results in an increase of $sp^2$ carbon content and formation of graphitic caps or fullerene-like reconstructions at their surface. Finally, at elevated temperatures spherical onion-like carbon (OLC) is formed by the transformation of all the surface $sp^3$ carbon to the one with $sp^2$ hybridization. During this process, one intermediate composite formed with a diamond core covered by a curved graphene outer shell is called bucky-diamond.

(b) **Introduction of the surface functional groups**
Introduction of surface oxygenated groups is normally carried out by oxidation with concentrated acid, hydrogen peroxide, annealing in oxygen containing gases, and by ozone treatment. The oxidation method mostly results in the enrichment of the surface with –COOH group. One of the effective ways to introduce hydroxyl functional groups is the use of Fenton reaction.[13] In this method, nanodiamond is treated with $H_2O_2$ and $FeSO_4$ under strong acidic conditions, and concentrated sulfuric acid is always used. Besides Fenton reaction, photochemical hydroxylation[14] can also be used to introduce the hydroxyl functional groups to the surface of nanodiamond. Fully hydrogenated nanodiamond can be prepared in a very short time period by the hydrogenation process in a microwave plasma reactor. Introducing the –OH or –COOH groups to the surface of the nanodiamond is relatively facile in comparison to other kinds of oxygen functional groups, which still represent a challenge in this field. Ozone treatment of NDs at increased temperature can lead to the formation of peroxide and nitroso groups, which are prone to form radical species.[15] The presence of such reactive species is an effective tool for their possible application as a radical initiator in certain organic reactions.

(c) **The doping or modification with heteroatoms**
Doping of nanodiamond with other heteroatoms such as N, P, and B is also one of the effective ways to specially influence its performance. However, in the case of nanodiamond, the stable $sp^3$ surface makes the doping with heteroatoms quite a difficult process. Low content of nitrogen has been found in

the core of NDs,[16] which probably gets trapped in them during their preparation by the detonation process when using nitrogen containing explosives such as TNT (2,4,6-trinitrotoluene). The location of nitrogen and boron impurities in the diamond nanoparticles has been examined by the density functional tight binding (DFTB) simulations.[17] It is believed that the existence of nitrogen at the extremum of a nanodiamond, bucky-diamond, or just below the fullerenic outer shell is energetically preferable. However, the existence of boron is energetically preferable at the sub-extremum sites for both bucky-diamond and nanodiamond. Nowadays, almost all doping is developed on the onion-like carbon (OLC) that is derived from the nanodiamond in order to get the benefits of doping. But the conversion of nanodiamond to OLC changes its surface and nature, and hence all the benefits offered by nanodiamond surface may change during its transformation to OLC.

## 18.2 APPLICATIONS OF NANODIAMONDS AS CATALYST OR CATALYST SUPPORT IN DIFFERENT CHEMICAL TRANSFORMATIONS

Nanodiamond has been employed in a variety of gas and liquid phase catalytic transformations. These transformations are categorized based on the common nature of the substrate or the type of reactions to which they belong. Table 18.1 summarizes different reactions with the possible active site(s) on the catalyst surface, where nanodiamond has been used as a catalyst with or without the supported metals. A general representation of different chemical transformations catalyzed by nanodiamond and metal supported on nanodiamond is also shown in Fig. 18.1.

### 18.2.1 DEHYDROGENATION OF ETHYLBENZENE

Catalytic dehydrogenation of ethylbenzene (EB) to styrene (ST) is one of the most important processes in the chemical industry and accounts for 90% of styrene production [Fig. 18.1(i)]. Potassium and iron based catalyst is generally used in the process that involves high iron loading. During this process, carbonaceous species are generated and deposited on the active sites of the catalyst, resulting in the drastic decrease of activity. In order to reduce coke formation during the process, a high amount of steam is added. Generation of steam requires large energy input that adds to the extra cost of the process along with the operational problems. Hence, a process that is resistant to carbon poisoning with high catalytic activity and without added steam is always beneficial. Metal free catalysts have been developed that are efficient under steam-free conditions; and nanodiamond is among them.

Detonation nanodiamond exhibited efficient catalytic activity for the dehydrogenation[19] of ethylbenzene to styrene at 550°C under atmospheric pressure without added steam. The reaction starts with the adsorption of ethylbenzene onto the surface of carbon, which then reacts with the carbonyl group generally located at the

**Table 18.1** Overview of reactions catalyzed by nanodiamonds

| S.N. | Reaction | Metal used in combination | Active sites | References |
|---|---|---|---|---|
| 1. | Oxidative dehydrogenation | Metal-free | Carbonyl/quinones | 18,19 |
| 2. | Nitrobenzene reduction | Metal-free or metal (Pt and Pd) | Carbonyl or metal | 20 |
| 3. | Reforming/Dehydration of alcohols | Metal-free or metal (Pt–Ni and Pt–Ru) | Acidic groups or metal | 21 |
| 4. | Aldol condensation | Metal-free | Nitrogen containing groups | 22 |
| 5. | Hydrodechlorination reaction | Metal-free or metal (Ni and Pd) | UDD surface groups or metal | 23,24 |
| 6. | Oxidation reaction | Au | Au | 25 |
| 7. | Hydroamination | Au, Pd, Pt | Au, Pd, Pt | 26 |
| 8. | Hydrogenation (C=C and C≡C) | Ni, Pd | Ni, Pd | 27 |
| 9. | Hydrogenating deamination | Pd | Pd/Acidic groups | 28 |

defects, leading to the dehydrogenation by the cleavage of CH bond.[18] The activity of the catalyst was also compared with conventional K-promoted $Fe_2O_3$ catalyst. The activity of conventional catalyst decreased drastically during initial 2 h and then remained constant during the course of reaction. Nanodiamond possesses 2.8 times larger steady-state activity in comparison to the K–Fe catalyst. The low activity of K–Fe catalyst was due to heavy coke formation on its surface. Restoring the activity of conventional catalyst is very hard, but in case of nanodiamonds, it can be achieved by heating them in the air flow at 400°C. This property of nanodiamond was helpful to regenerate them five times. The pristine detonation nanodiamond also displayed good catalytic activity in comparison to other carbon materials (Fig. 18.2), such as carbon nanotubes, nanographite, activated carbon, and mesoporous carbon. The activity of nanodiamond was up to 14 times higher than that of other samples. Also, all the nanocarbons provided higher selectivity than mesoporous carbon and activated carbon, which was probably due to their short or long range disorder with mixtures of $sp^2$ and $sp^3$ hybridized carbon. These disorders may provide a variety of surface functional groups and defects that further allow $\alpha$- or $\beta$-scission of ethylbenzene into side products, benzene and toluene. The spatial arrangement of spherical nanodiamond was also responsible for providing a better access for the reactants. High thermal conductance of the $sp^3$ diamond core may further lead to high activity of nanodiamond in the endothermic reaction. This core could improve the heat transfer efficiency near the active sites on the surface.

The chemical state at the working surface during this reaction was measured by in situ X-ray Photoelectron Spectroscopy (XPS) and diffusion reflectance IR spectroscopy (DRIFT) experiments. XPS analysis of nanodiamond in vacuum was used

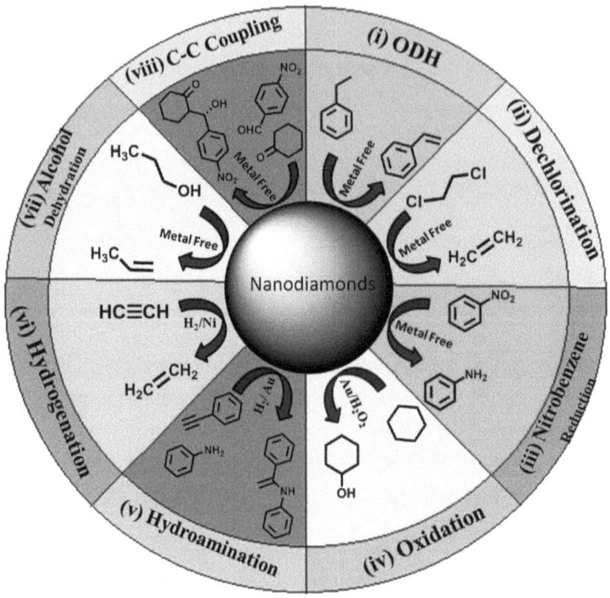


**FIGURE 18.1**

General representation of different reaction types performed on the surface of nanodiamonds. The reactions such as oxidative dehydrogenation (ODH) and nitrobenzene reduction are performed without any metal loading, whereas hydrogenation and hydrodechlorination (HDC) are mostly performed after loading a suitable metal. Metal loading on nanodiamonds can significantly improve their catalytic activity in comparison to conventional supports

to observe the surface functional groups, and it was observed that the fraction of C=O decreased at 450°C whereas the fraction of C–O increased from their initial concentrations. The change in concentration of both of these groups was probably due to the reduction of C=O to C–OH by the hydrocarbon. The ketonic C=O species could be correlated with the dehydrogenation reaction, and the decrease in activity agreed well with the consumption of C=O. The contact of ethylbenzene with the surface gives IR bands at 3470 $cm^{-1}$ due to –OH, indicating formation of hydroxyl-like intermediates. Also in the mid-range frequencies C=O stretching modes ($\nu$ C=O, 1700–1800 $cm^{-1}$) shows a significant red shift, indicating the activation of C=O bonds. Also, a significant change of the catalyst surface was observed due to hydrogen. Hydrogen passivation of the fresh and used catalyst was compared by the temperature programmed oxidation (TPO). The amount of water released by the used and fresh catalysts differs almost twice, indicating more coverage of the surface by hydrogen in the used catalyst after the reaction. This eventually leads to a decrease in number of exposed active sites that result in a drop in the catalytic activity.

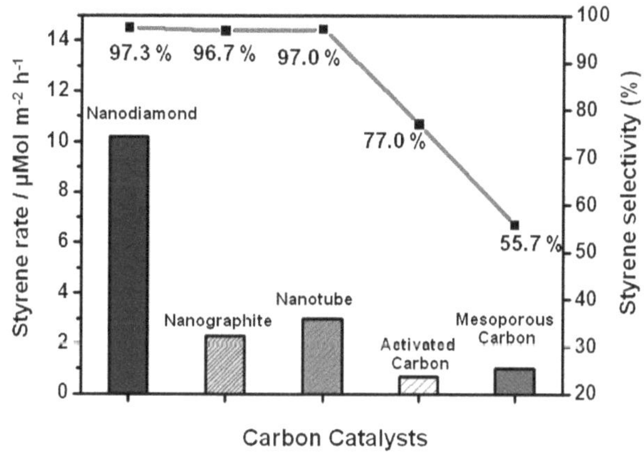


**FIGURE 18.2**

Steady state activity of various carbons during the ethylbenzene dehydrogenation reaction. Reaction conditions: 0.05 g, 550°C, 2.8% EB in helium, 10 mL $min^{-1}$. Reprinted with permission from: Zhang, J., et al. Angew. Chem. Int. Ed. 2010, 49, 8640–8644[19]

Powdered UDD is difficult to use in an industrial scale fixed bed reactor because it may cause a high pressure drop. Supporting UDDs on macroporous support can make them fit for the industrial use and ensure the operational safety. The dehydrogenation performance was higher when the NDs were dispersed on thermally conductive supports such as SiC in comparison to those supported on insulator supports. The catalytic activity of UDD was enhanced towards steam-free dehydrogenation of ethylbenzene by making a hybrid structure with a porous $\beta$-SiC foam matrix.[29] The catalytic activity of hybrid foam was excellent and stable. The dehydrogenation activity was about 3.8 times higher than that of powder-form NDs and 10 times higher with respect to the commercial K-Fe-based catalyst without a decrease in styrene selectivity ($S_{ST} = 98\%$). The increased activity of ND/SiC was probably due to the higher dispersion of UDD on the SiC surface, which might result in a higher exposure of active sites. The SiC foam structure can further generate microturbulence through the contact between the reactant and surface. The catalyst retains styrene selectivity of 98% even after 100 h of reaction and therefore it is a promising catalyst for an industrial process. The resistance towards deactivation could be attributed to the open porosity of the support and a connected pore that reduces the residence time of the products on the active site and hence in deposition of carbonaceous material.

Few-layered graphene (FLG) could also be used to make hybrid material with UDD for ethylbenzene dehydrogenation. The material displayed not only the individual properties of graphene and ND, but also new properties that arise from the synergism between the two components. NDs acted as a nano-spacer and also prevented the re-stacking of the FLG sheets. It was observed that catalytic performance [expressed in terms of styrene products per gram of nanodiamond per hour]

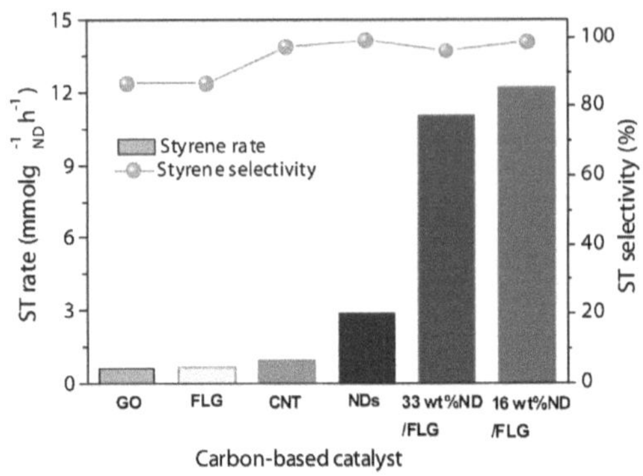


**FIGURE 18.3**

Catalytic performance of different carbon based catalysts under steady state during oxidative dehydrogenation reaction. Reaction conditions: 550°C, 2.8% ethylbenzene in helium, total flow rate = 30 mL $min^{-1}$, atmospheric pressure. Reprinted with permission from: Ba, H., et al. Catal. Today 2015, 249, 167–175[30]

on 33 wt.% ND/FLG composite was very high after 40 h in a stream at steady state (Fig. 18.3). Also, the styrene selectivity was slightly decreased from almost 99% at 550°C to 95% at 600°C. The decrease in selectivity was likely due to the formation of cracking products at high temperature. It should be noted that the ST formation rate on 16 wt.% ND/FLG was slightly higher than that on 33 wt.% ND/FLG, which could be ascribed to the better dispersion of 16 wt.% ND on FLG. K-promoted iron catalyst was also evaluated, and it showed an extremely high dehydrogenation activity at the start of the reaction, which gradually decreased and reached a low steady state activity at 1.00 mmol ST $g_{catalyst}{}^{-1}$ $h^{-1}$. A longer term catalytic testing was also performed on ND/FLG (33 wt.%) catalyst and it showed a relatively high DH activity even after 100 h of the test.

## 18.2.2 DECHLORINATION REACTION

Chlorinated products are a major cause of industrial pollution, and hence their removal by mild and renewable catalysts is always desirable. Carbon materials have also received considerable attention in this regard. Nanodiamond, being a relatively new class of carbon compounds, has attracted less attention. Chlorinated solvents such as 1,2-dichloroethane (DCE), hexachlorobenzene (HCB), chlorobenzene (CB), and 2,4,8-trichlorodibenzofuran (TCDF) represent a class of industrial chlorinated solvents. Such compounds are difficult to remove by incineration, and they form highly toxic dioxins during this process. Very high temperature up to 800°C is required for their complete oxidation to hydrochloric acid, and hence milder alternatives for this process are required. Disposing off the chlorinated solvents obtained as industrial waste can be achieved by three ways:

(a) **Dehydrochlorination.** The process is generally carried out at high temperature to facilitate the removal of two HCl groups from the molecule. For example, dehydrochlorination of 1,2-dichloroethane generates acetylene after removal of two HCl molecules.
(b) **Hydrodechlorination.** Removing the chlorine by gaseous hydrogen is called hydrodechlorination. This step requires hydrogen activation, and hence metal based catalysts are generally used. In this reaction 1,2-dichloroethane generates ethylene.
(c) **Dechlorination**. This process involves removal of chlorine directly from halogenated molecules. This reaction can be performed by a metal supported catalyst or in the presence of bacteria. 1,2-dichloroethane, when subjected to dechlorination, generates ethylene along with chlorine.

Dehydrochlorination is temperature-dependent and the last two routes in particular involve a transition metal based catalyst such as supported platinum-based bimetallic catalyst. Hence replacing the transition metals with carbon materials is a very good alternate in terms of economic and environmental benefits; and nanodiamonds are capable to perform this task [Fig. 18.1(ii)]. Several materials derived from graphene, nanodiamond, and carbon nanotubes have been tested for the dechlorination of 1,2-dichloroethane.[31] While evaluating the performance of different carbon materials during this process at 300°C in nitrogen flow, it was found that only detonation and synthetic diamonds ($D_{syn}$) prepared with other laboratory methods excluding the use of explosive and having average size of 0.1–1 μm were active in comparison to other carbon materials. 1,2-Dichloroethane conversion was reported to be 8–11 times higher on detonation nanodiamond surface than that on $D_{syn}$. The difference in the activity of both catalysts was most likely due to the difference in their surface area. It was also observed that the conversion for the catalysts decreased substantially during the initial stages and attained a steady state after 5 impulses. The initial decrease in the activity was probably due to the changing surface in the initial impulses, which later on attained a stability and steady state. Both dechlorination and hydrodechlorination were observed during the process. Hydrodechlorination was observed only during the initial stages of the reaction, and surface hydrogen present on detonation nanodiamonds was responsible for this. Once the chemisorbed hydrogen was consumed in the process, dechlorination dominated.

Not only metal-free nanodiamond but also the one supporting metal nanoparticles was used in this reaction. Ni was deposited on ND surface by thermal decomposition of nickel (II) formate under dynamic vacuum, and the material was tested for 1,2-dichloroethane decomposition. It was observed that 1,2-dicholoroethane conversion showed a decline during the initial stages of the reaction on pure nanodiamonds, whereas no such effect was observed on Ni supported on nanodiamonds. The reaction was also performed under the hydrogen and nitrogen flows at various reaction temperatures. The activity of NDs remained the same both in hydrogen and nitrogen flows and showed essentially the same activation energies of 73.5 and 74.0 kJ mol$^{-1}$, respectively. The same values for both streams indicated that gas-phase hydrogen did not react with 1,2-dichloroethane directly under the reported conditions. Ni-containing nanodiamond catalyst exhibited higher activity in the hydrogen flow, indicating that activation of hydrogen on the Ni surface was responsible for the reductive catalytic conversion. Nanodiamonds are capable of retaining the active hydrogen[32] on their surface during different hydrogenation reactions.

Besides 1,2-dichloroethane, dechlorination of other chlorinated substrates has also been investigated using a nanodiamond-based catalyst. One such investigated substrate is 1,3,5-trichlorobenzene,[23] and palladium nanoparticles supported on nanodiamonds were employed for this reaction. The catalyst was prepared by wet impregnation method using aqueous palladium nitrate solution followed by reduction at 280°C under the flow of hydrogen. Five different phases, including an aqueous alkali phase, gaseous hydrogen, a solid phase containing heterogeneous catalyst, an organic phase, and a low-solubility interphase transfer agent that forms an additional phase on the surface of the heterogeneous catalyst, were identified during this reaction. It was postulated that the presence of an interphase agent during hydrodechlorination[24] reaction can contribute to high rates and yields. Different catalyst containing variable weights of palladium from 0.5 to 5 wt.% were prepared and investigated for the hydrodechlorination of 1,3,5-trichlorobenzene (TCB) and hexachlorobenzene (HCB). The activity of the Pd/UDD catalyst was higher than the Pd deposited on activated carbon. It was hypothesized that the difference in the catalytic activity was due to the structural difference of the support, its porosity, and the presence of certain impurities.[33] Nanodiamonds are different from amorphous activated carbon in their structural configuration having small surface graphene islands, ordered crystal structure, and a large number of valence-unsaturated atoms on their surface. These unsaturated atoms can further stabilize the palladium nanoparticles. However, palladium on activated carbon (AC) did not have uniform metal distribution, and the size of palladium particles varied from 5 to 40 nm.

Carboxyl groups present on UDD surface have the tendency to form salts with heavy metal ions such as Pd, and, due to their constant concentration on the surface, they ensure regular distribution of the metal ions. Temperature programmed reduction (TPR) can be used to study the distribution of metal particles on UDD surface (Fig. 18.4). A narrow hydrogen adsorption peak is obtained in the spectrum of 5% PdO/UDD in comparison to PdO/AC which is characterized by a broader hydrogen absorption peak in TPR and hence with a broader particle-size distribution. Palladium and nickel supported nanodiamonds were also tested towards multiphase and vapor phase hydrodechlorination (HDC) of chlorobenzene[34] and polychlorobenzenes. Ni/UDD catalyst was found to perform the HDC of chlorobenzene at comparatively lower temperatures. The multiphase HDC of 1,3,5-trichlorobenzene, chlorobenzene, and 2,4,8-trichlorodibenzofuran on Pd/UDD occurred more rapidly in comparison to catalysts supported on activated carbon. 5% Pd/UDD was effective in bringing out the complete conversion of 1,3,5-TCB in about 40 min, and no appreciable difference in time was noticed for catalysts containing 1 and 2% Pd. Surprisingly, total degrees of conversion in the presence of 5% Pd/AC and commercially available 5% Pd/C were 12 and 5%, respectively. The catalyst with a lower Pd content such as 1

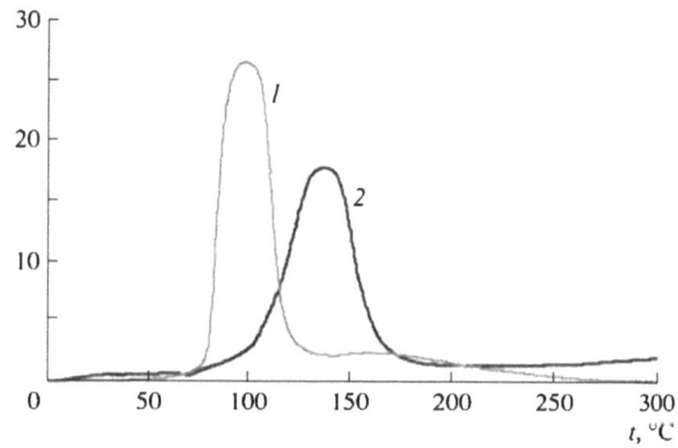


**FIGURE 18.4**

Temperature-programmed reduction spectra of 5 wt.% PdO/UDD (1) and PdO/AC (2). Adopted from: Kachevskii, S. A., et al. Russ. J. Phys. Chem. A 2007, 81, 866–873[23]

and 2% Pd exhibited almost the same activity in reactions with the participation of HCB. Benzene was the major product only when the reaction was performed with 5% Pd/UDD, and the complete conversion of HCB into benzene took place in the presence of 5% Pd/UDD within 80 min. This reaction did not complete on commercially available activated carbon supported catalysts even after a very long time of 500 min. Properties of the catalysts were found to be dependent on the surface structure of the nanodiamond support, which was highly ordered without the presence of micropores.

### 18.2.3 REDUCTION OF NITROBENZENE

Reduction of aromatic nitro compounds is an important method for the synthesis of corresponding aromatic amines during the preparation of fine chemicals and natural product syntheses. The reduction process can be achieved by pure carbon materials (without metals) as well as with the metal supported one [Fig. 18.1(iii)]. Carbon materials such as natural fullerene, graphite, carbon nanotubes, and reduced graphene oxide have the potential to reduce nitrobenzene into aniline. Nanodiamonds also have the capability to reduce the nitrobenzene with hydrazine,[20] but their activity was lower in comparison to carbon nanotubes. The main active site identified during the reaction was a carbonyl group present on the surface. The carbonyl groups are found to activate the hydrazine to reduce nitrobenzene. The activity of UDD for this reaction was improved by thermal treatment, and further oxidation with hydrogen peroxide makes them equally efficient to carbon nanotubes in the reaction. The degree of graphitization was increased after treatment at high temperatures, which might facilitate the adsorption of reactants.

ND supported metal catalysts have been found to be quite useful during the reduction of nitroarenes. Platinum and palladium nanoparticles loaded on the surface of nanodiamonds[35] have been efficiently used to catalyze the liquid-phase hydrogenation of nitrobenzene. Pt/ND samples were prepared containing 5, 10, 15, 20, and 25 wt.% of Pt, and similarly Pd/ND samples were prepared with 3, 6, 9, 12, and 15 wt.% of Pd. The catalytic property of the materials was determined towards hydrogenation of nitrobenzene. It was observed that hydrogen consumption was accelerated with the increasing metal content during the hydrogenation process by Pd/ND, whereas the increase was not proportional to the amount of palladium, thus indicating that some part of metal may not be involved in the catalytic process. The activity of the catalyst did not show an increase with very high metal content, thus the use of high metal loading may not be efficient in this process. The reaction rate and turnover number increased with longer catalyst activation, which may be due to the formation of Pd(0) that could further activate hydrogen. The activity of palladium supported on nanodiamonds was higher compared with the one supported on activated carbon.[36] Inductive coupled plasma (ICP) analysis was used to observe the metal leaching from the catalyst surface, and it was found that palladium supported on NDs was not easily washed off from the surface, whereas the metal loss occurred on the surface of Pd/AC. Nitrogen functional groups present on the catalyst surface were proposed to be an active functionality that might prevent the loss of metal nanoparticles from the catalyst surface. Also, on the activated carbon support (Pd/AC), a portion of the metal may have been located in micropores that were inaccessible to the reducing agent and reactants. Furthermore, ordered surface structure of ND may apparently cause the easy access of the activated sites to $-NO_2$ compared to activated carbon. Hydrogenation of nitrobenzene and its *p*-substituted analogs such as *p*-nitroaniline and *p*-nitrophenol was performed with both catalysts. The rate of the reaction was dependent on the type of the substrate used. The electron donating groups in the *para* position exhibited insignificant effect on the rate constant during hydrogenation of the nitro group; however, the introduction of an electron acceptor substituent such as carboxyl group decreased this parameter nearly two times.

### 18.2.4 OXIDATION OF ALKANES

Oxidation of functional groups such as alcohols, aldehydes, and ketones is used to perform different chemical transformation during fine chemical and drug synthesis. Lots of methods for the oxidation of alcohols and aldehydes (homogeneous and heterogeneous) are available today. But simultaneous oxidation of alkanes is quite difficult because of the stability of C–H bond and its reluctance to other chemical reactions such as nucleophilic or electrophilic reactions. The activation of the carbon–hydrogen bond in alkanes is very difficult and it prevents their use in the direct synthesis of value-added chemicals and products. Cyclohexanol and cyclohexanone are two important chemicals used by the polymer industry for the manufacture of nylon. In order to synthesize bulk quantities of these chemicals, their preparation from readily available cyclohexane is highly beneficial. Gold supported on nanodiamonds was used for the synthesis of cyclohexanol from petroleum derived cyclohexane [Fig. 18.1(iv)]. 1 wt.% gold based catalyst was prepared[25] by two different methods: the first was sol immobilization or colloidal method (COL), and the second was double impregnation (DIM). Sol immobilization involves dissolving the gold precursor [$HAuCl_4 \cdot 3H_2O$] in water, adding polyvinyl alcohol and $NaBH_4$, resulting in a ruby red sol, and finally adding the powder support. After a few days, a colorless solution is obtained that is filtered, washed and dried at 110°C overnight to get the Au/ND. Besides UDD, other supports were also used for the preparation of catalysts, which include activated carbon, carbon nanotubes, and carbon xerogels (mesoporous polymer based carbon materials). The catalysts thus prepared were used for the peroxidative oxidation of cyclohexane with $H_2O_2$ at room temperature to cyclohexanol and cyclohexanone via formation of cyclohexyl hydroperoxide. The best results were obtained with the Au/CNT samples, and the performance of nanodiamond-based catalysts was found to be superior in comparison to xerogel and activated carbon based catalyst. Au/CNT was most active because of its smaller Au particle size distribution on the surface. The end products of the reaction, i.e. cyclohexanol and cyclohexanone, are much more reactive than the starting compound cyclohexane, and hence, achieving high selectivity during this reaction was always challenging. In order to investigate the roles of the support, surface area, porosity, and surface functional groups were probed during the reaction, but no correlation was established.

The oxidation of cyclohexane by peroxide involves decomposition of $H_2O_2$ that generates oxygen-centered radicals $HO^\bullet/HOO^\bullet$ upon reduction of $H_2O_2$ by Au(0) or by oxidation with $Au^+$. Water could also possibly catalyze the shift of $H^+$ ions towards the formation of $HO^\bullet$ from $H_2O_2$. The cyclohexyl radical $Cy^\bullet$ is formed by H-abstraction from cyclohexane (CyH) by $HO^\bullet$ and a reaction of $Cy^\bullet$ with dioxygen generates $CyOO^\bullet$. This gets converted to CyOOH by H-abstraction from $H_2O_2$. Metal-assisted decomposition of CyOOH to $CyO^\bullet$ and $CyOO^\bullet$ would then yield cyclohexanol and cyclohexanone. A possible sequence of reactions is shown below:

**(a)** $Au + H_2O_2 \rightarrow HO^\bullet + Au^+ + HO^-$
**(b)** $Au^+ + H_2O_2 \rightarrow HOO^\bullet + H^+ + Au$
**(c)** $HO^\bullet + CyH \rightarrow H_2O + Cy^\bullet$
**(d)** $Cy^\bullet + O_2 \rightarrow CyOO^\bullet$
**(e)** $CyOO^\bullet + H_2O_2 \rightarrow CyOOH + HOO^\bullet$
**(f)** $CyOOH + Au \rightarrow CyO^\bullet + Au^+ + HO^-$
**(g)** $CyOOH + Au^+ \rightarrow CyOO^\bullet + H^+ + Au$
**(h)** $CyO^\bullet + CyH \rightarrow CyOH + Cy^\bullet$
**(i)** $2CyOO^\bullet \rightarrow CyOH + Cy_{-H} = O + 2O_2$

### 18.2.5 HYDROAMINATION OF ALKYNES AND ALDEHYDES

Nitrogen-containing organic compounds are widely present in nature and are a source of important compounds that provide biological and pharmaceutical activity. They can be used in cosmetics as well as intermediates in many agrochemical products and in industrial processes. Hydroamination of unsaturated hydrocarbons and imines is one of the important routes to synthesize such compounds. Various carbon supported catalysts including one based on nanodiamond are developed for the reaction. Hydroamination[26] of phenylacetylene with aniline was performed with carbon supported gold catalysts prepared by depositing gold on nanodiamonds [Fig. 18.1(v)], single or multi-walled carbon nanotubes, graphite, graphene oxide, graphitic cones, ordered mesoporous carbon, carbon black, carbon xerogel, and activated carbon. The gold nanoparticles (GNP) were synthesized by in situ reduction of chloroauric acid $H[AuCl_4]$ in water. The different catalysts exhibited good behavior for the targeted reaction, and 79% conversion was achieved by adjusting the gold loading amount on carbon support. Good catalytic activity was dependent on gold loading amount, structure and textural properties, and dispersibility of the carbon supports. The best gold–carbon catalyst performance was achieved by using graphene oxide, but nanodiamond supports also showed significant performance by suitably adjusting the desired parameters. Nanodiamonds were capable of showing catalytic activity due to their ability to have good dispersibility in water with high BET surface area with well-developed stacking pores.

Primary amines when reacting with aldehydes are capable of generating imines that can be further reduced to get secondary amines. This strategy can be used for the reductive amination of aldehydes. Hydrogenating amination[37] of propanal with 4-aminobenzoic acid was performed by Pd nanoparticles supported on nanodiamonds. The turnover number of catalyst decreased with increased palladium concentration. The activity of the catalyst was also compared with that of a commercially available Pd/C catalyst. It was observed that Pd/ND exhibited significantly higher catalytic activity in comparison to a commercial 1% Pd/C catalyst. The adsorption of substrates, formation of activated hydrogen, reduction of functional groups, and desorption of the product are also important for this reaction. Quantum chemical calculations were used to derive the values of total effective charges at the reaction sites of the substrates. The hydrogenation rates for the substrates were found to increase with increasing electron density at the reaction site (the oxygen atoms in the $NO_2$ group, the atoms of the >C=C< and >C=N– bonds). Steric effects also affected the hydrogenation rate, for instance, hydrogenation of nitrobenzene occurred more readily in comparison to 4-(propylideneamino)benzoic acid [4-$CH_3CH_2CH{=}N(C_6H_4)COOH$] probably because the >C=N– bond is shielded on both sides in the latter.

Platinum supported on nanodiamonds was also used for the hydrogenating amination[38] of propanal with 4-aminobenzoic acid along with the palladium counterpart. The activity of nanodiamond-based catalyst was more efficient in comparison to palladium or platinum-containing multi-walled carbon nanotubes, nanofibers, and fullerene black. During the reaction Pt/ND and Pd/ND were 2–18 times more efficient than the commercial Pt/C catalyst. Also, the activity of the catalysts decreased with the increase in the content of palladium on ND support. The dependence was opposite for Pt/ND catalyst, i.e. the activity of the catalyst increased with increasing content of platinum on the ND surface. The hydrogenating amination in the presence of Pd/ND was 1.3–2.3 times faster than that in the presence of Pt/ND.

### 18.2.6 HYDROGENATION OF UNSATURATED HYDROCARBONS

Hydrogenation is one of several very important processes used industrially for the manufacture and processing of petrochemicals, food stuff, fine chemicals, and drugs.

The reaction generally involves the dissociation of hydrogen gas in the presence of a suitable catalyst (Pt, Pd, Ni on different supports) to reduce compounds with a multiple bond (double or triple bond) [Fig. 18.1(vi)]. The hydrogenation processes basically have three reaction components that include an unsaturated substrate, hydrogen gas, and a suitable catalyst capable of dissociating hydrogen gas. Carbon based catalysts have been used for the hydrogenation reactions, and activated carbon has found many applications in this field until now. Many hydrogenation reactions require high temperature for completion and even a catalyst to operate at a higher temperature. $H_2$ is dissociated on the active metal that may "spill over" from the metal onto the catalyst support.[39] Nanodiamonds in this case make a good support for such a catalyst because they can perform well to keep this spilled hydrogen on its surface even at high temperature.

Nickel loaded on detonation nanodiamond was used for the C=C and C≡C bond hydrogenation.[40] It was found that oxygen-containing functional groups are presumably responsible for the hydrogen accumulation on the UDD surface. Various Ni-containing carbon catalysts (prepared on detonation nanodiamonds, sugar carbon, and synthetic diamond support) with different structures and surface chemical compositions were used to perform the catalytic hydrogenation of acetylene at elevated temperatures. Among all the tested catalysts, Ni supported on nanodiamond (Ni/ND) was capable of generating a sufficient amount of active hydrogen for the complete conversion of acetylene. The activity of the catalyst was highest at temperatures higher than 250°C and gave full acetylene conversion. The catalytic activity of all the materials was also tested by pre-adsorbed hydrogen, and Ni/ND displayed good activity during the hydrogenation of acetylene under the flow of nitrogen (Fig. 18.5). Therefore, the properties of carbon material are essential for the hydrogen accumulation on a surface. The changes on the catalyst surface were investigated by diffusion reflectance IR spectroscopy. All the catalysts were found to possess oxygen-containing groups in the region of 1700–1780 and 3700 $cm^{-1}$, but the overall absorbance for ND was significantly higher. The higher absorbance may have been observed due to the large surface oxygen content on ND surface, which may further act as active hydrogen acceptors and result in improved activity of Ni/ND. Similarly, the activity of nanodiamond based catalyst (Ni/ND)[27] was determined for the hydrogenation of acetylene and compared with its counterparts prepared by depositing Ni on quartz and coal ($C_{SUG}$) support. It was found that Ni/ND was an active catalyst for acetylene hydrogenation in comparison to Ni/quartz and Ni/$C_{SUG}$ and can hold active hydrogen on its surface. Though hydrogen accumulation can, in fact, also occur on activated carbon[41] during hydrogenation, the retained hydrogen is weakly bound to the surface and readily desorbed at room temperature.

Sometimes the conditions used for preparing the catalyst may also influence its activity towards hydrogenation of unsaturated hydrocarbon. The point was demonstrated during the catalytic hydrogenation of toluene.[42] While loading nickel metal on nanodiamond surface by using nickel acetylacetonate as a precursor, it was found that the gas environment and the temperature of the precursor's decomposition can influence the catalytic properties. At higher pressure, the amount of retained hydrogen

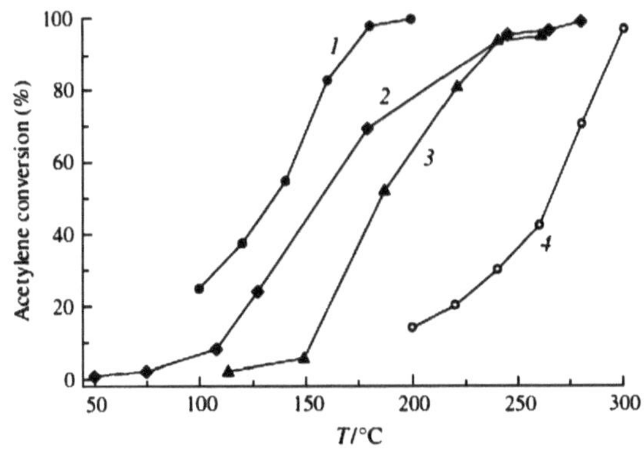


**FIGURE 18.5**

Temperature dependence of acetylene conversion in the flow of hydrogen gas on different carbon supported Ni catalysts Ni/ND (1), Ni/$D_{syn}$ (2), and Ni/$C_{sug}$ (3). Acetylene conversion by Ni/ND pre-adsorbed by hydrogen gas in a nitrogen flow. Figure adopted with permission from: Tveritinova, E. A., et al. Mendeleev Commun. 2013, 23, 249–250[40]

fairly increased on the catalyst surface. The ability of the three catalysts (Ni/DND, Ni/$C_{SUG}$, and Ni/quartz) to perform the reduction reaction was also determined in the stream of nitrogen after saturating them with hydrogen. From the study of diffuse reflection spectra of the three different catalysts, it was found that surface of detonation ND contained a considerable number of $sp^3$ C–O groups. On the other hand, quartz had a large number of OH groups, while $C_{SUG}$ had no hydroxyl or $sp^3$ C–O groups on its surface. The presence of these functional groups was correlated with their ability to retain hydrogen. $C_{SUG}$ failed to retain active hydrogen that had no hydroxyl groups, whereas the surface of quartz contained many –OH groups and hence could retain more active hydrogen. But due to smaller surface area of quartz, the overall content of adsorbed hydrogen was also low.

Palladium nanoparticles deposited on detonation nanodiamonds have also been used for the liquid phase hydrogenation of diphenylacetylene.[43] The activity of the same catalyst was simultaneously compared during hydrogenation of $\alpha$-phenylcinnamic acid and enamides. Palladium supported on ND displayed a high catalytic activity for all the selected unsaturated compounds. The reactivity of acetylenic substrate was higher compared to the ethylenic substrates; and catalyst also maintained the high activity during its recycling. Also, the activity of catalyst was compared with other palladium catalysts such as ($\eta$2-C70)Pd $(PPh_3)_2$ and Pd supported on carbon nanotubes during the hydrogenation of diphenylacetylene in methanol. In this reaction, Pd/ND exhibited very high performance in comparison to other catalysts, and the turnover number (TON) value was 6 for ($\eta$2-C70)Pd$(PPh_3)_2$, 2500 for Pd supported on carbon nanotubes, and 7200 for Pd/ND. The difference in the catalytic activity of Pd/CNT and Pd/ND was attributed to their structural features. Investigation of the detonation nanodiamond structure by various physicochemical methods has revealed[44] that its purification by conventional oxidation methods is unsuitable for the entire removal of impurities of fullerene- or graphite-like conglomerates. These materials form a thin shell on its surface and this fact may be beneficial during the preparation of the metal-containing catalysts. The $sp^2$-carbon centers of such surface fragments can enable more strong coordination to the metal clusters, compared to the non-modified surface containing only $sp^3$-carbon centers. On the other hand, palladium can form $\sigma$-bonds with $sp^3$-carbon atoms on surface. Moreover, the surface of natural diamonds contain small regions of the partially olefinic character,[45] hydroxyl or –COOH group that may further promote the formation of Pd–C $\sigma$-bonds and enhance the activity of the Pd clusters on the catalyst surface.

### 18.2.7 REFORMING OF ALCOHOLS

Catalytic reforming of alcohols such as ethanol is considered as one of the most efficient methods for hydrogen production. Use of ethanol is promoted as a substrate because of its low toxicity. The design of technologies for the production of different bio-fuels such as methanol and ethanol from renewable resources had gained worldwide attention. Therefore, steam reforming of bioethanol also provides a route to generate hydrogen from the renewable resources. Nanodiamonds themselves have a tendency to perform the steam reforming of ethanol. Different surface modified detonation nanodiamonds[21] were used for this process. The surface modification of ND was initially performed by hydrogen treatment that possibly resulted in the partial or complete removal of oxygenated functional groups from the surface. The activity of the modified ND was the same as that of the original ND and ethanol conversion remained the same in both cases. On the other hand, oxidation with nitric acid and hydrogen peroxide mixture increased the conversion at 400°C, and this effect almost vanished at higher temperature. The conversion of ethanol was increased for the entire temperature range for the air oxidized samples only. The main products of ethanol steam reforming were hydrogen and $CO_2$, side products such as CO, $CH_4$, and $CH_3CHO$ could also be observed. Methane was practically not generated at temperatures below 430°C, in the presence of all the catalysts and even at higher temperature little amount was formed.

Table 18.2 shows the different reactions during the course of ethanol reforming. The reforming of ethanol on nanodiamond starts with its adsorption on surface through the interaction of the hydroxyl groups. Decomposition of the adsorbed molecules with the formation of ethoxy fragments gave acetaldehyde after dehydrogenation. Re-adsorption of acetaldehyde at Lewis acidic sites forms hydrogen, carbon monoxide, $CO_2$, and $CH_4$. Surface carboxyl groups are believed to be the dehydration active sites, while Lewis acid and base sites are responsible for dehydrogenation. Almost every product can be formed on the surface of ND, and the difference in the product yields and selectivity depends on the way ND was modified. Air oxidation significantly improved the catalytic properties of the original nanodiamonds and a change in activity was generally caused by changes in the nature and quantity of functional groups on the surface. The most favorable condition was achieved when

**Table 18.2** Different reactions during the course of ethanol reforming and the heat effects. Data reproduced from: Mironova, E. Y., et al. Russ. Chem. Bull. 2013, 62, 2317–2321[21]

| Reaction pathway | $\Delta H^o$ 298/ kJ mol$^{-1}$ |
|---|---|
| $C_2H_5OH + 3H_2O \rightleftharpoons 2CO_2 + 6H_2$ | +173 |
| $C_2H_5OH \rightleftharpoons CH_4 + CO + H_2$ | +49 |
| $C_2H_5OH \rightleftharpoons CH_3CHO + H_2$ | +68 |
| $C_2H_5OH + H_2O \rightleftharpoons 2CO + 4H_2$ | +55 |
| $C_2H_5OH \rightleftharpoons C_2H_4 + H_2O$ | +45 |
| $CO + H_2O \rightleftharpoons CO_2 + H_2$ | −41 |

the surface oxidation was followed by the formation of Lewis acidic sites. Formation of Brønsted sites such as –COOH formed by the oxidation of mixture of nitric acid and hydrogen peroxide did not improve the catalytic activity.

Ethanol steam reforming (ESR) requires substantial energy consumption, and increasing the product yield requires higher temperature. This may lead to the formation of various undesirable by-products.[46] Detonation nanodiamonds have been used for the production of hydrogen by reforming of ethanol using Pt–M alloy (M = Ni, Ru) nanocatalysts deposited on ND in the presence of a commercial dense Pd–Ru membrane. The bimetallic Pt–Ru/ND and Pt–Ni/ND samples with 10 wt.% metal content were prepared and used for the production of hydrogen by ethanol reforming.[47] The use of Pt–Ni/ND catalyst at low temperatures provided substantially lower hydrogen yields in comparison to Pt–Ru/ND. However, methane yield increased considerably with the increase in temperature for this catalyst. Most likely, the decrease in hydrogen selectivity for Pt–Ru catalyst with temperature was due to the increase in selectivity of this catalyst for methane at high temperatures. The use of very high ratio of $H_2O$/ethanol [9:1 instead of 3:1] decreased the methane yield and simultaneously increased the hydrogen yield in the reaction that also increased with increasing temperature. The increase in hydrogen yield at higher temperatures was observed probably due to the fact that nanodiamonds were interacting directly with water vapor. Using a membrane for hydrogen removal shifted the equilibrium to the steam reforming products. When ~35% to 60% of hydrogen was selectively removed through the membrane during the catalysis by Pt–Ru/ND, it increased the hydrogen yield substantially. Very low conversion of ethanol was observed during the blank reaction (without a catalyst) in a membrane reactor at 450°C.

Detonation nanodiamonds were also used as a metal-free catalyst for the dehydration and dehydrogenation of $C_3$ alcohols (1-propanol, 2-propanol) along with ethanol [Fig. 18.1(vii)]. Nanodiamonds along with their modified forms obtained through heat treatment[48] with air or hydrogen were used in the process. The dehydration reaction proceeded with the formation of three products, namely ethylene (from ethanol), propylene (from 1-propanol and 2-propanol), and dipropyl ether (from 1-propanol self-condensation), whereas dehydrogenation involved the formation of acetaldehyde (from ethanol), propanal (from 1-propanol), and acetone (from 2-propanol). Oxidizing treatment of nanodiamond with air considerably increased the conversion of

2-propanol (from 28 to 100% at 330°C) along with the selectivity towards acetone formation (from 88.9 to 94.2%). The conversion of 1-propanol also increased on the oxidized nanodiamond, but the selectivity to propanal formation was decreased due to the formation of dipropyl ether. The conversion of ethanol and its selectivity to acetaldehyde were nearly the same on both oxidized and non-modified nanodiamonds. Reduced nanodiamonds obtained after the reduction process decreased the ethanol conversion, but simultaneously preserved the selectivity to acetaldehyde. Dehydration products were formed by the involvement of the carboxyl groups on carbon surfaces. On the other hand, the treatment of nanodiamond with hydrogen lowered its catalytic activity probably due to the decrease in the number of carboxyl groups. The oxidation treatment introduced the C–O–C linkage and reduction treatment increased the C–H bonds on the $sp^3$ hybridized carbon. The presence of these groups was verified by IR spectroscopic analysis where a band at higher wave numbers (1791 $cm^{-1}$) was observed after oxidation; and also bands at 2941 and 2877 $cm^{-1}$ appeared, indicating the formation of new C–H bonds.

### 18.2.8 COUPLING REACTION AND APPLICATION IN ORGANIC SYNTHESIS

Heterogenization of homogeneous catalytic systems is an important strategy for obtaining supported catalysts that retain their active catalytic sites of homogeneous analogues, and nanodiamonds have been utilized to make such systems. Furthermore, the ease of separation and reusability of the catalyst are the other major advantages provided with such a catalytic system. One such catalyst was prepared by conjugating a chiral ligand on ND surface for the asymmetric aldol reaction. Variable nature of functional groups on the nanodiamond surface makes them ideal for different surface modifications. Carboxyl, hydroxyl, carbonyl, and –CH bonds can be used for performing different reactions to modify their surface, and the resulting forms can further be applied in different chemical transformations. L-proline has been tagged on nanodiamond surface, and the resulting material was used as an enantioselective catalyst for the asymmetric aldol condensation[22] [Fig. 18.1(viii)].

For the surface functionalization of nanodiamonds, Boc-L hydroxyproline was first treated with sodium hydride and then reacted with nanodiamonds to yield chiral ligand-tethered nanodiamonds (Fig. 18.6a). Deprotection of the protecting group by acid treatment released the chiral heterogeneous ligand. The catalytic efficiency of the prepared catalyst was tested for the reaction of *p*-nitrobenzaldehyde with cyclohexanone under solvent-free conditions at room temperature. This reaction provided the anti-aldol product [Fig. 18.1(viii)] in 5% yield and in 73% ee. Longer reaction times lowered the enantioselectivity of the reaction with a slight improvement in the isolated yields. Furthermore, the enantioselectivities in different solvents were lower than that of the solvent-free reaction. A few additives such as $MgSO_4$, stearic acid, and benzoic acid increased the enantioselectivity, but the reaction did not complete. These modified nanodiamonds were further applied for the covalent attachment of drug molecules on the surface and used as a solid phase support in peptide synthe-

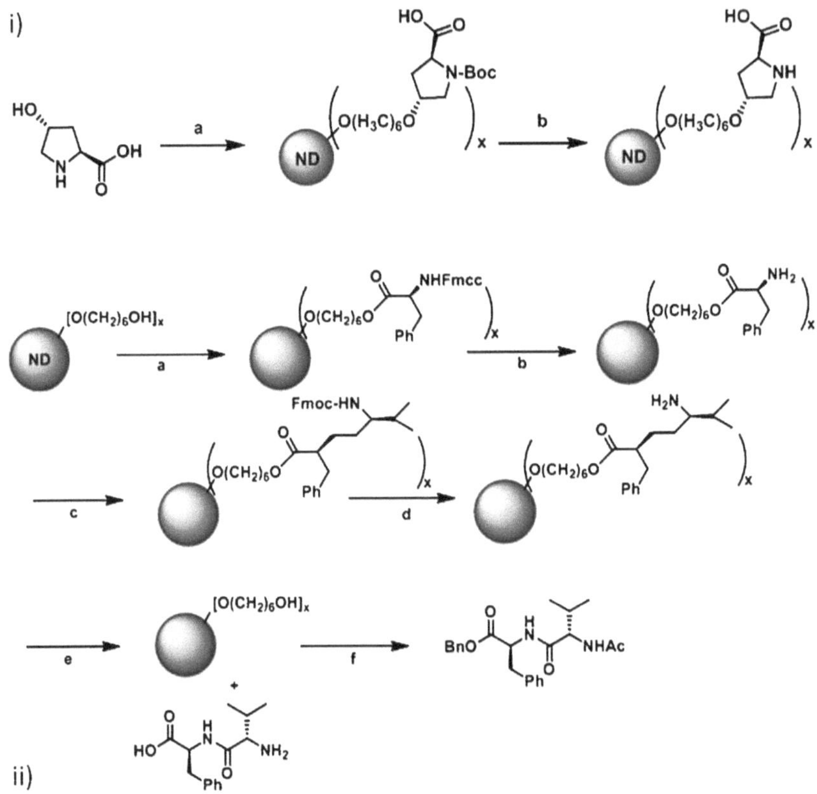


**FIGURE 18.6**

(i) Surface functionalization of nanodiamonds with L-hydroxy proline: (a) NaH, nanodiamonds, THF; (b) DCM, TFA (4:1). (ii) Use of hydroxylated nanodiamonds for peptide synthesis: (a) DCC, DMAP, N-(9-fluorenylmethoxycarbonyl)-L-phenylalanine, $CH_2Cl_2$; (b) piperidine in DMF; (c) DCC, DMAP, N-(9-fluorenylmethoxycarbonyl)-L-valine, DCM; (d) piperidine in DMF; (e) 1 M NaOH, rt; (f) $Ac_2O$, 1 M NaOH, benzylalcohol, DCC, DMAP, DCM. The figure was adopted with permission from reference: Zheng, W. W., et al. J. Mater. Chem. 2009, 19, 8432–8441[22]

sis (Fig. 18.6b). The terminal hydroxyl group of nanodiamond was esterified with N-(9-fluorenylmethoxycarbonyl)-L-phenylalanine, and DCC was used as a coupling agent. Subsequent deprotection of the N-Fmoc yielded free valine bond to the alkylated nanodiamond. This method was further used for the second coupling with Fmoc-L-valine to afford a dipeptide, which was finally isolated by saponification.

Suzuki coupling reaction can functionalize nanodiamond with trifluoroaryls and results in their increased solubility in organic medium such as ethanol and hexane. Also, ND functionalized by this method exhibits better resistance toward agglomeration. Recently, two routes for this purpose were developed.[49] In the first

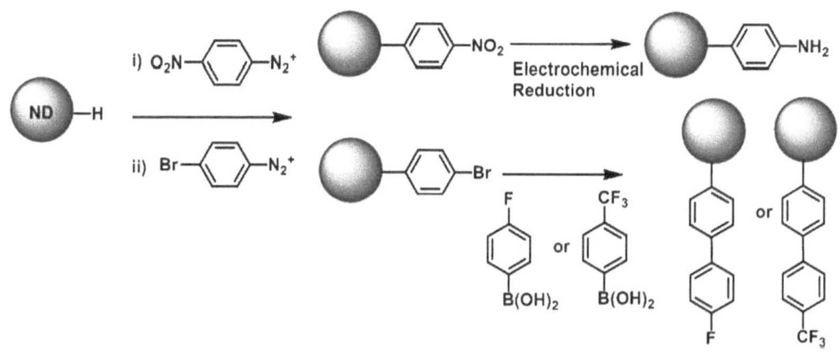


**FIGURE 18.7**

Coupling of diazonium salts on nanodiamond to generate the 4-nitrophenyl or 4-bromophenyl ND, which serves as a synthon for Suzuki coupling of 4-flurophenylboronic acid or 4-trifluorophenylboronic acid. The figure was adopted with permission from reference: Yeap, W. S., et al. Langmuir 2009, 25, 158–191[49]

route, hydrogenated nanodiamonds were derivatized with aryl diazonium salt to form the bromophenyl–nanodiamond complex, which upon subsequent reaction with phenyl boronic acid gave the biphenyl adduct. In the second route, the nanodiamonds were first derivatized with boronic acid groups to form the boronic acid–nanodiamond complex, which further undergo Suzuki cross-coupling with arenediazonium tetrafluroborate salts to yield biphenyl product. The preparative route for their formation is shown in Fig. 18.7. Different types of nanodiamond samples such as oxygenated ND, as-received ND, and hydrogenated ND (H–ND), were used in the reaction, but spontaneous reduction of diazonium salts was facile only on H–ND. Electrochemical activity of the nitrophenyl coupled nanodiamond was investigated by cyclic voltammetry to provide further evidence that the diazonium coupling can connect functional groups to the diamond surface. The nitrophenyl coupled nanodiamond was blended with ionic liquid 1-butyl-3-methylimidazolium hexafluorophosphate ($BMIMPF_6$) and then applied to the polished surface of the glassy carbon electrode. During this process an aryl nitro group undergoes a two-step reduction, giving a phenylhydroxylamine (–NHOH) intermediate before being further reduced to aniline. Two pronounced cathodic peaks were observed for the nitrophenyl coupled ND. These peaks were not present in the cyclic voltammograms of the control samples, and thus verified that they originated from the nitrophenyl groups on the nanodiamond surface.

Besides this, metal supported nanodiamonds have also been applied for the synthesis of amines and alkanes from organic nitriles. Palladium was loaded on nanodiamond support (Pd/ND), and its efficiency for the reduction and reductive deamination of benzonitrile was investigated.[28] The results were also compared with the analogous catalyst prepared by loading palladium on activated carbon (Pd/AC). Catalytic reduction of benzonitrile (BN) is a very important reaction used for the preparation of amines on large scale. The reduction process involves multiple hydrogenated or semi

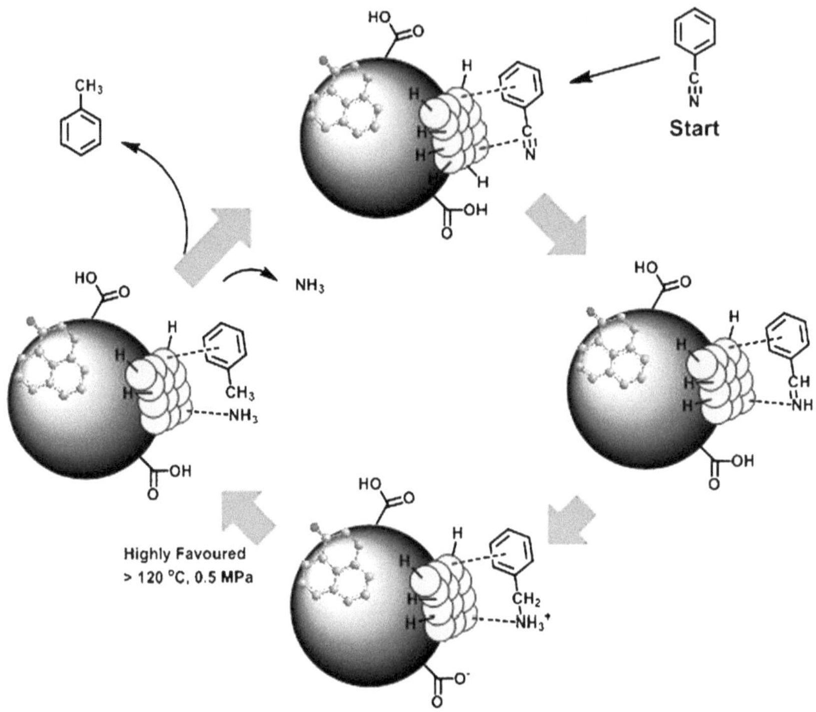


**FIGURE 18.8**

Plausible reaction mechanism for the hydrogenating deamination reaction of benzonitrile by palladium supported on nanodiamonds. Figure adopted after permission from: Gupta, N., et al. ChemCatChem 2016, 8, 922–928[28]

hydrogenated intermediates such as benzylamine (BA) and dibenzylamine (DBA). Using Pd/ND was beneficial in terms of providing high reaction rates and high selectivity towards bezylamine formation in comparison to conventional Pd/AC catalyst. The catalyst was also effective at lower temperature in comparison to the conventional catalytic system (up to 70°C compared to 100°C).

The Pd/ND based catalyst also provided exceptionally good tendency towards the hydrogenating deamination reaction of benzonitrile, and this tendency was used further to develop various synthetic applications. The methodology was used for the conversion of aromatic substrates such as nicotinonitrile, aromatic amines, and amines to their corresponding hydrocarbons. Also, the catalyst was demonstrated useful for the utilization of biomass by de-functionalizing the biomass derived aldehydes. Hydrogenating deamination of imine derivatives made a link to correlate aldehydes with alkane formation through this route. First, the aldehyde moiety was converted into imine by reacting it with benzylamine in the presence of activated molecular sieves. The obtained reaction mixture was then subjected to reductive deamination reaction to get the desirable product. This process did not result in removal of carbon atom constituting the aldehydic group and hence was good enough to provide "high atom efficiency". Functional groups on nanodiamond surface played an important role during the hydrogenating deamination reaction, and a mechanism involving surface acidic groups and active palladium metal was proposed (Fig. 18.8). The reaction starts with the interaction of benzonitrile with palladium nanoparticles, where the ring was probably activated by Pd. Well exposed Pd nanoparticles on the surface served as an active site for hydrogen absorption. Activation of aromatic ring then may lead to the hydrogenation of the –CN functional group, and acidic groups present on the surface might assist in the cleavage of the carbon nitrogen bond, leading to the deaminated product.

## 18.3 CONCLUSION

The goal of this chapter was to provide examples of catalytic applications of nanodiamond based catalysts. The spatial arrangement of spherical nanodiamond was responsible for providing better access for the reactants. High thermal conductance of the $sp^3$ diamond core may further lead to high activity of nanodiamond in the endothermic reaction. This core could improve the heat transfer efficiency near the active sites on the surface. Nanodiamond has a tendency to behave as a metal-free catalyst in different liquid/gas phase reactions. But the applications of nanodiamond as a metal-free catalyst are limited; and currently it finds application in ethylbenzene dehydrogenation, nitrobenzene reduction, alcohol dehydration, and hydrodechlorination. Surface modification of nanodiamonds with organic nitrogen containing compounds (organic bases) has led to its application as a metal-free catalyst in a few synthetic methodologies such as aldol condensation and solid phase peptide synthesis. Tagging non-polar organic molecules on nanodiamond surface can increase its solubility in organic solvents and make it suitable for organic syntheses and biological applications. Using nanodiamonds as a catalytic support for metals has extended their applications in reactions such as oxidation, dehydrochlorination, hydroamination, hydrogenation, and hydrogenating deamination. Metal supported on nanodiamonds can possibly make an efficient catalyst for the conversion of biomass derived aldehydes. Irrespective of the fact that nanodiamonds can be used as metal-free catalyst in the above mentioned chemical conversions and an excellent support for metals in a variety of chemical applications, their use as a catalyst is still limited. High cost of nanodiamonds and a few problems encountered during their operation (such as during ethyl benzene dehydrogenation) in chemical processes are possible limitations for their effective use. Such limitations can either be overcomed by making the synthetic process of nanodiamonds economical or by making nanodiamond based hybrids that can further minimize the amount of nanodiamonds required for the effective use in a chemical process. Moreover, we still need a lot of efforts for developing analogous applications of nanodiamonds, particularly as a metal-free catalyst to widen their scope and promote their use in different industrial applications.

## REFERENCES


1. Dai, Z. R.; Bradley, J. P.; Joswiak, D. J.; Brownlee, D. E.; Hills, H. G. M.; Genge, M. J. *Nature* **2002**, *11*, 157–159.
2. Danilenko, V. V. *Phys. Solid State* **2004**, *46*, 595–599.
3. Greiner, N. R.; Phillips, D. S.; Johnson, J. D.; Volk, F. *Nature* **1988**, *333*, 440–442.
4. Mochalin, V. N.; Gogotsi, Y. *J. Am. Chem. Soc.* **2009**, *131*, 4594–4595.
5. Maze, J. R.; Stanwix, P. L.; Hodges, J. S.; Hong, S.; Taylor, J. M.; Cappellaro, P.; Jiang, L.; Dutt, M. V. G.; Togan, E.; Zibrov, A. S., ; Yacoby, A.; Walsworth, R. L.; Lukin, M. D. *Nature* **2008**, *455*, 644–648.
6. Wang, R.; Sun, X.; Zhang, B.; Sun, X.; Su, D. S. *Chem. Eur. J.* **2014**, *20*, 6324–6331.
7. Mochalin, V. N.; Shenderova, O.; Ho, D.; Gogotsi, Y. *Nat. Nanotechnol.* **2012**, *7*, 11–23.
8. Pentecost, A.; Gour, S.; Mochalin, V.; Knoke, I.; Gogotsi, Y. *ACS Appl. Mater. Interfaces* **2010**, *2*, 3289–3294.
9. Xu, K.; Xue, Q. *Diam. Relat. Mater.* **2007**, *16*, 277–282.
10. Holt, K. B. *Phys. Chem. Chem. Phys.* **2010**, *12*, 2048–2058;
    Meinhardt, T. L. D.; Dill, H.; Krueger, A. *Adv. Funct. Mater.* **2011**, *21*, 494–500.
11. Slocombe, D.; Porch, A.; Bustarret, E.; Williams, O. A. *Appl. Phys. Lett.* **2013**, *102*, 244102.
12. Sakurai, H.; Ebihara, N.; Osawa, E.; Takahashi, M.; Fujinami, M.; Oguma, K. *Anal. Sci.* **2006**, *22*, 357–362.
13. Martín, R.; Heydorn, P. C.; Alvaro, M.; Garcia, H. *Chem. Mater.* **2009**, *21*, 4505–4514.
14. Girard, H. A.; Petit, T.; Perruchas, S.; Gacoin, T.; Gesset, C.; Arnault, J. C.; Bergonzo, P. *Phys. Chem. Chem. Phys.* **2011**, *13*, 11517–11523.
15. Agarwal, R. K.; Noh, J. S.; Schwarz, J. A.; Davini, P. *Carbon* **1987**, *25* (2), 219–226.
16. Dementjev, A.; Maslakov, K.; Kulakova, I.; Korolkov, V.; Dolmatov, V. *Diam. Relat. Mater.* **2007**, *16*, 2083–2086.
17. Barnard, A. S.; Sternberg, M. *Diam. Relat. Mater.* **2007**, *16*, 2078–2082.
18. Chen, D.; Holmen, A.; Sui, Z.; Zhou, X. *Chin. J. Catal.* **2014**, *35*, 824–841.
19. Zhang, J.; Su, D. S.; Blume, R.; Schlögl, R.; Wang, R.; Yang, X.; Gajović, A. *Angew. Chem. Int. Ed.* **2010**, *49*, 8640–8644.
20. Wu, S.; Wen, G.; Zhong, B.; Zhang, B.; Gu, X.; Wang, N.; Su, D. S. *Chin. J. Catal.* **2014**, *35*, 914–921.
21. Mironova, E. Y.; Ermilova, M. M.; Efimov, M. N.; Zemtsov, L. M.; Orekhova, N. V.; Karpacheva, G. P.; Bondarenko, G. N.; Zhilyaeva, N. A.; Muraviev, D. N.; Yaroslavtsev, A. B. *Russ. Chem. Bull.* **2013**, *62*, 2317–2321.
22. Zheng, W. W.; Hsieh, Y. H.; Chiu, Y. C.; Cai, S. J.; Cheng, C. L.; Chen, C. *J. Mater. Chem.* **2009**, *19*, 8432–8441.
23. Kachevskii, S. A.; Golubina, E. V.; Lokteva, E. S.; Lunin, V. V. *Russ. J. Phys. Chem. A* **2007**, *81*, 866–873.
24. Zinovyev, S.; Perosa, A.; Yufit, S.; Tundo, P. *J. Catal.* **2002**, *211*, 347–354.
25. Carabineiro, S. A. C.; Martins, L. M. D. R. S.; Avalos-Borja, M.; Buijnsters, J. G.; Pombeiro, A. J. L.; Figueiredo, J. L. *Appl. Catal. A: Gen.* **2013**, *467*, 279–290.
26. Seral-Ascaso, A.; Luquin, A.; Lázaro, M. J.; Germán, F.; Laguna, M.; Muñoz, E. *Appl. Catal. A: Gen.* **2013**, *456*, 88–95.
27. Tveritinova, E. A.; Kulakova, I. I.; Zhitnev, Y. N.; Kharlanov, A. N.; Fionov, A. V.; Chen, W.; Buyanova, I.; Lunin, V. V. *Russ. J. Phys. Chem. A* **2013**, *87*, 1114–1120.
28. Gupta, N.; Ding, Y.-X.; Feng, Z. B.; Su, D. S. *ChemCatChem* **2016**, *8*, 922–928.

29. Ba, H.; Liu, Y.; Mu, X.; Doh, W. H.; Nhut, J. M.; Granger, P.; Pham-Huu, C. *Appl. Catal. A: Gen.* **2015**, *499*, 217–226.
30. Ba, H.; Podila, S.; Liu, Y.; Mu, X.; Nhut, J. M.; Papaefthimiou, V.; Zafeiratos, S.; Granger, P.; Pham-Huu, C. *Catal. Today* **2015**, *249*, 167–175.
31. Tveritinova, E. A.; Zhitnev, Y. N.; Kulakova, I. I.; Cherkasov, N.; Maslakov, K. I.; Nesterova, E. A.; Ivanov, A. S.; Savilov, S. V.; Lunin, V. V. *Appl. Surf. Sci.* **2015**, *355*, 74–81.
32. Kachevskii, S. A.; Golubina, E. V.; Lokteva, E. S.; Lunin, V. V. *Russ. J. Phys. Chem. A* **2007**, *81*, 866–873.
33. Obraztsova, E. D.; Fujii, M.; Hayashi, S.; Kuznetsov, V. L.; Butenko, Y. V.; Chuvilin, A. L. *Carbon* **1998**, *36*, 821–826.
34. Lokteva, E. S.; Golubina, E. V.; Kachevskii, S. A.; Kharlanov, A. N.; Erokhin, A. V.; Lunin, V. V. *Kinet. Catal.* **2011**, *52*, 145–155.
35. Magdalinova, N. A.; Kalmykov, P. A.; Klyuev, M. V. *Russ. J. Gen. Chem.* **2014**, *84* (1), 33–39.
36. Kalmykov, P. A.; Magdalinova, N. A.; Klyuev, M. V. *Pet. Chem.* **2015**, *55*, 63–67.
37. Magdalinova, N. A.; Kalmykov, P. A.; Klyuev, M. V. *Pet. Chem.* **2012**, *52*, 299–304.
38. Magdalinova, N. A.; Klyuev, M. V.; Volkova, T. G.; Vershinin, N. N.; Bakaev, V. A.; Efimov, O. N.; Korobov, I. I. *Russ. Chem. Bull.* **2011**, *60*, 1085–1089.
39. Conner, W. C. Jr; Falconer, J. L. *Chem. Rev.* **1995**, *95*, 759–788.
40. Tveritinova, E. A.; Zhitnev, Y. N.; Kulakova, I. I.; Kharlanov, A. N.; Cherkasov, N. B.; Savilov, S. V.; Lunin, V. V. *Mendeleev Commun.* **2013**, *23*, 249–250.
41. Zieliński, M.; Wojcieszak, R.; Monteverdi, S.; Mercy, M.; Bettahar, M. M. *Catal. Commun.* **2005**, *6*, 777–783.
42. Mavrodinova, V.; Popova, M.; Mitev, D.; Stavrev, S.; Vassilev, S.; Minchev, C. *Catal. Commun.* **2007**, *8*, 1502–1506.
43. Turova, O. V.; Starodubtseva, E. V.; Vinogradov, M. G.; Sokolov, V. I.; Abramova, N. V.; Vul, A. Y.; Alexenskiy, A. E. *Catal. Commun.* **2011**, *12*, 577–579.
44. Osswald, S.; Yushin, G.; Mochalin, V.; Kucheyev, S. O.; Gogotsi, Y. *J. Am. Chem. Soc.* **2006**, *128*, 11635–11642.
45. Buriak, J. M. *Angew. Chem. Int. Ed.* **2001**, *40* (3), 532–534.
46. Bion, N.; Duprez, D.; Epron, F. *ChemSusChem* **2012**, *5*, 76–84.
47. Mironova, E. Y.; Ermilova, M. M.; Orekhova, N. V.; Muraviev, D. N.; Yaroslavtsev, A. B. *Catal. Today* **2014**, *236*, 64–69.
48. Tveritinova, E. A.; Kulakova, I. I.; Zhitnev, Y. N.; Fionov, A. V.; Lund, A.; Chen, W.; Buyanova, I.; Lunin, V. V. *Russ. J. Phys. Chem. A* **2012**, *86*, 26–31.
49. Yeap, W. S.; Chen, S.; Loh, K. P. *Langmuir* **2009**, *25*, 158–191.


## ACKNOWLEDGEMENTS

Authors (NG) thanks Chinese Academy of Sciences Fellowship for Postdoctoral and Visiting Scholars from Developing Countries, Grant number 2014FFGB0006 and also acknowledge financial support from the National Natural Sciences Foundation of China (21133010, 21473223, 51521091, 21261160487, 21411130120, 21503241, 91545119, 91545110), Strategic Priority Research Program of the Chinese Academy of Sciences, Grant Number XDA09030103, CAS/SAFEA International Partnership Program for Creative Research Teams for their work cited in this chapter.

# Index

## E

## R



## S

FSC
www.fsc.org
MIX
Papier | Fördert
gute Waldnutzung
FSC® C083411

Kolibri 360 GmbH
Ferdinand-Jühlke-Straße 7
99095 Erfurt, Deutschland
produktsicherheit@kolibri360.de